Synthese Library

Studies in Epistemology, Logic, Methodology, and Philosophy of Science

Volume 518

The aim of *Synthese Library* is to provide a forum for the best current work in the methodology and philosophy of science and in epistemology, all broadly understood. A wide variety of different approaches have traditionally been represented in the Library, and every effort is made to maintain this variety, not for its own sake, but because we believe that there are many fruitful and illuminating approaches to the philosophy of science and related disciplines.

Special attention is paid to methodological studies which illustrate the interplay of empirical and philosophical viewpoints and to contributions to the formal (logical, set-theoretical, mathematical, information-theoretical, decision-theoretical, etc.) methodology of empirical sciences. Likewise, the applications of logical methods to epistemology as well as philosophically and methodologically relevant studies in logic are strongly encouraged. The emphasis on logic will be tempered by interest in the psychological, historical, and sociological aspects of science. In addition to monographs *Synthese Library* publishes thematically unified anthologies and edited volumes with a well-defined topical focus inside the aim and scope of the book series. The contributions in the volumes are expected to be focused and structurally organized in accordance with the central theme(s), and should be tied together by an extensive editorial introduction or set of introductions if the volume is divided into parts. An extensive bibliography and index are mandatory.

Jan Voosholz

Towards a New Scientific Realism

Springer

Jan Voosholz
Institute for Philosophy and
International Centre for Philosophy
University of Bonn
Bonn, Germany

ISSN 0166-6991 ISSN 2542-8292 (electronic)
Synthese Library
ISBN 978-3-031-98885-1 ISBN 978-3-031-98886-8 (eBook)
https://doi.org/10.1007/978-3-031-98886-8

This Springer imprint is published by the registered company Springer Nature Switzerland AG
The registered company address is: Gewerbestrasse 11, 6330 Cham, Switzerland

For Noemi

Preface and Acknowledgements

This book is addressed to several different groups of readers. Depending on who you and what your interests are, you may wish to approach it in different ways. Here I hope to provide some guidance:

For readers primarily engaged in the contemporary debate about scientific realism—whether on semi-realism, ontic structural realism, perspectival realism, novel anti-realism or related positions—much of Chaps. 2–5 will be relevant, though many parts may already be familiar. The book is intentionally written so that even those without prior knowledge of the debate can follow the argument. If you are already well-versed in these discussions, I would particularly highlight the sections on the metalocal debate in philosophy of science (Sects. 2.3 and 2.4), my newly introduced concept of objective realism (Sect. 3.2), my interpretation of several oft-discussed arguments (Sects. 5.1.5, 5.1.6, and 5.2.4), and the addition of several new points or the consideration of less visible ones (Sects. 5.3–5.5).

The second part of the book develops ideas that are newly introduced into the scientific realism debate. If you are well acquainted with Quentin Meillassoux's concepts of correlationism and ancestrality, you may find much of Chaps. 6 and 7 familiar, although my particular interpretation, and substantial partial revision, of both concepts as analytical tools has not previously been published. If your primary interest lies here and not in philosophy of science, you may choose to read only these chapters.

For readers already acquainted with both the scientific realism debate and the debate concerning new and speculative realism in ontology and metaphysics, yet who are genuinely interested in my own proposal for a new scientific realism as presented in Chap. 8, I encourage reading the book from beginning to end, despite the effort. Of course, you are welcome to skip those parts with which you are already sufficiently familiar, or where you are willing to grant me the relevant conclusions.

In the case you are new to both debates and wish to work through the book to gain a broader understanding, I hope I have provided a readable introduction to each major topic, along with sufficient context and footnotes pointing to further sources, especially where I can only briefly mention a position or argument. My goal has been to present the complex arguments and diverse themes I engage with in a way that is both accessible to non-specialists and concise enough to remain useful for experts. I do not doubt that I have fallen short in some places, but I hope to have done a reasonable job overall. Since this investigation brings together approaches and debates that have, until now, remained largely separate, careful introduction is all the more essential—even more so than is already the case in philosophy.

This monograph results from an almost decade-long doctoral project, which I was able to carry out from summer 2016 to autumn 2021 at the Chair of Epistemology, Modern and Contemporary Philosophy at the Rheinische Friedrich-Wilhelms-Universität Bonn under the supervision of Professor Markus Gabriel, continuously supported by co-supervisor Professor Christoph Demmerling at the Chair of Theoretical Philosophy in Jena. The final book before you is an in a few places extended and substantially updated version of the doctoral dissertation that I submitted in 2021 (cf. Voosholz 2021a) and defended in 2022. Like many such projects, it began with an exposé outlining a clearly different endeavour. I wrote this exposé in 2015 and I primarily wanted to investigate the viability of Quentin Meillassoux's speculative realism as a critique of correlationism in ontology and philosophy of science, building on my master's thesis, submitted at the Free University Berlin in 2014. Some of my initial sketches and thoughts about the topic have made it into the final book in Chaps. 6 and 7, but otherwise this is a different monograph. Not only has the focus shifted considerably, especially in the early years of the project, but I have immersed myself more deeply in the existing realism-anti-realism debate in philosophy of science than I had previously thought. In particular, I only got to know Nancy Cartwright's, Anjan Chakravartty's, James Ladyman's and Steven French's philosophies more fully during my doctoral studies, and without them the result would have been completely different. The same applies to the developments in the debates surrounding new realism that have emerged in recent years, with which I was able to come into close contact in Bonn and during a research stay in Paris in 2017.

My deepest thank-you to Noemi Stelzig. Without Noemi's encouragement, I would never have dared to take the step from Berlin to Bonn, start a doctorate or ask complete strangers to become my supervisors. Equally crucial, however, were Noemi's always constructive criticism and our long conversations, without which my view of many topics such as feminist, intersectional and decolonial critiques of science would be a lot cloudier. Finally, I thank Noemi for her constant support since the first idea for a dissertation, which allowed me to write, translate, expand and publish this monograph without losing my mind or courage. Noe, thank you so much for all this and so much more! This book is dedicated to you for good reason.

I express immense thanks to my two supervisors, Markus Gabriel and Christoph Demmerling. Markus not only helped me write the doctoral thesis through many in-depth discussions and much substantial criticism, but also set a central landmark for my work with his philosophy, especially his pluralistic ontological realism. Without the philosophical exchange in countless colloquia, conferences and conversations in his office, this book would not be what it is. In addition, the interaction with the team that Markus has assembled at his chair in Bonn and into which he welcomed me in 2017, and in his international network of outstanding philosophers, natural scientists, humanities scholars and social scientists, was inspiring and valuable for me, far beyond the writing of this book. Christoph Demmerling gave me invaluable advice on the revision of the first version and the reorientation of the project, opened the door to the University of Jena and his chair, and decisively enriched my engagement with these topics through his own work on quietism, naturalism and anti-realism. Both of my supervisors invited me to participate in some of their publication projects, where I was able to gain wonderful experiences, and they always had an open ear for me, especially during the years of the pandemic, when academic exchange was fraught with unforeseen difficulties. Many thanks for your support.

I would also like to thank Jens Rometsch and Julia Petz. Jens went to Sisyphean lengths to comment on, criticise and discuss this work with me. Without his philosophical acumen, his patience and his humour, the chapters of this work—and I—would be philosophically much poorer. I am especially grateful to Julia for her precise and tireless philosophical, stylistic and orthographical corrections to the German manuscript, which would be much poorer and less comprehensible without them. In addition, the many conversations with Julia have always been a source of inspiration for me. Jens and Julia, thank you for your friendship, the many hours and the intellectual energy you have devoted to this work.

I would also like to express my gratitude to Florian Krockert. Florian, as my former flatmate, fellow student and closest friend from Berlin, has probably had more discussions with me about theoretical philosophy than anyone else. What I understand in philosophy and what I can express clearly is largely thanks to my conversations with you, Florian.

Finally, there is a long list of people who have helped me decisively, before and during the process of writing this, and I would like to thank them from the bottom of my heart: my godfather Thorsten Fischer has always supported me unconditionally, also decisively at the beginning of this endeavour; Thorsten, without you this doctoral period would not have been possible. My parents, Martina Fischer and Ulrich Voosholz, encouraged me in many ways to go my own way and even encouraged me to give academic philosophy a try. They have supported me many times during the writing process, thank you both! I would also like to thank Uli for helping me with typesetting questions and for initially teaching me how to write in LaTeX. I would like to thank Holger Keller for awakening my interest in physics and for the many discussions about the oddities of the cosmos. My families in Münster and my family of choice in Berlin: a thousand thanks to all of you for your support, both big and small, and for your forbearance and wonderful conversations over the years. I would also like to thank my friends in Bonn for their thoughtfulness, encouraging words and help. Thanks, too, to the staff at Café Mayras in Beuel for their considerable contribution in helping Noemi and me to get through the pandemic.

Alongside Noemi, Florian and Thorsten, it is thanks to my philosophical mentor from my time in Berlin, David Lauer, that I undertook my doctorate and started what would turn into this book in the first place: David, thank you for everything you taught me and for the important encouragement in autumn 2015 to set off for new shores. I would also like to thank my other philosophical teachers from Berlin (here in the order of the first event I attended with them): Holm Tetens, Sybille Krämer, Gunter Gebauer, Daniel Martin Feige, Hilge Landweer, Anne Eusterschulte, Jan Slaby, Mirjam Schaub, Robert Schmidt, Catherine Newmark, Mark Halawa, Julia Schiffers, Markus Rautzenberg, Georg Bertram, Rahel Jeaggi, Fabian Börchers and Susanne Lettow. I would like to thank my former logic tutor, David Löwenstein, for the time he spent answering dozens of questions on argumentation theory, as well as Philipp Wüschner, who spent hours talking to me about Meillassoux.

During my doctoral studies in Bonn, I benefited incredibly from the conversations and various initiatives at the Institute of Philosophy and beyond. My heartfelt thanks go to my doctoral brother Sergio Genovesi, who founded the student colloquium for French philosophy with me and spent many late nights discussing it; Diana Khamis, who convinced me of the deep difficulties of Meillassoux's ontology; and Marin Geier, who not only introduced me to Sellars, but also gave me crucial advice on more than one occasion. I would also like to thank my current and former, always helpful and bright colleagues and friends at the chair, IZPH and CST, Philipp Bohlen, Nick Dreikausen, Debora Eller, Alex Englander, Annette Feder, Philip Freytag, Charlotte Gauvry, Marin Geier, Mariya Halvadzhieva, Chelsea Haramia, Ana Ilievska, Eva Jeske, Julia Kastner, Tobias Keiling, Diana Khamis, Nicholas Kluge Corrêa, Johanna Koch, Joline Kretschmer, Lennard Landgraf, Virginia Lilley, Leila Mahmutovic, Ulf-G. Meißner, Timon Wade Meisner, Laura Michler, Tanja Mlinski, Julia Maria Mönig, Uwe Peters, Jens Pier, Huw Price, Jens Rometsch, Christiane Schäfer, Jerôme Schickschneit, Marlene Sieverdingbeck, Carla Spielmann, Apolline Taillandier, David Tain, Inga (Paul) Triebel, Victor Weisbrod, Barbara Wiehlpütz, Marie Charlotte Wolf and David Zapero. Special thanks go to Alex here, as he has been my office neighbour for a few years now, putting up with my complaints about this book with equanimity and humour as well as answering hundreds of translation questions I had when preparing the publication. I also cannot thank my collaborators Marlene, Charlotte and Debora enough for their fantastic support during the preparation of the book's publication at Synthese Library. Marlene holds a special place here, as she provided incredible support with a thousand details in the footnotes and in finding the right English translation of much of the literature, as well as being a dear friend. Thanks to my former colleague and friend throughout, Jens Pier, for many long evenings in the office and inspiring conversations (also, but not only, about this book project). Special thanks also go to Inga (Paul) Triebel not only for being most attentive and encouraging throughout the final phase of my PhD (especially my defence) but until the day I handed in the final proofs. I am indebted to all members of the Bonn Institute of Philosophy for their friendly welcome and collegial atmosphere. I would particularly like to mention Michael Forster, to whom (together with Markus) I owe my first assistant position in Bonn, and also Dennis Lehmkuhl, who discussed Michela Massimi's and Hasok Chang's more recent works, and his own cautious realism with me, from which I benefited greatly. I would like to thank the numerous members of Markus' research colloquium for their open critique and friendly suggestions over the years of writing.

Special thanks go to Jocelyn Benoist, who made it possible for me to spend a research stay in Paris in 2017 and warmly welcomed me to the Sorbonne, and to Quentin Meillassoux, who took the time to discuss the philosophy of science and his realism with me in private, as well as to Anna Longo, whose new ideas on the connection between emergence theory and nature in Meillassoux's work have enriched this book. I would also like to thank the Franco-German University for its financial support.

Markus kindly gave me the opportunity to benefit from his extensive network of excellent international partners. I also had the pleasure of meeting some incredible people through my position in Bonn. All of them have enriched my views. In particular, I would like to thank Zed Adams, Larissa Albantakis, Netaya Anbar, Neda Atanasoski, Jody Azzouni, Otávio Bueno, Anjan Chakravartty, David Chalmers, Kiyoshi Chiba, Lorraine Daston, Yasuo Deguchi, Shimon Dotan, Barbara Drossel, George Ellis, Maurizio Ferraris, Sarah Green, Sepp Gumbrecht, Graham Harman, Richard Healey, Alexander Kanev, Monika Kaup, Katerina Kolozova, Paul Kottman, Quill Kukla, Jocelyn Maclure, Greg Moss, Takahiro Nakajima, Denis Noble, Huw Price, Graham Priest, Dean Rickles, Jessica Riskin, Claude Romano, Jean-Marie Schaeffer, Umrao Sethi, Giulio Tononi, Charles Travis, Kalindi Vora, Jo Wolff, Jim Woodward and Aimee van Wynsberghe for their time and for their many discussions, lectures and texts. In Bonn, I would like to thank Tom Luu, Ulf-G. Meißner and Sach Mukherjee for exciting discussions on reducibility to the smallest physical level. I also thank my supervisors during my research stay during my Master's at IIT Bombay in winter 2014/2015, Siby K. George and P. R. Bhat, for their time and help with the first draft of a PhD synopsis.

I also have a number of people to thank for the publication of this book: Otávio Bueno, as Editor-in-Chief of Synthese and Synthese Library, encouraged me to produce this book from the PhD project and submit it for peer review. Otávio, your encouragement and your always enthusiastic manner were incredibly valuable. I have to thank Chris Wilby from Springer Nature for his always professional support, as well as Palani Murugasan from Strive. Last but not least (quite to the contrary, actually) I would like to thank my copy-editor Mary Newbould for her precise effort and her art in getting everything out of my sometimes very technical, Teutonic and dry sentences that was humanly possible. Any remaining errors and stylistic flourishes, of whatever kind, are mine alone.

Finally, I apologise to all those I have forgotten to thank here. It was a privilege to write this book, it was a wonderful time. However, I am also very glad that it is now over. Since this work about new scientific realism does not really seem finished to me—which according to Markus and Jens as is always the case with philosophical work—although I think I have answered the initial question exhaustively on the many pages that follow, I have the suspicion that new realism, its application to philosophy of science and the question of nature will keep me busy for a while longer. But here I have come to my senses, I won't let any more ink flow and I very much hope that the reader will enjoy one or two of the ideas contained in this work.

Bonn, Germany
May 2025

Jan Voosholz

Competing Interests The author has no competing interests to declare that are relevant to the content of this manuscript.

Contents

Part II New Realism for the Philosophy of Science

Acronyms

AN	Abstracta nominalism: Position according to which no abstract entities exist.
CN	Classic nominalism: Position according to which no general concepts and universals exist.
CPR	*Critique of Pure Reason*: Book by Immanuel Kant, Kritik der reinen Vernunft.
ECT	Positive epistemological tenet of systematic correlationism: A specific form of knowledge (in the extreme case: any possible knowledge) about the natural external world and/or certain entities (of a local, metalocal or global entity domain to be specified) consists only in a form of intersubjective consensus and/or is dependent on our interests and practices. Quietism, scepticism or agnosticism is therefore recommended as a philosophical attitude towards the natural external world and/or these entities.
ECTNS	Positive epistemological correlationist tenet about the natural sciences: Scientific knowledge about the natural external world, laws of nature, natural entities and/or some or all theoretical entities is only a form of intersubjective consensus and/or is dependent on our interests and practices; the philosophical stance advised in this domain is therefore quietism, scepticism or agnosticism.
ENSR	Epistemological thesis of new scientific realism: We can know some entities of nature in some of their properties and their local, structural dynamics by means of scientific methodology and theorising; our best theories about nature are partially true and should be understood literally.
EOSR	Epistemological thesis of objective scientific realism: We can know the subject-independent entities that together constitute the single real external world, at least in some of their properties and some of their connections to natural laws, by means of scientific methodology and theorising; our best theories about the external world are therefore approximately or partially true and should be understood literally.

EpiSR	Epistemic structural realism: Position according to which all that we scientifically know is structural knowledge of the world.
ESA	Epistemological scientific anti-realism: Position according to which the epistemological realist thesis in philosophy of science is false.
IBE	Inference to the best explanation.
MR	Metaphysical realism: Position according to which all subject-independent entities together constitute the world; there exists a complete and true description of this world; and parts of this description are expressed by us linguistically in the case of knowledge, i.e. true statements about the world correspond to it.
NR	New realism: Position according to which objective realism and systematic correlationism are false but both the ontological and epistemological realism theses are true for some local, metalocal or global context.
NRN	Non-reductive naturalism: Position according to which all tenets of naturalism are true but reductive physicalism and any other ontological reductionism are false.
NSR	New scientific realism: Position according to which both naturalism and systematic correlationism are false but both the ontological and epistemological realism theses are true in the context of philosophy of science.
OCT	Positive ontological tenet of systematic correlationism: The natural external world and/or certain entities (of a local, metalocal or global entity domain to be specified) are always ontologically dependent on their relation to subjects (in a sense to be specified). Conversely, there is also a form of dependence of subjects on the natural external world and/or the entities in question (often assumed as a causal, natural law-like or logical dependence).
OCTNEW	Positive ontological correlationist tenet about the natural external world: The natural external world and/or some or all theoretical entities are always ontologically dependent on their relation to the research community and/or our practices of theorising; conversely, there is also a form of dependency of the research community and/or our practices of theorising on the natural external world and/or the theoretical entities (often expressed as causal, natural-historical or natural law-like).
ONSR	Ontological thesis of new scientific realism: Our best natural scientific theories provide plural, correct domain ontologies of nature with their true parts, the entities of which are maximally modally robust and exist independently of their scientific-theoretical description.
OntSR	Ontic structural realism: Position according to which all that there is in the external world is structure.

OOSR	Ontological thesis of objective scientific realism: Our best theories provide in these true parts or in approximations a correct ontology of the single real external world, as the latter is subject-independent.
OR	Objective realism: Position according to which we know some aspects of some subject-independently existing entities of the single real external world.
OSA	Ontological scientific anti-realism: Position according to which the ontological realist thesis in philosophy of science is false.
OSR	Objective scientific realism: Position according to which we can know the subject-independent entities that together constitute the single real external world, at least in some of their properties and some of their connections to natural laws, by means of scientific methodology and theorising; our best theories about the external world are therefore approximately or partially true and should be understood literally; our best theories provide in these true parts or in approximations a correct ontology of the single real external world, as the latter is subject-independent.
PII	*Principium identitas indiscernibilium*, principle of the identity of indiscernibles: If two entities are indiscernible in principle, then they are identical.
PMI	Pessimistic meta-induction: Group of historical arguments against scientific realism.
RE	Epistemological realism: Certain entities can be registered directly or indirectly by systems (and we can in principle know some of them).
RO	Ontological realism: Certain entities have an independent existence and reality.
RP	Reductive physicalism: Position according to which all scientific ontologies can in principle be unified in a single scientific ontology, the best approximation of which is current physics as the leading and basic science; and every genuine advancement of knowledge can thus be expressed by reference to entities of the unified ontology of science, i.e. in the form of statements of physics. Epistemology not only coincides with philosophy of science; it can also in principle be reduced to physics.
SI	Subject independence: Independent existence from acting, conscious, mental, intentionally referring, personal, socially situated, knowing, speaking, discursive and/or culture-creating subjects.
UTE	Underdetermination of scientific theories by the data or evidence.

Chapter 1
Introduction

Global human society is currently facing a number of serious problems. These include an accumulation of various natural disasters such as floods, forest, field and bush fires, desertification, droughts, water pollution, a recent viral pandemic in the human population and its social consequences, and the loss of land mass to the oceans due to global warming. Linked to this is an exponential increase in species extinction, which is associated with a transformation of almost all ecosystems, and a global population growth for which our planet of origin does not appear to offer sufficient resources. Unsurprisingly, this work does not solve any of these problems. Nevertheless, I begin this introduction by listing them, as they are all based on the fact that we humans are part of nature. These problems also undoubtedly exist because humans are social and culture-creating beings who live out these characteristics collectively and individually. Ultimately, however, our survival will be determined by natural factors, with which we must come to terms through our actions in nature—from which we have never been detached and from which we have in truth never been removed. To achieve this, it is imperative to understand what is going on in nature, which factors are interrelated and what the future of nature looks like—from the smallest particles of water in our bodies to the stars.

It is precisely this knowledge that we expect to obtain from the natural sciences. How certain this knowledge is, what it means for our understanding of nature and what a currently appropriate perspective on the natural sciences looks like form the subject of this book. In order to answer these questions, I will introduce some core insights drawn from the relatively young philosophical current of new realism into the ramified debate about realism and anti-realism taking place in philosophy of science.

J. Voosholz, *Towards a New Scientific Realism*, Synthese Library 518,
https://doi.org/10.1007/978-3-031-98886-8_1

1.1 Objective

The aim of this book is to show why neither previously and currently predominant variants of *realism* nor the usual forms of *anti-realism* in philosophy of science are argumentatively sufficient and how this ongoing debate motivates a *new realism in philosophy of science*. I focus on the question of realism in the natural sciences. This scientific realism can provisionally be understood as the position according to which scientific statements and theories have a truth value regardless of whether we (can) establish it in our research, and according to which scientific entities described by theories exist and can be known how they are in nature, independently of us. Anti-realism, in relation to the natural sciences, denies this thesis.[1] As a method, it makes sense to develop a positive, new possibility from a critique of the current state of the debate in philosophy of science: new realism in philosophy of science, or new scientific realism for short. The aim of my work is not to elaborate such a philosophy in detail, but to show that a new solution must be sought and can be achieved by using ideas taken from the debate about new realism in ontology and epistemology. I do not claim here to follow the path of new scientific realism to its conclusion, but to take a first step in this direction. My aim is to chart a course that makes a new form of realist philosophy of science possible.

In doing so, I hope to make an innovative contribution to the current discourse on realism in philosophy of science, based on contemporary groundwork in the ontology of new realism. In particular, I will draw on the work of Quentin Meillassoux and Markus Gabriel, both of whom can be assigned to new realism in a sense that is still to be defined more precisely.[2] However, I will also analyse insights from other authors who contribute to the debates surrounding new realism, including Jocelyn Benoist, Peter Gratton and Anna Longo. I will spare the reader an introduction to the complex specialist discussion on realism and anti-realism in philosophy of science at this juncture, including its links to other areas of philosophy.[3] Importantly, I will not orientate my analysis in this book historically, but I will concentrate instead on the systematic level of the reconstruction of decisive arguments.

The claim to introduce an innovative proposal at this point may seem surprising in view of earlier and current voices that consider the debates on realism and anti-realism to have ended (cf. e.g. Blackburn 2002; Williams 2006). In their opinion, the major disputes have been conducted, the most important arguments have been worked out and criticised, and the essential positions elaborated. However, the representatives of current realisms in philosophy of science often discuss questions that directly result from the debate with anti-realism, such as: Are objects or rather structures fundamental to the best realist ontology of science? Which form of

---

[1] For more precise definitions, see my Sect. 2.1.1, p. 13.

[2] The following can be quoted as particularly central works: Gabriel (2015a), Gabriel (2016), Meillassoux (2008, 2011b).

[3] My entire Chap. 2 is devoted to a systematisation of the relevant debate, pp. 12–39.

realism has the right answers to the arguments of the other side? Can the entities of science all be reduced to minimal physical entities?

The partisans of anti-realist varieties of philosophy of science, on the other hand, often find themselves in other disputes: They focus on individual phenomena, such as the relationship between politics and science, or examine the (supposedly) ideological assumptions of disciplines such as evolutionary biology. The argumentative debate between realism and anti-realism has not concluded in philosophy of science either, despite enduring over many decades—and it is still unresolved (cf. e.g. Chang 2022; Massimi 2022). From this point onwards, I will use the metaphor of the stalemate, which originates from the game of chess, to describe this situation.[4] I am convinced that the argumentative stalemate can only be resolved if some of the basic problems of the situation are analysed. Drawing on the ideas of new realism, I argue that both conventional realisms and anti-realisms make problematic assumptions that are responsible for the stalemate.

This critical analysis allows me to show that future proponents of a new scientific realism can respond differently to anti-realist arguments if they do not cling to traditional variants of realism. To this end, I will derive a framework of conditions for such a new scientific realism from the preceding analysis at the end of this book. My research question arises from these goals.

1.2 Questions, Structure and Theses

In order to achieve the above-mentioned goals, ***two central questions*** must be answered: *Which arguments, premises, theses and philosophical positions belonging to the debate on realism and anti-realism in philosophy of science to date give rise to the stalemate in which none of the proposals are entirely convincing? And: Which concepts, arguments and insights derived from new realism can be introduced to philosophy of science and in what way, in order to resolve the stalemate?* Figuratively speaking, these two questions are the fixed stars of the ***two parts*** of this book.

To answer these two questions, it makes sense to break them down into a series of simpler questions. The realism/anti-realism debates in philosophy of science are inherently related to those in ontology and epistemology. My first two subquestions aim to clarify the central concepts and to determine the exact nature of

[4] I should note, for all the chess enthusiasts out there, that it would be more accurate to call it a "draw": because sometimes the opponents repeat argumentative moves, on some occasions it seems as though one side is cornering the other with rebuttals but is unable to secure its own position and win; sometimes the opponents agree that the opposite position is a rational possibility with its own merits, but in either case no side can finally win. Therefore, the metaphor of the draw would be most appropriate. My attentive copy-editor, Mary Newbould, and my colleague and friend, Alex Englander, both observed that "(argumentative) draw" sounds impenetrable and a bit weird, so I opted to disappoint the chess enthusiasts.

their discursive contexts. Therefore, in ***Chap. 2*** I will present ***A Map of the Realism-Anti-Realism Debates***. The goal here is to lay the foundation for further steps by defining the central concepts and defending them against potential objections. In particular, I argue for a principled separation of *epistemological* and *ontological* anti-realisms arising from various objections to the realist position. In addition to discussing current research literature of the realism/anti-realism debates, the chapter serves to classify various discourses on which I will draw throughout this book. I argue that, in addition to *local* debates about individual, clearly delimited areas of reality and the *global* discussion about reality as a whole, there are also *metalocal* debates about realism. They connect different local disputes in a systematic way. At the end of the chapter, I defend the thesis that the debate about realism and anti-realism in philosophy of science turns out to be significantly metalocal.

In the section above, I mentioned that I oppose many current realist positions. This raises the question of which positions these are, and of which theses, convictions and arguments they share; in short, what unites the first, realist group of my opponents. This is clarified in ***Chap. 3*** where I answer the question: ***What is Objective Realism?*** I develop this term specifically in the context of this thesis, and it serves to define a family of positions that are often equated with realism in the current debate. For this term to fulfil its purpose in my argument, it is important to clarify not only how it is defined, but also how it differs from related terms such as *metaphysical realism*. Throughout this chapter, I will reflect on whether Michael Dummett's thesis that there is a core to all realism/anti-realism debates is convincing, even in light of today's discourses and their proposed mapping. Here I will argue that there is such a core, but that it does not consist in linguistic-philosophical problems of this or that theory of truth and meaning, which merely point towards a deeper epistemological and ontological level. I conclude these considerations with an initial definition of *new realism* in contrast to *objective realism*.

Is objective realism advocated in philosophy of science or is it a straw man? Clarifying this question is a significant motivation behind ***Chap. 4: Objective Realism in Philosophy of Science***. I begin this chapter by discussing *naturalism*, aspects of which have already been criticised by Hilary Putnam, Christoph Demmerling, to name just two authors, alongside the new realists, to present a working definition that is as meaningful and fair as possible. This chapter shows that in this sense naturalism is represented in at least two systematic variants, one reductionist and one non-reductionist. The naturalism of Willard Van Orman Quine and many neo-Quinians provides an example of *reductionism*. I will explore various interpretations of Wilfrid Sellar's philosophy and Anjan Chakravartty's proposal of *semirealism* as non-reductionist naturalisms. My hypothesis regarding these reconstructions is that they always represent a form of objective realism, which explains why objective realism seems unavoidable to many debaters who defend realist theses in philosophy of science. In addition to proving this hypothesis, the chapter has other purposes. I will introduce two arguments that will turn out to be relevant for the future new scientific realism: the argument for the *disunity of the sciences* against reductionism and the *problem of pseudo entities*, which Huw Price calls the "placement problem"

(Price 2004, p. 7). The first concerns the fact that we are not able to explain everything by means of the theories of a fundamental or leading science. The second is aimed at the question of how such entities as the human mind, marital problems and numbers can appear to exist if only the concrete things of nature really exist. Chakravartty's idea of a *selective scepticism*, according to which scientific theories can be partially true and the scientific ontologies resulting from the considerations and assumptions of a natural science can be partially correct, also becomes central in the further course of my argument in this book.

The final chapter of the first part the book, ***Chap. 5***, represents the argumentative centrepiece: Here I reconstruct the main arguments against objective realism. In doing so, I prove my first thesis of the ***Double Stalemate of Objective Realism in Philosophy of Science***. This thesis states that the debate is paralysed by *two* argumentative stalemates: First, new- and anti-realist arguments ensure that objective realism in philosophy of science cannot be convincing without thereby properly motivating anti-realism. This is the *first stalemate*. Second, I show that a recent form of objective realism, *structural realism* as advocated in various forms by John Worrall, James Ladyman and Stephen French, tries in vain to overcome the first stalemate more successfully than object-naturalists such as Chakravartty. Between these principled versions of objective realism—(ontic) structural realism and (non-reductive) naturalism—persists a *second stalemate*: Neither option seems, in principle, more promising than the other, but both sides have strong arguments against the other side.

In order to demonstrate this argumentative double stalemate in the debate, I reconstruct the arguments against objective realism in philosophy of science from three families of arguments, alongside the structural realist objections, in Chap. 5: *Arguments from the underdetermination of scientific theories by evidence* oppose the idea that scientific theories can ever be proven to be true by empirical methods. *Historical arguments* take cases from the history of science to show that there is a constant change of theory that will not end with our best current theories and scientific domain ontologies. In addition to these well-known arguments, I point out the importance of *antitotality arguments*: Arguments belonging to this family oppose the idea, important to objective realism, that nature can be conceived as an external world that is fundamental to everything else, radically independent of us, and possessing a clear fundamental structure fixed by the laws of nature. The relevant arguments come partly from the new realism debate, such as Gabriel's arguments against the world, partly from those concerning relativism, such as Carol Rovane's arguments against unimundialism, and partly from debates in philosophy of science, which are often discussed separately from the realism question, such as Nancy Cartwright's argument against fundamental laws of nature. My analysis will suggest that antitotality arguments are crucial to the question of objective realism. My thesis is that the interplay of the arguments of all three families shows that objective realism cannot be maintained. However, since this neither suggests anti-realism, nor proves one particular objective realism to be better suited to the challenge than others, the double stalemate holds. With this conclusion, the *first of the two central guiding questions is (hopefully) answered by the* ***first part*** of this book.

The ***second part*** aims to resolve the double stalemate by introducing further new realist insights, concepts and arguments taken from debates in ontology and epistemology to philosophy of science. To this end, in ***Chap. 6***, I first examine Meillassoux's concept of ***Correlationism*** and develop it for my purposes. Roughly speaking, correlationism refers to the type of philosophy according to which there is a relationship of mutual dependence between, on the one hand, the natural external world, its natural laws and objects, and us subjects, our practices and languages on the other. Meillassoux believes that correlationism has been the predominant philosophical paradigm since Immanuel Kant's *Critiques* and argues in favour of dropping it. I attempt to separate Meillassoux's philosophical-historical hypothesis from the systematic considerations behind this endeavour and to sharpen the latter by means of working through the extensive critical research debate. In doing so, I develop correlationism into a tool that is clearly defined, albeit suitably complex, and can be used in the realism/anti-realism debates. In doing so, I defend the thesis that correlationism is not, as the research debate claims, firmly linked to anti-realism, but rather runs, as a categorisation, orthogonally to the dualism of *realism—anti-realism*. The use of correlationism as an analytical tool in philosophy of science reveals similarities between the realists Sellars and Anton Friedrich Koch and the anti-realists Bas van Fraassen, Brad Wray and Kyle Stanford. The analysis reveals that a new scientific realism must distinguish itself not only from objective realism, but also from all forms of realist and anti-realist correlationism. The subtitle of this book refers to the two main opponents of any new scientific realism, as the focus is on their refutation and on the search for an alternative path, not on a full-fledged formulation of a new philosophy of science.

How new realism refutes the approaches of correlationism is the question at the core of ***Chap. 7*** on ***The Problem of Ancestrality***. Meillassoux develops the thesis that correlationism, by assuming a relationship of mutual dependence between the objective and the subjective, torpedoes the possibility of properly conceptualising a time before any subjects. In this ancestral time, however, events would have taken place that made the factual existence of subjects possible in the first place. Meillassoux's ancestrality argument was heavily criticised in the debate surrounding his philosophy. In this chapter, I present a *reinterpretation of the problem* that identifies two levels on which it unfolds its effects: the *level of scientific discourse*, which is analysed by philosophers of science, and an *ontological level*, which scrutinises correlationism about the origin and the conditions of the possibility of correlation. I argue that rephrasing the ancestrality problem has value in arguments against the various correlationisms presented in my previous chapter.

In ***Chap. 8***, I conclude my answer to the central question of part two by bringing us on the way ***Towards the New Scientific Realism***. In doing so, I present a framework for any such future position against the background of the critical insights of my previous chapters. In a first step, I will argue that objective realism in philosophy of science has had its day, while against correlationism I will put forward the reformulated argument from ancestrality. This situation justifies the need for a new scientific realism. My thesis in this chapter is that such a philosophy is possible and that many of its basic features have already emerged throughout my analysis of the debates. I argue that a *new philosophy of nature* needs to be considered, since nature cannot be equated with the external world of humans. New realism and my reflections suggest that nature is *radically plural* and that this is often misjudged from a realist perspective in the debate to date. Here, it becomes clear that new scientific realism does not denote a single position, but a family of positions, as there are different ways of adequately philosophising about nature within my framework. Furthermore, I introduce the concept of a *natural-scientific interrelation*, according to which nature, properly understood pluralistically, is connected to us and our research. In the next step, I point out various ways in which a new scientific realism can answer questions that arise in this context, with regard to the criticism of the natural sciences, to the interpretation of theories, of models and of scientific domain ontologies and also to the transdisciplinary exploration of reality. With this outlook, the initial question of the second part of my book is answered.

I will conclude by stating the main features of a new form of philosophy of science that is motivated by a critical analysis of the current debate and that can make use of many of the arguments and achievements of both objective realism and correlationism. The results of this work are briefly reviewed and summarised in ***A Short Conclusion***, which constitutes ***Chap. 9***.

My limitation to the systematic level of philosophical argumentation in the context of a contemporary debate—which is placed in its historical context at appropriate points during the argument—is obvious. Methodologically, I will make use of formal reconstructions where they highlight important points or bring new clarity to an argument or serve the purpose of clear definitions of philosophical positions and their categorisation terms. I have deliberately tried to avoid predicate-logical, modal-logical or other formalisms in order to ensure the readability of the book. In its philosophical approach, here my work is close to new realism, even if its premises and arguments are never adopted uncritically. This explains why no fundamental distinction is made between *analytic* and *non-analytic* (*Continental, ...*) philosophy. I am convinced that this, at most, marks a sociological or historical difference that is of no assistance in answering my systematic questions, should it be justified at all. In the following work, I assume that it is philosophically honest to draw on the arguments of different authors of academic philosophy and for us to enter a dialogue with each other—beyond supposed or actual affiliations according to tradition.

Both parts of this book are preceded by their own introductions. The reason for the decision to introduce the contents and the procedure in both parts separately is due to the length of the present book and the complexities resulting from the different debates. I hope that the following remarks will present a fair critique and a helpful further development of the realism-anti-realism debate in philosophy of science. As always, the reader may decide whether they are convincing.

Part I
Objective Realism and Philosophy of Science

Introduction and Structure of Part I

In this book, I examine the consequences of the critical perspective of new realism for philosophy of science. By *critical perspective* I mean, on the one hand, the negative philosophical findings that can be summarised under the terms correlationism and ancestrality, which arise from the arguments against any form of totality and for ontological pluralism, and on the other distinguishing new realism from other realist positions such as naturalism. By presenting the critical perspective of new realism in an organised way, I want to make visible the main consequences that apply to the realism/anti-realism debate in philosophy of science.

The tasks assigned to this first part build on each other: first, *how* the multifarious terms of the debate—realism, anti-realism, naturalism—describe *which* positions must be established. The first task is therefore to clarify these central categorisations and to show which concrete philosophical positions they are linked to. Only by maximising clarity can an analysis avoid dealing with straw men, thereby forfeiting a major part of its value. This requires categorising the most important contributions in the field.

The second task is to assess the state of the debate: For this I will introduce a new label, *objective realism*. Positions of objective realism are those that insist on an absolute reality of the *external world*, which is understood as being *independent of subjects and subjectivity*. At first glance, it seems obvious that this is also a minimum requirement for any realist position in philosophy of science and that objective realism should therefore also be accepted by opponents of the more demanding *metaphysical realism*. This new concept, objective realism, is necessary to prove my first hypothesis in this first part of my book: Even the most promising positions of objective realism in philosophy of science are in an argumentative *double stalemate*—on the one hand against each other, on the other against anti-realist critique. The conclusion to this argumentation provides the insight required to go beyond objective realism to seek a more convincing *new scientific realism*.

My approach to achieving this argumentative goal consists of four steps: First, in Chap. 2, I will provide conceptual clarity by means of a *map of the debates* and an explanation of fundamental terms. I argue that realism contains an ontological and epistemological thesis (2.1). Furthermore, I use examples from current scholarly debates to show that a specific realism/anti-realism debate can either be limited to a local area of entities (2.2), combine several local areas, or even be concerned with reality globally (2.3). I conclude by drawing the outlines of metaphysical and scientific realism on my map, since the related debates play a central role (2.4).

Chap. 3 takes the next step by introducing the new concept of objective realism. To this end, I pursue Michael Dummett's idea that there is a single core to all realism/anti-realism debates (3.1). This allows me to introduce objective realism and to present its essential characteristics (3.2).

In the next step, in Chap. 4, I apply the newly acquired concept of objective realism to the metalocal debate within philosophy of science. First, I follow the trail of why, in philosophy of science, objective realism often takes the form of reductive naturalism, as it was and is advocated in the tradition of W. V. O. Quine—despite the problems associated with it (4.1). I then turn to the question of what alternatives have been developed within objective realism in philosophy of science to resolve these difficulties, thus characterising non-reductive naturalism (4.2). Here, special attention is paid to Anjan Chakravartty's semirealism, as it provides an example in the discussion that follows of an objective realism in philosophy of science that is argumentatively as strong as possible.

I prove my thesis of the argumentative double stalemate of objective realism in philosophy of science in Chap. 5. To this end, I reconstruct the anti-realist critique of objective realism, starting with the arguments from the underdetermination of scientific theories by empirical evidence that form a first family (5.1). These objections are classically supplemented by historical arguments from the history of science (5.2). I then outline several arguments against totalities and, in particular, against the concept of the external world, of which objective realism makes use (5.3). These include arguments from related debates in philosophy of science, such as Nancy Cartwright's discussion of fundamental laws of nature, but also arguments from new realism, such as Markus Gabriel's arguments against the world. I then argue that even the more recent proposal of structural realism is not significantly preferable to Chakravartty's semirealism in overcoming all these counter-arguments (5.4). At the end of the chapter, I draw my conclusions and summarise why objective realism in philosophy of science is in an argumentative stalemate with anti-realist counterarguments on the one hand, and why, on the other, its best options, semirealism and ontic structural realism, cannot prevail against the alternative position, resulting in yet another argumentative stalemate (5.5).

Chapter 2
A Map of the Realism-Anti-Realism Debates

The present chapter prepares the ground for proving the main thesis of Part I: *Different forms of objective realism and anti-realist critique in philosophy of science are in a double stalemate*. It is first necessary to understand what kind of debate this is before I can be in a position to develop my concept of objective realism in the subsequent Chap. 3. Clarifying the terms realism and anti-realism and mapping the debates around both terms helps me to identify their distinctive features in philosophy of science. To develop them further, it is crucial to consider their differences with comparable discussions about realism and anti-realism. In the chapters that ensue in Part I, I introduce my concept of objective realism, and I analyse the arguments between realism and anti-realism in philosophy of science.

Here, the *debates about realism and anti-realism* in various areas of philosophy are initially analysed from a bird's eye view. My starting point is to conceptualise the fundamental concept of *realism* and its opposite of *anti-realism* as generally as possible (2.1). I argue that realism in the full sense consists only in the combination of two positive theses: an *ontological thesis about independent existence* and an *epistemological thesis about the knowability* of a certain entity or entities. Individually, however, these positive theses are also put forward as *ontological realism* and *epistemological realism* or rejected as *ontological anti-realism* and *epistemological anti-realism*.

These distinctions prove relevant in a whole series of debates. I demonstrate the robustness of my definitions by mapping several related realism/anti-realism debates. The structure of these debates is modelled by distinguishing between *local*, *metalocal* and *global* forms of discourse: Debates that concern a clearly delimited area—e.g. unobservable scientific-theoretical entities—are local (2.2). Those that thematise several local, interrelated areas I call metalocal. These debates are more complex than local debates. When the ontological and epistemological question is all-inclusive, however, one enters the global realism/anti-realism debate (2.3). In doing so, I argue that the questions concerning both metaphysical and scientific

J. Voosholz, *Towards a New Scientific Realism*, Synthese Library 518,
https://doi.org/10.1007/978-3-031-98886-8_2

realism are often falsely characterised as global or local, respectively, when in fact they both are metalocal (2.4).

To this end, realism and anti-realism as concepts will first be defined based on the acceptance or rejection of an ontological-epistemological thesis. I will then motivate my definition in order to illustrate it with reference to concrete, systematically delimitable, but related research discussions.

The categorisation of philosophical positions has a long tradition and has repeatedly been the target of fundamental criticism. Without being able to delve deeply into that debate at this point,[1] I will allow myself a brief preliminary remark: The reduction of a piece or even the entire work of an author to a more or less complex concatenation of *-isms*, "Simone de Beauvoir is a feminist, post-Heideggerian existentialist ...", does not in itself provide deep systematic insights, nor does it harbour philosophical-historical profit. If this categorisation is carried out with a primary text-based justification and the categories are underpinned by explanatory definitions and surrounding reflections, it can provide greater clarity in a complex and lengthy philosophical argumentation with a multitude of presented positions and systematic classifications, such as the present work. That is the aim of this chapter.

2.1 Ontological and Epistemological Realism

In order to differentiate the twin concepts of realism and anti-realism more precisely and to be able to address the debates in which realists and anti-realists compete in philosophical disputes, it is necessary to present clear working definitions of the core concepts. My proposal for such a conceptualisation makes use of another philosophical concept, which I must first obtain:[2]

In the following, I call ENTITY everything that exists or appears in the broadest possible sense. Of course, this statement requires ontological clarification. My use of *entity* is very similar to Graham Priest's deployment of *object*[3] and Theodore Sider's of *entity*.[4] Anything that can be the real or counterfactual referent of a linguistic expression or variable is an entity—any target of a possible and real reference.

---

[1] This approach is already problematised in introductory literature. Cf. e.g. Craig 2002, Chap. 6 "Of "isms"", pp. 61–73.

[2] In order to ensure the consistency of the present work in this book, I present working definitions to which my use of language will adhere in what follows. For the sake of clarity, they are set in CAPITALS unless they appear in quotations. In some cases, these definitions will be further sharpened in the course of my argumentation. This is unavoidable as some definitions are criticised, expanded and specified. These points are all specifically highlighted.

[3] See Priest 2014, p. 27: "Now, to be an object is to be something." Priest also argues that *objects* can be quantified over, that they are nameable and that they can become the object of an intentional mental state.

[4] See Sider 2020, p. 46: "It will be convenient to have a predicate whose interpretation is broad in a way that mirrors the broad interpretation of the quantifiers. I will use "entity" to play this role. Thus absolutely everything is an entity, on my usage."

Entity serves me as an expression for an ontologically general something. However the fundamental concepts of ontology are understood exactly: objects, things, stuff, items, properties, relations, events, processes, systems and structures are all entities in my sense.[5] This remains true regardless of the modality in which the entities exist. The cat on the mat is an entity, just like the Big Bang, world history, justice, the empty set, stomach aches or this morning's potential accident, which was narrowly avoided but still scared me. The relation of two objects being connected, the property of redness and the process of an object shrinking are all entities. Even the non-existent, the contradictory and the impossible, such as the round square, are entities in my usage.

This definition suffices for my definition of realism and anti-realism. Similarly generalised definitions can be found implicitly or explicitly among many contemporary realists and anti-realists. But how do I operate in the context of this book?

2.1.1 Definitions for Ontological and Epistemological (Anti-)Realism

A complete REALISM regarding an entity domain consists in the combination of two philosophical theses:

RO Certain entities $E_1, \ldots, E_n$ have an independent existence and reality.

Independence paradigmatically refers to E's independence from all perceiving observers and acting agents, i.e. *subject independence*. This is to be interpreted more broadly to include all registering systems, such as measuring devices.

RE These particular entities $E_1, \ldots, E_n$ can in turn be registered directly or indirectly by systems (and we can in principle know some of them).[6]

The first thesis RO—certain entities exist independently, including: independently of our knowledge, cognition and access to them—is called *ontological realism*. The second thesis RE—the entities can in turn be directly or indirectly registered, and can (at least in principle) also be known by us humans—is called *epistemological realism*. *Realist* in the complete sense are all those positions that affirm both theses (i.e., advocate RORE).[7]

---

[5] This is not to say that realism or anti-realism would be linked to a particular form of ontology such as a process, event or object ontology, or a systems or structural metaphysics or the like.

[6] This definition is based on common definitions such as those in Miller 2019; but I borrow the term registering systems from: Gabriel 2016, p. 152–158, which is used to replace common formulations of subject or mind independence. The choice of this working definition will be justified by the conceptual reflections in this and the next two Sects. 2.1, 2.2 and 2.3, the usefulness of another concept of independence compared to subject independence will become apparent in Part II of this book.

[7] A list of abbreviations and acronyms that applies to the entire book can be found on p. xvii.

RO also states that existence attributions and reality attributions are always made in unison. This means that it is not possible to separate existence and reality. *E* exists independently if and only if *E* is independently real. If one is denied, then the other is also denied.

As an antonym to this, a definite use of the concept of ANTI-REALISM with regard to certain entities necessarily results: The denial of RO with respect to certain entities, i.e. the denial of their independent existence and reality, means the advocacy of *ontological anti-realism*. If, on the other hand, their fundamental registrability and thus also recognisability, perceivability and knowability is denied, and consequently RE is rejected, an *epistemological anti-realism* is assumed with regard to these entities.[8]

In my view, this nuanced understanding of anti-realism has a decisive advantage over many other definitions of the term: It makes it possible to capture the opposition to realism but without limiting the theoretical possibilities of anti-realism *per definitionem*. This danger exists if anti-realism is merely understood as a thesis of subject-dependent recognisability or knowability of an entity domain. According to this alternative view, anti-realism would be the position that argues that an entity or domain of entities is recognisable to us and knowable while it constantly depends on us.

Under both definitions, for example, an error theory about an entity domain can be classified as anti-realism, which explains how the illusion of the independent existence of these entities comes about. In such an error theory, an ontological anti-realism is defended, and RO is consequently rejected. The knowability of the entities *dependent* on our practices is not denied, i.e. a variant of RE is assumed. I see the advantage of my more nuanced definition of anti-realism elsewhere: Only in this view is it correct to call anti-realism any position that considers the knowability of, and even any statements about the potential registration processes of an entity to be impossible—therefore rejecting RE—but which holds on to its independent existence for some other reason—therefore advocating RO. Thus, certain sceptical and solipsistic positions can be identified as anti-realism. Anti-realisms may therefore have more in common with certain forms of realism than with other anti-realisms that opt for opposite preferences in RO and RE.

2.1.2 Different Forms of Anti-Realism

Irrespective of whether only RO, RE or RORE are denied, a variety of different anti-realist positions can be adopted—ranging from a radical social constructivism that declares all entities to be products of social discourse to a scepticism that declares

[8] I would like to thank Markus Gabriel for helpful criticism of a first version of my definition of anti-realism, which led to the double variant I am presenting here, and Jens Rometsch for his smart objections to an earlier version of the independence thesis of realism.

all knowledge to be impossible. The same applies to the affirmation of both theses, since distinct realisms understand and justify RO and RE completely differently.

As already mentioned above, the nuanced differentiation into 1. ontological, 2. epistemological and 3. ontological-epistemological anti-realisms can always be made. The category of anti-realism is therefore even more heterogeneous than that of realisms. However, due to the rejection of at least parts of RORE, anti-realisms have the common feature that they bind foundational concepts such as existence, reality, knowledge, necessity, possibility, object, property, etc. solely to "our formative use of these concepts" (Gabriel 2016, p. 13, my translation), as Markus Gabriel states. This also remains true in my nuanced conception of anti-realism. Thus, as Christoph Demmerling rightly puts it, fundamental ontological registers such as existence, reality or world are understood as "dependent on the human mind, on our thinking, our possibilities of cognition and our language" (Demmerling 2004, p. 30, my translation)—if one can speak of them in a meaningful way at all. For they either have "no definite structure" (Demmerling 2004, p. 30, my translation) without this relation (*ontological anti-realism*) or it remains unknowable (*epistemological anti-realism*).

Since the components of RORE can be separated, mixed forms of realism and anti-realism are also conceivable on an ontological and epistemological level. If it is conceded that there are independently existing entities, but that these are unknowable, we have a combination of ontological realism and epistemological anti-realism. These intermediates can be described as forms of scepticism.[9] These mixed positions represent an over-all anti-realism, since they leave the entities in the ontological picture, but it is no longer possible even in principle to clarify what their status is. Such epistemological anti-realisms argue, for example, that as soon as cognition attempts to latch onto the entity, it does not grasp objective properties of the entity, but only subjectively moulded properties. The properties inscribed in the cognitive relationship are thus no longer objectively attributed to an independent entity, but are inseparable from subjective access.

At first glance, the reverse hybrid forms of ontological anti-realism and epistemological realism simply appear to be contradictory: According to my definition given above, independent entities do not exist, but they can be registered or be known by us. We would therefore arrive at objective knowledge of something non-existent. However, this difficulty can be avoided by qualifying the rejection of the first part of the thesis: No independent entities exist, but subject-dependent ones do, and it

[9] Some interpretations of the *Critique of Pure Reason* and the *Prolegomena* suggest a special form of this position for Immanuel Kant: While the thing-in-itself must exist, since we have intuitions and these must always be intuitions of something for analytical reasons (otherwise this would lead to a contradiction), the thing-in-itself is completely unknowable. This is shown by the fact that no properties can be ascribed to it (apart from existence, which for Kant is not a real property), that it does not stand in knowable relations and so it cannot be apparent to us whether it is involved in any events, processes or structures. Cf. my Chap. 7, pp. 263–268, for realist and anti-realist interpretations of Kant in the context of my discussion of the ancestral argument.

is those that we can know.[10] The error theories already mentioned, which explain how the illusions of certain entities arise for us, are striking examples of such a position. The apparent contradiction outlined at the beginning of this paragraph can thus be avoided. These positions are susceptible, at least on a global level, to the fact that although almost everything can be conceived ontologically as subject-dependent, ultimately at least one subject with full ontological existence must be assumed, which is either independent or registers itself.

This last point can be generalised: Being a realist always with respect to certain entities means that it is possible to argue for or against realism regarding a certain single event (e.g. the moon landing) or a single certain object (e.g. the god Odin), a certain domain of entities (such as mathematical, scientific, mental, moral or fictional entities) or even all entities in general, by denying its or their independent existence or by doubting in principle the knowability of any real properties of it or them.

An obvious objection to the two-part definition of *realism* and *anti-realism* presented here is that the ontological thesis should actually only consist in the existence statement (*X* exists). Then, obviously, alleged realists and their opponents would agree in almost all areas, since (at least it is to be assumed) questions of existence are often unanimously answered superficially. The question of independence is therefore one of the real central points of the debate. From this it could be concluded that one deals essentially with an epistemological problem and that ontological questions are not affected, as Demmerling suggests in an overview of several debates.

From an examination of anti-realist arguments and positions, Demmerling draws the conclusion that they argue primarily epistemologically and incidentally in terms of philosophy of language, here mostly the theory of truth and meaning (cf. Demmerling 2004, pp. 40–41). This seems to provide a reason not to refer exclusively to the question of existence when it comes to justifying realism. I take this argument into account here, first by anchoring *independent existence* centrally in RO, and second by insisting on the epistemological thesis, contrary to a tendency to deal only with ontological questions.[11] These two linked aspects can be demonstrated in a series of related debates, as the following subchapters show.

[10] George Berkeley's so-called subjective idealism can be read along these lines. Berkeley can be interpreted in such a way that all objects are dependent on their "registration" in god's mind, but by god's grace these in turn appear to us as objectively given. Cf. Brock and Mares 2007, pp. 49–59. Such a reading is developed in more detail in: Winkler 1989.

[11] Cf. for a restriction of the definition of realism to an ontological thesis, e.g. Searle 1995, pp. 149–160 and also Brock and Mares 2007, pp. 2–3.

2.2 Local (Anti-)Realisms

Although philosophical realism/anti-realism debates about the independent existence and knowability of individual entities (e.g. the Christian god) can be ascertained, debates about larger domains of entities predominate. The ontological and epistemological levels are not always sharply separated. Several important debates can be identified in specific, quite clearly delineated domains, which are, however, often intertwined with each other through arguments. In these discourses, the respective positions can be characterised as a form of *local (anti-)realism*, in contrast to *global (anti-)realism*, which refers not to one specific but to all entity domains.[12]

One aspect of the local-global distinction is often overlooked: Between local (anti-)realisms and global (anti-)realism, discussions remain that are not limited to a clearly delineated local entity domain but do not concern all entities either—and thus, strictly speaking, nor do they concern the global level. I will call this intermediate class *metalocal (anti-)realism*. Additionally, I will provide an argument that the debate concerning scientific realism also belongs to this class, since it concerns several local disputes *in their connection*. This allows me to develop an understanding of the interconnectedness of multiple local issues that is not possible if, for example, the debate in philosophy of science is conceived of as solely local (e.g., concerning only unobservable entities postulated by scientific theories).

This does not mean that local debates are irrelevant in the context of the present work. After first giving some prominent examples of local debates in this subchapter, I will particularly focus on the debate about *temporal* and *modal realism*, which will later prove relevant for the philosophies of science of new realism.

But first I would like to start with a classic example of a local debate: In relation to mathematical entities such as sets, numbers and geometric shapes, positions of *(anti-)realism regarding mathematical entities* are advocated. I have two reasons for picking out this particular discussion. First, it is a very clear example of a local debate, as it only concerns the most fundamental elements of mathematics. The question of what they are (only sets, only mereological wholes, only numbers, ...) is again part of the debate. It thus provides a good archetype for a local debate. Second, it will become apparent through the course of this book that, from some perspectives, this local debate directly belongs to the metalocal debate in philosophy of science (e.g. in Quine's realist naturalism) and is clarified there, while other authors consider this to be a completely separate field, since their positions in either area have no implications for the other. This local debate is therefore a prime example of the fact that the different positions are not only to be distinguished in their theses, but also entail different understandings of the context of the debate at the local level.

[12] For lexicon articles and handbook chapters reflecting this, cf. Brock and Mares 2007; Loux 2003; Miller 2019; Tetens 2017b; Voosholz et al. 2024, 2025, esp. on the question if this pertains to problems in metaphysics, ontology, epistemology or philosophy of language.

In this discussion about the status of mathematical entities, realist positions such as different variants of mathematical Platonism, which understands (some) mathematical entities as abstract, real objects[13] and mathematical naturalism, which understands (some) mathematical entities as properties of concrete objects,[14] argue against anti-realist alternatives. The latter include constructivism and intuitionism, which regard mathematical entities as mental constructions,[15] structuralism, which considers structures to be the fundamental mathematical entities, although these do not necessarily occur independently of the subject[16] and nominalist formalism and instrumentalism, which regard simple mathematical entities as symbols for concrete objects and higher mathematics as a mere formal game.[17] A particularly strong anti-realist position that has developed from this last theory option is nominalism regarding abstract entities in general.[18] This attitude towards abstracta will be further examined in my later subchapter on *global and metalocal realisms*. Subsequently, fictionalism was developed in philosophy of mathematics, according to which mathematical entities have the same non-existent status as fictional entities from literature.[19]

This discussion is therefore clearly a local debate since, as foundational entities of mathematics, the entities in question—sets, numbers, merelogical wholes, geometric forms, etc.—form a relational structure and thus a definable, delimitable domain within ontology. They can thus be investigated by a science using its methods. Of course, numbers or geometric shapes also appear in other debates; no type of entity is isolated in this sense. This fact does not detract from the locality of the question of realism here. It can be answered for this domain only, without necessarily having decided on the status of all abstract objects, let alone all entities.

For different classes of abstract entities, as in philosophy of mathematics, a separate local research discussion is conducted, which is connected via the metalocal debate concerning all abstracta. The latter summarises all the local debates, but remains below the global level with respect to all entities. Therefore,

---

[13] Here the starting point of the most recent debate is the logicism of Gottlob Frege; cf. Frege 1953, 1988, 1998, 2013, whose ideas are experiencing a renaissance in: Wright 1983, Other main variants of Platonism are represented by: Balaguer 1998b; Gödel 1947, 1951; Putnam 1971, 1979; Quine 1953, 1969b; even though there are also interpretative reasons to include Quine and Putnam among naturalism, although some scholars argue against this, cf. for this Linsky and Zalta 1995.

[14] Cf. e.g. Maddy 1990, 1997.

[15] Cf. the classical starting point in Brouwer 1908, 1919, 1981, and more recent contributions such as: Moore 2019; Wright 1982.

[16] The starting point here is Benacerraf 1965, 1973, cf. the important work of; Carter 2008; Hellman 1989, 2001; Landry 2011; Resnik 1997; Shapiro 1997.

[17] In addition to preliminary work by H. E. Heine and Johannes Thomae, the classic source here is Hilbert 1990, 2015, cf. also: Azzouni 2004, 2009; Curry 1951; Field 1980; Gabbay 2010.

[18] The formalist-nominalist work Goodman and Quine 1947, can be counted as the beginning, particularly influential is: Field 1980, for an overview of the development of philosophy of mathematics afterwards cf. Burgess and Rosen 1997; Whittle 2021.

[19] Cf. Bueno 2009; Colyvan 2010; Field 1989; Leng 2010; Yablo 2005, 2012.

common arguments and symmetrically constituted positions emerge in all these local discussions.

Other local debates are not central to my present endeavour, but the implications of these issues about mathematics and time will become clear in several places. Among the other local debates, the most important are mental, moral and fictional realism: metaethical or *moral realism* argues in favour of the objective existence of certain values, good and evil, moral commandments or universal human rights. On the one hand, these appear to be historically and culturally variable, but on the other, a universal core of moral standards can be identified. *Realism regarding fictional entities* concerns characters, actions or places from literature and is based primarily on the realisation that sentences such as "Sherlock Holmes is admired for his deductive abilities by many members of the police force" seem intuitively true and meaningful. However, this seems to presuppose that Sherlock Holmes is known and recognised as having deductive abilities, which in turn presupposes his existence, which one would intuitively deny. *Mental realism* evidently concerns the existential status of, for example, thoughts, feelings, imaginations and intentions. Here again, several even more detailed individual discussions, e.g. on colours, emotions and intentions, can be distinguished, so that only a closer examination of the debates would shed light on whether it is a local or a metalocal realism debate in philosophy of mind.

2.2.1 Temporal (Anti-)Realism

Two further debates about local realisms are of interest in this context on account of their ontological and philosophical implications: the debate about the independent existence of time and the debate concerning modalities.

In the context of *realism about time, past, present and future*, the realist thesis refers to different sub-questions. The most fundamental question is whether time exists independently at all. The modern debate on this was initiated as early as 1908 by J. McTaggart Ellis McTaggart (1908), who advocates GENERAL TEMPORAL ANTI-REALISM: According to McTaggart, time is an illusion that owes itself exclusively to our subjective perspective, against which he offers a two-step argument whereby the so-called A-series of time (i.e. past, present and future) is logically necessary for the B-series (the ordering of events and points in time as *later than/earlier than*). However, according to McTaggart, the consequences of the A-series lead to logical contradictions. From this he concludes that time is unreal.[20]

This argumentative strand is taken up by ETERNALISM, although the generality of the conclusion is resisted. Eternalism takes physical theories seriuosly, most notably special and general relativity, and McTaggart's critical stance towards the A-

[20] Today this position is defended with partly different, often empirical arguments, e.g. by Baron et al. 2021, as a form of fictionalism.

series. From this, eternalism rejects both the order of past, present and future as part of reality and ascribes it to subjective perspectivity; eternalism also seeks to treat events and objects independently of their temporality, as merely temporally ordered along their world lines.[21] Eternalism can be represented in the form of a staunch realism regarding time as an order of events in the sense that it often considers all events and points in (space-)time to be equally real.[22] It is often chosen as a position when the further study of time is treated as an empirical question and placed mainly in the hands of physics. Eternalism can be reformulated in such a way that it assumes a non-growing block universe:[23] All events and objects are located somewhere in space and time (or spacetime), without it making any ontological difference where and when exactly. An important and much-discussed recent form of eternalism is Theodore Sider's four-dimensionalism, which sees concrete objects as composed of spatial and temporal parts.[24] As an alternative position closely related to eternalism, GRADUALISM lends itself if the non-existence and thus particular openness of the future is to be assumed as something ontologically robust, while still incorporating scepticism or subjectivism towards the A-series. A gradualist argues that our cosmos is a growing block universe: While past and present are equally real and form a block universe as described by general relativity theory, greater numbers of future events and objects are added to that block.[25]

The largely realistically understood position of PRESENTISM argues against general temporal anti-realism, gradualism and also eternalism.[26] It consists in the ideas that not only time, but (at least) the present also has objective reality, and that the present is importantly ontologically superior, and in some sense *more real* than the past and future.[27] According to most varieties of presentism present events and

---

[21] The *locus classicus* of an argument against the passing of time, which has given eternalism its still common form today, is: Smart 1949, cf. for the most important positions of eternalism that follow suit: Lewis 1986b,d; Oaklander 2004; Price 1996; Putnam 1975; Quine 1950; Smart 1963a, 2008.

[22] For a reformulation of eternalism, which trivially affirms the passage of time, see Mozersky 2013, esp. p. 181 more recent contributions, which in various respects speak out in favour of eternalism, gradualism or B-theories of time and against presentism include: Baron and Miller 2013; Miller 2018; Oaklander 2012; Skow 2009, 2011a,b, 2012a,b, 2015; Tooley 2012.

[23] Cf. on this and the following Stoneham 2009, here pp. 201–202.

[24] Cf. esp. Sider 1997, 1999, 2001, in contrast, there are attempts to transform four-dimensionalism into a presentist position: Brogaard 2000, as well as attempts to negotiate four-dimensionalism and the question of the persistence of concrete objects relatively independently of the realism question about time and the eternalism-presentism debate: Hawley 2001; Moss 2012.

[25] Two well-known formulations of this metaphysical position, combined with partly semantic justifications, can be found in Adams 1986; Tooley 1997, for a more recent variant cf.: Ismael 2012.

[26] For an overview of important publications in this debate, see Haslanger and Kurtz 2006; Le Poidevin 1998, also helpful for an overview on eternalism and presentism in general are: Dyke 2013; Maudlin 2007, here esp. chap. 4; Paul 2010; Miller 2013; Zimmerman 2005.

[27] Arthur N. Prior, the developer of temporal logic, must be mentioned as a pioneer of realist presentism: Prior 1957, 1967, 1968, from this especially "Changes in Events and Changes in

objects receive their reality, the reason for their existence, based on their present position in the dimension of time. While the past is usually assumed to be fixed and no longer real, the future is (at least in part) still open and does not yet exist, so that the present enjoys an ontological primacy and superiority. In most cases, presentism is even based on the hypothesis that only the present and therefore only present objects exist.[28] Consequently, many presentist positions can be reformulated in such a way that they represent a realism regarding the present, but an anti-realism regarding the past and future. Presentism is often identified as the position that accounts for the flow of time, the A-series and our everyday relation to time as the only theoretical valid option.

More recent positions, some of which, such as Quentin Meillassoux's (Meillassoux 2014), can be categorised as new realisms, also adopt a realist position in the debate about time. However, they differ in important aspects from the standard positions of realist presentism and realist eternalism, as will be shown later (cf. Longo 2015a). There are nonetheless also attempts within the debate to undermine the basic assumptions of general temporal anti-realism, eternalism, gradualism and presentism, which are often treated as exhausting the veriaty of options with regard to the ontology of time.[29] In addition, there is scepticism as to whether part of the debate is not too ignorant and distanced from the actual scientific debates about time on the one hand, while another part of the debate cites the (supposed) results of physics as arguments on the other.[30]

Things", pp. 1–14; "On Spurious Egocentricity", pp. 15–25; "Time, Existence, and Identity", pp. 78–87; "The Logic of Ending Time", pp. 98–11, 1972; 1996; in addition, George Schlesinger may also be considered a realist presentist: Schlesinger 1980, 1982; whereas Dummett 1978, esp. in "Bringing about the Past (1964)", pp. 333–350, "A Defence of McTaggart's Proof of the Unreality of Time (1960)", pp. 351-357, "The Reality of the Past (1969)", pp. 358–374, integrates his presentism into his general linguistic–philosophical anti-realism. In his more recent works on the philosophy of time, however, he no longer seems to advocate presentism, but instead leans towards a general temporal anti-realism. Cf. Dummett 2004, esp. pp. 73–92.

[28] The current leading figures of presentism include Markosian 1993, 1994, 2004, 2012; Cf. for other positions of presentism that exhibit major internal differences: Bourne 2006; Brogaard 2006; Chisholm 1990; Crisp 2007, 2008; Eker 2023; Fitch 1994; Hinchliff 1996, 2000, 2010; McKinnon 2013; Merricks 1995; Monton 2006; Sullivan 2012a,b; Zimmerman 1996, 1998, 2008, 2011.

[29] Cf. for a critique of the current debate between eternalism and presentism as failing in their common basic ontological convictions: Dolev 2007, esp. chap. 1; some authors consider the question whether the positions are not actually metaphysically or explanatorily similar; cf. on this Callender 2012 further criticisms can be found in: Dyke 2012; Savitt 2006.

[30] For a contribution that is strongly based on interpretations of quantum gravity, see Baron et al. 2021; criticism formulated Dolev 2007, pp. 25–36.

2.2.2 Modal (Anti-)Realism

Modal realism concerns itself with the questions of whether, in addition to the actual ones, possible objects, events and alternative states of affairs exist independently, whether the possible entities are knowable, and whether this also applies to necessities. Since David Lewis's *On the Plurality of Worlds* (1986), this debate seems to be inextricably linked to the question of the existential status of possible worlds.[31] Here, too, new realisms attempt to propose an alternative form of realism.

Lewis's realist analysis states that all modal expressions and many other, e.g. counterfactual statements can be translated into the semantics of possible worlds, so that their translated equivalents consist in simpler, non-modal, non-contrafactual propositions. In this way, modalities can be reduced to possible worlds. Lewis offers a consistently ontologically realist position: These possible worlds exist in an ontologically robust sense and thus serve as truthmakers for the propositions in question. Stuart Brock and Edwin Mares offer an insightful characterisation of Lewis's ontology, which at its core consists of five ontological and two conceptual theses. The ontological theses are:

> *Minimal realism.* There are other worlds and other individuals that occupy those worlds.
> *Mind-independence.* The other worlds and individuals are not of our own making. They do not depend on us, our minds, our linguistic practices or our conceptual schemes. They would exist no matter what we thought or said.
> *Parity* The other worlds are of the same kind of thing as our universe. There is an ontological parity between this world and the other worlds. The individuals that occupy the alternative worlds vary in kind, just as they do in this world.
> *Isolation* No single world has spatiotemporally disconnected parts. No two worlds are spatiotemporally continuous with one another. Two individuals are part of the same world if and only if there is some spatiotemporal relation that holds between them.
> *Plentitude* For any collection of individuals, or parts of individuals, from any number of worlds, there is a single world containing any number of duplicates of each, and in any arrangement (so long as the individuals are not too large, or too numerous, or shaped too awkwardly to fit). Brock and Mares 2007, p. 175.

With this ontology, which in itself has counterintuitive theses, Lewis can now offer an analysis of modality and related topics that is reductive and elegant. To do so, he only needs to introduce two further conceptual theses to explicate what he means by *possible world*, and how he reduces the first modality (the others then follow from the interplay of the theses):[32]

---

[31] Cf. Lewis 1986c, Although the term of course goes back to Leibniz and earlier contributions by Lewis and others have already used this terminology, since 1986 there has rarely been an attempt to provide philosophical information about modal facts without reference to possible worlds. For a systematic overview, see Divers 2002; on the connection to modal logic, see e.g. Barcan Marcus 1993.

[32] With this characterisation, Brock and Mares are close to other standard interpretations of Lewis, cf. almost all contributions in: Loewer and Schaffer 2015, as well as: Coggins 2003.

> *Possibility* The worlds are *possible* worlds. The worlds of which we are not a part are the other ways the world might have been; the *merely possible* worlds. All individuals that occupy those worlds are *counterparts* of things in other worlds.
> *Actuality*. This world is the *actual* world. The other worlds are non-actual worlds. But every world is actual at itself. "Actuality" is an indexical term that serves to pick out the world in which its utterance occurs. Brock and Mares 2007, p. 176.

Correspondingly, *possible* can be defined in such a way that a fact is possible precisely when it is the case (so is actual) in at least one possible world, and a fact is *necessary* precisely when it is the case in all possible worlds. Consequently, modalities *de re* can also be clarified: Hypatia is essentially a philosopher iff she is a philosopher in all possible worlds, i.e. iff all her *counterparts* are also philosophers.

The idea that there are other possible worlds that are causally and spatiotemporally completely isolated from our world, which are (at least for the most part) also concrete, i.e. spatiotemporally structured, has provoked significant objections, even if it may be advantageous from an explanatory point of view. Of the many counterarguments, that which is one based on Ockham's razor stands out:[33] In contrast to other possible explanations, Lewis's theory of innumerable possible worlds, all of which must be assumed separately, is characterised by its complete rejection of scientific and ontological parsimony. Lewis addresses this objection in several places in his work:

> I subscribe to the general view that qualitative parsimony is good in a philosophical or empirical hypothesis; but I recognize no presumption whatever in favour of quantitative parsimony. My realism about possible worlds is merely quantitatively, not qualitatively, unparsimonious. You believe in our actual world already. I ask you to believe in more things of that kind, not in things of some new kind. Lewis 1973, p. 87.

Lewis thus recognises that it would be problematic to have to accept countless qualitatively novel individual things as really existing, but believes that, since his possible worlds are by and large similar to our world, e.g. concrete, only quantitatively numerous *counterparts* must now be assumed. Even if it remains open whether quantitative parsimony is a theoretical value in ontologies, Lewis still has a problem: as Joseph Melia rightly points out, Lewis not only assumes many possible worlds, but also the existence of an uncounted number of individual things that populate these worlds (cf. Melia 1992). Many among them that are radically qualitatively novel, as he himself states: "[T]here are differences of kind between things that are parts of different worlds—one world has electrons and another has none, one has spirits and another has none" (Lewis 1986c, p. 2). According to him, there exist as many qualitatively novel individual things as possible. This raises the question of whether there are other forms of modal realism that have similar advantages to modal realism concerning concrete possible worlds and individual things.

[33] Cf. e.g. Forrest 1982.

A popular candidate is so-called Ersatz modal realism.[34] Although it is represented in various forms, all forms of Ersatz modal realism have one thing in common: They reject the *parity* thesis, according to which other possible worlds are of the same kind as ours. Ersatz modal realists do not agree on *plentitude* and *isolation*: both theses divide the field into advocates and opponents.[35]

As far as the conceptual theses are concerned, most Ersatz modal realists claim that there are no *counterparts* in Lewis's sense, but that the identity of individual things is possible across worlds (cf. Brock and Mares 2007, p. 181). As far as the thesis on the definition of actuality is concerned, Ersatz modal realists consider all worlds to be actual or real. However, only one possible world (as an abstract object) represents our universe as it is, while all others deviate more or less strongly from it: According to the Ersatz modal realists, only this possible world—as an abstract representation—is thus actualised.

Lewis and others have provided individual counter-arguments against various families of Ersatz modal realism. To pick just one against Armstrong as an example: If possible worlds are conceived as a set of properties or propositions, then it is necessary for the Ersatz modal realist to ensure that this set contains no contradictions. Consistency, however, is a modal concept—a set of properties or propositions is consistent if and only if they can all exist or be true at the same time. The notion of possibility is therefore needed to obtain that of consistency, so that this variant of Ersatz modal realism cannot reduce the use of modal language to the theory of possible worlds, but must assume primitive modal expressions, thereby losing a key advantage.

A realist alternative to the possible-worlds picture in modal realism is modalism as introduced by Forbes (1985). Here, a local anti-realism regarding possible worlds is combined with a fundamentally realist view of modalities.[36] Modalism essentially consists in the thesis that modal operators are logical primitives, i.e. that they are not further reducible. This does not mean that the talk of possible worlds or of possible individuals or facts cannot have a heuristic value, as with Leibniz; but, according to

---

[34] Cf. as the main representative of combinatorialism esp. Armstrong 1978a,b, 1986a,b, 1989, 1997, 2004a,b; other important contributions to Ersatz modal realism are Berto and Jago 2019; Melia 2008.

[35] A good reason to reject *plentitude* is that the world of all worlds is neither conceivable as a set nor as a mereological whole without contradiction. Cf. the argument in Forrest and Armstrong 1984. Lewis's attempt to escape the paradox by introducing an unknown upper limit for the number of elements as well as geometric and topological restrictions in Lewis 1986c, pp. 102–104, however, seems rather arbitrary. Herein lies a second strong counterargument to Lewis, while Lewis (1992) raises the possibility that he, for his part, can adopt Armstrong's own answer to the paradox, consisting of a restriction on the range of possibilities themselves, as can any other modal realist. This might exploint problematic anthropic assumptions, as Gómez-Torrente 2024 hints at.

[36] Cf. Forbes 1985, 1989, it should be noted that due to the self-designation of his theory as modal realism, which Lewis also chose for his proposal but later regretted, Forbes's theory is often referred to as modal anti-realism, although this is a misnomer under my definition of the term. This term generally only expresses that Forbes opposes Lewis and the Ersatz modal realists with respect to the existence of possible worlds, not possible or necessary entities.

Forbes, in these cases no assertions of existence can be made for these as special *possibilia*. One problem, however, is that in order to translate modal expressions into natural languages and possible-world semantics, a large number of logical primitives must be introduced as existent, which runs counter to Forbes's motivation (cf. Brock and Mares 2007, pp. 190–192). Modalism is thus an instrumentalism with regard to possible worlds and possibilia: Forbes proposes to start from modal facts in the world instead of possible worlds.

An anti-realist alternative, which reduces modal expressions to possible worlds but treats them as decidedly not real, is Gideon Rosen's so-called prefix fictionalism (cf. Rosen 1990, 1993, 1995, 2002): Modal expressions can thus be translated into statements about possible worlds, and are capable of truth, but should not be taken literally, just as fictional speech needs to be interpreted. According to Rosen, the idea behind this is that the sentence "A detective lives at 221B Baker Street" is true if an implicit operator ("... in the stories of Arthur Conan Doyle") is always added. For Rosen, a parallel fictionalism can be introduced for any modal expression (cf. Rosen 1990, pp. 332–333). For Lewis, modal expressions are true insofar as they can be translated into statements about possible worlds and supplemented by "... under the assumption of the hypothesis of many possible worlds". According to Rosen, our modal statements thus correspond most closely to fictional, hypothetical stories with alternative courses of action and not to processes in causally isolated or abstract worlds. One difficulty with this approach is that many (if not all) of the arguments against realism also apply here.[37]

What motivates modalism, prefix fictionalism and related positions?[38] Some general arguments speak against realism about possible worlds, be it Lewisian, Ersatz modal realism or otherwise. It should be noted here that these arguments are not directed against modal realism *per se*, insofar as it is considered possible to offer a realist conception of modalities that does not make use of possible worlds as a background theory. The following are the two most compelling arguments in my view:

1. Saul Kripke argues against Lewis that the latter uses a theory of *counterparts* that is not only counterintuitive but even incomprehensible, thus vehemently opposing the *possibility* thesis:

> The counterpart of something in another possible world is never identical with the thing itself. Thus, if we say "Humphrey might have won the election ..." we are not talking about something that might have happened to Humphrey but to someone else, a "counterpart." Probably, however, Humphrey could not care less whether someone else, no matter how much resembling him, would have been victorious in another possible world. Thus, Lewis' view seems to me even more bizarre than the usual notions of transworld identity that it replaces. Kripke 1980, p. 45.

[37] According to Rosen, a further problem of hidden necessity arises of its own accord. Cf. on this Rosen 1993.

[38] An anti-realism that attempts to resolve the difficulties of both Forbes and Rosen can be found in: Chihara 2001, esp. Chap. 4, Modal Anti-Realism.

It is therefore a problem of the attitude we have towards modalities, especially that of possibility. For our concept of possibility, it seems relevant that the possible concerns our own actions and thoughts: we have possibilities and must make appropriate decisions. Lewis now explains this by the fact that our *counterparts* realise other possibilities than we do and make other decisions.[39] However, we have a completely different attitude towards the options and decisions of our *counterparts* that are causally completely isolated from us, even if they resemble us in many respects, than towards our own options and decisions. In this respect, Lewis's theory seems too focussed on reducing the language use of modal expressions to seemingly clearer non-modal expressions about facts in other possible worlds. He is not taking into account that a blatant attitudinal difference occurs.[40] This argument can also be applied to other theories of possible worlds that take them to be real: The attitudinal difference also occurs between abstract possible worlds in Ersatz modal realism.[41]

2. Another, ever more serious problem is the *epistemic counterargument*, as presented here by Brock and Mares:

> For the [possible world; J.V.] realist of any stripe, possible worlds and individuals are causally impotent with respect to individuals in the actual (concrete) world, perhaps because they are spatiotemporally isolated from us, or perhaps because they are abstract objects occupying no region of spacetime. But, runs the objection, it is a necessary condition of someone's having knowledge about a certain domain that there be a causal connection of an appropriate kind between the knower and that which is known. The realist violates this condition, at least when it comes to our knowledge of the merely possible truths. We might express the worry as follows. [...]
>
> (P1) We have abundant knowledge of the modal facts (including knowledge of mere possibilities).
> (P2) We do not have knowledge of other worlds and merely possible individuals.
> (C) Therefore, modal facts are not reducible to facts about other worlds and individuals.
> Brock and Mares 2007, p. 186.

This argument can only be rejected by realists about possible worlds if (P2) is refuted, since (P1) is to be defended. Here, however, the burden of argument rests on the side of the realist about possible worlds, since common empirical knowledge fails as a candidate both for causally and spatiotemporally completely isolated concrete worlds and for abstract worlds. Here we find parallels to the local realisms in the philosophy of mathematics and metaethics, which also have to struggle with epistemic arguments regarding their abstract entities.

New and speculative realists now often have their own reasons, such as in the cases of Gabriel, Boghossian and Meillassoux, for advocating a modal realism

[39] Deviating from the slight misrepresentation above by Kripke, who claims that Lewis explains this by what *would have* happened to Humphrey's *counterpart* and not by what *has* happened to this counterpart, which is a relevant deviation from Lewis's conception. If one removes this misinterpretation from the objection, as I do in my reconstruction, Kripke's argument retains its force.

[40] For a deeper analysis of this argument: De 2018.

[41] This insight goes back to: Lewis 1986c, p. 196.

without positing possible worlds. In doing so, they are aware of the state of the discussion described above, ensuring that they do not also run into the problems raised by the two counterarguments previously mentioned.[42]

* * *

There are other related discussions about *various other local (anti-)realisms*, such as (anti-)realism with regard to perceptual content, and in principle it would be possible to introduce any number of local realisms. However, this concludes the overview of the local realisms relevant to my current project. Before I embark on my qualification of global versus metalocal realisms and my consideration of metaphysical and scientific realism respectively, it should already be anticipated that the diagnoses of the relevance of various local realisms for new realists such as Meillassoux or Gabriel will be disparate: Meillassoux's ontological project represents a combination of local realisms concerning concrete objects, time (including past, present and future), and modalities as well as (some) mathematical entities, and, thus far, Meillassoux has only indirectly committed himself to a scientific realism, as I will go on to elaborate. Gabriel's ontological project attempts to represent *almost* a global realism: With respect to all the local realisms mentioned above, he advocates at the least the unrestricted existence of the respective entities, even if not independence to the same degree and in the same sense for every case.[43] In the debates that are particularly relevant here in philosophy of science, philosophy of mathematics, philosophy of time and concerning modalities,[44] however, he explicitly advocates a position of registration-independence, i.e. independence from any conceivable registering system. I will argue in favour of these findings in Part II in close discussion with the negative core insights of new realism and with the aim of a broader outlook on its prospects in philosophy of science.

2.3 Global and Metalocal (Anti-)Realisms

In addition to the discussion about metaphysical realism, which I deal with in the next subchapter, at least five divergent debates can be distinguished historically and systematically within the framework of *global and metalocal (anti-)realism*, in which the respective realism is confronted with a counter-thesis: The first debate concerns the independent existence of general concepts or universals, where the antagonist is classical nominalism. Since the existence of concrete objects is not in question here, it is not a global realism debate in the strict sense, but as it does not

[42] Related reservations from the debate and corresponding projects are not only pursued by new realists, however. Cf. e.g. Krebs 2019; McDaniel 2004; Roy 1993.

[43] I defer here the important question of why Gabriel replaces subject-independence with registration-independence. I will return to this in Part II of this book in the context of philosophy of science.

[44] However, with the decided restrictions that these are not obtained via possible worlds, cf. Gabriel 2015a, p. 207 and pp. 263–368.

only concern a single entity domain, it is not strictly speaking a local realism either. This debate thus falls into the intermediate class of *metalocal (anti-)realisms*.

This first, longstanding debate about universals must be clearly separated from the above-mentioned meta-discussion of ontology, metaphysics, philosophy of mathematics, science, morality and mind, which revolves around abstracta generally. This meta-discussion is the second of the five discussions that can be identified. Here, a clear distinction is made between classical nominalism (CN) and abstracta-nominalism (hereafter AN), yet both research debates concern metalocal realisms. This also applies to the third debate concerning (anti-)realism in the philosophy of language: here, the main focus is on realist and anti-realist theories of truth and meaning, with which various of the local realism debates can be bundled.[45] For this reason, this concerns also a metalocal realism.

This is different for the remaining two debates, which are closely related to each other. The fourth debate concerns the distinction between global realism and global idealism—both are extreme positions—in which independence is called completely into question. Put simply, global realism assumes that all existing entities must also be independent, whereas in global idealism existing entities are always understood as dependent on an instance of the subjective or the ideal. In the fifth debate, global realism can be conceived as the antithesis of global scepticism. There is clear evidence of such a separate debate: Here, the epistemological part of the realism thesis is radically under attack, in that the global sceptic attempts to question all knowledge of reality, whereas the global realist holds the radical counter-thesis that all, or (less radically) that at least many, existing entities are minimally in principle knowable and thus fall within the realm of possible knowledge. I will now turn to these five debates in turn.

The debate between CN and realism regarding the existence or non-existence of general concepts or universals has one of its roots in the ancient philosophy of Plato.[46] The terms *universal* and *general concept* have been used differently over the centuries. But a core feature of both, according to realists, is that different entities can fall under a general term (all concrete tables, for example, fall under the general term of artefact) and that different entities likewise exemplify a universal (all individual hares exemplify the mammal family of *Leporidae*). Classical nominalists, as opponents of realism in this debate, are by no means global ontological anti-realists and do not doubt the existence of all entities or objects, but they limit themselves to the non-existence of universals and only consider individual objects, processes, events or relations to exist independently, which can be either concrete or abstract. As the references to universals extend across all subject domains, this discussion is structured differently to local debates and shows parallels to the debate in the philosophy of language.

The CN-realism debate reached a climax in the context of medieval scholasticism, in which Petrus Abaelardus, William of Ockham and others argued in

---

[45] Cf. my Sect. 3.1.1, pp. 42–46.

[46] For an overview see: Amerini and Cesalli 2017; Klima 2017.

favour of CN with a series of arguments. While Abaelardus characterises universals as man-made concepts, Ockham, for example, argues that no new, unobservable entities such as universals should be introduced to explain observations if this could also be done by individual objects.[47] This principle is known as Ockham's razor. In a radical form, John Scotus Eriugena and, more moderately, Thomas Aquinas, John Duns Scotus and others on the side of realism argued that universals must exist, for example by claiming that without universals the human mind would never be able to grasp the nature of a single object.[48]

In modern days, the universals debate is continued in a modified form in so-called analytic metaphysics and philosophy of language, for example by David Lewis, Keith Campbell, Gonzalo Rodriguez-Pereyra and Douglas Ehring for nominalism against David M. Armstrong, John Bigelow and D. H. Mellor for universal realism.[49]

The second, closely related debate is not about the (non-)existence of universals and general concepts, but about *abstract entities*. Abstract entities, or abstracta, are not localised in spacetime.[50] Examples of abstracta are mathematical objects (numbers, sets, etc.), logical objects (modus ponens, existential quantifiers, etc.), fictional objects (literary figures and places, etc.), and can include propositions, properties, types, values, thoughts or laws of nature. The discussion about these objects is thus clearly the meta-debate for all the local realism/anti-realism discussions about these entity domains, but raises the question in its most general form and thus touches on every local manifestation. The core of this debate also goes back to Plato's theory of ideas, which is realist in the sense that it argues in favour of the existence of abstracta. In contrast to CN, any form of AN claims that only concrete entities exist independently and denies this status to all abstracta (including individual abstracta): AN always denies their existence or their independence generally.

A frequently used argument against the existence of abstracta is that they are in principle unrecognisable, imperceptible and unknowable due to their apparent lack of causal power (cf. Benacerraf 1973). In contrast, the abstracta realist would have to assume a special kind of abstract intuition,[51] against which the AN often makes use of Ockham's razor: If all phenomena that are attributed to abstracta in abstracta realism can also be explained from the AN position, then the assumption of

---

[47] Cf. on the last point e.g. Keele 2010; Michon 1994; Panaccio 2004, 2015; Pelletier 2013; Spade 1999.

[48] Cf. e.g. Beierwaltes 1994; Carabine 2000; Cross 1999; Dewan 2006; Goris 2012; Ingham and Dreyer 2004; King 2002; Moulin 2016; O'Callaghan 2003; Ward 2014; Wood 2011.

[49] Cf. on CN: Campbell 1990; Ehring 2011; Lewis 1986a; Rodriguez-Pereyra 2002, whereby the first two contributions, as trope ontologies, represent both an abstracta and a universal nominalism and thus also position themselves in the second debate presented below. On realism about universals: Armstrong 1978a,b, 1997; Bigelow 1986; Bigelow and Pargetter 1989; Mellor 1997, 2012.

[50] For an alternative definition of the difference between abstracta and concrete objects: Burgess and Rosen 1997, pp. 13–25.

[51] Such an intuition offers Gödel 1947.

such a mysterious intuition could be avoided. The realist counterarguments mostly focus on a certain type of abstracta and often show either that AN cannot explain all phenomena,[52] that the enigmatic abstract intuition can be replaced by a clearer concept[53] or can attempt to bind certain abstracta to concrete entities.[54]

Both nominalist positions, CN and AN, are independently defensible. Armstrong, for example, argues for the existence of universals, which he also considers to be spatiotemporally localised and rejects abstracta outright (cf. Armstrong 1978a,b, 1997). Quine, on the other hand, advocates—at least in some of his writings (cf. Quine 1953, 1981b)—mathematical Platonism and thus the existence of abstracta, but at the same time rejects the existence of universals.

The *(anti-)realism in philosophy of language* is closely linked to the question of theories of truth and meaning.[55] In a sense, this metalocal debate and especially Michael Dummett's work can be seen as the starting point of the modern overall debate, as the term *anti-realism* owes its creation to him. In his article *Realism* and other works that he later published in the collection *Truth and Other Enigmas*, Dummett attempts to uncover a common core of all realism debates (cf. Dummett 1978, mainly "Realism (1963)", pp. 145–165, here esp. pp. 145–149): In this article, Dummett asks what the debates in philosophy of science, philosophy of mathematics, philosophy of mind, etc. have in common and whether the individual positions that speak out against the realist option do not also have more substantial things in common than the mere rejection of their respective realism. His idea is that in every single debate the antagonism concerns a class of statements, and the realist claims that they are true or false, regardless of whether we can ever correctly determine their truth or falsity. Thus, realists would always advocate the principle of bivalence: propositions are always either true or false. According to Dummett, the realist leans towards the correspondence theory of truth, according to which truth consists in the correspondence of a statement with the facts, or towards other strong

---

52 So-called *indispensability* arguments can be found e.g. in: Quine 1951, 1981b; Resnik 1985, 1997, this last article also presents a structuralist argument, as in: Shapiro 1997.

53 In Balaguer 1998a, he argues, for example, with regard to Platonism, that for all existing abstract entities, it is replied to this epistemological argument that this problem can be avoided if it is assumed that every consistent mathematical or semantic system exists, since consistency proofs are not mysterious.

54 For example, Penelope Maddy, in her early work, attempts to argue for a naturalism of mathematical objects, especially sets, that conceives of them as properties of concrete objects: Maddy 1990.

55 A good overview of this debate is provided by: Hale et al. 2017, esp. in the contributions: *Michael Morris*, "Metaphysics, Philosophy, and the Philosophy of Language", pp. 3–26; *John Skorupski*, "Meaning, Use, Verification", pp. 73–99; *Bob Hale*, "Realism and its Oppositions", pp. 493–531; *Ralph C. S. Walker*, "Theories of Truth", pp. 532–555; *Patrick Shirreff/Brian Weatherson*, "Relativism", pp. 787–803; *Bob Hale/Crispin Wright*, "Putnam's Model-Theoretic Argument against Metaphysical Realism", pp. 703–733, classic contributions to this debate are: Davidson 1984, 2001, the two aforementioned volumes contain many of Davidson's central articles from the 1960s to the 1990s; Devitt 1983, 1991; Haack 1976; Putnam 1978, 1981, 1994b; Quine 1960; Wright 1992, 1993.

theories of truth and meaning. In contrast, he sees the anti-realist as holding that the truth of a statement consists merely in *warranted assertability* (cf. Dummett 1976, pp. 85–88) or another, weak criterion of truth.

Consequently, contrary to many of his critics, Dummett considers the debate about the theory of meaning and truth in philosophy of language to be of central importance. Nonetheless, this and related positions are a metalocal (anti-)realism and not, as Dummett suggests, a global debate: Dummett reliably attempts to subsume all (or at least almost all) local debates. However, some of the local disputes convincingly resist integration by means of the question of theories of truth and meaning: The debate about temporality, as already explained above, can hardly be reduced to the semantic level. The debate on metaphysical realism can only be reduced to it if, like Putnam, one is prepared to solely focus on the semantic thesis in the definition of this position, as will be shown in my next section.

Another, even more important reason against identifying the discussion of the theory of meaning and truth with a global debate, and which Dummett himself recognises, is the fact that the question of realism or anti-realism can be negotiated piecemeal, local entity domain by local entity domain. Thus, it is relatively easy to advocate the correspondence theory of truth in relation to the physical external world and scientific entities (i.e. to be a realist according to the principle of bivalence-definition) and to reject it for the domains of moral values and fictional objects (i.e. to insist on anti-realism in the same sense). Dummett's attempt at a method of integrating all debates therefore provides the blueprint for my proposed characterisation of metalocal (anti-)realism.[56] In my next Sect. 3.1, I will explore the question of whether Dummett's idea of a single core for all debates is convincing and whether an alternative to philosophy of language could be identified.

The *debate about global realism and global idealism*, which also goes back to antiquity but which was only conducted in its most radical forms in modern times, is closely linked to all these debates about truth, CN and AN. GLOBAL IDEALISM can be characterised as the position that questions independence in general by assuming that everything that exists can only exist because it is in principle intelligible.[57] If strict independence is impossible, then *any* ontological realism fails, which is why global idealism—conceived of in this way—is tantamount to global anti-realism.

---

[56] Apart from metaphysical and scientific (anti-)realism as my examples for metalocal debates (see the next subchapter) and this exemplar of the one in philosophy of language, one could argue that the discussion about deflationism in contrast with substantive metaphysics could also be phrased in this manner, even though this possibly shifts with the stance that is taken in that debate itself (cf. Hirsch 2010; Schaffer 2017; Sider 2011; Thomasson 2007, 2015, 2020b; Thompson 2023). For my present purpose, however, this question can be ignored.

[57] However, already this characterisation is highly controversial, I am borrowing a formulation here from Brock and Mares 2007, p. 48. The idealism debate is characterised by the fact that many of its participants introduce the opposing positions in a way that often leads to straw man accusations. This debate is therefore only outlined very briefly here. I would like to thank Jens Rometsch for pointing out that actual idealists would rather claim that everything is intelligible, but that this does not necessarily entail the thesis that all entities owe their existence to the presupposed possibility of being intelligible, recognisable or knowable.

Thus, this debate is distinct from those described above because it is indeed global: here the second part of the thesis of ontological realism, that of independence, is fundamentally at issue.

The frequently occurring qualification of *global idealism* into specific forms comes in several guises. George Berkeley's so-called subjective idealism is frequently associated with this form of global anti-realism in some readings (cf. Priest 1995, pp. 65–77; Winkler 1989; Brock and Mares 2007, pp. 49–59), just as certain interpretations suggest this for Immanuel Kant's so-called transcendental idealism[58] and Georg Wilhelm Friedrich Hegel's so-called absolute idealism.[59] First, I find it questionable which of these or other authors really represented global idealism in the sense described above; all of the three positions mentioned are completely different. In addition, it is highly doubtful whether the interpretations of these three philosophers as ontological global anti-realists are fair, and also whether the fundamental opposition of realism and idealism outlined here is credible at all, which becomes apparent in the context of the second global debate.

A similar problem characterises the *debate about* GLOBAL SCEPTICISM: Representatives of this position doubt that reality can be recognised, i.e. that knowledge about it is possible.[60] An opposition to realism is often established here, which professes the fundamental knowability of reality in the epistemological realism criterion. Global scepticism is therefore the most radical form of epistemological anti-realism.

Even if we leave aside the question of whether anyone has ever advocated global epistemological anti-realism in the sense of global scepticism, this characterisation alone presents a problem: The global realist who, in maximal opposition to global scepticism, claims that *everything* is knowable in principle, represents the position of global idealism in the other instance of the global debate.[61] This is hardly surprising, since it is impossible to characterise two truly *global* debates that are not necessarily related to one another. Between global scepticism and global idealism, then, realism exists only as an intermediary position, asserting the existence of a reality that is partly knowable and partly registration-independent, with an overlap between these two parts.[62]

---

[58] Cf. Chiba 2012; Strawson 1966; Willaschek 1997, argued against by e.g.: Langton 1998.

[59] Cf. Bertram 2017; Pippin 1989, 2018, argued against by e.g.: Emundts and Horstmann 2002; Fulda 2006.

[60] Cf. for an overview Gabriel 2008a; Heil 1998; Perler 2009; Zalabardo 2012.

[61] Rometsch 2018, convincingly argues in pp. 228–244 for the fact that René Descartes already submits a more nuanced response to the problem of global scepticism.

[62] Arguing that this was already the constellation of the debate in ancient philosophy: Gabriel 2009, esp. §8.

2.4 Metaphysical and Scientific (Anti-)Realism

In what follows, I will discuss two debates that represent problem cases. Metaphysical realism is often presented as global realism. After setting out the classical view of metaphysical realism, I propose an argument for why it is not a global realism at all, but a metalocal realism. One consequence of this is that metaphysical realism contains more assumptions than are necessary for a convincing realism. This motivates the development of my concept of objective realism.

I then address the second problem case: philosophical realism in the sciences or scientific realism is often defined as the assumption that for unobservable theoretical entities of natural sciences both theses of realism—the ontological and the epistemological—are affirmed: quantum particles, prehistoric ecosystems, genes or the Big Bang thus exist independently and we can gain knowledge about them—according to realism. I argue below that this only amounts to a truncated view. The status of unobservable theoretical entities gives us an important local question, but it is linked to other questions in philosophy of science, which also turn out to be vital for the issue of scientific (anti-)realism. Thus, I argue, one should regard scientific realism also as a metalocal realism.

2.4.1 Metaphysical (Anti-)Realism

In METAPHYSICAL REALISM proponents argue firstly for the independent existence of a world (or even several worlds),[63] especially a physical and causal external world, secondly for the presence of one complete and true description of this world, and thirdly for our linguistic expressibility of parts of this description in the case of true judgement and knowledge. Hilary Putnam characterises metaphysical realism in what has become a classical point of reference as follows:

> One of these perspectives is the perspective of metaphysical realism. On this perspective, the world consists of some fixed totality of mind-independent objects. There is exactly one true and complete description of "the way the world is." Truth involves some sort of correspondence relation between words or thought-signs and external things and sets of things. I shall call this perspective the *externalist perspective*, because its favourite point of view is a God's Eye point of view. Putnam 1981, p. 49.

[63] On the one hand, David Lewis, as shown above, argues that the causally isolated but independent existence of all possible worlds is given. Cf. Lewis 1986c, esp. pp. 1–5. On the other hand, the Everett interpretation of quantum theory, according to which different worlds exist in parallel at the same spacetime points, and the various subsequent multiverse hypotheses, according to which our universe is only one among many, lead some representatives of certain *metaphysics of science* to claim that there are independently existing external worlds. For an overview see: Vaidman 2018; Wallace 2012.

Putnam's definition thus corresponds to the definition I share.[64] Due to the immense influence that Putnam and his interlocutors exerted on the debate, using their take as the starting point for my consideration is an obvious choice. Of course, there are also problems with Putnam's definitional approach, which I will address later in connection with my concept of objective realism.[65] There are two slight differences between my working definition and Putnam's version:

Firstly, Putnam depicts metaphysical realists as inferring from a possible, complete, and true description of the world that we could take a *God's Eye point of view*. I have formulated my definition somewhat more defensively by making it clear that we can only ever express parts of the overall description. But Putnam does not necessarily have to assume that, according to the metaphysical realist, we can grasp the complete description and thus adopt the *God's Eye point of view* at will. He clearly exposes a weakness in the position in this characterisation: On the one hand, the principled knowability of a complete description requires that the entire world is completely intelligible, which would introduce a certain idealist flavour that, in my view, metaphysical realism does not necessarily have in each of its instances.[66] On the other hand, Putnam's arguments against a *God's Eye point of view* are of course rather renowned.[67]

Secondly, Putnam requires that the metaphysical realist advocates for a correspondence theory of truth. This is an obvious way in which the third criterion I give above can be satisfied: In the case of veridical perception and cognition, our statements correspond not only possibly but actually to a part of the true and complete description of the world. Consequently, there is a form of correspondence. Of course, this does not determine how this overlap is understood in detail. For in addition to the usual correspondence theory of truth, which cites the correspondence of a proposition with the objective matters of fact as the criterion, a deflationist view of truth, for example, would also fulfil Putnam's requirements. According to deflationism about truth, "Snow is white" is disquotationally true if and only if snow is white. In order to take the place of the correspondence theory in metaphysical realism, the disquotational theory would have to conceive of facts as metaphysical facts, but would otherwise be free to detach itself from the idea of correspondence.

[64] Cf. for an excellent analysis of Putnam's changing treatment of metaphysical or external realism Button 2013, in which he shows why we should not take this position.

[65] Cf. my Sect. 3.2, p. 58.

[66] Cf. for a related view of metaphysical realism as the thesis that the world "potentially diverges completely from our beliefs about the world": Gabriel 2008a, p. 25, my translation; which habours the seed of scepticism.

[67] Elsewhere Putnam is fairer and argues thus: "[Metaphysical realism] traditionally included the idea that there is a definite totality of all objects [...] and a definite totality of all "properties." [...] It follows, on this picture, that there is a definite totality of all possible knowledge claims, likewise fixed once and for all independently of language users or thinkers. The nature of the language users or the thinkers can determine which of the possible knowledge claims they are able to think or verbalize, but not what the possible knowledge claims are." In: Putnam 1994a, p. 466.

At first glance, it is unclear whether metaphysical realism should be categorised as local, metalocal or global. On the one hand, metaphysical realists do not necessarily have to take a realist position with regard to moral, fictional or mathematical entities, which is a good indication that we should not speak of global realism here. On the other hand, many forms of metaphysical realism assume that the independently existing parts of the world—e.g. matter and energy distributed in spacetime, forming complex systems and undergoing processes governed by laws of nature—have given rise to humans. And in turn, we humans, as a biological species, brought moral, fictional and mathematical entities into existence through our cultural history. Such socially and culturally constructed entities would be inscribed in a relationship of historical dependence on our own existence. In this respect, metaphysical realisms often claim to clarify the question of the existence (or non-existence) of all conceivable entities. This is why metaphysical realism is often equated with global realism. From a comparative perspective, however, this categorisation must be resisted, as metaphysical realism can be wonderfully combined with mental, moral or any local anti-realism and need not fall neatly between scepticism and idealism.

The question of the metaphilosophical localisation of metaphysical realism is further complicated by Gabriel's ontological position. As described above, he apparently represents a global realism through cumulative local realisms, but on the other hand is fundamentally critical of metaphysics. According to Gabriel, the world does not exist because indefinitely many fields of entities would have to be conflated into a unified whole. Since such a totalisation is not possible, it cannot be equated with the world or the *external* world (cf. Gabriel 2015a, pp. 187–207). All fields of entities therefore do not form a whole that could be equated with the world and certainly not with the external world. In rejecting the world, Gabriel thus declares himself a metaphysical anti-realist in ontology, if one understands metaphysics as the subdiscipline that deals with the philosophical description of the world or all there is. However, he is far from arguing for a global anti-realism.

However one wants to mediate between Gabriel's position and metaphysical realism, this is not the only worry: I have shown that metaphysical realism does not equate with global realism, but neither does a metalocal realism seem to be clearly present. This is because metaphysical realism bundles certain local discourses, but in a notoriously unreliable fashion: One metaphysical realism may argue only for concrete, directly perceptible objects and the mind-independent external world, thus ignoring almost all local realism debates; the next treats concrete objects only as entities on the level of scientific theories, but justifies mainly logical objects and quantifiers as the cornerstones of the external world as a mereological fusion.

Does metaphysical (anti-)realism therefore represent a special case that does not fit into the local/metalocal/global realism distinction? This question must be answered in the negative—it is possible to categorise it. Firstly, it could be argued that Gabriel's position is not a global realism, since almost all entities exist according to him, but not *the* decisive entity of metaphysical realism, the external world, which convenes all other entities. However, this does not resolve the first contradiction. It is easy to imagine being a metaphysical realist and a realist in the philosophy of mathematics, science, morality and mind, but not in relation to fictional objects, e.g. literary figures such as Bilbo Baggins. In that case, there would clearly also be no global realism in the sense of the previously discussed objection, but it seems nonsensical to speak of a local realism, since the external world here comprises quite disparate entities such as numbers, electromagnetic fields, the prohibition of killing children and the feeling of loneliness. Even when asked about Bilbo Baggins, one can answer from this position that Bilbo Baggins does not exist in the external world, but is only variously imagined by hearing, seeing or reading *The Lord of the Rings*. In short: metaphysical realism falls outside of the categories of local and global realism.

According to the definition given above, however, the existence of at least one world, in particular a physical and causal external world, must be represented in every metaphysical realism. This means that it is always a metalocal realism, since a group of local realisms must always be claimed for the validity of the metaphysical thesis: It is not sufficient to take only one single form of concrete entities of the external world, be they everyday objects or entities of the natural sciences, as independently existent. The objects of the natural sciences do not in themselves constitute a world. By this I do *not* mean that in metaphysical realism several forms of *concrete* entities must always be assumed—the objects of physics are already sufficient as concrete entities. Rather, these do not yet form a world, a unified whole. Unification of a whole must always be achieved by means of an *integration method*. For example, the concrete entities of the external world can be understood as forming a logical (mereological, set-theoretical, etc.) whole, which requires abstracta. Another option is that the concrete objects of physics form a whole of an (external) world, insofar as we (or idealised observers) can grasp these objects in a comprehensive description (cf. Dyke 2023). For this, however, there must be (idealised) subjects that cannot themselves belong in every respect to the known types of concrete entities. In all cases, metaphysical realists must necessarily take a realist position in several local debates, thus metaphysical realism is always metalocal.

One objection that could be made against this argument can be expressed with the example of an eliminative metaphysical realism.[68] Someone holding this position would argue that there are only the concrete, fundamental objects (either strings or particles of the Standard Model of quantum theory) and supervenient properties of systems composed of these objects, with the external world being the maximally largest of such systems. This would be a metaphysical realism that represents only one local realism and not several. The problem here is that there is no reason to assume that there is a complete description of this overall system, because this would also have to arise from the concrete fundamental objects as a supervening system. It seems to me questionable what an argument for this could look like, or even an explication of what a description is if not a linguistic, mental or abstract object. Moreover, I will later show that this position leads to self-contradictions or

[68] I thank Jens Rometsch for pointing out this objection to me.

to the impossibility of formulating one's own position under the assumption of its truth. Thus, the objection made by the eliminative metaphysical realist is based on a hidden contradiction, since nothing more is done than to use this position as a counterexample.[69]

The counterargument outlined above, which denies the categorisation of metaphysical realism as metalocal, disappears when it is understood that not every metaphysical realism must represent *the same* local realisms; yet one nevertheless deals with a metalocal realism in every single case. My argument shows that this is indeed so. The in-depth analysis of metaphysical realism in its various forms could demonstrate this judgement by breaking down examples, but this falls outside the scope of my present investigation.

2.4.2 Scientific (Anti-)Realism

Similar to metaphysical realism, *scientific realism* is controversial in terms of definition, yet the various versions are even more heterogeneous. The corresponding position often refers to unobservable entities referred to in theories of the natural sciences such as physical fields, elementary particles, species or the Big Bang.[70] THEORETICAL ENTITIES are entities the existence of which requires a scientific theory to be true. *Unobservable* means that there is no direct method to perceive them.[71] This definition appears to indicate a local realism. It could be extended to include unobservable theoretical entities in the humanities and social sciences, such as market forces, intertextual dependencies, diplomatic relations or the behavioural preferences of groups of people. The question would still be a local one: Do these unobservable theoretical entities exist independently of postulating the theory or not? Do the scientific ontologies that are assumed for the theories have only a functional, instrumental character or do they deal with real entities, i.e. do the corresponding variables denote something? This is an important part of the question of (anti-)realism in philosophy of science, but not the only one.

The overall debate in philosophy of science concerns not only (allegedly) unobservable theoretical entities, but at least all theoretical entities, including (allegedly) observable ones: Biological sex, genders, genes, cells, species, the planets, tectonic plates, the sea level, casualty figures in wars, human reactions, such

[69] Cf. in my Sect. 4.1 the sections on eliminative materialism, esp. p. 77.

[70] To give just one example of this definition: Psillos 1999, p. xv.

[71] Part of the debate concerns where exactly the boundary between observable and unobservable lies: Does "observable" mean perceivable with bare, healthy sensory organs, magnifying glasses, amplifiers, light microscopes, electron scanning microscopes, etc.? In the present context, it is sufficient to assume that both predicates are vague and yet understandable. Perceptual aids for observations are permissible, but they become more and more theory-laden with increasing complexity, i.e. the observations made are only reliable and veridical if certain scientific theories are true.

as fear, and many more entities besides also play a role in scientific theories.[72] The ontological or epistemological question of realism is raised for all of these. This also concerns the extend in which ordinary objects appear in scientific theories—the trees outside my window appear in everyday references as well as in a possible urban ecology study.

However, the ontological question of realism does not end with the aforementioned entities, which can possibly be interpreted as concrete objects or their properties. It also refers to laws of nature and their status.[73] Another field of discussion concerns measurements.[74] These problem must nonetheless be decided on a different level to that of the question of theoretical entities. Here, due to the claim to universality that laws of nature usually entail, questions of scope and dependence on our formulaic language and our epistemic-historical approaches become relevant in a different way.

Another consideration of scientific realists, which is often, but not always, combined with a realistic attitude towards theoretical entities and laws of nature, corresponds to Anjan Chakravartty's dictionary definition of the term: "Scientific realism is a positive epistemic attitude toward the content of our best theories and models, recommending belief in both observable and unobservable aspects of the world described by the sciences. This epistemic attitude has important metaphysical and semantic dimensions, and these various commitments are contested by a number of rival epistemologies of science, known collectively as forms of scientific antirealism." (Chakravartty 2017c). Chakravartty also recognises ontological implications in this epistemological attitude: He claims that scientific realists are committed to a form of metaphysical realism that incorporates the ontological determinations of the best scientific theories. Even if this definitional consequence can be contradicted, this definition is widespread. Accordingly, the epistemological dimension often takes centre stage.[75] In addition, there is also talk of realism with regard to certain scientific methods. This is about determining which methods of scientific measurement, data collection and (research) logic are best suited to enabling scientific models to represent their respective target systems or phenomena as successfully as possible and to deliver reliable, true results (cf. e.g. Frigg 2023; Mets et al. 2024; Olsen 2010; Potochnik 2022).

At the beginning of Chap. 4, I will discuss several definitions of realism in philosophy of science, followed by a detailed discussion of their anti-realist

[72] A lengthy exploration could be pursued here of the extent to which individual local realism-anti-realism debates are conducted about these (supposedly) observable theoretical entities in each case. As examples of the debate in the feminist critique of biological sex and social gender that is explicitly concerned with the concept of realism, suffice it to refer here to: Ásta 2011; Mikkola 2006.

[73] Cf. two exemplary anti-realist contributions to this ramified discussion: Cartwright 1983; van Fraassen 1989.

[74] Cf. for a realist and an anti-realist perspective Tal 2018; Wolff 2023.

[75] Sometimes with a link to discussions about general epistemological realism, cf. e.g. Sangeetha 2024.

critique.[76] Therefore, the complex structure of the overall debate in philosophy of science will not be presented here. It will be a focus of the present work, especially in the next chapter and in the entire third part. However, it can already be said that the impact of logical empiricism and the Vienna Circle in the 20th century and their critics made it increasingly common to view the debate on realism in philosophy of science as largely congruent with the general ontological-metaphysical debate. This is hardly surprising, as opponents such as Carnap and Quine agreed that metaphysics had become obsolete as a discipline *foundational* to the empirical sciences. In logical-empirical anti-realism, common metaphysical statements do not refer to any method of verification or are otherwise statable solely with recourse to well-founded scientific language, and are therefore nonsensical (cf. e.g. Carnap 1998, 2003, and the overview in: Voosholz 2017) whereas in Quine's realist-naturalist programme metaphysical and ontological questions are answered (or at least could be answered) by the natural sciences without help from any foundational discipline (cf. e.g. Quine 1953, and the overview in: Hylton and Kemp 2020).

Based on my exposition of the scientific realism debate above, it is easy to argue that various local realism/anti-realism debates are conducted around individual questions that stand in a complex relationship to each other and to the debate about metaphysical realism, meaning that scientific (anti-)realism should be understood as a metalocal discussion. There is nothing wrong with the abbreviated definition I quoted above, as it makes it clear on which aspect the article focuses.[77] For the context of this book, however, the entire debate is relevant.

---

[76] Cf. Chap. 4 and 5 of this book, pp. 69–189.

[77] Cf. considerations that go in a similar direction to my characterisation of the debate as metalocal: Chakravartty and van Fraassen 2018.

Chapter 3
What Is Objective Realism?

The main thesis of Part I of this book is that there is a double stalemate: on the one hand, between different variants of objective realism in philosophy of science, on the other, between these variants and anti-realist counterarguments. In order to take a further step in my argumentation for this thesis, I must clearly define my concept of objective realism. That is the task of this third chapter. Its solution is essentially orientated towards my map of the realism-antirealism debates that I presented in the previous chapter.

In the *first subchapter* (3.1), I pursue Michael Dummett's idea that the various debates that deal with the concept of realism and its opposites, whether these discussions concern numbers, colours, possible worlds, scientific theories, unobservables or the external world, not only show family resemblances, but that in all cases a *single philosophical core question* is articulated in them. Considering whether this idea is correct, and what that question may be, helps us to better understand more fully the systematics of the numerous related discussions on a philosophical level and to identify the problems that characterise the debate as a whole.

In doing so, I reflect on whether there is a common philosophical core to the local, metalocal and global questions, and why, although its connection with considerations belonging to philosophy of language and theories of truth must be recognised, the focus here is on the ontological and epistemological levels (3.1.1). I argue in favour of Dummett's thesis of a core of these discussions, but against his one-sided identification of this *focal point of the theory of truth*. Furthermore, the possible follow-up objection that the pair of terms *realism* and *anti-realism* is then too broad to do meaningful philosophical work is refuted (3.1.2). Finally, I discuss a consequence of the commitment to an ontological-epistemological core of the debates (3.1.3): If I am correct in articulating the realism problem in this way, can we not identify two aspects within the ontological definition of realism, an *axis of existence* and an *axis of independence*?

This reflective stage in the first subchapter makes it possible to introduce a fundamental differentiation in the debate about realism: In the *second subchap-*

J. Voosholz, *Towards a New Scientific Realism*, Synthese Library 518,
https://doi.org/10.1007/978-3-031-98886-8_3

ter (3.2), I distinguish between *objective realism* and—at first purely negatively—*new realism*.[1] The former, objective realism, consists roughly in an ontological-epistemological realism that focuses on three core characteristics of the real: the existence of an entity in the *single real external world*, the *knowability* of the entity and the *independence of the external world from subjects* (3.2.1). Objective realism thus delineates fundamental areas of reality that exist genuinely from those that are merely subject-dependent. According to objective realism, the genuinely existing part of reality can be characterised as coherent and uniquely real.

I then counter the possible objection that I am only giving the well-known and well-researched *metaphysical realism* a new facade with my concept of objective realism. By analysing Hartry Field's arguments against Hilary Putnam's conception of metaphysical realism, I refute this doubt (3.2.2). A *look at the consequences* resulting from three basic tenets of objective realism (3.2.3) shows that objective realism tends towards the thesis of a fixed structure of the external world and suggests naturalism. The chapter concludes with a *distinction of new realism* that differentiates it from objective realism on the one hand and from anti-realism on the other (3.2.4). Both anti-realism and new realism oppose objective realism—without calling it that—but draw opposite conclusions from this opposition.

3.1 Dummett's Idea of a Core of All Debates

The aim of this subchapter is to show why my conception and thus definition of realism is based on the combination of an ontological and an epistemological thesis. The subchapter also explains why I agree with Dummett that we can identify a core of all realism debates, but that, contrary to his assertion, this core does not lie in the philosophy of language. Finally, the thesis of ontological realism is examined in closer detail in order to render two aspects of it visible, against both of which anti-realists can set themselves apart in different ways.

3.1.1 Philosophy of Language, Ontology or Epistemology as Core Concerns?

> I shall take as my preferred characterisation of a dispute between realists and anti-realists one which represents it as relating [...] to a class of *statements*, which may be, e.g., statements about the physical world, statements about mental events, processes or states, mathematical statements [...]. This class I shall [...] term the "the disputed class". Realism I characterise as the belief that statements of the disputed class possess an objective truth-

[1] *New realism* is intended and used as a systematic categorisation and not a historical one, which would refer to how old or young a particular realist philosophy is.

> value, independently of our means of knowing it: they are true or false in virtue of a reality existing independently of us. The anti-realist opposes to this view that statements of the disputed class are to be understood only by reference to the sort of thing which we count as evidence for a statement of that class. [...] [T]he dispute thus concerns the notion of truth appropriate for statements of the disputed class; and this means that it is a dispute concerning the kind of *meaning* which these statements have. Dummett 1978, p. 146. Emphasis as in the original.

This passage from Dummett's *Realism (1963)* places the question of theories of meaning and truth at the centre of all debates about (anti-)realism. On the one hand, this stance is understandable, since—as already shown in the previous chapter—a large number of local and global debates can be organised around this question. Dummett's proposal of a core of the different discussions has two welcome implications: (1) Who is a realist and who is not can now, at least superficially, be easily decided by merely determining the respective theory of truth. The desire behind this seems to be that more complex questions about local metaphysical and ontological commitments can be replaced by a, supposedly, simpler question about the meaning of a class of statements. (2) If the characterisation is correct, and the bivalence principle (*any statement must be either true or false*) and a strong theory of truth are *the* signs of realism *par excellence*, then their refutation or proof can settle *all* local debates at once.

> Philosophers such as Michael Dummett who attempt such a recharacterisation of the traditional divide, suggest that realists about a domain will claim that distinctive statements about that domain are either true or false, independent of our means for coming to know which. This way of marking the boundaries of any realist–anti-realist dispute suggests that realists and anti-realists have different conceptions of truth itself. Realists, for example, tend to embrace a correspondence theory of truth, according to which a proposition is true if and only if it corresponds to the facts. Antirealists, on the other hand, are thought to hold a contrary theory: perhaps a theory that equates truth with warranted assertability, practical utility or something else altogether. [...] While contemporary philosophers such as Thomas Kuhn, Hilary Putnam, Dummett and Crispin Wright all give an account with a semantic spin, other equally important figures working in the area, such as David Lewis, Michael Devitt and Paul Horwich, vigorously deny any connection between the problem of realism and the problems of truth and meaning. Brock and Mares 2007, pp. 7–8.

This description, presented by Stuart Brock and Edwin Mares in their introduction to the overall debate, clearly draws a line between philosophy of language and the ontological (and possibly epistemological) fundamental question. Brock and Mares clearly take a side: In their view, there is no connection between the problem of realism and the problem of truth and meaning in the philosophy of language. It seems, however, that Dummett can base his proposal to identify a core of the discussions on similarities between them and has the two advantages mentioned above on his side.

Brock and Mares now argue that Dummett has correctly recognised that the different elements of the debates have something in common. According to Brock and Mares, or critics such as Michael Devitt (cf. 1983, 1991), Dummett is mistaken in taking bivalence and a problem in the philosophy of language concerning the meaning and truth of propositions to be the common denominator. My initial definition of realism already reveals that I follow Brock, Mares and Devitt in this respect. This raises two questions: firstly, why is the question of theories of truth not at the core of all debates and, secondly, if it is not the philosophy of language that determines the shared core, which modes of philosophy do?

In his argument against Dummett's version of anti-realism in philosophy of language, Devitt has put forth the core argument against reducing the realism-anti-realism debates to the theory of truth. He first objected that the traditional definitions of realism as *X* "objectively exists independently of the mental" (Devitt 1983, p. 76) does not contain a hidden semantic thesis, nor does a semantic thesis follow from it together with an unproblematic further premise. Devitt then ascertains whether it follows from a realist theory of truth or the correspondence theory of truth that its proponents must be in favour of the above-mentioned traditional definition—a variant of my ontological realism. He summarises *realism about truth in the philosophy of language* as the following thesis:

> Physical statements are true or false in virtue of: (i) their objective structure; (ii) the objective referential relations between their parts and reality; and (iii) the objective nature of that reality. [. . .]
>
> Does Realist Truth entail Realism? It does not. Realism (as I have defined it), requires the objective independent existence of common-sense physical entities. Realist Truth concerns physical statements and has no such requirement: it says nothing about the nature of the reality that makes those statements true or false, except that it is objective. An idealist who believed in the objective existence of a purely mental realm of sense data could subscribe to Realist Truth. He could believe that physical statements are true or false according as they do or do not correspond to the realm of sense data, whatever anyone's opinion on the matter: we have no "incorrigible knowledge" of sense data. [. . .] Analogously some nominalists in the philosophy of mathematics accept a doctrine of realist truth for mathematical statements by taking them to refer to linguistic items. And an operationist can accept realist truth for statements apparently about unobservables by taking them to be really about observables. In sum, mere talk of truth will not yield any particular ontology. Devitt 1983, p. 77.

Here Devitt makes it clear that proponents in various debates arguing for mutually exclusive positions on the fundamental ontological and epistemological questions raised can defend the *same* theory of truth. Global idealists, sense-data epistemologists, abstracta nominalists, and operationalists with respect to unobservable theoretical entities can also argue for realism about truth in the philosophy of language, and in some cases even for a correspondence theory. Devitt's argument, of course, does not show—and explicitly does not claim—that the question of ontological and epistemological (anti-)realism in the local, metalocal and global debates is fully *independent* of the debate about truth. Whenever a position is taken in one of the debates, this has an impact on the possible philosophical stances that can be adopted without contradiction and convincingly in other areas. However, this only shows that we are dealing with a genuine, philosophical core problem in the question of realism. Devitt, on the other hand, proves that the position "pro realism" in the theory of truth does not necessarily imply a *particular* stance or position in all local debates.

Furthermore, it can be noted at this point that the two supposed advantages of Dummett's proposal—(1) the simple assessment of philosophers as anti-realists or realists and (2) the possibility of solving all local debates at once—are deceptive. With regard to (1), it could be argued that this task of reliable categorisation can also be be effected using my initial definition of (anti-)realism, and that it can subsequently be illustrated using elements of the debates. Contrary to (2), and against the background of the Devitt argument, it must be admitted that not all local debates can be resolved immediately using the philosophy of language either. Ultimately, Dummett cannot deliver on this tempting prospect in terms of philosophical labour economics.

On the other hand, I consider Dummett's first insight from *Realism* to be completely correct (cf. Dummett 1978, p. 145): There is a philosophical core that connects all local debates. In my view, there is much to be said for placing the ontological questions of the independence and existence of entities at the centre and supplementing them with the epistemological question of their recognisability and knowability: The various realism debates I mentioned can be entirely mapped in this way, as these two levels are prominently discussed in each of the debates. The metaphysical question of the world and its complete describability as well as the question of philosophy of language regarding theories of truth and meaning both presuppose those two aspects, as I have shown in the previous chapter. Furthermore, as Dummett argues, some of the archetypal positions (naturalism in philosophy of mathematics, naturalism in philosophy of science, naturalism in metaethics, ...) and arguments (model-theoretical arguments in philosophy of science, philosophy of language, metaphysics, ...) feature in many different debates: The concerns are too similar for these related questions to be treated disparately and, consequently, to dispense with a unifying metaphilosophical perspective.

Thus, the interconnections between the problems of truth and meaning and the problem of realism become clearer: The *reason* to move from an anti-realist position, such as logical positivism, to a realist naturalism in philosophy of science[2] may well consist in a rebuttal of theories of truth and meaning, such as verificationism and its successors. However, this only means that there is always an epistemological level to the problem of realism, which also plays a central role in theories of truth and meaning. The question of whether and how (in-)dependent entities can be known is always directly relevant if a theory of truth and meaning draws immediately on entities as truthmakers, and indirectly relevant if it cites facts, states of affairs or data instead, since these refer or link back to entities in most conceptions. The question of realism also arises when a philosophy of language instead ties meaning and truth back to the epistemic method. Dummett's idea of reducing these questions to one another may fail, but the connection between the questions is neither trivial nor accidental. The epistemological dimension links these

---

[2] Naturalism as well as some of the arguments in its favour are the subject of Chap. 4 of this book, pp. 72–102.

debates, and it cannot be reduced to matters of ontology or philosophy of language alone.

The critical discussion of Dummett's idea of a core of all realism/anti-realism debates in this section motivates my definitions of realism and anti-realism, as introduced at the beginning of Chap. 2. Even if the proposed definitions of these opposites are largely shared in the debate in similar form, the following section will deal with an objection that is easy to make: realism and anti-realism might be concepts that are too broad and worn out to be useful in the context of philosophical research today.

3.1.2 Realism—Too Broad a Term?

If one considers how broad the philosophical research fields that the concept of realism addresses are, one can doubt whether *realism* or *anti-realism* do any meaningful philosophical work at all. Various internal differentiations of the two opposing sides have already been touched on. The question of whether anti-realism as a stance is taken up at all goes deeper than the question of its usefulness. Jocelyn Benoist aptly remarks: "The real seems evident in such a way that one must ask oneself how we can even be brought to question this evidence philosophically." (Benoist 2011, p. 45, my translation.) A critic may object here that, rather, the binarism creates confusion, since it groups together concerns that should each be located in their own sub-discipline of theoretical philosophy. The binary opposites would be thus unsuitable due to their excessive generality. This is made more difficult by their long histories and their use as philosophical *labels*. Is there more left than just a combative term and an empty shell? Throughout this book, at least two other terms will emerge that must also face similar criticism: *idealism* and *naturalism*.

Philosophy is dependent on the discussion of different positions and stances. In view of the unbridled diversity and long history of our discipline, it is impossible to dispense completely with the categorisation of stances and positions. Even if the project of banishing "-isms" from philosophical language were feasible, the problem of identifying common ground in the form of shared theses, commitments or arguments remains. Within an enquiry such as mine, philosophical categories are useful, as already described at the outset.

However, is the scepticism towards realism, anti-realism, idealism and naturalism unjustified? After all, these categories have already been assigned a disarray of heterogeneous positions. The critic's main argument is that the terms make systematic work in philosophy more difficult and imprecise. One example of such a critic is Jens Rometsch. In *What is a Realistic Concept of Reality*, he argues against the indirect definition of an (epistemologically) undemanding realism in Paul Boghossian's *Fear of Knowledge*. Boghossian argues against relativism and constructivism. In the context of Rometsch's argument, one passage is particularly relevant, in which Boghossian opposes a form of constructivism, so-called factual constructivism, which assumes that facts only exist because people construct them through their descriptions: "[It must be] assumed that there are some basic facts—the basic worldly dough—on which our redescriptive strategies can get to work" (Boghossian 2007, pp. 37–38). Boghossian thus insists that reality cannot consist only of descriptions and redescriptions. Rometsch now sees the problem that Boghossian demands nothing more of a realistic conception of reality than a commitment to *basic facts*, his *worldly dough*. Rometsch shows that even an intelligible solipsist, who assumes that only her own mental inner life and her descriptions of it exist, can have a realistic conception of reality in Boghossian's sense. He thus argues that Boghossian's seemingly realistic concept of reality is too weak (cf. Rometsch 2019, pp. 112–117). Such broad and self-evident conceptions of reality and such unintelligible counter-positions are problematic for Boghossian, if Rometsch's reading is correct, because his own position then remains so diffuse that a commitment to such *realism* counts for nothing.[3]

As I have already noted with regard to Demmerling's objection concerning overly broad definitions of realism—those that are only interested in the existence of entities—, the reflection of the realism-anti-realism problem that I propose here establishes the deep connection between the ontological and epistemological levels of consideration. In this way, Rometsch's concerns are also taken into account. The versions of both concepts defended here allow for intelligible and philosophically sophisticated positions on both sides: Both realisms and anti-realisms can be put forth that deserve to be taken seriously. Neither an anti-realist nor a realist picture of reality is necessarily *unrealistic* or contradictory. The same applies to the concepts of idealism and naturalism,[4] which are likewise clearly delimited and defined, but not over-stringently fenced in or unintelligibly constructed from the outset.

I therefore share the above-mentioned concern about overly broad terms, but only up to a certain point: of course their use is divergent and often diffuse. This can be remedied by clearly defining these terms on the basis of shared theses, assumptions, premises and arguments. Precisely *these* terms of realism, anti-realism, idealism and naturalism are used here, as each has two characteristics: firstly, these terms are already effective in the debate, and secondly, they target relevant focal points of the debates discussed in this book.

Since the terms cannot be removed from the debates in ontology, epistemology and philosophy of science, one does not resolve their sometimes confusing or misleading roles simply by introducing new terms and trying to displace the old ones. It is better to define these terms for one's own use and to point out similarities with and differences to other meaningful definitions.

---

[3] Accordingly, Rometsch 2019, p. 110, initially argues against the charge that anti-realism, according to its own aspiration, is hostile to reality and unrealistic: "No theoretical position (not even a declared anti-realism, if there is such a thing) would want to label itself as *unrealistic*. Constructivist, relativist, post-modern stances of whatever kind happen to convince themselves of their own capacity to provide the best theoretical or discursive approach for describing what they take to be the reality of matters."

[4] For my definition of idealism, see this book, p. 31, for that of naturalism, see p. 72.

The criticism that the four terms mentioned would intersect with problematic areas of different sub-disciplines is not only unfounded, at least in this case, but intersecting different fields is actually a strength of the terms. Even if ontology, epistemology and philosophy of science can be distinguished, it seems naïve to want to resolve the debates completely separately. Each of the terms picks out a philosophical family of attitudes belonging to a core problem that appears in different guises in the sub-disciplines of theoretical philosophy. Even if specialisation is necessary in order to keep up with highly focused systematic discussions, these concepts nevertheless make it possible to illuminate the central concerns of philosophy as such. Finally, Benoist answers the question he raises as to why we can question the real: "Where the question of reality arises (what is "real", what is not?), we must give ourselves the means to *recognise* reality" (Benoist 2011, p. 72, my translation). Here too, the ontological and epistemological levels form a team that must be deployed to achieve any possible answer to the realism-anti-realism problem.

The reflection in this section also shows that the thesis of ontological realism deserves a deeper consideration. It makes sense to do so based on the two axes along which it is organised: the axis of existence of entities and the axis of their independence.

3.1.3 Axis of Existence, Axis of Independence

The thesis used in the definition of realism and anti-realism—*certain entities have an independent existence and reality of registering systems, which in turn can be registered directly or indirectly by systems*—has already been broken down into a first, ontological and a second, epistemological part. The ontological aspect can be analysed further: it states that (1) the entity exists and (2) it is independent of registering systems and events of registration. The registering systems also include us as perceiving observers. Two questions can now be posed: Why is it not enough to claim that an entity exists in order to be a realist with respect to that entity—so does one necessarily need (2)? And: Can anti-realisms be qualified according to whether they deviate on the axis of existence (i.e. reject the partial ontological thesis (1)) or on the axis of independence (corresponding to partial thesis (2))?

The question of whether (1) and (2) are necessary for realism is assessed divergently in the literature. Stuart Brock and Edwin Mares argue in favour of their definition as follows:

> Realism about a particular domain is the conjunction of the following two theses: (i) there are facts or entities distinctive of that domain, and (ii) their existence and nature is in some important sense objective and mind-independent. [...] But what sense is the important sense? The sense-data of the phenomenalists, the ideas of the idealists, the monads of the Leibnizian, the phenomenal world of the Kantian and the verifiable world of the logical positivists have all been alleged to exist. But such entities are all, to a greater or lesser extent, subjective and mind-dependent, and so lack reality in our fundamental sense. For a domain to be objective and mind-independent in our sense it must not exist in any of these ways. Facts about the domain must be out there to be discovered rather than constructed;

> entities within the domain will be detected rather than constituted by our mental states. This is, very roughly, what it means for a realm of facts and things to exist objectively and mind independently. Brock and Mares 2007, pp. 2–3.

The opening of the argument that typical examples of anti-realist positions also grant existence to their entities in the sense of (i) is, of course, weak. Simply wanting to exclude someone from the privilege of being able to call themselves a realist cannot justify why (ii) necessarily belongs to the definition. The last part of the quotation is more important. Facts that exist with regard to an entity should exist regardless of whether someone grasps this fact in order to be able to say that it stands, i.e. that it exists. If this were not the case, facts would depend on the respective subjective access and would be socially or otherwise constructed, thus they would be an expression of the subjective and not simply concern a state of affairs in which an entity is located. The same applies to entities themselves. If these were constitutively dependent on mental states, then they would not be robust enough to appear as elements of reality for Brock and Mares, since they disappear when the constitutive subjective episodes are concluded.[5]

Furthermore, it is noteworthy in this context that (i) refers not only to entities but also to *facts*. This is because Brock and Mares argue that it is possible for facts about a state of affairs to be true without asserting the existence of entities. The examples they give include a scientific realist who considers it a fact that everything that is thrown up falls down without believing in gravitational fields and a fictional realist who accepts the fact that in the novel *Wuthering Heights* Heathcliff is chased across the moor by a ghost without being bound to the existence of Heathcliff or ghosts (cf. Brock and Mares 2007, p. 2). I do not agree with this definition, as I do not share the view that both cases must involve realists. An anti-realist with regard to fictional objects can also consider facts about the plot of a story to be facts; an anti-realist in the philosophy of science will hardly believe that the everyday fact described does not have full empirical validity. Rometsch's argumentation on Boghossian, presented above, emphasises this point. My assessment of the situation, as well as the two ontological sub-theses (1) and (2), are shared by many authors who participate in the debate, such as Devitt or, in his lexical definition, Alexander Miller:

> Realism [is] a claim about what entities exist and a claim about their independent nature. Devitt 1991, p. 14.

> There are two general aspects to realism, illustrated by looking at realism about the everyday world of macroscopic objects and properties. First, there is a claim about *existence*. Tables, rocks, the moon, and so on, all exist, as do the following facts: the table's being square, the rock's being made of granite, and the moon's being spherical and yellow. The second aspect of realism about the everyday world of macroscopic objects and their properties concerns *independence*. The fact that the moon exists and is spherical is independent of anything anyone happens to say or think about the matter. Miller (2019).

[5] For mental entities, Brock and Mares give a decidedly different characterisation of the corresponding realist position. Cf. Brock and Mares 2007, pp. 25–27.

Does this fully answer the question of why it is necessary for realism always to advocate partial thesis (2) in addition to the ontological partial thesis (1)? There is still no real argument for or against it. If one affirms (1) and denies (2), one restricts the validity of (1). Entity *E* does exist, but only based on the fact that *E* is registered, whereby a registration is usually thought of as a (human) subject that helps *E* into existence by referring to or accessing *E*. Such a position is often referred to as MINIMALISM.[6] The term minimal realism and, more rarely, the term minimal anti-realism can also be found in the literature. Both are distinguished from an ontologically robust realism that subscribes to both sub-theses.

Minimalism obviously attempts to occupy an intermediate position in relation to ontological realism and anti-realism, as I have defined it, since the ontological thesis (consisting of (1) and (2)) is clearly accepted in part and clearly rejected in part. Together with the two strategies of QUIETISM of rejecting an answer to the ontological question altogether or categorising the question as nonsensical,[7] minimalism is an attempt to *escape* the dynamics of realism and anti-realism. If minimalism and quietism are understood in this way, they cannot be categorised as realism or anti-realism. Rather, they reject the dynamics of the latter two terms or want to prove their binarism to be irrelevant.

However, without going into whether minimalism is possible and convincing with respect to certain local or metalocal domains, two qualifications must be made. First, I present an argument for why global minimalism collapses into anti-realism or realism. Then I explain why local or metalocal minimalism is only possible if it clearly argues in favour of (1) and against (2), since otherwise minimalism in turn becomes anti-realism or realism. Then I come to the second question raised above about the classification of anti-realisms by (1) and (2).

Any supposed minimalism that accepts (1), rejects (2) and applies this to *all* entities is faced with a problem, namely that of a self-application that cannot be redeemed: If the *registration*, e.g. through the *discourse* of human subjects, helps every entity to exist in the first place and thus every entity depends constitutively on its registration, then at least the registration process—in the example, the discourse—seems to establish all forms of existence. But the discourse is an entity, which leads to a circular argument. Now there are two options open to global minimalism: Either the preferred registering system is not supposed to be an entity in the sense of the other entities and does not exist in the same way. In this case, (1) is called into question for the registering system, which would be a form of global anti-realism. Or the circular argument is accepted and contained by postulating the existence of something outside the registration process that provides a reason or cause for it but is not affected by the process. This thing-in-itself would then make global minimalism an ontological realism, even though almost all entities depend on registration except for this mysterious thing outside of it. This mystical thing would

[6] An example from philosophy of science for such a middle position is: Saatsi 2019a,b.

[7] Cf. the illuminating analysis of such local ontological quietisms in Spiegel 2017, pp. 20–21.

cause the registration. Therefore, global minimalism collapses in any case—either into a particularly minimal form of ontological realism or into anti-realism.

The same is true if (1) is not clearly accepted, while the other thesis (2) is clearly rejected. It quickly becomes apparent when we consider how this would play out in any local minimalism that is defective in this way. If someone clearly assumes that everyday objects only exist insofar as they occur in our practices of use and insofar as they depend on their constant registration in practice, (1) is restricted to such an extent that the concept of existence can no longer be maintained in this case as it is in other areas. In all conceivable local cases, it is clear that this is by no means minimalism, but anti-realism. It would be going too far at this point to explore whether the clear rejection of (2) ultimately restricts (1) in all local cases to such an extent that minimalism is generally impossible, and so we must speak of a form of deferred anti-realism.

Brock and Mares also make a suggestion regarding the second question of whether it is possible to distinguish anti-realists according to their deviation in relation to sub-thesis (1), the axis of existence, and sub-thesis (2), the axis of independence:

> [I]dealism, verificationism, social constructivism and response-dependence on the one hand, and the other forms of anti-realism [fictionalism, instrumentalism, error theory, etc; J. V.] on the other, have less in common with one another than they do with the realist theories they oppose. According to idealism, verificationism, social constructivists and response-dependence theorists, there are positive facts about the relevant domain; individuals of the relevant kind exist, but they are not objective—they depend on us. According to other forms of anti-realism, however, there are no individuals or facts of the relevant kind, and their non-existence does not depend on us; they would still fail to exist even if there were no minds at all. [...] The existence thesis and the mind-independence thesis are not independent of one another. In order for an individual or fact to be dependent on our minds, it must exist; mind-dependence therefore guarantees existence. Brock and Mares 2007, pp. 4–5.

Regardless of what exactly these terms mean for different philosophical positions, and whether the grouping proposed by Brock and Mares is correct, it appears that ontological anti-realisms cannot reject sub-thesis (1) and (2) at the same time. It makes no sense, at least at first sight, to claim that something does not exist but is dependent on the human mind. Of course, an anti-realist could try to argue that certain objects have a form of being that is weaker than ontologically robust existence (which is what is meant by sub-thesis (1)), which can then be dependent on an instance of the subjective: If someone is an anti-realist regarding scientific theoretical entities, she could argue that unobservable theoretical entities (photons, gravitational waves, phlogiston) do not exist robustly, but merely have a lesser form of being as long as subjects use the corresponding theories, i.e. there is a relation of transitive dependency from theoretical entities via theories to subjects.

Many variants of neo-Meinongianism or noneism—such as Richard Routley's (Richard Sylvan's)—depend on such a doubling of existence and being (cf. Routley (Sylvan) 1980). These duplications pose a difficulty for evaluating a position in the debate. This also applies to the options of treating existence as a gradual variable, whereby things can be more or less existent, or the qualification of existence into *simply exist* and *robustly exist/actually exist*. It can already be stated here that I will classify many debaters operating in this way as anti-realists, since the rejection of (2) leads to a clear restriction or reinterpretation of (1), so that the entire ontological thesis is negated. This already suggests an argument that can possibly be used against minimalism, which attempts to interpret (1) entirely in the sense of realism and to reject (2) entirely, as anti-realism does. I will address this later, as it would take the discussion too far afield at this point.

It remains to be said that Dummett's thesis of a core of the discussions is fundamentally correct, even if his restriction to philosophy of language and its theories of truth and meaning is ultimately misguided. It is not just a matter of one central question, but of *three*: Firstly, the question of ontology regarding the existence of entities and entity domains, and secondly, their (in)dependence on the (also epistemic) level of their registration by a system. Thirdly, depending on the answers to the first two questions, the issue is either about the epistemological question arising from an ontological-realist perspective as to whether and how these entities are knowable, or about the epistemological-ontological question arising from an ontological-anti-realist perspective, concerning how non-existence or dependence can be proven. These three complex and intertwined questions lie at the heart of all realism-anti-realism debates—and with them the philosophical core problems of existence and knowledge, ontology and epistemology.

3.2 The Features and Consequences of Objective Realism

The aim of this subchapter is to introduce a new concept that is essential for my main thesis of the first part of this book: *Different forms of objective realism and anti-realist critique in philosophy of science are in a double stalemate*. As I will now show, the concept of objective realism (OR) emerges from the map of the realism-anti-realism debates presented in my previous chapter. In the ensuing paragraphs, I will first explain why in my view the analysis of the debate suggests a connection between the assumption of an external world and a particular conception of the independence criterion. Then, in the subsequent section, I present the two core ontological features and the core epistemological feature of OR. In the section that then follows, I defend the concept against the objection that OR is merely a reformulation of metaphysical realism. In doing so, I show that OR advocates these that are less demanding, and that it can seem obvious to see it as only a minimum requirement for *every* possible realism. By examining some consequences of OR in the third section of this subchapter, I aim to show that this is not the case. Finally, I provisionally define *new realism* as those realist positions that reject OR without advocating any form of anti-realism.

Through the reflection of the previous Sect. 3.1, it becomes possible to make a further fundamental differentiation of the concept of realism: Every ontological realism can be characterised both by how it spells out the ontological criterion of independence and by whether all (ontologically robust) existing entities jointly constitute a single, real external world.[8] There is a connection between these two steps, the commitment to a certain characterisation of the criterion of independence and to a single real external world:

Let us say that a philosopher for some reason advocates a *metaphysical realism*: If such is presupposed, then she must at least assume an external world to which a single, complete and true description applies.[9] In addition, it is then clear that parts of this description can be expressed by us linguistically in the case of a successful knowledge claim. This lends itself to certain theories in epistemology (and, subsequently, certain theories of truth and science) that focus their attention primarily on objects of the external world and perceiving, theorising subjects:

The objects of the external world are thereby always knowable in principle, since a single and complete description of them exists. Now there are two options: On option one, the objects of the external world are then represented passively in the perceptual-cognitive process, as they exist independently and the subjects have to do work in order to arrive at the only correct description. This means that subjects, as the active agents of the perceptual-cognitive process, are nevertheless able to access these entities directly or indirectly, albeit with difficulty. On option two, the objects of the external world are set as the active element. The subject is then primarily receptive, passively dependent on the outside world acting on it and appearing to it.[10] Both are two descriptions of the cognitive knowledge-acquisition process: the registration, active or passive, of objects of the external world by passive or active subjects of perception and cognition becomes the paradigm of knowledge in general. In both cases, there is a clear difference between subject and object. It is only through their radical independence from speaking subjects of perception and cognition that objects of the external world are guaranteed to be knowable through a single, complete and true description. If metaphysical realism is therefore assumed, *subject independence* imposes itself as the criterion of independence. If the relation is understood in this way, the interesting but problematic question always arises as to how knowing subjects fit into the external world and the corresponding ontology.

But even if we do not begin with the assumption of the metaphysical realism of the external world and its describability, but with the ontological characterisation of independence as the independence from subjects, a similar picture emerges: Let us assume a realist who defines the independence criterion as independence determined by whether or not we, or more generally subjects, exist. For example, if there were no subjects—so this realist could judge—then there would be neither good nor evil, but there would be our solar system. Therefore, good and evil do not exist

[8] How exactly this external world is defined varies. Before I analyse this notion more precisely later, for now it can be defined as the domain that encompasses all robustly or objectively existing entities. This is closer to the everyday understanding of *external world* than, for example, the well-known and similar, but stronger definition by Adrian W. Moore. He understands the external world as the "absolute conception of reality", and thus follows Bernard Williams' understanding of it as the set of everything "what is there anyway". For this: Moore 1997, p. 68; Williams 1978.

[9] For the following, see also my discussion of metaphysical realism, Sect. 2.4.1, p. 33.

[10] I would like to thank Jens Rometsch for this point.

ontologically, but the solar system does. Whatever entities are assumed for these reasons—for our realist it follows from this version of the independence criterion that all independently existing entities can be conceived either as a set ("the set of all ontologically existing entities") or as a single entity ("the independent/the objective/the whole"), which in turn has various identifiable individual parts. In both cases it is possible to say that all entities together form the single real external world or are parts it. It is then also an open question how subjects can occur in this independent external world.

The assumption of the external world and subject independence as a criterion go hand in hand. Both give rise to a similar problem. For example, if independence is defined as the independence of an entity from acts of the human mind (and is thus a variant of subject independence, however exactly *mind* may be defined), then for metalocal or global realists a problem arises with the idea of a single real external world, which Gabriel in a similar context has labelled "'spooky' [...] *world without spectators*" (Gabriel 2015a, p. 40): If "robust" existence of *E* presupposes that *E* is e.g. independent of the existence of the human mind, and the external world is formed solely from "robust" existing entities, how can the human mind itself be part of the external world or appear in it? Does this dualism not make the outside world unobservable in principle? There are various strategies for solving this problem, but they will not be pursued here. What is decisive for my argument at this juncture, however, is that the assumption of a single real external world and subject independence are linked and lead to similar constellations of problems, although a metaphysical realism is not necessarily assumed in these cases. I do not want to claim that a very specific definition of the independence criterion (e.g. independent of acts of the human mind) also *necessarily* entails a specific position on the single real external world (e.g. fusion of all mind-independent entities). I maintain, however, that there is always a connection between the two. This connection means that if an ontological realism leans towards either subject independence or a single real external world, the other assumption seems obvious and thus the same complex of problems arises. My characteristics of OR derive from this analytical result.

3.2.1 Three Core Features of Objective Realism: Subject Independence, Single Real External World and Knowledge

Every ontological realist must explain what she understands by *independence*, as she states that certain entities have independent existence and thus reality from registering systems. This is most often defined as independent *of us* or of any conceivable, knowing observer or acting agent. But what exactly does that mean? Independence can then still be realised in very different ways: It can be defined as independence from acts of consciousness or actions, mind-

or *Geist*-independence,[11] independence from intentional reference by a person or independence from the institutions, practices and discourses of a linguistic, knowledge or cultural community. I refer to all such or similar variants of the criterion as SUBJECT INDEPENDENCE (SI). What all these specifications have in common is that they presuppose, in a minimal form, acting, conscious, mental, intentionally referring, personal, socially situated, knowing, speaking, discursive and/or culture-creating subjects. All variants refer to and assume subjects. This does not yet say whether the respective SI implies independence from the individual, independence from the intersubjective community or both. SI, as I will use the term from now on, includes all three options. SI is the predominant realisation of the criterion of independence, precisely because it is directly accessible for the everyday world of our experience—the forest is there, whether I am there or not, whether we are there or not.

This long list of deeply and hotly debated philosophical concepts, of course, cannot be treated at this point. Adult, awake, healthy people are subjects, but so are children, the unconscious, the sick, perhaps some non-human animals and the aliens so popular in philosophical thought experiments. Some of them, such as unconscious people do not fulfil the entirety of the list contained in the last paragraph. However, this does not contradict the fact that they are subjects. The criterion of SI offers room for many, and especially very minimal, conceptions of what subjects are.

The overwhelming majority of the local, metalocal and global ontological realisms cited so far represent SI one way or the other—and this across large internal differences within the individual debates.[12] It is precisely because of this apparent lack of pretension that SI is suitable as the first criterion for objective realism. OR always employs (a variant of) *SI as the ontological criterion for independence*, this is its *first ontological core feature*.

In addition to SI as the execution of the independence criterion, there is a *second ontological core feature* of OR, which results from the analysis carried out above: All (ontologically robust) existing entities together form a *single real external world* or are parts of a single real external world. The EXTERNAL WORLD is an open concept, which, however, always requires that it can at least be formed as a set, sum, system, fusion or whole of all (ontologically robust) existing entities. This includes all concrete, i.e. spatiotemporal, entities, usually understood as physical objects. In this sense, often all robustly existing, concrete objects form a single system, the external world as a causal and relational network or structure. Usually, but not in every OR, abstract, ontologically robust entities also play a role—often

[11] It should be noted here that the German word *Geist* and the English word *mind* denote overlapping, but not congruent or even coextensive areas of the mental. From here on, when I write of (in)dependence from the mental, I always mean both (in)dependence from the mind and *Geist*.

[12] The debates on mental and social (anti-)realism seem to be special cases in this respect, since the usual formulation of SI seems to fail here at first glance. However, this only appears to be the case superficially; here, too, SI formulations are ubiquitous. Cf. e.g. Demeter 2009, p. 65: "Mental entities exist independently of what we think about them."

these are mathematical and logical entities as well as laws of nature. In all cases, the external world is either regarded by OR as ontologically primary, in the sense of the really existing whole, of which everything else is always only a part or a (partial) property, or it is constituted as a maximum single set of the ontologically primary or robust individual entities. The external world is also frequently associated with an "absolute conception of reality" (Moore 1997, p. 68), as Adrian W. Moore advocates following Bernard Williams (1978), and defines as a set of everything "what is there anyway". However, this additional step is not necessary in order to represent the existence of a single real external world in my sense; the mere assumption of a comprehensive set or a constitutive whole of all (ontologically robust) existing entities is sufficient.

A further aspect is included in this second core feature of OR, which goes beyond the assumption of the existence of the external world: It also requires that this is the *single real* external world. This means that, whatever entities it may comprise in different conceptions of OR, the external world is always singularly that which exists completely independently of subjects and subjectivity. This can be understood, as already mentioned in the case of Moore, as *absolute* existence. Others, as shown above, use the label *ontologically robust* for the entities of the external world. Neo-Meinongians and noneists reserve *existence* as a predicate for the entities of the external world in contrast to the mere *being* of other entities.[13] This list could be supplemented by any number of further qualifications. The core of the second ontological feature of OR is thus the assumption that entities that belong in the single real external world are ontologically clearly distinguished from those that do not. Both ontological criteria are usually combined in OR in a form such as: *The entities of the single real external world exist independently of the subject.*

The second ontological core feature of OR, the existence of a single real external world, is also held as a thesis by every *metaphysical realism*, and is even a component of its first thesis: This has already been defined here, following Putnam, as the thesis that a world (or many worlds), especially a physical and causal external world, exists independently, a complete and true description of it exists, and parts of that description can be expressed by us linguistically in the case of knowledge. This already shows that metaphysical realism obviously goes beyond the assumption of a single real external world. In the next section, I will argue that, firstly, the two concepts cannot be equated with each other, secondly, OR involves less argumentative ontological, epistemological and linguistic theses than metaphysical realism, and therefore, thirdly, that OR is a more plausible form of realism.

The third core feature is epistemological, and builds on both ontological features: The entities of the single real external world exist subject-independently and *are known in part by us subjects*. The OR thus adopts a *minimal epistemological optimism*. It does not merely consider knowledge about external entities to be possible in principle, but also states that we are actually in a position in which we

[13] Cf. e.g. Priest 2014, pp. xxi–xxii.

know some of these entities in some of their properties in such a way that we know how they are subject-independently. I call this thesis minimal, because not every OR posits that our knowledge claims about the subject-independent external world and its entities are predominantly true. It is true that there are objective realists, such as John McDowell, who argue in favour of this stronger thesis. He does this by means of his form of semantic externalism. According to this view, the propositional content of singular thoughts—that is, those that are directed towards a concretely present entity such as *my coffee cup here*—and our linguistic practices are only possible if the entities are predominantly what we think they are in our mental and linguistic references (cf. McDowell 1998c, esp. p. 228 and pp. 240–255, 1998d, esp. p. 204, as well as the helpful presentation in: Liptow 2014). But other representatives of OR are more sceptical about our epistemic situation, even if they must argue against the spectre of global scepticism. Therefore, only a minimal epistemological optimism is necessary for OR.

These reflections on the three core features now make it possible for me to summarise the title concept of the present subchapter: All forms of realism that

1. use *subject independence* as a criterion for independence and
2. assume the existence of a *single real external world* and
3. assume that some entities of the external world are known by us subjects in some of their properties,

I will call OBJECTIVE REALISM (OR).[14] This position claims that *we know some aspects of some subject-independently existing entities of the single real external world.* OR is thus a very comprehensive "ism": Many of the local, metalocal and global realisms already presented in this and the previous chapter represent a form of OR. Indeed, I assert that OR is an obvious position to take when attempting to defend the realist position: SI imposes itself when one wants to say that something

[14] To my knowledge, there is only one extant discussion in which the term *objective realism* occurs: in physics and the philosophy of physics. First, I would like to emphasise that this is a mere duplication of names for two completely different concepts. In the context of this book, I will not refer to the concept of physics outside of this footnote, which is why there is no danger of equivocation. In physics and the philosophy of physics, *objective realism* and *locality* denote the two intuitive assumptions, one of which must be abandoned if Bell's Theorem is correct. An overview is provided by Myrvold et al. 2019. While some authors assume that *objective realism* is a straw man assumption by Bell or his interpreters, cf. e.g. Maudlin 2014, others consider the objective realism thesis to be central to classical physics cf. e.g. Zukowski and Brukner 2014. The objective realism thesis in this special physical sense roughly states that an experiment only shows properties of the experimental system that have already existed before the experiment cf. e.g. Brukner 2018, p. 2. Each measurement result determines actually existing properties of quantum objects present before the measurement, which are either part of the observed target system or of the measuring apparatus. In my sense, this physical objective realism is therefore a local realism with reference to the properties of quantum objects or quantum structures. My concept of OR and that which pertains in physics only connect what any arbitrary local realism and metalocal realisms always connect—and additionally in this case the name—, but quantum theory or its philosophical reflection does not necessarily imply OR in ontology, nor does an attitude towards Bell's dilemma derive from the latter position.

exists in a demanding sense. The assumptions of a single external world and knowledge about it lend themselves to metalocal or global realist claims in ontology and epistemology. On the face of it, OR seems like default realism.

In the following section, I address the objection that OR is just another name for metaphysical realism and that its conceptualisation is therefore superfluous here, before discussing some of the consequences of accepting OR.

3.2.2 Is Objective Realism Only Metaphysical Realism?

At this point, an objection to the introduction of the concept of OR suggests itself: Is this not merely a cumbersome way of expressing the old and already sufficiently discussed metaphysical realism? The discussion of this objection will allow me to explain in more detail what makes OR particularly attractive, how individual positions can be qualified as OR, and where some of its difficulties originate—all three topics will be taken up again in my next Chap. 4 and deepened for the context of the philosophy of science. I begin by contrasting the two in harmonised formulations. OR states:

OR1 An entity E_1 exists independently iff it exists (objectively, actually) subject-independently.
OR2 All entities $E_1, \ldots, E_n$ that exist subject-independently together constitute the single real (ontologically robust, primary, absolute) external world.
OR3 We have knowledge of some of these entities in some of their properties.

Metaphysical realism asserted:

MR1 All independent entities $E_1, \ldots, E_n$ together constitute the (ontologically robust, primary, absolute) world, in particular a single real external world.
MR2 There exists a complete and true description of this world.
MR3 Parts of this description are expressed by us linguistically in the case of knowledge, i.e. true statements about the world correspond to it.

OR2 and MR1 are identical except for the fact that OR2 already specifies SI as the independence criterion, whereas MR1 does not directly demand this. However, I have already argued above why it makes sense to choose SI as a criterion if the assumption of a single real external world is made. This is the connection between OR and metaphysical realism. OR1 contains nothing more than that this obvious assumption is explicated.

But the other theses are clearly different. While OR3, as already explained in the last section, is a minimal epistemological thesis, MR2 and MR3 are extremely heavyweight and problematic semantic-epistemological theses. Putnam and other opponents of metaphysical realism use them to refute it. For this reason, I consider OR to be less demanding, easier to justify and more obvious as a position for realists. My concept of OR therefore serves to characterise a position that is as strong as possible in terms of argumentation.

In order to substantiate these conclusions, I will now engage with a discussion between Hartry Field and Putnam on metaphysical realism to illustrate how the weakness of metaphysical realism presents itself and relates to the position of OR.

Taken together, the three core features of OR suggest that the subject-independent external world must have a fixed constitution for its (partial) knowability, i.e. that it needs a way of being structured. This raises the question of whether OR implies a "*God's Eye point of view*" (Putnam 1981, p. 49), as Putnam calls it in *Reason, Truth and History*. Putnam decisively rejects such a view from "Nowhere" (Putnam 1990, p. 28):

> This is one reason I feel justified in having taken the "metaphysical realist" to be a philosopher who accepts what Hartry Field calls "metaphysical realism [MR1; J.V.]," (the world consists of a fixed totality of mind-independent objects), *and* accepts "metaphysical realism [MR2; J. V.]," (there is exactly one true and complete description of the way the world is), and also accepts "metaphysical realism [MR3; J.V.]," (truth involves some sort of correspondence). These doctrines have been held by philosophers of every historical period, and one can think of a rich filigree of ideas, doctrines, and detailed arguments which flesh out these abstract theses in different ways. Putnam 1990, p. 30.

Hartry Field has argued against Putnam that MR1 and MR2 can be represented without MR3: one can argue for Field's "redundance theory" (Field 1982, p. 554), which amounts to disquotational deflationism in the theory of truth: ""Snow is white" is true iff snow is white." (Field 1994, p. 405) Putnam summarises this as: "[T]o say *"P"* is true is merely to affirm *P*." (Putnam 1990, p. 31).

In addition to MR1/OR2 and MR2, Field also argues for OR1 and OR3. In the nomenclature I present here, he can therefore be assigned to OR, but not to metaphysical realism. However, he advocates a relativism (i.e. anti-realism) with regard to justifications (cf. Field 1982, pp. 562–567). This relativism states that epistemic values are dependent on subjects, and that therefore the justification and substantiation of knowledge claims are relative to the subjects who express them. It is not a relativism about facts that exist about entities of the external world, otherwise Field would also contradict his assumption of MR2. Moreover, Field also advocates an AN, i.e. is an anti-realist with regard to abstract entities, which Putnam omits in the passage quoted above, but later in the same text takes up and reflects on (cf. Field 1980). Field agrees with Putnam that relativism follows from MR1 and MR2, but that it is not harmful.

Putnam counters this objection by arguing that deflationism about truth casts doubt on the status of truth as a property (cf. also for the following Putnam 1990, p. 32). Interestingly, Putnam cites Richard Rorty to argue that this stance on truth is mistaken. Accordingly, for Putnam, either an ELIMINATIVE MATERIALISM OR PHYSICALISM, or a form of anti-realism such as CULTURAL RELATIVISM, would need to be advocated. In the first case, it would be assumed that ontologically robust properties exist only in spacetime; in the second case, that truth is reducible to assertibility from a culturally dependent individual perspective. For Putnam, both are forms of reductionism with regard to truth, which ultimately cannot be maintained.

Conversely, Field accuses Putnam of being a global anti-realist (cf. Field 1982, pp. 555–556) and takes up the commitments of OR himself. Putnam, on the other hand, believes that he himself is a realist and that Field is not (cf. Putnam 1981, pp. 20–31). This seemingly confused situation can easily be clarified with the distinctions made here: Field simultaneously adheres to OR and AN. He is thus a realist with regard to the single real, physical external world, but an anti-realist with regard to all abstracta and attempts to advocate a deflationism with regard to truth. Putnam rejects (at least in the context of *Realism with a Human Face*) MR1 and MR2. Since MR1 essentially implies OR, he is both an objective anti-realist and a metaphysical anti-realist. Of course, this does not mean that he could not explicitly take a realist position in certain local debates, such as those in philosophy of science, as it would be possible to argue. Putnam's *internal realism* is neither an OR nor a metaphysical realism. I leave the question of whether it is a realism at all unaddressed here, as it goes beyond the scope of my argument. I consider the fact that neither Field's nor Putnam's positions are ineligible, but that both are philosophically intelligible via my proposed framework, even if they contradict one another, to be a strength of the analysis of concepts introduced here.

Is it correct, as Field and Putnam both suggest, that from MR1—which can be associated with OR2 in this context—then MR2 follows? And if so, only with the aid of demanding additional premises or by itself? For MR2, the position that there is a single, complete and true description of the constitution of the single real external world is what Putnam criticises as a *God's Eye view*. Similar positions are attacked as a "view from sideways on" (McDowell 1998b, p. 214) by McDowell and a "view from Nowhere" (Nagel 1986, pp. 25–27) by Thomas Nagel.

MR1, as Putnam also notes, can be realised in different ways. OR2 offers a still open but clearer formulation: All entities $E_1, \ldots, E_n$, which exist independently of the subject, together constitute the single real (ontologically robust, primary, absolute) external world. Does it now follow from its core characteristics that there is a single, complete and true description of the constitution of this single real external world? The constitution of the external world, since it is constituted exclusively by subject-independent entities, is fixed precisely when the real entities display stable patterns of relations and interactions that are independent of space and time. All external world entities that we know of exhibit stable patterns and the external world as such appears stable. This is classically conceptualised as natural causality. In any case, a stable external world has *some sort of fixed structure*. Since the external world is thought of as a set or whole, this premise is obvious. Therefore, it is usually assumed to be physically and causally ordered. I refer to this as the STRUCTURE THESIS and will come back to it later. It should be noted here that at least this aspect of thesis MR2 follows from the unproblematic premises set out here: The external world has a fixed structure in OR.

It is not so simple with the single, complete and true description. If a description is the linguistic formulation of a knowledge claim regarding states of affairs, then even if one does not claimed that we as a finite subject community can find the single, complete and true description, it must at least be possible to conclude that the condition of its possibility has been met. Knowledge claims are only made by epistemic subjects. OR3 asserts that some of our knowledge claims are true and that we thus possess knowledge. Some of the subject-independent entities of the external world are (at least partially) knowable.

Even with this premise, the second aspect of MR2 still cannot be concluded without assuming more: it is not only the individual entities of the external world, but their total constitution, "the way the world is" (Putnam 1990, p. 261), which is alleged to be knowable. This is not a trivial premise that would result from what has been shown so far. Furthermore, the structure should not only be knowable in parts, but as a whole, which assumes its complete intelligibility for subjects. Here, too, we are not dealing with an assumption that necessarily follows or is undisputed, but is initially far from the minds of many objective realists: Why shouldn't there be things in the cosmos that are unknowable to us subjects in principle? It can therefore be said that the path from OR to the complete thesis MR2 is not only long, but anything but compelling for most. This does not rule out the possibility that Putnam's suspicion is correct that—contrary to Field's assumption and that of many others—the only course of OR worth pursuing leads from OR2/MR1 to MR2 and on to MR3, only ultimately to fail. I refrain from passing judgement here on this issue. However, this consideration makes it clear that OR offers more options than committing oneself to the perfect describability of the structure of the external world and defending a metaphysical realism.

I have thus shown that OR and metaphysical realism do not describe the same position. OR allows metaphysical realism as one option among many. However, Field's position, which includes AN, is also an OR. Based on this argument reconstruction, I also see my thesis confirmed that OR is easier to maintain than metaphysical realism and seems more obvious.

3.2.3 Consequences of the Core Features of Objective Realism

In this section, I will discuss some consequences that follow from the three core features of OR. I will show that although OR is a broad category that allows for a variety of philosophical options, variants of realism are also conceivable that cannot be categorised as OR because they do not share some or all of the basic assumptions. I will refer to these variants—at first in this purely negative way of not being OR—as new realism, although this term will be further sharpened in the discussion of correlationism in Chap. 6.

The previous section demonstrated that the *structure thesis* follows from the two ontological core theses, which include SI as the criterion of independence and the existence of a single real external world: The external world has a *fixed structure*. The argument in favour of this assumption requires only two further premises:

P1 If the external world is constituted only by subject-independent entities, then the structure of the external world is fixed precisely when these SI-entities all exhibit stable, space- and time-independent patterns of relations and interactions (e.g. all concrete entities are subject to natural causality or laws of nature; all logical entities are subject to logical laws; etc. for everything that is considered part of the external world, depending on the variant of OR).

P2 All entities of the external world that we know of, as well as the external world as such, as far as we know it, exhibit stable patterns of relations and interactions.

From P1 and P2 follows together with the three core features of OR:

C The (subject-independent) external world has a fixed structure (*structure thesis*).

Since both premises P1 and P2 are indisputable in OR, all representatives of the OR are committed to them. What conditions must this structure fulfil? Firstly, it must itself be subject-independent, as otherwise the external world would essentially be co-determined by an instance of the subjective. However, OR denies this. Secondly, it must extend to everything that exists independently of subjects, the entire external world and all its entities, and in this sense be independent of space and time. Thirdly, as a structure it must guarantee and regulate the stability of the external world; a reference to it must therefore be able to serve as an explanation of the events and processes of the external world. It is this third point that then serves many variants of OR to attribute this function of an external world structure to natural causality, natural laws, logical laws or comparable concepts. This step itself is not compelling without other premises and arguments, but it is an obvious one to take at this stage. However, the third point of structure as a guarantor of the stability and regularity of the external world is always implied.

It is now possible to give a definition of realism according to OR. This results when the three core features of OR are brought together with the structural thesis and my definition of realism as a combination of the ontological and epistemological theses:

Realism according to OR: All really (ontologically robust, primary, absolute) existing entities are those that exist independently of the subject and are contained in the single real external world. The external world, in turn, has a fixed structure into which all really existing entities are integrated. Most entities can in principle be known directly or indirectly by subjects. In some cases, we succeed in our knowledge claims about them and therefore have knowledge about some subject-independent entities of the external world in some of their properties.

Overall, the definition is a commitment to both ontological and epistemological metalocal realism with respect to at least the concrete entity domain of objects with which we are familiar through our everyday lives. OR does not have to be a global realism, as Field shows. He advocates AN as metalocal anti-realism with respect to abstract entities simultaneously with OR. In many cases, these nested differentiations are not made by OR proponents or they remain implicit, but this does not detract from their consequences.

All variants of OR invite two further conclusions, each of which follows with only one further premise for OR:

Together with the premise

A. not all entities for which we have descriptions are parts or elements of the (single real) external world.

follows from OR that we only refer to facts or real entities with statements about objects of the external world, depending on our preferred philosophy of language. For those parts of our vocabulary that are not obviously about objects of the external world, such as fictional characters, numbers, the God of Islam, state institutions or the human mind, the question now arises as to what their status is. For each such object, or the entire object domain, each OR initially has the option of choosing realism or anti-realism. Either the objects do belong to the external world in a non-obvious way (which this particular OR must then justify) or they do not (whereby it must be explained why this is nevertheless mentioned).

Many of the local and metalocal debates are dominated by the fact that objective realists, among whom there is agreement on, for example, concrete everyday objects, mainly discuss the finer issues from their divergent respective positions, such as the exact reasons why anti-realism is untenable and whether local realism or anti-realism is appropriate for particular cases. They and their critics further discuss whether OR is better justified by local naturalism or pragmatism, or whether anti-realism is better justified by a theory of supervenience or a theory of error.[15] Through the lens of OR plus (A.), many structures of local debates become better to grasp.

If, in addition to the OR's definition of realism, the premise

B. the best methods for the knowledge of the external world do not consist in philosophical thinking, but are practised by the empirical (natural) sciences

is accepted, a clear division of scientific labour results from OR: Philosophy does not have to provide a substantial and empirically informed metaphysics by itself, nor a generally binding epistemology, but receives its information on reality from the empirical individual sciences upstream. This by no means entails that OR, by accepting thesis (B.), would no longer allow metaphysical realism or that metaphysics could no longer be pursued within the framework of this division of labour: Such a metaphysics, according to (B.), then, can and must build on the ontologies of the individual sciences and integrate them, which emerges, for example, in the expression of a *metaphysics of science*.[16]

However, OR does not have to take this metaphysical path, but can limit itself to acting within the sciences as a self-reflexive authority that can provide a meta-theory and methodological description of the human acquisition of knowledge. Further, it can act as a defender of the sciences against perceived anti-realist philosophical, religious or other undermining. Following this line of thought in philosophy, the world-related mind, social space or the results of empiricism form

[15] Cf. also the discussion on this point in the present work on naturalism and supervenience, pp. 72–102.

[16] Cf. exemplarily the introduction by Healey 2016, as well as the overview presentation in: Schrenk 2017.

the starting point of the philosophical endeavour, and not the external world with its ontological structure of existence itself, for the exploration of which philosophy is *not* responsible.

With (A.) *and* (B.) together, the three greatest tasks for OR do not seem to lie in ontology, metaphysics and epistemology, but in philosophy of science, mind and language. It demands a theory of science,[17] as a special form of *philosophy of science*, to provide the aforementioned reflexive instance of the sciences, which pursues OR on the path of naturalism with explicating a meta-theory and methodological reflection. Against the background of the considerations following (A.) and (B.), it seems clear that philosophy can only ever be a meta-level evaluator of the structure of the external world (metaphysics) and of different types of existence (ontology) second-hand; empirical science deals in real, first-hand information. The role of philosophy is on the one hand that of argumentative support, and on the other that of an interpreter. I will pursue the idea of the path of naturalism in Chap. 4.

One puzzle that must consequently be solved from this point of view, which brings one from philosophy of science to philosophy of mind, is the question of how seemingly mental objects, especially consciousness, are possible in the external world and why beings with consciousness can have phenomenal experiences. David Chalmers has called this the *hard problem of consciousness* (Chalmers 1995). Here OR with (A.) and (B.) creates a problem in philosophy of mind. From this, it is but one step to the linked problem in philosophy of language. This is because how our language can refer to the real at all also becomes problematic, i.e. how reference to the external world is possible. Language does not obviously seem to be a part of the external world, so how can we guarantee that it grasps it correctly? The question of the right kind of epistemology is transformed into questions of the right meta-theory of the empirical sciences, their methods and their languages (which OR poses as a task). Only after these questions are answered do longstanding questions of ontology, metaphysics and epistemology arise again, such as in the aforementioned research discourse of metaphysics of science.

The three core features of OR were introduced in the first section (3.2.1), as well as the structural thesis that follows from them as a conclusion. The two additional premises (A.) and (B.), on the other hand, are not among those core features of OR. However, since the latter assumptions easily lend themselves to proponents of OR against the background of OR's definition of realism, this explains the structure of many of our discussions about realism. Indeed, jointly they argue in favour of a form of naturalism, which is the most dominant form of OR. In the following Chap. 4 of this study, I take a closer look at its varieties in relation to philosophy of science. From the preceding considerations, however, it should already be clear that there is no lack of unresolved disputes within OR: Premises (A.) and (B.) are non the less controversial, and even more so, the views of OR's definition of realism allow for strongly divergent perspectives. It could be argued

[17] This also includes anti-realistic versions of such a project, which then take OR and naturalism as a negative starting point.

that Paul and Patricia Churchland, Ruth Millikan, John McDowell, Daniel Dennett, Philip Kitcher, Michael Williams, Chalmers and Priest are all objective realists.[18] However, from the inconsistency of these positions with respect to basic ontological and epistemological questions, it can be concluded that there can be no further agreement on the correct form of realism in OR either. The arguments that were and are exchanged between these proponents of OR are also interesting because they indicate that none of these OR positions have been able to establish themselves as a common standard.

3.2.4 New Realism in Contrast to Objective Realism

As a final step in this chapter, I will distinguish OR from NEW REALISM (NR), which does not follow OR in its three core features and in the structural thesis, but which also does not aim to advance a metalocal or global anti-realism. This is a preliminary, purely negative determination of NR, which is supplemented in Chap. 6 of this work by a further criterion, the simultaneous rejection of correlationism. Even though I develop a positive definition of NR for philosophy of science in the course of Part II, I do not claim to have defined the philosophical core of New Realism in general.[19]

A possible misunderstanding should be avoided at this point: The NR agrees with OR that we have (among other things) knowledge of independently existing entities, i.e. it agrees with a neutral variant of its epistemological thesis. Moreover, not every NR rejects *both* basic ontological features of OR (SI and the single external world). Also, there are different positions within NR concerning how to understand (or completely reject) (in)dependence as a reality criterion and how to relate to the (external) world. By minimally rejecting one of the two basic ontological convictions of OR, however, the structural thesis about the fixed structure of any subject-independent external world no longer follows in *any* New Realism. This provides a good criterion for testing whether a given realism is a NR. After analysing correlationism, a further criterion is added to this first one.

Categorisations related to OR are also widespread in the NR literature. For Gabriel, *old realism* is central, and is defined by SI of OR:

---

[18] Without being able to justify this categorisation in detail in the context of my definition of OR at this point, I would at least like to point out the self-assignment to realism in very similar definitions here: Chalmers 2009; Churchland 1981, 1986; Dennett 1991; Kitcher 2001; McDowell 1981, 1998a; Millikan 1984; Priest 2014; Williams 2006.

[19] Cf. the influential contributions: Gabriel 2015a, in the extended, German version, 2016; Benoist 2011, 2017, 2021; Ferraris 2014; Gabriel 2015b, 2024a, speculative realism (SR) must be mentioned as an importantly related current, some works of which must also be counted as NR under the definition used here, esp. Brassier 2007; Meillassoux 2007, 2008, 2013a, 2014, 2016, the classification between NR and OR is questionable in: Boghossian 2007, cf. also the overviews in: Alloa and During 2018; Benoist 2018; Gabriel 2014.

> Old realism is the assumption that there is a reality independent of our thought, speech, action, consciousness and *Geist*. The reason given for this is that thinking, speaking, acting, mental and conscious beings have not always existed. This motivates the conclusion that reality cannot be identical with reality as it appears to us. For there was reality before us, and there will be reality after us. The old realism presents this as the gold standard of reality. Realism in the old sense is therefore the commitment that there is such a reality that is independent of our thoughts, actions and consciousness and that this reality decides what really exists. This conceptual definition automatically gets the old realism into hot water. [. . .] If we define reality using the concept of independence from consciousness, the question immediately arises as to what the state of consciousness, language, thought and mind is. These qualities then appear to us through the conceptual definition as unreal, as not really belonging, as mysterious. Eckholdt and Gabriel 2019, p. 9; my translation.

First, it can be noted that the two ontological features of OR explicated above overlap with features of old realism according to Gabriel,[20] however, with a strong emphasis on the first feature of SI. According to Gabriel, this also allows NR to formulate an objection, since the two features of old realism directly impose the following problem on it: How do subjects, and with them consciousness, mind, language, intentionality and humans in general, fit into the single real external world? As much as our bodies belong in a causal, physical world, now old realism needs an explanation as to how the subjective can appear in the objective external world, the "world without spectators" (Gabriel 2016, p. 36, my translation). Gabriel and other New Realists, but also (anti-realist) critics of metaphysical realism, cite a number of arguments that arise from this PROBLEM OF THE SUBJECT. But the second ontological feature of OR is also the target of arguments by Meillassoux, Gabriel and other anti-realists, correlationists and new realists, especially against the totalising, absolute concept of a single real external world, which is defined as the whole or set of all real entities. In addition, there are independent, more far-reaching reasons to argue against OR. I will take a stand on all these points in the context of analysing the arguments against OR in philosophy of science.[21]

New, old and possibly middle realism can usually be traced back to a connection between temporal epochs with different philosophical paradigms, as can be seen from the temporality of the designations—this goes far beyond the debates of ontology.[22] In addition to aesthetics, media theory and cultural philosophy, terms of the same name can especially be found in the fields of literary studies, anthropology,

---

[20] If it has not yet become clear from the previous quote that Gabriel also understands the second ontological feature of OR as one of old realism, the following passage should be pointed out: "The old realism is not free of cost, yet many still believe that it is correct. Even many philosophers still associate the term "realism" with precisely this assumption of a consciousness-independent external world." Eckholdt and Gabriel 2019, p. 10.

[21] Cf. my Sect. 5.1–5.3, pp. 109–158.

[22] This can range from contributions in the (anti-)realism debates of philosophy in the usual sense, the only ones considered up to this point, but also beyond. An example of this categorisation from aesthetics, media theory and cultural philosophy can be found in: Schweppenhäuser 2018, pp. 9–70 on old realism, pp. 71–139 on middle realism and pp. 141–206 on new realism.

art history and political science.[23] These concepts of realism, which are widespread in wider philosophical, humanities and social science discourse, do not merely exhibit family resemblances, but are part of a discursive network with the terms used in ontology, epistemology, metaphysics and philosophy of science, which are at the core of my work. In this context, Gabriel, Benoist and Maurizio Ferraris in particular speak formatively of an old and a new realism. However, the historical dimension of the terms is put aside at this point in favour of a systematic consideration. New Realism is used in my context as a first approximation only as a systematic definition in contrast to OR and anti-realism, it does not imply a historical judgement.

At present, there are certainly many more ORs than NRs, and it is by no means clear which current will prevail in future. It would be legitimate to choose other names for these positions in order to avoid any misunderstandings, but—as has already been explained above with regard to *realism* and related terms—I would then abandon the goal of sharpening the terms that haunt the debate. The positions of SPECULATIVE REALISM, if one includes here the stances adopted by Quentin Meillassoux, Iain Grant and Graham Harman, as well as the early Ray Brassier and their followers, belong in part to NR, namely in the cases of Meillassoux and early Brassier. However, the concept of speculative realism is not chosen here as the counter-concept of OR: On the one hand, not all NR positions, such as those of Gabriel, fall under the definition of speculative realism that Meillassoux, Grant, Harman and Brassier have jointly put forward. They conceive of speculative realism as a form of realism that permits ontological speculation and opposes the fundamental anti-realist tendencies of postmodernism and post-structuralism, whatever is meant by this exactly.[24] On the other hand, there is an opposition to correlationism in speculative realism, but not necessarily to OR. NR and speculative realism share a subset of positions, but the essential term for my consideration is NR. For the moment, the negative definition of NEW REALISM as *any realism that rejects the first, the second or both ontological features of OR—and thus the structure thesis—without falling into metalocal or global anti-realism* suffices.

* * *

This concludes my analysis of the debates about realism and anti-realism and the introduction of my concept of objective realism. I can now turn to the metalocal debate in philosophy of science and show how the positions of OR attempt to distinguish themselves from each other and to defend themselves against anti-realist criticism. Through this consideration, I will then show why the various objective realisms are in a stalemate with each other and with anti-realism in philosophy of science.

[23] For the discussion in literary studies, cf. Lavocat 2016; Strowick 2019; Thanner et al. 2018 and in anthropology Danowski and de Castro 2017; Holbrad and Pedersen 2017. Overlap between litary theory and speculative realism exists, such as in Longo 2015b; Meillassoux 2012. The terms used in art history and political science are much older. Cf. here for an overview Bark and von Nayhauss 2015; Korab-Karpowicz 2023; Morgenthau 1954.

[24] This is the original definition of speculative realism, cf. Meillassoux et al. 2007, followed by a lively discussion on how this can be better and more precisely defined, cf. Bryant et al. 2011; De Sanctis and Longo 2015.

Chapter 4
Objective Realism in Philosophy of Science

In this chapter and the next, I will substantiate my main hypothesis in Part I of this book: *Even the most promising positions of objective realism (OR) in philosophy of science are in an argumentative double stalemate—on the one hand against each other, on the other against anti-realist critique*. In this chapter, I will outline the best-known and most promising forms of OR in philosophy of science and explain how they differ from each another in terms of their arguments for OR. In the following chapter, the core of the double stalemate reveals itself through an exploration of the main (counter-)arguments in this debate.

The *first subchapter* introduces the most important current of OR in philosophy of science, *naturalism in its various forms*. From the debate surrounding this term, I will argue for a definition of a realistic *naturalism* that must be categorised as OR in the context of philosophy of science (4.1.1). I go on to show that this definition of naturalism is valid and not a straw man. To this end, I analyse a selection of the central ideas of one of the most important naturalist of the 20th century, Willard Van Orman Quine (4.1.2). These include the idea of ontological commitments acquired through the acceptance of scientific theories, of the unity of science, and of reductive physicalism as constituting core features of what I call *reductive naturalism*. My aim is not to present a comprehensive account of Quine's position on science, but to show that he advocates a naturalism as encompassed by my definition and to clarify the over-all form he gives it through the additional theses of the unity of science and reducibility to physics. The persistent lack of empirical unity in the natural sciences and the problems of reduction and theories of supervenience are addressed in the following step (4.1.3). They motivate the search for alternatives within objective realism and naturalism.[1]

[1] This is not only motivated by the internal mechanism of any philosophical debate, but leads to real differences of how we report on, use and value scientific theories and results, cf. Pils 2025; Pils and Schoenegger 2024.

J. Voosholz, *Towards a New Scientific Realism*, Synthese Library 518,
https://doi.org/10.1007/978-3-031-98886-8_4

Two influential variants of objective realism in philosophy of science are reconstructed in the *second subchapter*. First, I consider a proposal for how naturalism can be advocated without reductionism. To this end, I will analyse the naturalist work of Wilfrid Sellars, who, like Quine, is still extremely influential in today's debate (4.2.1). Here, too, my aim is not to develop a full-blown interpretation of Sellars, but to demonstrate his approach to a *non-reductive naturalism* in philosophy of science, which, as with Quine's naturalism, still finds many adherents today. By providing a glimpse into the interpretative debate surrounding Sellars, I strive to show that different readings of Sellars fashion different varieties of OR in the metalocal debate in philosophy of science. By looking at the scope for interpretation, it also becomes clear that a lucid philosophy of science is needed in order to introduce non-reductive naturalism as an alternative to reductive naturalism. To this end, I go on to discuss Anjan Chakravartty's attempt to clarify the question what kind of metaphysics *any* realism in philosophy of science must and should presuppose (4.2.2). This opens a route towards Chakravartty's minimal position of realism, his so-called semirealism. Here, too, Chakravartty's position will not be comprehensively reconstructed. Instead, I show that Chakravartty's take leads to a lucid non-reductive naturalism and therefore counts as a promising OR in the context philosophy of science. This step also shows that naturalism is still the standard form of OR in philosophy of science today. In conclusion, the reconstructions and discussions will allow me to specify the hypothesis underlying philosophy of science's OR in today's debate. This prepares the ground on which, in the concluding Chap. 5 of Part I, I can prove my hypothesis of the double stalemate on the basis of the most important arguments concerning OR in philosophy of science.

4.1 Objective Realism I: Reductive Naturalism

OR was introduced in the previous chapter as the position that defends the idea that *subject-independent (SI) entities* exist that together constitute a *single real external world* and which we can recognise in at least some of their properties. Consequently, we have knowledge about some SI entities and the single real external world. I subsequently argue that this leads to the *structure thesis* of OR. This is the assumption that the SI external world has a fixed structure, i.e. that the external world entities have stable patterns of relations and interactions.

In Chap. 2, I argued that the debate about (anti-)realism in philosophy of science is a *metalocal* one: It combines the debate about observable and unobservable theoretical entities (i.e. those described in scientific theories) with other debates. This includes the question of whether individual scientific ontologies, which must be assumed for all theories of the respective sciences, have only a functional or instrumental character or whether they comprise real entities. Other closely related realism disputes revolve around the status of laws of nature, natural causality and the question of the claims to truth of scientific results, methods and models.

If the insights from both chapters are combined, the plausibility of advocating an OR in philosophy of science becomes apparent. This is particularly true with regard to the empirical natural sciences. If, as an objective realist, one assumes that the best methods and theories are predetermined by the single real external world and, in a second step, that the best ontology is not introduced by speculative philosophical reflection in metaphysics and epistemology, but instead by the empirical natural sciences, I represent the basic ideas of *naturalism* (in a broad sense of the word). This already foreshadows why naturalists are always objective realists in my use of these terms. If, in the debate of philosophy of science, one insists that the (natural) sciences[2] grasp entities as they are, objectively and independently of our interventions, that they are always governed by laws of nature, and that this is the privileged way of understanding the (external) world in terms of knowledge, then all the core elements of OR are implied.

4.1.1 What Is Naturalism?

First, I would like to clarify the concept of naturalism, as it diverges in similar ways to that of realism; naturalism often remains vague as a self- and external designation. For example, Putnam complains "a further very common feature of [naturalist writings; J. V.] is that, as a rule "naturalism" is not defined" (Putnam 2004, p. 59), after he compares the affirmation of naturalism in current Anglophone philosophy with the affirmation of the compatibility of one's own opinion with that of Stalin during the period of the Stalinist Soviet Union. In the article "Naturalism" in the *Stanford Encyclopedia of Philosophy*, David Papineau claims in the very first sentence: "The term "naturalism" has no very precise meaning in contemporary philosophy." (Papineau 2020.) Papineau does not develop a definition in what follows, and only outlines various theses that are often associated with naturalism. He notes that the majority of English-speaking philosophers today identify as naturalists.

In the context of the methodology of the present book, it is necessary to define naturalism more precisely, as otherwise my criticism is easily lost in a "swarm of naturalisms", as Demmerling (2008, p. 241, my translation) calls the multitude of heterogeneous uses of the term. Since naturalism as a strong realist position in philosophy of science will be in my focus, I will define the term in such a way that it corresponds to the widespread idea of an ontological-realist naturalism that banishes everything supernatural from the world, and, following Quine, aligns itself with the ontologies of the natural sciences. I combine this with an orientation towards

[2] Naturally, the empirical social sciences, technical sciences and mathematics, and possibly also the humanities, could also be counted among the core disciplines that are in contact with the external world. As I will show later, naturalists often either limit themselves to a single leading science (usually physics) or at least draw a line under the non-mathematising sciences.

a methodological naturalism that associates epistemology with the philosophy of science and scientific methods of knowledge acquisition. Naturalism understood in this way aligns its own philosophical methodology, in the tradition of logical positivism and Karl R. Popper, with Francis Bacon's scientific method of hypothesis generation, logical testing and verification or falsification. My aim is to unite these frequently mentioned basic aspects in my definition and to identify naturalism as an attractive, obvious, but ultimately inadequate position that also belongs to OR.

For the purpose and context of this book, NATURALISM means any form of realism that holds at least the following assumptions:

N1 A reliance on scientific methodology, especially that of the natural sciences.

N2 The thesis that our best or future or ideal scientific theories are true, especially those of the natural sciences.

N3 The thesis that our best or future or ideal scientific ontologies (presupposed by the true theories) are correct, especially those of the natural sciences.

N4 The thesis that these scientific ontologies contain all entities of reality/the external world that exist in a sophisticated sense and independently of any theorising and subjectivity, especially ours.

N5 The thesis that there is a strong criterion of truth and in particular that the verificationist theory of truth is incorrect, according to which the meaning of a proposition consists in the method of its verification, and also that the consensus theory of truth is incorrect, which defines truth as the agreement of beliefs within an intersubjective community. Scientific theories are true in a strong sense, as suggested by the correspondence theory of truth or the disquotational theory of truth.

N+ Following N4, naturalism also generally assumes that our best or future or ideal scientific ontologies fulfil the task of ontology as *prima philosophia* and essentially settle the question of the existence of things, thus rendering dogmatic metaphysics superfluous.[3] It follows from this that individual sciences and philosophy are continuous and that the theories of philosophy should therefore be evaluated according to the same criteria. N+ also excludes the influence of something supernatural (gods and goddesses, spirits, a supernatural human soul, etc.) on nature and the external world.[4] This often also includes the assertion that subjectivity or the human mind is generated by nervous systems. N4 is often associated with the term *ontological naturalism*, but the assumptions in N+ do not commit to naturalism in my sense. Thus, one only needs to accept N1-5 to be a realist naturalist in my sense.

Realist naturalism, thus defined, is closely related to logical positivism, which advocates an anti-realist/empiricist twist on these assumptions by introducing a verificationist theory of truth for the interpretation of assumptions 2 and 3 or proposing

[3] An illuminating idea of the naturalist as *second philosopher* is elaborated in: Maddy 2007, esp. Parts I and II.

[4] Cf. for a naturalism in which N+ is decisively rejected Ellis 2014.

empiricist equivalents to assumptions 4 and 5 in which scientific ontologies are reduced to scientific languages that are accepted or rejeted for pragmatic reasons. Quine's demarcation from Carnap's variant of logical empiricism can be seen as groundbreaking. Accordingly, realist naturalism (from here onwards I will abbreviate to "naturalism") is not necessarily committed to the correspondence theory of truth. However, representatives of naturalism usually use it to explain the truth of scientific theories. According to the correspondence theory in philosophy of science, a theory is true if it correctly represents its target system or the part of nature that it describes. This explains its ability to make predictions, act as an explanation for phenomena or make the target system or part of nature technologically controllable. Alternatively, some naturalists follow Quine in his disquotational and deflationary theory of truth. However, this leads to the same consequences as the correspondence theory in the following respect: In both cases, naturalists are committed to the fact that scientific theories presuppose certain scientific ontologies that are understood as systems of theoretical entities (e.g. electrons, galaxies) and laws of nature (e.g. the law of gravity) and to whose elements the theory refers. This basic insight of Quine's is shared by all naturalists in my sense.

N1 is often given as the only criterion for the ascription of naturalism in a broad sense, as so-called "methodological naturalism".[5] In this respect, the term is only of limited use in my view, especially in the context of my study. A stricter methodological commitment, for example of philosophy itself to the methods of the natural sciences, could appear to be a solution to the problem of overgeneralisation. However, this would have the disadvantage of possibly excluding classical figures of naturalism such as Wilfrid Sellars.

I do not want to deny that reliance on scientific methodology, especially of the natural sciences, played and still plays a central role in naturalism. It has contributed significantly to the understanding of philosophy in naturalistic circles and beyond. However, this does not ensure realism. What is more, Amanda Bryant even argues that naturalised metaphysics can be had without a commitment to (scientific) realism (cf. Bryant 2024). As will be explained in the next chapter, there are indeed anti-realist positions that regard science as a futile enterprise and that are therefore sceptical about its methodology. Other anti-realisms do not consider the scientific approach to reality to be privileged. It simply stands alongside other "ways of accessing the world", such as religion, art, poetry or the "everyday" reference to the world. Categorising naturalism simply as accepting science as one way of accessing the world is not sufficient because all variants of OR, NR, SR, but also constructive empiricism, scientific fictionalism and logical positivism would then all have to be considered as naturalism. Contrary to this version of methodological naturalism (or comparable "liberal naturalisms" such as Putnam's; cf. for this term Caro 2025), the

[5] Cf. e.g. Forrest 2000; Maddy 2011, p. 120: "The leading theme of methodological naturalism [...] is roughly that science is the best way we have of finding out about the world. If we could define "science", this rough version would be enough [...]." Similarly broad definitions are widespread, e.g. in Williamson 2007, p. 18: "everything is part of the natural world, and should be studied as such".

term I use is a narrow one: My definition combines methodological and ontological naturalism with a commitment to an OR.[6]

However, different variants of naturalism can still be presented, some of which show significant differences in the resulting philosophical positions. This applies not only to the philosophical understanding of individual branches of science or science as a whole, but also to ontology and epistemology. Quine, probably the most prominent naturalist, describes it as follows:

> [Naturalism; J. V.] sees natural science as an inquiry into reality, fallible and corrigible but not answerable to any supra-scientific tribunal, and not in need of any justification beyond observation and the hypothetico-deductive method. [...] [A] source of naturalism is unregenerate realism, the robust state of mind of the natural scientist who has never felt any qualms beyond the negotiable uncertainties internal to science. [...]
> The naturalistic philosopher begins his reasoning within the inherited world theory as a going concern. He tentatively believes all of it, but also believes that some unidentified portions are wrong. He tries to improve, clarify, and understand the system from within. He is the busy sailor adrift on Neurath's boat. Quine 1981a, p. 72.

Neurath's boat is a metaphor that Quine likes to use: "Neurath has likened science to a boat which, if we are to rebuild it, we must rebuild plank by plank while staying afloat in it. The philosopher and the scientist are in the same boat." (Quine 1960, p. 3.) This already indicates that Quine neither believes that we have reached the end of scientific inquiry, nor that all current scientific knowledge claims are true, but that many of them are. This also makes the naturalist a workmate of the natural scientist. At the centre, however, is the assertion that the natural sciences are concerned with the single real external world, which Quine here simply calls *reality*, and that no *prima philosophia* or metaphysics is necessary to justify it.[7] Quine remains true to this conviction of Carnap and logical positivism.[8] He stands in this tradition with his argument that it is necessary to accept a certain domain ontology if a scientific theory is held to be true:

> When we want to check on existence, bodies have it over other objects on the score of their perceptibility. But we have moved now to the question of checking not on existence, but on imputations of existence: on what a theory says exists. The question is when to maintain that a theory assumes a given object, or objects of a given sort—numbers, say, or sets of numbers, or properties, of points. To show that a theory assumes a given object, or objects of a given class, we have to show that the theory would be false if that object did not exist,

[6] This is still often assumed as a background condition of scientific realism *per se*, cf. Park 2022a; I will take up some of the problems laid out by Park in his Chap. 11 in my Chap. 8.

[7] Quine 1969c, p. 26: "I hold that knowledge, mind, and meaning are part of the same world that they have to do with, and that they are to be studied in the same empirical spirit that animates natural science. There is no place for a prior philosophy."

[8] Quine's rejection of anti-realist verificationism, which characterised the early Vienna Circle and some of the empiricists of logical positivism (cf. Friedl 2021), as well as his criticism of Carnap, especially in his rejection of the analytic-synthetic distinction, cannot be discussed in detail here for reasons of brevity. Cf. e.g. Carnap 1998, 2003; George 2000; Price 1997; Richardson 1997; Surovell 2017; Thomasson 2015.

> or if that class were empty; hence that the theory requires that object, or members of that class, in order to be true. [...]
> To show that some given object is required in a theory, what we have to show is no more nor less than that that object is required, for the truth of the theory, to be among the values over which the bound variables range. Quine 1969b, pp. 93–94.

Quine thus claims that if it is necessary for the truth of a theory that there is something—an entity—to which the values of certain bound variables of this theory must be able to refer, we accept the existence of the entity or class of entities in question. If someone asserts the truth of this theory, then this person is ontologically obliged and committed to assume the existence of the entity or class of entities in the subject-independent, single real external world. On this point, as can be infered from my definition above, all naturalists in my sense follow Quine. What does the ontological commitment of scientific theories mean for naturalism and how is it spelt out if we follow Quine's reasoning?

4.1.2 Reductive Physicalism, Quine and the Unity of Science

For Quine and other naturalists, the crux of the ontological commitment to a certain SCIENTIFIC ONTOLOGY is that we assert the existence of certain entities when we assume the truth of any (natural) scientific theory: "Our ontology is determined once we have fixed upon the over-all conceptual scheme which is to accommodate science in the broadest sense [...]." (Quine 1953, p. 36.) If it is claimed that Newton's theory of gravitation is true, then there must, for example, be extended bodies with a fixed mass and velocity per point in time for which the axioms of the theory apply. As can be seen from the discussion of naturalism above, it is now assumed that the best or future or ideal scientific ontologies are correct (thesis N3) and that all the best/future/ideal scientific ontologies together contain all the entities of the subject-independent, single real external world (thesis N4). Two questions arise at this point: First, are our best current scientific ontologies, those of the future, or even merely ideal ones, *de facto* unattainable ontologies that correctly determine the external world? Second, can the scientific ontologies of different true theories and different sciences be unified? Both questions are not answered unanimously in naturalism.

In what follows, I will postpone the first question:[9] An examination of the various answers to the second question will show that all attempts run into problems. Thus, how the answer to the first question is resolved for different naturalists is not of central import for my argument. Quine, as can be seen from the quotations in the previous section, thinks that some of our current theories are true and the associated scientific ontologies are correct. These are our best theories today. Others are most likely false, and it is the task of the philosopher and of the scientist to find out

[9] For a critique of *ideal scientific ontologies* combined with a defence of naturalism, cf. Schulte 2010.

together which those are. In a finite amount of time, we must not hope to ask *all* scientific questions and to answer them correctly. This optimistic, but not hopelessly naïve view of the sciences rests on Quine's answer to the second question:

Quine believes that the different domain ontologies of science can be unified because there is a basic science into which the theories of all other sciences, including naturalistic philosophy, can be translated—physics. He thus combines his naturalism with a reductionism in philosophy of science, epistemology and ontology: all true theories can in principle be reduced to physical theories, as physics investigates the fundamental elements of the external world, the universe. This means that the human cognitive process can also be theoretically reduced to these. The ultimate ontology of the universe, the outside, the single real external world, is written by physics. I refer to this cluster of theses as REDUCTIVE PHYSICALISM (RP).

Whereas Carnap and Neurath's *physicalism* was still based on a language-theoretical idea—*all meaningful propositions can be translated into the mathematical-physical universal language of science*—it is now based, through Quine and the dominance of naturalism, on the ontological, epistemological, and philosophy of science idea described above (cf. Keil 1993, p. 33). Naturalists must answer, on the basis of their ontological commitments, whether it is possible to unify all ontologies of science. For from this answer follows whether it is possible to arrive at a single, empirically-scientifically grounded ontology, which will also be the ultimate ontology of the external world. It is no coincidence that Quine chooses physics, which investigates the smallest and largest entities of the universe and its most fundamental laws.[10] Physicalism is often understood in such a way that it accepts *concrete entities*, which are spatiotemporally localised and material-energetic, as the only subject-independent entities of the external world. This view, already introduced above as ELIMINATIVE MATERIALISM, can go hand in hand with the position I call RP, but it does not have to. Instead, my definition given above allows that there are also abstract entities, especially mathematical entities, since, depending on the prefered interpretation of physics, these also occur in the scientific ontology of the leading science of physics.[11]

[10] On the discussion of the reduction of biological theories to physical theories, which continues to this day, cf. e.g. the criticism of Nagel 1961, by: Kaiser 2012.

[11] This leniency with RP is also appropriate because, firstly, it demands an abstracta-nominalism, which is contrary to the realistic impetus of many self-declared naturalists, and, secondly, the stricter variant threatens to slip into epistemological scepticism due to its (apparent) inexpressibility: If there were only material-energetic entities, it would be questionable what discourse or a linguistic statement exactly is and how it should be possible for one part of it to have the property of truth. In accordance with the principle of charity, RP should therefore not be understood in this way.

In his middle and late writings, Quine is a representative of RP in my sense, as he ascribes existence to some abstract mathematical entities. He originated the so-called indispensability argument for certain abstract mathematical entities:[12]

(P1) Some mathematical theories are indispensable for our best scientific theories and cannot be eliminated.
(P2) We should accept everything that is indispensable for our best scientific theories and cannot be eliminated. (Follows from N3, N4.)
(P3) We should accept those mathematical theories that are indispensable in the said sense. (Follows from P1, P2.)
(P4) These mathematical theories are committed to realistic assumptions of the existence of certain abstract mathematical entities. The scientific ontology of mathematics requires at least sets or numbers as entities.
(P5) We should accept the existence of those entities to which our best theories are committed. (Follows from N3.)
(K) We should accept the existence of those abstract, mathematical entities, at least sets or numbers.

If (P1) is accepted, it appears as if a strict, ontologically monistic, eliminative materialism is untenable, according to which there are only concrete, physically describable, material-energetic entities. This would mean that PR cannot be eliminative materialism after all. Eliminative materialists must therefore work through (P1). They must show that the supposedly most fundamental theories of physics and their own meta-theory (according to which the most fundamental theories of physics completely describe the single real, subject-independent external world and the subject-independent entities) can be formulated with these entities only. No concept of the truth of this theory can be developed, nor can a formulation of these theories be given that is (not only in principle but in fact) reducible to the fundamental concrete entities (strings, particles of the Standard Model, etc.) on pain of self-contradiction. In my view, it seems simply contradictory to claim that anyone could show that one's own assertion is actually reducible to these concrete entities. I think Quine saw this and did not try argue for eliminative materialism, but clearly argued instead for a version of RP with the indispensability argument in mind.

The mapping I undertook in Chap. 2 proves helpful here: By advocating a metalocal realism in philosophy of science, Quine is able to formulate a new argument in favour of the realist position in the local debate about mathematical entities and the metalocal debate about abstracta. This succeeds because at least Quine and Putnam assume that mathematical entities are necessarily abstract, as (P4) explicates here.[13] According to RP, the external world consists of the physical,

[12] Cf. e.g. Quine 1953, esp. pp. 15–19, 1951, esp. pp. 36–42; 1966; the argument was supplemented by Putnam 1971; cf. to the following presentation Colyvan 2001.

[13] Maddy 1990, tries to argue for a naturalism that turns mathematical entities into properties of groups of concrete entities.

concrete universe and possibly some abstracta. Thus it can be stated that RP is a naturalism that includes these two theses in addition to N1–N5:

RP1 All scientific ontologies can in principle be unified in a single scientific ontology, the best approximation of which is current physics as the leading and basic science.

RP2 Every genuine advancement of knowledge can thus be expressed by reference to entities of the unified ontology of science, i.e. in the form of statements of physics. Epistemology not only coincides with philosophy of science; it can also be reduced to perceptual psychology, neurobiology and furthermore to organic physics.

4.1.2.1 The Unity of the Sciences

RP is not the only variant of reductionism within naturalism, but it is on the one hand so dominant and on the other hand so similar to other reductionisms in its internal difficulties that I feel justified in focussing on it for the purpose of this book. RP implies a unity of all the sciences and this hypothesis is problematic.

The assumption of naturalism with RP1 and RP2 leads to the following conviction: If the sciences recognise the truth about some entities and the sciences C and B can be reduced to science A in the sense that all true theories of C and B can be fully expressed by the more fundamental theories of science A, then our C-knowledge and B-knowledge is, in truth, A-knowledge. Moreover, this relationship is transitive: if science D can be reduced to C under the same conditions, then D can also be reduced to A. If all sciences can be reduced to a single science A in this sense, then there is a *unified science* A with which all scientific, i.e. from the point of view of naturalism, objective knowledge can be expressed.

Most reductionists admit that this unified science has not yet been fully achieved, but is either an ideal that can only be reached theoretically or a state of science that lies in the future. This is where the question of unity touches on the other question posed above. Physics is often assumed to be the most fundamental science, as in the classical unified scientific formulation of RP by Paul Oppenheim and Putnam (1958) in *Unity of Science as a Working Hypothesis*.[14] In this article, the authors claim that the thesis of unified science is empirical and not an ideal. They argue that the most elementary building blocks of the universe are explained by physics, meaning that in the future more and more knowledge of all empirical sciences—including the empirical social sciences and humanities—can be understood physically.

For Oppenheim and the early Putnam, the sciences can be roughly summarised as follows: Sociological and socio-psychological theories can be reduced to the theories of psychology, which in turn can be reduced to those of (neuro-)biology. The biological theories can be reduced to those of (organic) chemistry. Chemistry,

[14] Important precursors can be identified in the Vienna Circle, for an example of the universal language of physics for all sciences cf. Carnap 1931.

in turn, can be reduced to physics. Therefore, at the end of research, every theory of social psychology on buying behaviour on the art market could eventually be reformulated in a (presumably quite complicated) physical expression. While this basic structure of RP only seems to make statements in philosophy of science, there are ontological consequences of RP:

Together with the naturalistic theses N1–N5, it follows that the scientific ontology of the unified science of physics is the fundamental ontology of the external world. If all empirical knowledge can be reformulated in such a way that it is physical, one is committed solely to the entities of the best or future physics. This is ontologically parsimonious and is usually further justified with a reference to Occam's razor. Moreover, the physical entities are (at least, when superficially viewed with the gaze of the general philosopher of science who has turned away from practice) *tame in terms of epistemology and ontology* because they are accessible through precise definitions of mathematical formal language and are numerically measurable.

Sometimes, though not always, this results in the convenient ontological *monism* described earlier: Everything in the external world, which turns out to be the universe of theoretical physics, is material-energetic. This is the already well-known thesis of eliminative materialism, in which the human mind is also considered to be a material-energetic system or its effect or property by complete reduction (cf. e.g. Churchland 1981; Churchland 1986). However, RP does not have to follow this idea, but can be extended to an almost equally convenient ontological *dualism*. On the one hand, the external world consists of the universe of theoretical physics, i.e. material-energetic objects. On the other hand, it contains some (maybe even only a few) abstract objects (e.g. the empty set), from which, according to the dualists of RP, mathematics and thus the formulaic language of physics can be fully derived (for instance, by the construction of all numbers from the empty set). Furthermore, alternative theories of subjectivity and the laws of nature that involve abstractions can be developed. This means all entities of the external world are either concrete or abstract. A great advantage of RP in both cases of dualism and monism is that the old dogmatic metaphysics, which sees itself as the basis of scientific investigation, can be rejected, just as a strong, science-friendly realism can be advocated against scepticism and idealism.[15]

4.1.2.2 Quine's Holism and the Principle of Ontological Relativity

On the basis of the provisions given above, Quine's naturalism can be understood as a dualistic form of RP (cf. other receptions e.g. Verhaegh 2018). Further aspects of his naturalism can be summarised as follows:

Firstly, Quine pursues the project of naturalising epistemology (which essentially corresponds to philosophy of science) itself. In this way he aims to assign it a place

[15] Cf. a critique of Quine's refutation of scepticism Bernecker 2003.

among the natural sciences—or at least next to them in the same boat. Philosophy and natural science appear in a continuum with each other.

Secondly, he develops a theory of truth that is, on the one hand, clearly realistic and distinguishes itself from the verificationism advocated by early Carnap, Neurath and representatives of logical positivism. On the other hand, Quine wants to avoid the presuppositional correspondence theory of truth advocated by many other reductive physicalists. This is the trajectory from which Quine's deflationary, disquotational theory of truth comes.

Thirdly, Quine advocates a Platonism regarding mathematical entities as abstracta. With the indispensability argument, however, Quine binds this abstract level to the true theories of the natural sciences.

In addition, there is another central idea of Quine's of importance here, as he deems it necessary for his form of RP. This is the hypothesis of a *holism of knowledge*. It is complemented by Quine's hypothesis of *ontological relativity*. Both raise immense problems. I focus here on the first aspect, Quine's holism, because it illustrates how much his RP is attached to the idea of a unified science of physics. Ontological relativity roughly states that our ontology is always coupled with our human perspective due to our epistemic and linguistic limitations (cf. esp. Quine 1969c). The various tensions that this hypothesis causes within his naturalism—commentators even see a contradiction here[16]— I will not go into further detail here, without wishing to skim over this further problem of his position. Whether Quine's stance can be upheld in this case, however, is no longer relevant if the criticisms in the next sections and the general anti- and neorealist counterarguments against OR apply.

Quine asserts a HOLISM of knowledge, human cognition and scientific hypotheses. According to his position, these three things are inextricably intertwined: We possess knowledge when we hold a true theory or hypothesis. Knowledge is particularly relevant when we recognise connections between the subject-independent external world, i.e. when we formulate a correct theory about one of its aspects or domains. Since there is a unified science that can describe the universe as a whole in its most basic elements and most fundamental laws of nature, all true theories are necessarily interrelated. Because our knowledge is thus structured like a network, it can only be described holistically, according to Quine.

What holism has in common with the hypothesis of ontological relativity is that our epistemic access to the subject-independent external world becomes problematic from the perspective of Quine's naturalism. Quine uses both concepts to address, among other things, the concern that our limited perspective as animals with contingent episodes of perception is in tension with the global truth claims of our comprehensive production of theories. This is to take account of the fact that humans sometimes err and that in the past scientific theories were accepted as true when they were false, such as Newton's celestial mechanics.

[16] Cf. e.g. Fogelin 1997, for the convincing conclusion that ontological relativity and Quine's holism are not compatible with his strong naturalism, as I have also characterised it here.

Quine's holism of knowledge is derived from the understanding of epistemology that I have already discussed, according to which it is a reflection of the best natural sciences and is simultaneously assigned to them.[17] Epistemology and natural science are in a reciprocal relationship in Quine's opinion, but in different senses: Quine's epistemology is *qua* its subject matter—the research processes of the human species—a part of social psychology and thus, under the assumption of reductionism, part of the natural sciences. However, it is precisely these natural sciences that become the object of research in this discipline, whereby Quine reduces epistemology to philosophy of science. Quine also believes in the unified science of physics.

If these views are supplemented by the fact that ontological commitments depend on theories that are accepted as true, a problematic consequence arises: It is not immediately clear which theory should be accepted as true if there are two or more theories that fit equally well with the observations, evidence and data, but cannot be reconciled with each other—either for epistemic reasons (e.g. because they make different, non-validatable predictions) or for ontological reasons (e.g. because they presuppose different entities). From this problem, Quine derives the argument of *holistic underdetermination*.[18] Quine begins with Pierre Duhem's insight that a single hypothesis or a single sentence of a theory does not yet result in a prediction that can be empirically tested. He thus argues against the idealised scientific procedure that verificationists have in mind:

> It is that the typical statement about bodies has no fund of experiential implications it can call its own. A substantial mass of theory, taken together, will commonly have experiential implications; this is how we make verifiable predictions. [...] But the failure falsifies only a block of a theory as a whole, a conjunction of many statements. The failure shows that one or more of those statements is false, but it does not show which. Quine 1969a, p. 79.

From this, Quine concludes that the language of science cannot be reduced to simple observation sentences and mathematical statements. Verificationism is therefore untenable. However, the implications of this idea threaten Quine's realist-naturalist convictions: there is no *experimentum crucis* for a theory, i.e. no way to specifically test an individual scientific hypothesis empirically through experiment or observation in isolation. Only the totality of all hypotheses of a theory and other background assumptions of an ontological nature—the stability of all regularities in the part of nature under consideration and often also those of other presupposed theories—serve as a holistic network that enables predictions about and testing of

[17] Quine 1969a, begins with the statement, p. 72: "Epistemology is concerned with the foundation of science." Quine continues in the same text, pp. 82–83: "Epistemology, or something like it, simply falls into place as a chapter of psychology and hence of natural science. It studies a natural phenomenon, viz. a physical human subject. This human subject is accorded a certain experimentally controlled input—certain patterns of irradiation in assorted frequencies, for instance—and in the fulness of time the subject delivers as output a description of the three-dimensional external world and its history."

[18] For more details on underdetermination arguments and this very specific point, see Sect. 5.1 of this book, pp. 109–122.

empirical reality. According to Quine, this explains the holistic underdetermination of these theories established by the different types of evidence. This is where Quine concludes his justification for the holism of our knowledge:

> The totality of our so-called knowledge or beliefs, from the most casual matters of geography and history to the most profound laws of atomic physics or even of pure mathematics and logic, is a man-made fabric which impinges on experience only along the edges. Or, to change the figure, total science is like a field of force whose boundary conditions are experience. A conflict with experience at the periphery occasions readjustments in the interior of the field. But the total field is so underdetermined by its boundary conditions, experience, that there is much latitude of choice as to what statements to reevaluate in the light of any single contrary experience. No particular experiences are linked with any particular statements in the interior of the field, except indirectly through considerations of equilibrium affecting the field as a whole. Quine 1951, p. 39.

In Quine's holistic theory of knowledge and science, this means that we are free to hold some beliefs, even if data or evidence seem to contradict them, provided that we are willing to change other parts of the field of knowledge claims, that is, to abandon other beliefs and adopt new ones. Only in the fringe areas that are supported by a *substantial mass of theory* can we validate findings. Moreover, every theory is related to every other theory via the unified science. By abandoning and changing some parts, which has consequences for other parts, our knowledge can be better adapted to the circumstances and data. However, conversely, this means that we can, if we wish, hold on to any single hypothesis, no matter what observations we make, if we are only willing to adjust other parts of the web sufficiently.

In the light of Quine's naturalism, this fact is problematic. If it is true, then the door is open to sceptical doubts about individual areas of theories. This seems to conflict with the methodological optimism of naturalism (thesis N1). If no *experimentum crucis* is possible and any hypothesis can be brought into agreement with the evidence, is it not likely that many scientists will go wrong in practice? Does this lack of certainty not sow doubt that certain theories will be defended regardless of their aptness if they can only be supported (or invalidated) with the help of a broad system of our beliefs?

Besides N1, there is another, more serious problem: As Robert Fogelin notes, holism and ontological relativity are at odds with the "hard" (Fogelin 1997, p. 543) part of Quine's naturalism, especially the hypotheses of RP. Several of Quine's critics emphasise this point. Berry Stroud draws on knowledge holism and Quine's understanding of underdetermination when he says that Quine's RP version of naturalism is caught up in a variant of scepticism:

> Countless "hypotheses" or "theories" could be "projected" from those same slender "data", so if we happen to accept one such "theory" over others it cannot be because of any objective superiority it enjoys over possible or actual competitors [...] our continued adherence to our present "theory" could be explained only by appeal to some feature or other of the knowing subjects rather than of the world they claim to know. And that is precisely what the traditional epistemologist has always seen as undermining our knowledge of the external world. Stroud 1984, p. 248.

Penelope Maddy summarises this point succinctly when writing about Stroud's criticism of Quine: "Because Quine's line on [ontological; J. V.] positing and underdetermination is supposed to have resulted from scientific inquiry, Quine's science has undermined itself from within." (Maddy 2007, p. 90.) Indeed, when holism is brought together with thesis N3,[19] a picture emerges that is fatal both for the claims to truth of (natural) science and its theories (i.e. thesis N2) and for the second hypothesis of the RP about the reducibility of all true knowledge to physical statements. Quine's holism forces us to recognise that only blocks of theories can be tested on the basis of evidence. Any single hypothesis can be reconciled with any evidence by a sufficient number of other adjustments in the web of beliefs. If both are true, then it is unclear why it is inappropriate to be sceptical about all knowledge claims about the external world. This is what Stroud and Maddy attest to. If the unified mathematical-physical ontology continues to hang on the truth of *all* theories, it seems presumptuous to think that even a substantial proportion (let alone the majority) of our current scientific ontology of external world entities is correct. But how then is RP's thesis that epistemology/philosophy of science can be reduced first to perceptual psychology, then to neurobiology and finally to (organic) physics supposed to be convincing?[20]

We can draw the interim conclusion that Quine's attempt to represent naturalism together with RP runs into a series of problems that are difficult to resolve. In addition to the unfortunate introduction of the idea of a holism of knowledge, there are further difficulties that affect not only Quine's renowned variant of RP and naturalism, but also his intellectual heirs who attempt to represent an alternative variant of these positions.

4.1.3 Quine's Legacy: Pseudo Entities and the Disunity of Science

The problem that holistic underdetermination poses for Quine's naturalism is obvious. RP in particular is threatened by it. It faces two further challenges: the *explanatory problem of pseudo entities* and the *problem of the disunity of science*. I will briefly discuss both of the resulting counterarguments and explain why they, together with the problem of holistic underdetermination, make RP an unattractive position. This makes it possible to understand why, in addition to defenders of RP

[19] It should be noted that thesis N3 states that the scientific ontologies presupposed by our best theories are correct.

[20] Sometimes, commentators doubt if Quine was indeed a realist. Thomas Bonk points out that Quine presupposes realism for naturalism, but cannot justify it effectively from his holism: Bonk 2006, whereas Haack 1993, argues that Quine vacillates between two variants of naturalism, which are differently compatible with RP on the one hand and with traditional philosophy on the other.

and forms of reductive naturalism, there are also those who accept non-reductive variants of naturalism, which I discuss in the next Sect. 4.2.

4.1.3.1 The Explanatory Problem of Pseudo Entities

First, I present both problems, beginning with the *explanatory problem of pseudo entities*: PSEUDO ENTITIES in the sense of RP are entities that are neither abstracta grounded in our best scientific ontologies (like some mathematical entities), nor are they concrete entities. Examples include such diverse things as mental entities—beliefs, feelings, plans—, moral values, social institutions, or the human mind. They do not all fall obviously into one of the two categories of a dualistic RP. They pose a problem for Quine and other proponents of this position (let alone eliminativist materialism) because they occur in our language and practices, and reference to them is (apparently) successful.

From RP's point of view, there are three options for how to deal with pseudo entities:[21] The first option is that we are mistaken about the respective pseudo entity in our practice of reference, and so it lives up to its name. We then only think that the entity exists, but in reality we are not talking about anything, so we are only producing verbal, unscientific nonsense. Here RP would have to provide an error theory for any type of pseudo entity. The second option of RP is to conceive the pseudo entity as in truth *identical* with a concrete system or one of its properties. In this case, discussion of the pseudo entity can in principle be eliminated and replaced by scientific language. The third option is the assertion that the pseudo entity exists ontologically only secondarily and is dependent on a concrete system. This means that the entity in a sense exists but is constantly dependent on certain ontologically primary and robust entities, namely concrete, material-energetic systems. RP then states that the pseudo entity supervenes on the concrete entity. SUPERVENIENCE here means that a set of properties A of an entity x supervenes on another set of properties B of the entity x exactly when no entity y can exist that has other A-properties than x, but identical B-properties.

The first option, the presentation of a convincing *error theory*, may be a viable route for some of the examples above to adopt, such as values or morality. For the entire metalocal domain involved in the case of RP, any attempt fails: It not only seems evident that I am not mistaken about the existence of my own feelings in the present; more importantly, the thesis that thinking, mind, hypothesising and scientific discourse are merely errors negates the very enabling conditions of science and thus runs counter to basic naturalistic conviction.

Similar objections can be levelled against the second option, *eliminativism*. It describes the hypothesis that a pseudo entity is identical with a concrete system or one of its properties. In relation to the mind and thinking, it is prominently

[21] I am not aware that the theoretical possibility of tracing the typical candidates for pseudo entities back to ontologically robust mathematical abstractions has been pursued, so I neglect it here.

represented—e.g. in variants of eliminative materialism by Patricia and Paul Churchland or as so-called strong illusionism by Keith Frankish and Dennett—but it is also strongly criticised.[22] This debate cannot be reconstructed here in its breadth.

To cite just one striking counterargument, I would like to highlight the problem of multiple realisability (cf. e.g. Shoemaker 2007, Chap. 2; Francescotti 1998): That an entity is MULTIPLE REALISABLE means that it can be realised in completely different systems. The desire to go dancing again after the corona pandemic was realised in completely different biological systems in different ways, so it supervenes differently. According to eliminativism, however, this would mean that two individuals cannot possibly share this desire, as they consist in two completely different states, since the conscious mental content of the desire does not exist, but merely similar properties of two nervous systems. However, these two highly complex systems are completely different. Therefore, this does not seem to explain what is to be explained.[23] This argument, together with the self-application argument against the global thesis of eliminative materialism already set out in the previous section, shows that this option is very difficult to defend.

The third option, on the other hand, the advocacy of *supervenience theories*, especially of mind and thought, enjoys great popularity among (RP) naturalists.[24] Mind, mental contents and other pseudo entities are assumed to exist, but are determined in constant ontological dependence on concrete systems, since they supervene on their properties. Mental properties thereby supervene on physical properties.[25] As a weak hypothesis, however, this does not explain enough: it would only say that there can be no physically identical bodies that possess different mental properties. Assuming sufficiently strict criteria (there are *de facto* no two exactly physically identical human bodies or persons with mental content), the hypothesis appears trivial, but does not solve the problem of pseudo entities. The

[22] Cf. e.g. the defences in: Churchland 1981, 1986; Dennett 2016, 2019; Frankish 2016; Poslajko 2019, and the criticism in: Chalmers 2020; Corns 2016; Pitman 2003, I treat strong illusionism regarding consciousness here as a special form of eliminativism, since Frankish and Dennett claim more than *mere* error theory, namely that all mental pseudo entities can be eliminated in favour of a set of neuronal states.

[23] The same applies to scientific theories themselves: Quantum theory can be presented equally well on a digital computer, in a printed handbook and in a series of lectures by an expert with a blackboard. The concrete systems are completely different in each case and no common properties can be specified that are identical to "quantum theory".

[24] In philosophy of mind, RP is also represented in the variant of realisation theories, which are closely related to supervenience theories, but which are ignored here because usually they do not claim to justify RP as a philosophy of science or ontology. Cf. e.g. Shoemaker (2007). For supervenience and against such realisation theories argues Kim (2011), noting that the objections I raise against supervenience-RP as a global theory of science can also be raised against realisation-RP.

[25] I should note here that these supervenience theories of RP are reductionism in my usage of the word. In the self-designation of these positions, especially in philosophy of mind, these positions are often called *non-reductive physicalism* to distinguish themselves from what is referred to as reductive physicalism, which is labelled *eliminativism* in my nomenclature.

supervenience theorist with RP in mind must go further and advocate a strong or global supervenience thesis.[26] She may, for example, claim that the same mental properties of two persons supervene on the same physical properties, in which case supervenience is associated with ontological dependence.[27] RP is even often defined in terms of supervenience, as by Michael Roche (2016, p. 670): "(Supervenience Physicalism) Any possible world of type T that is a minimal physical duplicate of our world is a duplicate of our world simpliciter. "

However, several aspects of this third option seem extremely questionable to me and many proponents of the debate. There are a number of strong counterarguments in the literature: Robert Francescotti and Ronald Endicott criticise that it is not clear how such a strong variant of supervenience as RP assumes should be possible for the mental realm if mental properties are not simply physical properties: "[I]t is a mystery why the supervenience required by physicalism should be thought to obtain if mental properties are not physical." (Francescotti 2014, p. 29.) Then the third option would in truth either be just a veiled variant of the second option, and the same counterarguments would apply; the realm of the mental would be completely reducible to the realm of the physical and thus eliminable. Or else, RP allows that mental properties are not physical. In that case, however, according to Francescotti and Endicott the strong variant of supervenience would not be tenable.

This second route leads to the core of Francescotti's criticism (cf. Francescotti 2014, pp. 29–43): Supervenience-RP must (1) allow, for example, that mental properties cannot be completely reduced to physical properties, (2) assume a strong supervenience relation with ontological dependence of mental on physical properties, (3) prevent that mental entities must therefore be postulated that exist independently and outside the physical universe. In order to fulfil these conditions, the physical systems would have to evoke the mental properties and thus be their carriers. But this process of evocation is impossible, since physical systems do not have to have mental properties for logical reasons, nor for physical reasons (the known natural laws of our universe). According to Francescotti, logical or conceptual reasons cannot exist, since in the first instance only physical conclusions are possible with purely physical premises and in the second instance physical vocabulary would have to be clearly translatable into mental vocabulary, which is not the case. Physical reasons would imply that the laws of nature concern not only physical properties but also mental properties, which—due to the ontological commitment of naturalism to the entities of our best theories—would entail that mental entities exist.[28]

[26] On the distinction of supervenience theses cf. Bennett and McLaughlin 2018.

[27] Cf. on the point that supervenience does not necessarily imply ontological dependence McLaughlin 1995 and on the point that for RP an *middling global supervenience* with ontological dependence would be necessary, cf. Bennett 2004.

[28] I have greatly simplified the research discussion here for the sake of brevity. Francescotti supports this conclusion with further arguments, e.g. that it is possible for all physical properties of natural history to remain unchanged while a mental property changes. If this scenario is possible, it proves that no physical or logical law governs the supervenience relation between the physical and

Endicott formulates it similarly: "[A] total theory of the world must have the resources to show that the ontology of all non-fundamental theories is explainable in a physically acceptable way." (Endicott 2016, p. 2232.) According to Endicott, the task of RP is either to explain which physical processes ensure that supervenience relations exist in every area of the apparent entities, or why the apparent entities and supervenience relations belong to the ontological presuppositions of physical theories themselves, i.e. to the scientific ontology of physics. Neither alternative seems feasible: There are no known *physical* reasons *why* certain constellations of matter and energy must be correlated with the occurrence of certain pseudo entities. The integration of pseudo entities such as justice, fraud, the plan to go to a party and magistrates' courts into the scientific ontology of physics (i.e. depriving them of their "pseudo" status and making them fully-fledged physical entities) is not only implausible, but also contrary to the intentions of RP.

Terry Horgan and Kevin Morris give a related but more direct argument against supervenience-RP as a thesis of metaphysics and philosophy of science, as cited above in Roche's variant (cf. the arguments in Horgan 1993, 2006; Morris 2018, and the discussion in: Wilson 2005): For such a global thesis to be true, even the strong supervenience described would not suffice; instead, a "superdupervenience" (Horgan 1993, p. 577.) would have to be assumed. That is a metaphysical principle *sui generis* that explains the properties falsely ascribed to pseudo entities. This would not only generate the ontological dependence, but could also be expressed in a purely physical theory. In their view, however, this implies a direct self-contradiction: According to RP, such a metaphysical principle, the validity of which must be presupposed, would have to come directly from physics. But there is no such principle in physics. Philosophically charged discussions in physics seem more to be conducted between eliminativists and emergentists.[29] Morris even suggests that in a purely physical world such metaphysical principles *sui generis* do not apply on principle (cf. Morris 2018, p. 277).

To summarise, it is not clear how supervenience theories catch up with *all* apparent entities to be explained via physical properties. Supervenience denotes a relation between two sets of properties of *one* existing, concrete (or mathematical) entity according to RP, the bearer of the ontologically primary and secondary (supervening) properties. If the RP claims that all mental, social, etc. entities supervene on physical entities in the sense of an ontological dependency relationship, there is a need for explanation that seems hard to give. A significant problem in such an explanation is that a mere correlation between a physical entity and a pseudo entity is not sufficient to show that in the second case we deal with a merely secondary property of the physical entity. Especially for non-localised pseudo entities, such as

the mental. Cf. Francescotti 2014, pp. 41–42 and his remarks on the incompatibility of RP and basal supervenience in Francescotti 1998. The status of supervenient mental properties is controversial in the current debate in the philosophy of mind: some argue in favour of irreducibility like Paull and Sider 1992, others doubt that such a solution still has the desired ontological dependence and argue in favour of reducibility: Kim 1993a,b.

[29] Cf. e.g. the discussion between Ellis 2016, and Luu and Meißner 2021.

justice or a price slump on the stock market, it is difficult to specify the physical entity that is a substitute and bearer of this pseudo entity. At least all aspects that are necessary for scientific practice, to which RP refers, would have to be explained in this way by supervenience theory. However, the burden of explanation seems unbearable.

Thus, it can be said that RP seems to have no options that have any prospect of success in solving the problem of pseudo entities.[30] Perhaps a new option could be proposed here that escapes all counterarguments. The burden of the argument lies clearly on the side of RP. However, there is another, equally serious problem for RP: the disunity of the sciences.

4.1.3.2 The Problem of the Disunity of the Natural Sciences

The last central problem of RP that I would like to address is the fact that the hypothesis of a unified science of physics stands on feet of clay. The empirical evidence from the sociology and history of science speaks against it. It seems as if a proliferation of disciplines is taking place, resulting in the plurality of sciences that we are currently observing.[31] This applies in methodological, discursive and ontological terms. Arguments in favour of this idea about the sciences and supporting empirical studies come, for example, from the so-called Stanford School in philosophy of science. However, Quine already admits that the current sciences are not "monolithic" (Quine 1975, p. 314) in their holistic constitution.

Of course, there are attempts to link different theories with each other through inter- and transdisciplinary endeavours. However, in such cases it is not clear that this must lead to a *unification* of the sciences or theories involved. This applies to the practical-organisational level as well as to the choice of methods and with regard to existing, accepted theories or their scientific ontologies. Demmerling, for example, argues that a one-sided recourse to the methods of physics does not appear to be "particularly promising for reasons of meaningful methodological pluralism. A one-sided diet leads to deficiency symptoms. This also applies to the sciences." (Demmerling 2008, p. 243; my translation.) He therefore rejects this form of physics-scientism.

Furthermore, in the present day not even the physics of condensed matter can be reduced to seemingly more fundamental areas of physics, such as particle

[30] To be sure, I have not considered many arguments from the over-arching debate here. A damning argument, if valid, was made by Merricks 1998, according to this, obvious versions of supervenience RP, such as the microphysical supervenience theory—the thesis that any two concrete entities with the same microphysical composition must have identical intrinsic properties—are self-contradictory. Merricks argues by assuming the intrinsic property of a person to be conscious, conceiving of the person as a concrete entity with this supervenient property, and then introducing physical variations that microphysical supervenience theories cannot handle.

[31] For philosophical and historical reflections on this hypothesis, cf. Galison and Stump 1996.

physics.[32] Chemistry or even biology, including evolutionary biology, ecology and neurobiology, are out of the question with regard to the actual reduction of theories altogether.[33] The reference to the future, which some reductionist physicalists attempt, seems problematic, since from an anti-dogmatic position—think here of the early Putnam and Oppenheim—it would have to be a purely empirical, thus open question as to how science will develop in the future. For the reductionist physicalist, this leaves only the option of claiming that there are compelling, meta-theoretical reasons for the future existence of a unified science and fundamental scientific ontology.

A classical reason for assuming the unity of science(s) is given by the *generality argument*: All events that fall under the laws and within the structures of any science are generally also physical events that fall under physical laws of nature.[34] Thus runs one justification of RP's unity hypothesis. Jerry Fodor contends against this argument and the unity of the sciences: bridge laws are needed that enable the transformation of a law from the so-called special sciences (e.g. history or psychology) into physical laws. These bridging laws state that some historical or psychological event necessarily goes hand in hand with a certain physical event according to the laws of nature. The background assumption is that *all* events are also always physical events. For Fodor, this seems difficult to justify without committing a circular argument with RP. Even if such an argument were possible, it would not show that any scientific property must also be a physical property. For even if a certain e.g. psychological or historical event is necessarily accompanied by a certain physical event (as the bridge law requires), this does not mean that both have the same properties.[35] Even if we grant that this is not a problem for RP, the generality argument still does not lead to RP's conclusion that all scientific ontologies can be unified into a single ontology by which all genuine knowledge can be reformulated. According to Fodor, the reason for this is that the premises of RP would then imply that every conceivable *natural kind* would also be a *physical kind*.[36] All certain psychological or historical events about which lawful generalisations can be made would demand this consequence. Fodor cites Gresham's law from economics as an example, according to which one type of

[32] This is the subject of a longer debate in physics and philosophy of physics: Clark and Lancaster 2017; Drossel 2021; Kiefer 2007; Luu and Meißner 2021.

[33] This debate about the life sciences is also complex. From a philosophical perspective, see for example Dupré 1993, Chap. 5./6.; from that of theoretical biology, see i.e. Noble 2012, 2016, 2021.

[34] Cf. also for the following Fodor 1974, pp. 97–107. I do not analyse here Fodor's attempt in the second half of his paper to preserve the unity of the sciences in a non-reductionist way, since, first, it no longer supports RP and, second, is not very convincing in light of the further arguments.

[35] The property of my toothache to be unpleasant does not imply that the state of my nervous system is unpleasant, even if the toothache necessarily depends on the state of my nervous system by the laws of nature.

[36] Here I use *natural kind* as a *terminus technicus* of the debate, since Fodor does not offer a definition either. In this context, it can be understood to mean any natural category of entities into which they fall of their own accord.

money displaces another from circulation under certain economic conditions. He thus shows that it would be surprising if all conceivable forms of money—cash, cheques, electronic data—were also of the same physical kind. It would be even more surprising if the objects of this physical kind fell under a certain physical law that predicted their circulation. Yet, this is exactly what is claimed not only by the generality argument for all theories in all sciences, but also by the sum of the background assumptions of RP.

Other central counterarguments have been made by members of the Stanford School in philosophy of science, which includes John Dupré, Ian Hacking, Peter Galison and Nancy Cartwright, as well as Patrick Suppes. The latter aptly summarises the group's approach to unity and disunity: "The rallying cry of unity followed by three cheers for reductionism should now be replaced by a patient examination of the many ways in which different sciences differ in language, subject matter and method, as well as by synoptic views of the ways in which they are alike". (Suppes 1978, p. 9.) What the proponents of the Stanford School have in common is above all a commitment to the pluralism of scientific methods and different scientific cultures between disciplines and paradigms.[37] However, other important objections to the unity of the sciences also stem from their ranks, which are derived from considerations of scientific ontologies, laws of nature and the empirical plurality of theories.

Dupré argues for the disunity of the sciences by emphasising that the classification and theory systems of different sciences subdivide reality into other natural kinds, each of which can prove real causal efficacy at their respective level of observation with a plurality of methods (cf. also for the following Dupré 1993, chap. 1–4 and 8–10 and the commentary on kind realism in Chakravartty 2023). From this he draws the conclusion, among other things, that reality could neither be completely described with the domain ontology of the microscale in particle physics; nor is reality causally closed on this scale, since influence can be exerted from other scales. Elsewhere, Dupré argues that laws at the macrolevel are only stable if the factors at the microlevel are not resolved too finely (cf. also for the following Dupré 1983, pp. 331–336). He uses the predator-prey relationship from ecology as an example: In populations in which the relationships at the population level hold up empirically excellent, more difficulties arise the more detailed the attempt to take into account other factors such as the constitution of the individual hunters, the specific environment of the respective hunt, the circumstances of the prey, etc. Dupré sees this as proof of the independence of the population level from the individual level in ecology. He argues that even in these cases, there is irreducibility and inconsistency, but that that is not a problem. Following Dupré, a unity of the sciences can only be established if the sciences are viewed as an open, historical process. However, this view carries no weight for an imagined ontological,

[37] Cf. e.g. Galison 1998; Hacking 1996; Kusch 2010 although I will not discuss these complex historical and meta-theoretical points separately here, as they have provoked many objections even among anti-realist relativists, as the last article suggests.

methodological and epistemological unity; on the contrary, this view supports their historically necessary plurality. In this sense, Dupré argues for a plurality of ontologies of science and a pluralistic epistemology, which he recommends philosophy of science should recognise and explore (cf. Dupré 1993, pp. 242–243).

Cartwright rejects any attempt to save the uniformity of science by arguing that, despite the different domain ontologies, methods and disciplinary cultures, at least the goal of the sciences is uniform, namely to formulate universally valid laws that provide explanations for phenomena (cf. also for the following Cartwright 1983, 1999, esp. pp. 23–95). From a purely naturalistic perspective, there is no reason to assume that the world is so homogeneously structured that even laws of nature that are assumed to be fundamental always apply, everywhere and under all conditions. Cartwright counters this uniformity view of the laws of nature with a *dappled world*, a pluralistic view of reality, which in turn has become the target of fundamental criticism (cf. e.g. the strong criticism of Ruphy 2016, pp. 58–65) But even this criticism does not call into question the fact that all sciences cannot possibly be committed to the goal of law-like explanations for phenomena in the same way. The unity of the sciences in the sense of RP cannot be justified by such a manoeuvre.

A unity of the sciences—or even only of the natural sciences or physics—can be neither empirically established nor meta-theoretically proven. It is true that the excerpts from the debate cited here have not proven that the unity of science is a fundamentally untenable idea. However, they have shown that the current distribution of the burden of proof is clear: RP seems to be based on a questionable empirical assumption. Referring to the future does not help the empirical facts. With the current gaps between quantum and relativity theory, there are even non-reducibilities in physics because, first, there is no unified basic theory in sight for all fields—unlike in the time of Newton's mechanics—and because, second, subfields such as condensed matter physics formulate models and laws that cannot in fact be fully expressed in quantum theory and relativity physics. Since Quine advocates a broad concept of "science" that includes at least empirical social research and historical science, it seems unlikely that Quine's RP programme will ever be realised. The empirical study of the scientific landscape suggests instead that science will always be plural, and the meta-scientific considerations outlined above also speak in favour of this view. However, this would mean that the monism or dualism envisaged by reductionists lack any foundation: a reference to *the sciences* does not justify that only mathematical and physical entities should be included in our scientific ontologies, since this move already presupposes all possible other entities, including minds, mental entities and social institutions.

To avoid this difficulty, of course, a narrower concept of science can be devised that includes, for example, only quantum theory (or a future development of it) within physics. But even this position cannot be fully justified. Anti-realist arguments such as Larry Laudan's pessimistic meta-induction—in which it is inferred from the empirical fact that fundamental physical theories have been shown to be false in the past that our present theories are likely to be proven false in the future—are an almost unmanageable challenge for an RP that accepts only current or future quantum theory as science, for this RP could only make use of empirical findings of (future) physics and the scientific methodology of abduction (cf. Laudan 1981, and my chap. 5.2). But even if these historical-empirical challenges from PR were answerable, the question would remain open as to why precisely this one scientific ontology of quantum theory should prove to be *the* ontology of the external world—and not any other scientific ontology. The attempt to justify this by claiming that quantum theory describes the most fundamental entities—quanta—from which all other entities are constructed, must fail. This is clearly a circular argument, since a monistic, reductionist, object-centred ontology of RP is already presupposed, and which is to be justified. In my view, actual reducibility and the unity of the sciences cannot be demonstrated empirically or deduced meta-theoretically; there are even good reasons to assume the opposite. Neither physics nor any other (natural) science can be shown to be the absolute basis of all sciences, which means that there is not just one, but that there are many ontologies of science. This is not to say that there cannot be unity in the sciences in other respects, but the sciences are not unified ontologically.

* * *

Given the problems of pseudo-entities and the disunity of sciences, it is questionable whether RP-naturalism is a tenable position. This also applies to any other reductionist naturalism. It is therefore hardly surprising that some naturalists do not want to see themselves as Quine's heirs and turn away from reductionist projects. In the context of my argument here, this raises the question of whether non-reductionist naturalisms have a better chance of representing an OR in philosophy of science. However, the analysed counterarguments to reductive naturalism do *not* disprove the Quinian idea that for a realistic interpretation of a certain scientific theory, the associated scientific ontology must also be assumed to be correct. The alternative varieties of naturalism also remain true to this idea of ontological commitment in a broad sense, as I will show in the following subchapter.

4.2 Objective Realism II: Non-Reductive Naturalism

The internal difficulties of RP, such as the compatibility of Quine's holism with naturalism, and external problems, such as that of pseudo entities and the disunity of the sciences, are overwhelming. Nevertheless, the appeal of naturalism remains unbroken, as the number of its current proponents confirms. It is therefore hardly surprising that realists in philosophy of science look for alternative positions within the spectrum of naturalist OR. This subchapter deals with such alternatives.

The positions can be categorised with reference to the map of the debates that I drew in Chap. 2 of this book. For most realists, OR undoubtedly appears to be the most minimal position that can be defended. If the failure of RP is accepted, but the attractiveness and philosophical stance of naturalism is convincing, then the realist will try to defend a NON-REDUCTIVE NATURALISM (NRN). By this I designate all positions that defend the theses N1–N5 of naturalism but reject both theses of RP. I will show that an important reason for the attractiveness of NRN is that it clarifies not only the question of unobservable theoretical entities, but also the *whole metalocal* realism-antirealism debate in philosophy of science. The positions of Wilfrid Sellars and Anjan Chakravartty will serve as my examples of NRN in this subchapter.

There are also alternatives to NRN: If an objective realist is not convinced by naturalism's understanding of philosophy and science or its core ideas, she can try to advocate a form of anti-naturalistic pragmatism. This can be outlined as a position according to which human action is inextricably linked to our knowledge and scientific insights. Without analysing them in more detail in the present context, it can be said that some of these positions threaten to slip into anti-realism, as is often claimed in the debate for Richard Rorty's pragmatism (cf. Backmann et al. 2005). Other pragmatist positions, such as Chakravartty's semirealism, which are decidedly realist, I will examine with regard to the question of whether they are a form of OR. Another alternative to NRN is structural realism, which introduces scientific ontologies based on structures that differ from classic naturalism. I will elaborate on these further options of an epistemic or ontic structural realism in my Chap. 5.4, following the anti-realist and neo-realist critiques of naturalism, since strucural realism is essentially a reaction to the poor state of OR in philosophy of science.

Overall, my aim in this subchapter is to show that the OR's alternatives to reductive naturalism are subject to a double critique: On the one hand, they threaten to slide into an ontological or epistemological anti-realism; on the other hand, they have to commit themselves to a daring interpretation of OR that is closely tied to the subjects' epistemic access to the external world.[38] In the next chapter, it will be shown that the OR is not sufficiently equipped to meet the challenges posed by critical arguments, even if the anti-realist alternatives are not convincing.

There is a wide range of NRNs in the landscape of philosophy of science. My task here is not to analyse the full range of positions available. Instead, I will focus on two influential examples the basic convictions of which are taken up or shared by many other representatives of NRN. The first example is Sellars' paradigmatic attempt at NRN. Alongside Quine, he can probably be considered the most influential naturalist of the 20th century and continues to shape the debate. The second position is pragmatist naturalism and so-called semi-realism as represented by Chakravartty. My task is, firstly, to show that both positions represent the naturalistic theses N1–N5 in their basic convictions and clearly reject RP. Secondly, I will show why each position is dependent on a subject-centred interpretation of OR in philosophy of science if it does not wish to be accused of representing a form of anti-realism.

[38] I will later expose this dichotomy as a conflict that stems from the fact that the alternative positions attempt to represent OR and correlationism simultaneously. This is fundamentally problematised in Part II of this book.

4.2.1 Sellars' Two Images

Non-reductive naturalism is not necessarily best understood as an invention to be intruduced through the failure of RP. Quine and his current reductionist heirs have, on the contrary, always received objections, especially from the ranks of naturalism. It would also be short-sighted to think that Sellars' philosophy is a reaction to Quine or primarily a negative counter-proposal to reductionism. Sellar's form of naturalism can be read as both, but it is based on independently productive and far-reaching considerations.

In the following, I would like to show that Sellars, like Quine, affirms all assumptions of naturalism (N1–N5) if it is assumed in the interpretation that he falls back on the subject independence (SI) criterion of OR. However, he argues against the idea that the external world, grasped by science with subject-independent certainty, is the only relevant domain of reality, even if—and here the readings of Sellars already vary[39]—the only one of which we can provide objective truths. What the various interpretations have in common,[40] however, is that Sellars distinguishes the objective, "scientific image of the world", from a "manifest image" (cf. Sellars 1956, 1961, 1962, 1963, 1968). The latter is roughly the concept of the world of everyday reference (other interpreters prefer the term *pre-theoretical*), in which we can locate ourselves self-reflexively. Sellars describes the *manifest image* as "the framework in terms of which man came to be aware of himself as man-in-the-world" (Sellars 1962, p. 42). In this respect, subjects are the most important objects in the ontological drafts of the manifest image, which is why, according to Sellars, subjects are *the* subject of classical philosophy. However, by conceiving of ourselves as perceivers and knowers, it is possible to develop scientific methodology and theory construction, and thus to elaborate the other image, the *scientific image*, according to which the world is an objective and subject-independent external world.[41] The subject-independent external world has, among other things, causally produced our environment and humans as a biological species. Therefore, the scientific image is already as complete as manifest image and contains the human being, so that a tension arises between the two images (cf. Sellars 1962, pp. 56–57).

The interpretative debate on Sellars is extensive and heterogeneous. I will not be able to represent it fully, nor will I be able to clarify all the contentious points of that discussion in my reading here. My aim is more modest: I merely claim that *if*

[39] Cf. for a proof of naturalism in Quine and Sellars and for their differences Rosenberg 2007, Chap. 2, *Sellars and Quine: Compare and Contrast*, pp. 33–46; as well as an argument in favour of naturalism and realism, also in the sense defined in my present inquiry deVries 2005, pp. 15–22, and on Sellars' opposition to the anti-realist forms of empiricism Brandom 2009.

[40] A good overview of different approaches is provided by the essays in deVries 2009, esp. by McDowell, Brandom, deVries, O'Shea and Rosenberg.

[41] Some interpretators would object here. In my Sect. 6.3.3, p. 241, I will present an account of a reading of Sellars as a *correlationist realist* that does not categorise him as part of OR (and, thus, classical naturalism as defined in this chapter).

Sellars is read as an ontological and epistemological *objective realist* who advocates SI as an important criterion of what is real, then such readings of Sellars as a non-reductive naturalist—in the sense of my map of the debate developed in my Chaps. 1 and 4—make good sense. I am explicitly not presenting a fully fledged interpretation of Sellars, nor will I elaborate if anti-realist or correlationist interpretations of his writings may also be justified.

Sellars represents a clear naturalism above all through his famous *scientia mensura* dictum: "In the dimension of describing and explaining the world, science is the measure of all things, of what is that it is, and of what is not that it is not." (Sellars 1956, §41, p. 303.) Jay Rosenberg convincingly shows that Sellars is to be read realistically on this point:

> This understanding of the epistemology of scientific inquiry is robustly realistic. On Sellars' view, stories that postulate "theoretical entities" are not merely manageable second-class surrogates for more complicated and unwieldy stories about entities that we have good, i.e., observational, reasons to believe actually exist. Theoretical entities, rather, are those entities we warrantedly believe to exist for good and sufficient theoretical reasons. The results of scientific inquiry, in fact, are ontologically definitive. Rosenberg 2007, pp. 14–15; here Rosenberg refers mainly to Sellars 1956, 1961.

Unlike Quine, Sellars does not claim to be able to make a reduction to a single scientific ontology and does not advocate a supervenience theory of the mind. The irreducibility in Sellars' naturalism is not limited to assuming a plurality of sciences and methods. For Sellars, the logical space of reasons of the manifest image and our epistemic practices are not fully reducible to the causal processes in space-time.[42] Against this background, his insistence on the *manifest image* is significant at least as a historical precondition of the *scientific image*: It is only through the further developed self-reflexive practices in which we understand ourselves as knowing, perceiving and acting—i.e. practically influencing the world—that it is possible to develop the objective *scientific image*. Sellars thus attempts to represent a NRN, to avoid the problems of supervenience theory and the disunity of science and at the same time to find a way from an object-centred to a process-oriented form of ontology. It can already be seen here that the subjective approach to the objective external world must play a key role in Sellars' philosophy read as an OR.

Depending on Sellars' reading, this consideration of the irreducibility of the epistemic, as in McDowell's interpretation below, leads to a variant of philosophy that asserts a reciprocal, indissoluble entanglement of both images—and thus to a co-dependence. Contrary to this idea, Rosenberg and deVries think that Sellars understands the task of contemporary philosophy as a "fusion"[43] of both pictures.

[42] See Sellars 1956, §5, p. 257: "[T]he idea that epistemic facts can be analysed without remainder—even "in principle"—into non-epistemic facts, whether phenomenal or behavioural, public or private, with no matter how lavish a sprinkling of subjunctives and hypotheticals is, I believe, a radical mistake—a mistake of a piece with the so-called "naturalistic fallacy" in ethics." and the commentary in: O'Shea 2009.

[43] An important passage for such an interpretation can be found in Sellars 1962, pp. 40–41: "the philosopher is confronted not by one complex many-dimensional picture, the unity of which, such

In this way, we could also successfully refer to the subject-independent external world in the manifest discourse in accordance with our non-abandonable, manifest practices (cf. Rosenberg 2007, pp. 9–32; deVries 2005, pp. 12–13 and 276–282). These interpretations by McDowell, Rosenberg and deVries are not exhaustive, and possibly a synthesis of these views could even be developed. But that is not my aim here.

Rather, in my view, this interpretive debate shows the following: Rosenberg and deVries read Sellars as a solid realist who assumes a SI assertion for the current scientific worldview and a future fused picture of reality. If this reading is to be prioritised over McDowell's, then Sellars' philosophy can also be attributed to all assumptions of naturalism. The scientific method is one means of finding out truths about reality (N1).[44] Some theories of science are true in a strong sense (N2 and N5) and they provide ontological knowledge about the domain of the world they treat (N3). Finally, the scientific worldview is complete in the sense that it contains all entities of the single real external world and thus also humans who exist subject-independently (N4).[45]

In the deVries-Rosenberg reading, Sellars represents an OR, since all naturalisms assume a strong OR, as has already been shown. Thus, the arguments against OR in philosophy of science, which will be developed in my next chapter, also apply to Sellars understood in this way. But what about McDowell's Sellars? It is not immediately clear whether we should attest realism if both images are interdependent and indissolubly intertwined.[46] Assuming that scientific ontologies depend on the existence of our scientific practices, then the following assertion is possible: In the *scientific image*, i.e. viewed through a scientific lens, entities such as the solar system or the tree outside my window are independently existent and knowable in a realistic sense.[47] In the *manifest image*, i.e. seen through the everyday

as it is, he must come to appreciate; but by two pictures of essentially the same order of complexity, each of which purports to be a complete picture of man-in-the-world, and which, after separate scrutiny, he must fuse into one vision."

[44] Cf. for proof of these assumptions: Sellars 1965; deVries 2005, Chap. 1, 6 and 10, whereby esp. on pp. 142–157 it is emphasised that Sellars' understanding of the theories and laws of nature deviates precisely as outlinded from the standard naturalistic conception.

[45] Manifest objects, such as the everyday tree outside my window, are thus subject-dependent according to deVries-Rosenberg's Sellars. Cf. e.g. Sellars 1962, p. 26: "Manifest objects are "appearances" to human minds of a reality which is constituted by systems of imperceptible particles." Cf. Sellars 1968, p. 49, 1961, p. 52; and the reading of these passages in: deVries 2005, pp. 158–159.

[46] McDowell seems to have changed his reading of Sellars over time. The interpretative approaches developed in the following can be found in: McDowell 1999, 2009b,g,h, whereby my explanations in the following will sharpen the interpretation to a radical and a realist variant in order to show what a consistent philosophical position would be.

[47] The choice of these examples, which suggest an object ontology, is due to the ease of understanding, at the cost of suppressing the aspect of the processuality of everything that exists in Sellars. Nevertheless, I believe that I am presenting a problem here that would not be resolved even if his process ontology were taken into account.

or pre-theoretical lens, it makes no sense to speak of the solar system. Here only the moon appears, larger than the sun, but less bright, and at night the stars (cf. Rosenberg 2009, p. 290). The tree, on the other hand, receives its meaning as part of the ordinary environment. Which things have existence is essentially determined by the relatedness of language acts, ideas and everyday practices (cf. McDowell 2009b,e). This can be understood as a draft for a *radical* Sellars reading.

McDowell's radical Sellars seems to undermine the distinction between realism and anti-realism by linking the question of the status of any entity to an—in the end almost arbitrary—choice of our world views (cf. McDowell 1999, 2009f, pp. 208–210). Some entities appear in both images, like the tree outside my window, but they conceive the entity ontologically and epistemologically different. Other entities appear only in one image, like the solar system, and disappear in the other.

Here I would like to highlight a problem that arises when my map of the realism-anti-realism debates is brought into play: It is not that easy to get rid of the realism problem. Much depends on presenting the choice between the two images as *almost arbitrary*. If it were completely arbitrary, then a contingent act from a subjective perspective would decide whether, for example, the solar system or genes exist after all. However, this implies a blatant violation of Sellars' *scientia mensura* dictum. According to Sellars, the sciences really do make *discoveries*, they do not invent the world. They describe and explain reality. Therefore, there can be no completely arbitrary choice here.

How then can McDowell's Sellars claim that our understanding of the world determines the ontological and epistemological status of entities? The answer is: only if he assumes a higher level of ontological anti-realism. According to this, both images would be partly contradictory, but indispensable and therefore not literally true. They would relate to each other in parts, but would lead us to different assumptions about the world. This radical Sellars recommends a strong realism and naturalism for the scientific image, but not for the manifest image. This tension could never be resolved, as one image is crucial for explaining our external world, while the other is irreplaceable for opening up our minded inner world. McDowell argues for example, that this radical Sellars would, from a scientific perspective, be an ontological anti-realism regarding everyday objects.[48] This dialectic would be dominant. However, ontological realism would be ruled out even for external world entities such as the tree in front of the window, since the scientific description and explanation is the dominant and more objective one, but it only makes sense within the framework of the scientific picture, which in turn is not fully true and stable.

McDowell's radical Sellars seems easy to reproach. McDowell does not wish to impute such a position to Sellars; he is primarily concerned with Sellars' Kantian

[48] To this end, McDowell first quotes a passage from *Empiricism and the Philosophy of Mind* and comments on it: ""*[S]peaking as a philosopher*, I am quite prepared to say that the common-sense world of physical objects in space and time is unreal—that is, that there are no such things." According to this doctrine, the red cubes and so forth that are, apparently, immediately present to us in intuitions do not really exist." McDowell 2009g, p. 41. McDowell then immediately criticises this idea as "misguided" (ibid., p. 42).

philosophy of perception and intentionality and his critique of the Myth of the Given (cf. McDowell 2009a,d,h). For McDowell's reading, this means that such a higher-level anti-realism must be ruled out. This is also important in order to preserve Sellars' scientific realism, as James O'Shea convincingly argues (cf. O'Shea 2009).

If this is true, then likewise McDowell's more moderate Sellars does not seem to undermine the realism-anti-realism debate in philosophy of science. Instead, he appears to hold a strong realist position that amounts to naturalism. This seems appropriate, if Sellars is interpreted with this McDowellian realist reading: "[A modern philosopher is; J. V.] confronted by two conceptions, equally public, equally nonarbitrary, of man-in-the-world and he cannot shirk the attempt to see how they fall together in one stereoscopic view." (Sellars 1962, p. 41.) A realistic stereoscopic view would be the position that represents OR for *both* images, but holds that the philosophical *conceptions of the ways of looking at reality* are in a relationship of tension, not the different levels and entities of reality. For philosophy of science, McDowell's moderate-realist Sellars, like Rosenberg and deVries, represents OR, but in contrast to his Rosenberg-deVries mirror image, McDowell's Sellars is convinced that the *manifest image* will continue to exist because it describes a part of reality that cannot be captured by the scientific level at all, but that it also requires a realistic description.

At least deVries, on the other hand, interprets Sellars as a realist naturalist in philosophy of science, who represents an ontological anti-realism with regard to abstracta (i.e. AN) and norms (cf. deVries 2005, pp. 18–22). However, supplemented by the idea of a future fusion of both images, this means nothing more than that Sellars also represents a strong OR and naturalism. He expects the fused, synoptic picture to be capable of integrating the necessary normative and minded elements into the plural scientific ontologies using scientific methodology, without reducing or eliminating these elements (cf. also deVries 2005, Chap. 10; O'Shea 2009; Rosenberg 2009). What the very different interpretations have in common, then, is that Sellars' NRN must be a form of strong OR in order to be convincing. At the same time, Sellars must arm himself against the accusation of anti-realism in every reading by defending a realistic view of the immensely central, epistemic access of the subjects.

4.2.2 Chakravartty's Semirealism

Partly due to the difficulty of interpretation, Sellars' position is represented directly in philosophy of science less frequently than Quine's. However, there is no shortage of followers who advocate a version of non-reductive naturalism. At this point, one of the most promising current positions of NRN in philosophy of science will be discussed in order to avoid operating with a straw man in my following critique of the double stalemate: I am referring to Anjan Chakravartty's semirealism. I will first show that semirealism corresponds to my definition of naturalism. In the next chapter, I repeatedly return to Chakravartty's position by showing how the criticism reconstructed there applies to his philosophy.

In *A Metaphysics for Scientific Realism*, Chakravartty attempts to show why the question of what kind of metaphysics any form of realism in philosophy of science *presupposes* is important for both realists and anti-realists (Chakravartty 2007). In the course of this investigation, which was continued in further articles and the monograph *Scientific Ontology*, he develops his position of a semirealism (Chakravartty 2013, 2017a,b, 2018). His argumentational approach consists in the idea that metaphysical questions in philosophy of science are often tied to individual concepts (spacetime, causality, laws of nature, quantum entities, etc.) and remain systematically isolated from the realism-anti-realism debate in practice. He considers this to be a mistake: "For there is a sense in which the metaphysics of science is a precursor to its epistemology. One cannot fully appreciate what it might mean to be a realist until one has a clear picture of what one is being invited to be a realist about." (Chakravartty 2007, p. 26). I agree with the quoted argument, as my map of the debate shows: it is a metalocal debate that does not only concern the ontic and epistemic status of unobservable theory entities, as is often claimed, but which includes the question of the ontology and epistemology of the sciences and their central concepts (encompassing those enumerated above).

In contrast to Chakravartty, however, I see from this conclusion no compulsion to develop a metaphysics for realism in philosophy of science. He understands metaphysics to be the fundamental science that attempts to formulate a theory about the world as such (cf. Chakravartty 2007, p. 18). This definition already makes it clear that the realisms he analyses are always ORs that refer to the single real external world. In the next chapter, however, the antitotality argument is used to argue against this concept of the external world and thus against metaphysics understood in this way in philosophy of science. I consider it more important to deal with the ontological and epistemological level of the problem of realism in philosophy of science than to present a theory of the world as such. I will argue for this in detail in the second part of this book.

Chakravartty names the *underdetermination of theories by evidence, i.e. data, proof and observations* (UTE), the *pessimistic meta-induction* (PMI) and the *argument about the inference to the best explanation* (IBE) as the three main difficulties for realism. While the IBE and the UTE point out that the truth of our best theories is questionable, the correctness of our scientific ontologies uncertain, and the scientific method prone to error, the PMI (and other historical arguments) negates realism: probably our best scientific theories are wrong. According to Chakravartty, the aims of the argument are different. UTE and IBE are invoked to sow doubt about the central tenets of realists, especially naturalistic assumptions, and to emphasise realist arguments as dismissible. Meanwhile, PMI and other historical-relativist arguments attempt to refute realism directly.

Chakravartty defines realism in philosophy of science in three steps: an ontological thesis (there is a subject-independent world or reality), a linguistic thesis (theory statements have a literal meaning, are either true or false) and an epistemological thesis (theories give us knowledge about the world or reality): "That is, predictively successful (mature, non-ad hoc) theories, taken literally as describing the nature of a mind-independent reality are (approximately) true. The things our best scientific theories tell us about entities and processes are decent descriptions of the way the world really is." (Chakravartty 2007, p. 9.) Chakravartty would thus be an objective realist in my sense, if he were additionally to claim that the subject-independent world or reality in the relevant sense contains only scientifically graspable entities. However, he is not a reductionist in any sense, since he neither assumes a reducibility of the sciences to a basic science, nor that smaller or larger structures have priority.

This leads me to the question of whether Chakravartty's semirealism can be interpreted as an NRN. To do this, I must first explain some key points of his theory in order to then show whether they fulfil the naturalistic theses N1–N5. In doing so, I cannot present them in all essential points, but this is not necessary in order to answer the question.

It follows from the quotation above that Chakravartty assumes that scientific theories ontologically commit realists to certain assumptions of existence. This applies to both observable (i.e. directly accessible to the senses) and unobservable entities and properties that are assumed in the respective theory. Chakravartty's semirealism, however, is supposed to be a *selective scepticism*: As a realist, he wants to grant existence to certain concrete structures of nature that are truthfully described by the theory, but he recommends a sceptical, agnostic attitude to other parts of the theory, thus remaining an anti-realist in this respect (cf. Chakravartty 2013, p. 41). This is the origin of the self-designation of semirealism.

Chakravartty believes that the scientifically relevant structure of nature is identical with the relations between first-level properties of natural entities.[49] Chakravartty thus tries to avoid Bertrand Russell's abstract structures, which are a higher-level, formal property of arbitrary (abstract or concrete) systems—and are used by today's so-called epistemic structural realism, as I will argue in the next chapter. Instead, Chakravartty ties his concept of structure to concrete structures of first-level properties of concrete entities. When he speaks of structures, these are the concrete structures present in the various concrete systems in nature, which consist of concrete individual entities. The maximum overall structure of all individual entities is presented by him as the only subject-independent external world. Since this overall structure must in turn be understood concretely, and concrete entities are at least in principle accessible to the natural sciences, I can answer my question: Chakravartty is an objective realist, since he believes that the single real external world contains (in principle) only scientifically ascertainable entities.

Chakravartty now characterises these first-level properties as causal. By this he means that they are dispositions for the interaction of concrete entities in a system and thus for the relations that these relata maintain within it. They form the concrete

[49] Cf. also for the following Chakravartty 2007, pp. 39–40, especially: "the relata I have in mind here are generally, in the first instance, quantitative and determinate properties (such as masses, charges, volumes) of particulars (such as objects, events, processes)".

structure (cf. Chakravartty 2007, p. 41 an elaborate defense and overview of his view until today can be found in Chakravartty 2024). This presupposes the existence of both the first-level causal properties and the entities that bear them—otherwise there would be no concrete structures. Chakravartty agrees with this conclusion.

Now critics might object that semirealism reverts to realism about objects or to a form of reductive naturalism. Thus, selective scepticism would no longer provide a way out of the historical-relativist arguments of anti-realism (cf. Chakravartty 2007, p. 44). Chakravartty opposes this by claiming that semirealism opens up a method that makes it possible to distinguish between those structures described by theories in relation to which we should be realists and those to which this does not apply—and not merely *post hoc*:

"The structures to which realists should commit, echoing the most persuasive insight of ER [entity realism; J. V.], involve properties and relations that are essential to describing our causal connections to the world. Realists should commit to concrete structures that are detected and described in scientific theories." (Chakravartty 2007, p. 47.) According to Chakravartty, the properties that we use for manipulation and measurement are necessarily causal properties that connect us to the respective structures. He calls those *detection properties*.[50] He calls the other properties that we ascribe to certain structures in our theories *auxiliary properties*. Some of these are causal, others are not. In the course of scientific development, some of these are discarded, some become detection properties and some remain auxiliary properties. Semirealism is ontologically committed only to the detection properties and thus a selective scepticism. This shows that Chakravartty differentiates himself from anti-realism by pointing out that the epistemic access of us subjects is central (this is what defines detection properties), but makes objective knowledge about the single real external world accessible to us.

Chakravartty identifies detection properties as properties that are necessary to give a minimal interpretation of a scientific equation or theory when it comes to predictions and measurements related to the equation (Chakravartty 2007, p. 48). His example is Fresnel's system of equations for the refraction of light, which, Chakravartty argues, only contains the intensity and the direction of propagation as detection properties. The ether (and, after Fresnel, electromagnetic fields) are only theoretical entities with solely auxiliary properties of the structure, with regard to which Chakravartty is a sceptic, i.e. an epistemological anti-realist. Light intensity and the direction of propagation are causally linked to our practices of measurement and manipulation so successfully that future theories cannot do without them—here a strong ontological assertion of existence is required. In this way, Chakravartty

[50] Also for the following Chakravartty 2007 p. 47. cf. also p. 48: "Detection properties are connected via causal processes to our instruments and other means of detection. One generally describes these processes in terms of mathematical equations that are or can be interpreted as describing the relations of properties."

means to consider both the PMI and other historical-relativist concerns as well as the realist no-miracle argument.[51]

Thus, semirealism can be identified as NRN: Clearly, Chakravartty represents N1, N4, and N5. He restricts N2 and N3—the assumptions that our best theories are true and the ontologies they presuppose are correct—by his selective scepticism, but makes it clear that detection properties are to be understood ontologically as properties of existing and knowable entities in certain concrete structures of the single real external world. Chakravartty thus represents a maximally cautious variant of naturalism.

Various problems—such as the dispositional conception of causal properties (cf. Spehrs 2024), the separation between detection and auxiliary properties, the question of a minimal interpretation of mathematical formulae and the status of non-mathematising sciences—I will pass over at this point. I have two reasons for this. In order to deal with them adequately, I would need to present Chakravartty's position in much more detail. Semirealism serves as a topical and satisfactory example, why every OR in philosophy of science is in a double stalemate. In the next chapter, I show why, in my view, even his carefully formulated proposal does not hold up. This brings me to my second reason: the points of criticism mentioned address specific difficulties of semirealism. Dealing with and formulating them would be an unnecessary digression.

It remains to be said that semirealism is an OR and a cautious non-reductive naturalism. This should not come as a surprise, since semirealism explicitly aims to be a minimal and well-defended fusion of current realisms.[52] By reconstructing the anti-realist and neo-realist arguments against the OR in philosophy of science in the next chapter, it becomes clear why semirealism also has a difficult standing as a realism.

* * *

With this, I have achieved my aim for this chapter: it is now possible to summarise what constitutes OR within philosophy of science. Before embarking on this task, I would like to relate the result of this chapter back to the map of debates in my second chapter and the introduction of OR in my third chapter. The question arises as to why the two OR positions of Quine and Sellars, as well as Chakravartty's OR of semirealism as a current and tentative position, are all naturalisms. From its beginnings in the middle of the last century to the present day, the metalocal debate in philosophy of science has been more strongly characterised by naturalism than by any other variant of OR. The reason for this can be deduced from the previous analysis:

[51] For an example from chemistry arguing for selective scepticism, cf. Penner (2021), as opposed to "functional realism" Ochiai (2024).

[52] Cf. e.g. Chakravartty 2013, pp. 39–40.

The feature analysis of OR revealed that this form of realism does make some clear ontological-metaphysical presuppositions, but these seem to be minimal. As I have shown in Chap. 3, many authors consider the definition of OR to be a definition of realism in general. From my map of the realism-anti-realism debates and the fact that the debate in philosophy of science is a metalocal one, it follows that positioning oneself within this debate has implications for several local debates (e.g. those about time and about mathematical entities, etc.) and other metalocal debates (for instance, those about abstracta). It is also easy to move from a metalocal debate to a global debate in ontology and epistemology. The advantage of advocating naturalism in philosophy of science is that a universally valid metaphysical position that can be applied to many local debates can be easily established. This applies to all three positions mentioned in more detail. It is a clear advantage that the additional assumptions that have to be made beyond OR are readily accepted by many objective realists. They do not seem outlandish or daring, but consist in an optimism about scientific methods and scientific progress, a great confidence in science to provide knowledge about the fundamental nature of the external world and the central importance of theories for scientific knowledge acquisition. In my view, this explains on a systematic level why the variants of naturalism dominate OR in philosophy of science.[53] It can now be summarised which epistemological thesis (EOSR) and which ontological thesis (OOSR) the OR IN THE CONTEXT OF PHILOSOPHY OF SCIENCE, MEANING OBJECTIVE SCIENTIFIC REALISM (OSR) at least represents:

EOSR We can know the subject-independent entities that together constitute the single real external world, at least in some of their properties and some of their connections to natural laws, by means of scientific methodology and theorising. Our best theories about the external world are therefore approximately or partially true and should be understood literally.

OOSR Our best theories provide in these parts or in approximations a correct ontology of the single real external world, as the latter is subject-independent.

In addition, there is usually the assumption that there is scientific progress; theories and models can therefore gradually describe, explain and control the external world better and more comprehensively. The EOSR and OOSR theses are based on the idea of a subject-independent, single real external world that is governed by fixed structures, in particular fundamental laws of nature, which are responsible for the emergence and properties of concrete entities, their structural relationships and the development of the external world in general. This is usually defended as natural causality.

[53] Of course, there are also philosophical-historical and institutional reasons for this, which I will not go into further at this point.

OSR is thus a seemingly obvious minimal assertion for realism in philosophy of science.[54] The position is less daring than any form of naturalism, whether reductive or non-reductive naturalism, but is encompassed by it. Standard definitions of scientific realism such as those of J. J. C. Smart, Richard Boyd, Bas van Fraassen or Brian Ellis, and many others, imply OSR (Smart 1963b; Boyd 1983, p. 8; van Fraassen 1980; Ellis 2005, esp. p. 374).

In the following chapter, I will argue that the OSR is in a argumentative double stalemate: on the one hand, against anti-realist counterarguments and proposals; on the other hand, in relation to different variants of the OSR, namely semirealism and ontic structural realism. This double diagnosis explains *why*, in my view, neorealist alternatives to OSR are to be sought—namely to resolve both stalemates. The analysis also reveals which elements of OSR are problematic and which challenges need to be overcome by a new scientific realism.

[54] Among the few exceptions to currently represented philosophies of science that reject the OSR but at least want to represent a form of realism are, in my view, Ian Hacking's and Nancy Cartwright's entity realism and Huw Price's pragmatic naturalism. While I discuss why entity realism has significant problems in Chap. 5 (see pp. 119–120 in this book), I will show in Chaps. 6 and 7 that realistic pragmatisms designed as correlationisms run into difficulties (see for Price as a correlationist alternative to OR, p. 240 in this book).

Chapter 5
The Double Stalemate of Objective Scientific Realism

In this fifth chapter, I will argue for the main thesis of the first part of this book, according to which objective scientific realism (OSR) is bound up in a double stalemate. I will proceed as follows:

First, I reconstruct the main arguments against OSR. This shows why different families of neo- and anti-realist arguments put OSR in irresolvable difficulties. I mainly refer to Chakravartty's semirealism as a relevant, topical contribution to show how it fails despite its subtle construction. It also becomes obvious that the counter-arguments do not establish a better position in the form of an anti-realism in philosophy of science. This is the first stalemate between anti-realism and OSR in philosophy of science.

Second, I show why the more recent solution of structural realism, which partly takes into account the basic constellation of the first stalemate, ends up repeating the manoeuvres of the original debate. Since what I will call antitotality arguments, which constitute one of the three central challenges to OSR are not fully comprehended, structural realism itself falls back into the patern of the first stalemate. I demonstrate this by reconstructing the argumentation of Steven French's ontic structural realism and comparing it to Chakravartty's semirealism. My analyses of Chakravartty's attempt to justify OSR as well as French's attempt to overcome it provide reasons why both are able to raise legitimate objections to each other without being able to refute the arguments against their own position. This demonstrates the second stalemate: That between subtle and structural realist OSR. Thus, I conclude by drawing a summary from the entirety of the first part of this book for the case of the metalocal debate in philosophy of science by arguing for the necessity of a New Realism in philosophy of science, a *new scientific realism*. In my view, this is the only way to resolve both stalemates.

Previously, Chap. 4, I showed that central proposals for OSR advocate a form of naturalism that is the predominant form of OR in this metalocal debate. This is particularly understandable in light of the results of the debate map and feature-analysis of OR: naturalism combines the basic assumptions of OR with an optimism

J. Voosholz, *Towards a New Scientific Realism*, Synthese Library 518,
https://doi.org/10.1007/978-3-031-98886-8_5

about scientific methods and scientific progress. This also applies to Chakravartty's semirealism as a moderate, cautious NRN. In this chapter, he provides me with a topical, argumentative example for a subtle OSR. My aim in the first three subchapters is to present the three families of the main arguments against any OSR.[1] An examination of these main arguments based on the research literature shows how they seem to oppose any form of realism in philosophy of science and hit OSR hard:

The first challenge is posed by the group of *underdetermination arguments*. In my analysis in the first subchapter (5.1), I show that these impose a defensive manoeuvre on OSR. *Underdetermination of theories by the data or evidence* (UTE) is a massive problem for OSR in general and naturalism in particular. This is because if UTE is assumed, the (approximate) truth of the theories can no longer be assumed. With Quine's holism and Chakravartty's selective scepticism, I have already presented two strategies in the last chapter to counter underdetermination arguments. This mainly concerns classical formulations of *experimental and holistic UTE*—the idea that no single hypothesis is ever tested in isolation, but that our theory testing must always be theory-laden itself—by authors such as Pierre Duhem, Larry Laudan or Quine (5.1.1). An extreme reading of these forms of UTE amounts to *radical UTE*, which I will show to be excluded from the discussion entirely, however, since those arguments belong to a global scepticism that must (and can) be dealt with not in philosophy of science but in the global debate of epistemology (5.1.2). The form of *contrastive UTE*—based on the question of whether the set of theories we review contains sufficiently good theories—gives rise to a refutation of J. C. C. Smart's abductive argument for OSR based on the *inference to the best explanation* (IBE) (5.1.3). Other arguments, especially Bas van Fraassen's bad lot and model argument, question epistemological realism about unobservable scientific external entities altogether (5.1.4). UTE arguments thus impose defensive manoeuvres on OSR, as I show using the example of Chakravartty and with an eye on alternative realist strategies (5.1.5). I highlight the complexity of the consequences of UTE at the end of the subchapter (5.1.5).

I discuss the second family of *historical arguments against OSR* in the second subchapter (5.2). Incommensurability arguments, the so-called pessimistic meta-induction and other considerations derived from the history of the sciences can be divided into two groups: While weak historical arguments sow doubt about central theses of OSR, strong historical arguments attempt to prove them wrong. A weak historical argument central to the debate is Thomas S. Kuhn's and Imre Lakatos' incommensurability argument concerning different paradigms, which aims to characterise the truth of the best theories as unprovable and improbable (5.2.1). The so-called pessimistic meta-induction (PMI), which is influential in the current debate, is intended to support the conclusion that the theses of OSR are false and therefore ranks among the strong historical arguments. In the reconstruction, however, I show that the label "PMI" conceals various arguments, some of which are

[1] These are mainly anti-realist arguments. Some of the arguments cited are (also) formulated by objective or new realists, but the anti-realist critique should be emphasised.

false and others of which are weak historical arguments that can be turned against certain optimistic assumptions of classical positions of OSR (5.2.2). Kyle Stanford's influential argument of *unconceived alternatives* combines contrastive UTE and historical considerations to refute OSR (5.2.3), and is thus a strong historical argument. I then show how Chakravartty's semirealism nevertheless manages to overcome the hurdle posed by these arguments, although some concessions must be made to the anti-realist (5.2.4). The question of why the critique of the sciences from a philosophical perspective takes on a new meaning against this background is part of this consideration.

While all these arguments are well known in the debate in philosophy of science, the third major challenge of OSR is usually not fully recognised: I will argue in the third subchapter (5.3) that the group of *antitotality arguments* is an unavoidable problem, even for those OSR positions that can overcome the first two challenges. I demonstrate this again using the example of semirealism. Antitotality arguments question whether the concept of the single real external world with a fixed structure can be as all-encompassing and stable as OSR demands: The pseudo entities argument, already discussed in the previous chapter, belongs to this family, as it emphasises that reductionist naturalisms are not successful in covering all of reality ontologically. These arguments are either—in their first type—directed against the ontological construction of a totality of the subject-independent external world. They include Alain Badiou's set paradoxes and Nelson Goodman's reflections on the indescribability of a total external world (5.3.1), but also Markus Gabriel's no-world arguments (5.3.2) and Carol Rovane's relativistic critique of unimondialism (5.3.3). The second type of antitotality argument is directed against the idea that the fixed structure of the single real external world is determined by necessary, all-encompassing natural laws. This includes examples in philosophy of science, such as Nancy Cartwright's argument against fundamental laws of nature, as well as in the ontological debates of New Realism, such as Meillassoux's thesis of the necessary contingency of all concrete entities, causal relations and laws of nature (5.3.4). Here I focus on Cartwright's argument. Finally, I show why the antitotality arguments hit OSR hard under the conditions in which UTE and historical arguments placed it, and why its prospects for a solution are not promising (5.3.5). This establishes the *first argumentative stalemate between anti-realist criticism and objective scientific realism*. While the criticism is sufficient to reject Chakravartty's subtle position, it cannot sufficiently motivate anti-realism as a position: UTE and antitotality arguments do not suggest such a stance *per se*, whereas the ostensibly strong historical arguments prove to be too weak.

The second part of the first main thesis is demonstrated in the fourth subchapter (5.4): The subtle defence of OSR and the structural realist overcomings of the object-centred OSR, *Chakravartty's semirealism and Steven French's ontic structural realism, also find themselves together in a second argumentative stalemate*. With regard to the construction of their position and the handling of the three main challenges, both make mistakes: While Chakravartty's mistake has already been shown by this point, French's ontic structural realism must first be analysed. To this end, I first trace John Worrall's attempt at epistemic structural realism, which tries to reconcile PMI and the no-miracle argument, but which cannot counter the criticism of his resulting position (5.4.1). In response to this failure, James Ladyman and French develop two different variants of ontic structural realism, in which structures are recognised as the only ontologically fundamental, subject-independent entities of the external world (5.4.2). In the context of the analysis, French's variant proves to be superior and an OSR. Finally, I can also show that this theoretical option does not escape its pitfalls, especially the antitotality arguments (5.4.3). This justifies the second stalemate: Semirealism and ontic structural realism ultimately fail in the face of the same arguments.

The over-all aim of my argument is to show that the different variants of OSR do not give an adequate answer to the three challenges when they are considered in combination. The best arguments of these three families together show that any OSR may provide a strategy to counter both underdetermination arguments and historical arguments, but not one that refutes all three families of arguments together. Chakravartty's semirealism is motivated precisely by the attempt to stand up to the two classical challenges of underdeterminism and historical arguments. But even such a balanced OSR is not able to also evade the antitotality arguments. With this, I am also demonstrating a thesis in the field of philosophy of science that is advocated by critics of naturalism such as Gabriel, Demmerling and Price in metaphysics and philosophy of language: Naturalism is unable to counter anti-realist criticism by referring to the scientific ascertainability of external-worldly facts.[2] Strictly speaking, however, my thesis for philosophy of science is even stronger: In view of all three main objections, not only any naturalism, but also the less presuppositional, ontologically minimal OSR cannot be convincing.

Finally, in the brief fifth subchapter (5.5), I summarise the two conclusions of the first part: the first argumentative stalemate between objective realism and anti-realism in the philosophy of science; neither position is fully convincing. I explain why OSR is not compelling in the face of criticism and why the motivation for realism in philosophy of science still remains despite the objections. This partly explains why, despite the many critical arguments, anti-realist philosophies of science are not particularly popular today. The second argumentative stalemate between Chakravartty's semirealism and French's ontic structural realism is explained, and the aspects of their philosophies of science that have proved problematic are clarified. I conclude Chap. 5 and part I of the present book with a conclusion arguing for the necessity of a new scientific realism in the metalocal debate in philosophy of science based on the thesis of the double stalemate.

---

[2] In his discussion of Millikan's and Papineau's philosophy of mind and language, Demmerling writes: "Scientific theories are incapable of thematising phenomena of meaning: nature has no meaning, there is no meaning in nature." Demmerling 2002, 120, my translation; while Gabriel objects to naturalism as a metaphysical theory option: "Although NONT (the naive ontology) provides an ultimately incoherent framework of premises, it underlies the naturalistic metaphysics that is customary today." Gabriel 2016, p. 89, my translation; further criticism can be found in: Spiegel 2020.

5.1 Underdetermination Arguments

The idea of underdetermination arguments is that any serious overall theory about the external world or nature allows for more theories and models, and allows us making predictions for more cases, than can actually be observed or otherwise verified. This surplus of theories, models and predictions is not only due to the seemingly contingent circumstance that we sometimes observe phenomena (e.g. because we happen to be on site and awake) and sometimes not. Rather, the UNDERDETERMINATION OF THEORIES BY EVIDENCE, I.E. DATA, PROOF AND OBSERVATIONS (UTE) is firmly anchored in the core of science itself. A single hypothesis or theory can only be tested by evidence if we take other theories and assumptions as given. It is always possible to formulate more than one well-supported theory to our limited body of evidence; trivially, because an additional premise can be added to any theory as long as it does not contradict the evidence and has no other consequences for possible measurements, observations or evidence to be provided; non-trivially, because it can never be ruled out that an alternative theory can be formulated that explains the evidence equally well (or better) but has significant differences. Significant differences between theories are given, for example, when different predictions are made, different explanations for phenomena are given, different scientific ontologies are assumed or different theories about a different field are supported, although both refer to the same donain of the external world or nature.

5.1.1 Experimental and Holistic Underdetermination

The first form, *holistic* underdetermination, was dealt with in this book already by reference to Quine's holism of knowledge in the last chapter. Duhem already points out that a single hypothesis or a single proposition of a theory cannot be tested empirically: It alone does not yield predictions, nor does it demand the existence of very specific observations.[3]

> A physicist decides to demonstrate the inaccuracy of a proposition; in order to deduce from this proposition the prediction of a phenomenon and institute the experiment which is to show whether this phenomenon is or is not produced, in order to interpret the results of this experiment and establish that the predicted phenomenon is not produced, he does not confine himself to making use of the proposition in question; he makes use also of a whole group of theories accepted by him as beyond dispute. The prediction of the phenomenon, whose nonproduction is to cut off debate, does not derive from the proposition challenged if taken by itself, but from the proposition at issue joined to that whole group of theories; if

[3] Historically, precursors of this train of thought can be found in Descartes' *Genius malignus* arguments, cf. Descartes 1996a, 1st Meditation, pp. 17–23, and ancient scepticism. Stanford 2017 finds a first reference to the underdetermination of scientific hypotheses in Mill 1867–1900, p. 328, but the *locus classicus* remains Duhem 1914.

> the predicted phenomenon is not produced, the only thing the experiment teaches us is that among the propositions used to predict the phenomenon and to establish whether it would be produced, there is at least one error; but where this error lies is just what it does not tell us. Duhem 1914, p. 185.

Duhem and other proponents of fictionalist instrumentalism—roughly the thesis that scientific theories and models linguistically function analogously to fictional speech with the particular characteristic that they are useful tools and instruments for understanding the world—conclude that there can never be an *experimentum crucis* that decides success or failure of a theory. Thus, it seems impossible to specifically test a single hypothesis empirically through experiment or observation. Duhem thus advocates a proto-form of holistic underdetermination, experimental underdetermination. According to this theory, a single hypothesis cannot be falsified in an experiment, since further hypotheses and theories must always be assumed for the experimental context, and nor can a hypothesis be verified definitively by a *reductio ad absurdum* of all alternatives in any experimental science, contrary to logics or mathematics, because a complete differentiation of all cases is impossible. In my view, a holistic idea already follows from the first point. Only the totality of all hypotheses of a theory and other background assumptions with an ontological nature, such as the stability of all regularities in nature, and often also of other theories, serve as a holistic network that enables predictions about and tests of empirical reality. If the prediction is not correct or the test is negative, it is never clear without further ado which hypothesis or assumption must be abandoned.

This tendentially anti-realist, instrumentalist thesis was taken up by Quine in order to turn it against logical positivism, in particular (earlier) Carnap and indirectly Karl Popper, whose verificationist or falsificationist understanding of theory required observation sentences, i.e. simple experimental predictions, to be derived from a theory, which can then be tested atomically (or at least can independently fully confirm or refute a given theory).[4] However, Quine argues that not even an entire theory, such as Newtonian mechanics, can do this job. Only when a substantial amount of further theory, ontological assumptions and contextual descriptions are added is it possible to arrive at experimental implications, verifiable and falsifiable predictions. If these predictions then fail, according to Quine, it remains open which parts of the tested theory, other background theories, ontological assumptions or contextual descriptions are falsified and which remain. Quine aims to use this situation to justify his holism of knowledge and the argument initially appears valid. However, it has already been shown that Quine's naturalism as a variant of OSR leads to serious difficulties.

What remains besides the valid criticism of verificationism of early Carnap and falsificationism of Popper? The argument establishes that more than one theory is always necessary for experimental implications. Conversely, this in turn means that a rival theory can always be constructed that also agrees with the evidence

[4] Cf. Quine 1969a, p. 79. It has not been conclusively proven in the history of philosophy whether Quine developed his underdetermination argument completely independently of Duhem or not.

by introducing an empirically unimportant additional hypothesis. Larry Laudan characterises Quine's thesis as follows: "for any theory, T, and any given body of evidence supporting T, there is at least one rival (i.e. contrary) to T that is as well supported as T" (Laudan 1990, p. 271).

The challenges posed by Duhem's and Quine's approaches give rise to two tasks for any OSR: On the one hand, the realist conception of theories must be compatible with the insight that for a given theory it can never be shown to by itself describe fully and truly a part of the external world alone. On the other hand, every OSR must be able to deal with the fact that hypotheses and theories can never be clearly tested without making certain background and contextual assumptions. This enables, but at the same time limits, any scientific methodology. In the case of Chakravartty's semirealism, it has already been shown that it attempts to fulfil this task. The basic idea of selective scepticism with regard to theories is intended to do precisely that. By admitting that parts of theories and their underlying scientific ontologies contain only auxiliary properties or purely postulated entities that we should not include in our ontology of the single real external world, selective scepticism takes into account the fact that theories are not completely true. The idea of semirealism here is precisely to recommend a sceptical attitude for the questionable part of our theoretical assumptions, in other words: ontological anti-realism. Chakravartty admits that hypotheses and theories cannot be tested without further theories and assumptions, without seeing this as a contradiction to his proposal. In scientific practice, this is unproblematic and the proposal of semirealism consists precisely in postulating auxiliary properties without having to make assumptions about their existence.

5.1.2 Radical Underdetermination

In the essay cited above, Laudan characterises a second idea often ascribed to Quine: Every theory is just as well supported by the evidence as every conceivable rival theory (cf. Laudan 1990, p. 271). This thesis is called global, strong or *radical underdetermination*. On the one hand, it can be argued for in a similar fashion to holistic underdeterminism: Any theory can be reformulated into any rival by adding and removing individual hypotheses. Without further assumptions, each of these hypotheses is neither provable nor disprovable if sufficient other adjustments are made in the holistic web of beliefs of our knowledge. As a result, the rival can be considered to be just as well supported as the initial theory.

This idea is the starting point for the discussion of whether the rival is a genuine alternative and whether this argument is true for all theories (cf. Kukla 1996, esp. pp. 145–158). From the point of view of many contemporary anti-realists such as Stanford and K. Brad Wray, these discussions lead away from the narrower topic of scientific realism and towards general sceptical scenarios. I agree with this analysis (cf. Stanford 2006, pp. 11–13; Wray 2008, 2018, p. 38): Radical underdetermination leads to a global scepticism in which the metalocal debate of philosophy of science is abandoned and the global realism-anti-realism debate is reached. This possibility is easily explained by the debate map I have developed. It is simple to run into a global realism problem with the arguments of a metalocal debate, such as the philosophy of science-debate. This is one reason why I deal with radical underdetermination only to the extent that it is used as an argument against OSR.

In addition, the thesis of radical underdetermination is not very convincing, as it already presupposes a global scientific scepticism. This is because in order to be able to formulate the thesis, the discursive situation in the sciences must be completely ignored. The question of radical underdetermination is equal to that of whether the sciences are groping in the dark and any scientific discourse is devalued as a consequence if radical underdetermination is accepted. A third reason that neither OSR needs to respond to radical underdetermination, nor anti-realists should use the argument, is that there are serious anti-sceptical objections to any such global scepticism.

Most authors do not consequently take on such radical anti-realism and scepticism from UTE in philosophy of science. Laudan argues, for example, that the significance of all UTE arguments is already exaggerated by Quine (cf. Laudan 1990, p. 271). He reaches the conclusion that there are usually arguments and examples from science for weak versions, such as experimental underdetermination, but that consequences are drawn from the unsupported, strong versions. Thus Quine may have shown that it is psychologically possible for people to hold any conviction despite striking evidence to the contrary, but that this is indifferent for the discussion of (theoretical) knowledge: "[It is not the case that; J. V.] any theory can be rationally retained in the face of any recalcitrant evidence." (Laudan 1990, p. 277).

5.1.3 Contrastive Underdetermination and the Inference to the Best Explanation

The second powerful argument that can be derived from UTE aims to refute a classical justification of OSR. This so-called abduction argument uses the IBE to prove that OSR is the best explanation for the practices and the success of science (cf. Shook 2021). Anti-realists use *contrastive* underdetermination in trying to undermine this and related realistic arguments.

Smart's abduction argument for OSR is a particular variant of the no-miracl argument. However, as I will show in Sect. 5.2, improved variants of the latter can be given. Smart argues as follows (cf. Smart 1989, Chap. 3): Our best current theories allow accurate predictions about many observable phenomena. These theories postulate certain theoretical entities, laws of nature, etc. The observable phenomena behave as if the theoretical entities, laws of nature, etc. postulated by the sciences qua theories exist. The best explanation for this is that these theoretical entities, laws of nature, etc. do in fact exist. Since in the meta-theory of science, as in the sciences, the inference to the best explanation should be followed, Smart concludes that these theoretical entities, laws of nature, etc. exist.

Similarly, Sellars states: "As I see it, to have good reason for holding a theory is *ipso facto* to have good reason for holding that the entities postulated by the theory exist." (Sellars 1963, p. 97, fn. X). According to Smart and Sellars, if the best explanation for an observed phenomenon, an experimental result or the functioning of a technology consists in a scientific theory, it is not only permissible but required to assume the truth of this theory as well as the independent existence of the entities postulated therein (cf. also the prominent presentation in Boyd 1983, for an alternative variant of the same argument). The argument that already lies behind this conclusion in Sellars and becomes explicit in Smart is that good reasons for the former are necessarily good reasons for the latter. This is solely due to the fact that an inference to the best explanation always combines usefulness and correctness: An explanation is good if it is true. And it can only be true if the entities assumed by the explanation exist independently, and for Smart this means SI. This applies at least in cases of scientific theorising in which the explanation is directed towards the subject-independent external world. If, for example, the entities were purely fictitious but useful entities that only exist within the explanation, then there would be no explanation of a real phenomenon, a veridical observation or a functioning technology.

Arthur Fine has criticised the abduction argument, especially in the variants of Smart and Boyd (cf. Fine 1986a, esp. pp. 152–156).[5] One part of the theses defended by the proponents of OSR—and, without Fine mentioning this, also by the positivists—is the fundamental trust in scientific methodology. On closer inspection, it must be explained in more detail what exactly this scientific method, which by no means appears uniform in scientific practices, consists of. However, according to Fine, abduction and induction are always central parts of scientific methodology. Anti-realists attack the reliability of abduction and induction, for example with the help of the problem of induction. Therefore, abduction's and induction's usefulness should never be assumed in an argument against anti-realists. However, this mistake is made by Smart, Sellars, Boyd and other realists. The abduction argument in favour of OSR is circular, according to Fine, due to the inference pattern used. Fine concludes that the instrumental success of the sciences can be explained quite simply by an instrumentalist thesis that sees theoretical entities as instrumentally useful fictions that are pragmatically reliable.[6]

In addition to this obvious circularity objection, Stanford provides a more profound objection to Smart's argument (Stanford 2000, 2001). He criticises Smart's

[5] For an argument against Fine's take, cf. Lipton 2004.

[6] Fine attempts to conceptualise this fictionalist instrumentalism not as anti-realism, but in line with the *natural ontological attitude*. Cf. on this Fine 1986b, p. 9 and on the variant of the counter-argument pp. 114–117. I cannot elaborate here why I think that this leads to an ontological anti-realism in my sense. The same holds in my opinion for Mauricio Suárez proposal Suárez 2010, esp. p. 225, of stating a broad fictionalism about models and general scientific discourse as neutral with regard to scientific realism, as Gentile and Lucero 2024 convincingly show.

third premise with a UTE argument: several alternative theories with different entities etc. could be constructed that also fit the observable phenomena. So, Stanford asks, isn't it more likely that, in choosing from many empirical equivalents, we have deliberately chosen a theory that is wrong rather than the right one by chance?

Van Fraassen also criticises the IBE as a reason for OSR (van Fraassen 1980, Chap. 1). The IBE as an inference pattern is inductive or abductive. Thus, it is not truth-preserving from the premises to the conclusion, but knowledge-enhancing (cf. Tugendhat and Wolf 1986). If the evidence/data B is available and the explanations E^1 to E^n can be given for B, then E^1 should be favoured over all other E^n iff E^1 is a better explanation than all other E^n. The criteria that constitute a better explanation are discussed in philosophy of science in a debate about epistemic theory values and epistemic norms. Simplicity, compatibility with other theories, generality, elegance and many other epistemic values are considered, also depending on disciplinary issues. Consistency with B, internal consistency and compatibility with other accepted explanations in the same natural domain are usually considered to be of central importance (cf. e.g. Chakravartty 2017b, p. 110). In line with this, van Fraassen raises the following two points against the realist argument:

First, the observation that we apply the IBE in everyday situations does not yet form a basis for the realist argument. After all, it does not have to be assumed that because explanation E^1 best explains the empirical data, it is also true. "Here is a rival hypothesis: we are always willing to believe that the theory which best explains the evidence, is empirically adequate (that all the observable phenomena are as the theory says they are)." (van Fraassen 1980, p. 20). This is now the basic idea of van Fraassen's constructive empiricism: the aim of science is not to find the ultimate truth about the nature of the external world, but to provide empirically adequate explanations that are compatible with the observable phenomena (van Fraassen 1980, p. 12).

Second, realists require a further premise for their argument, even if one is not inclined towards constructive empiricism, but rather in favour of the realist alternative. The IBE supposedly makes it possible to choose the hypothesis that is closest to the truth from a finite set, regardless of the specific theoretical norms and epistemic values in which it is spelt out. Realists thus always imagine a choice between different hypotheses that explain the regularities of nature. Anti-realists, on the other hand, only want to evaluate which theories are empirically adequate and empirically strong. "So the realist will need his special extra premiss that every universal regularity in nature needs an explanation, before the rule will make realists of us all. And that is just the premiss that distinguishes the realist from his opponents." (van Fraassen 1980, p. 21). In fact, almost any form of anti-realism, be it instrumentalism, historical relativism or constructive empiricism, would reject that *every* regularity in the external world can be explained by a causally explicable reason in the form of a theory.

A third reason, which van Fraassen does not mention, but which also plays a role, is the fact that it sometimes happens in the sciences that two rival theories, which are *not* empirically adequate or adequate in terms of any possible evidence, are nevertheless *equally well* confirmed by the *existing* evidence. An example of this phenomenon, which Larry Sklar calls "transient underdetermination" (Sklar 1975, p. 380), is the evidence situation in 1912 he cites with regard to the special theory of relativity with Lorentz covariance and that with Galilei covariance. In Sklar's view, it cannot be ruled out that such a situation might occur permanently for two theories.

In my view, these three points are sufficient to show that the abduction argument for OSR is not viable. However, arguments from contrastive UTE are not only suitable to counter this justification of OSR. They can also be used to cast doubt on the truth of the two theses of OSR.

5.1.4 The Bad-Lot and the Model Argument from Contrastive Underdetermination

Already by the end of the 1980s, van Fraassen offered two positive arguments that are not intended to refute the abduction argument in favour of realism, but which argue independently in favour of anti-realism. The so-called *bad-lot argument* emphasises that epistemic optimism is necessary for the assumption of OSR, which he deems to be unfounded. It thus shifts the burden of proof to the disadvantage of OSR. With the *model argument*, on the other hand, van Fraassen wants to show that OSR is overwhelmed in recognising what counts as a theory and how a debate situation in the natural sciences can present itself. To show realism's overextension, van Fraassen argues in *Laws and Symmetry* as follows:

> [The IBE] selects the best among the historically given hypotheses. We can watch no contest of the theories we have so painfully struggled to formulate, with those no one has proposed. So our selection may well be the best of a bad lot. To believe is at least to consider more likely to be true, than not. [...] For me to take it that the best of set X will be more likely to be true than not, requires a prior belief that the truth is already more likely to be found in X, than not. van Fraassen 1989, pp. 142–143.

The crucial point in this passage is similar to Stanford's critical question: Even if the anti-realist concedes the methodological and metaphysical presuppositions of the abductive argument for OSR, it is never a certainty that the external world is captured by the historically actually represented theories as it really is. This *bad-lot* argument is a central starting point for recent anti-realisms such as those of van Fraassen, Kareem Khalifa or Wray (van Fraassen 1989; Khalifa 2010; Wray 2018). Stanford (cf. Stanford 2001, 2006, pp. 27–48), in discussing and extending the bad-lot argument, brings out what was already anticipated by Nelson Goodman with the *new riddle of induction* in relation to hypotheses in general,[7] namely, that this problem can be radicalised if it is taken into account that it applies not only to

[7] Goodman 1983, gives the well-known argument that the evidence supports the conclusion "that all emeralds are grue", p. 74.

historical theories, but to all conceivable theories and also to alternative models of a theory (Schultz 2018). This difference can be understood as follows:

In van Fraassen's historical version, the bad-lot argument states that the best theory T_1 from the set of historically proposed theories on a given phenomenon and the best theory T_x from the set of all meaningful theories on the same phenomenon need not be identical. According to the bad-lot argument, the fact that OSR assumes that these are identical or at least sufficiently similar is based on the assumption that it is more likely that T_x is contained in the set of historically proposed theories than that this is not the case. However, it is precisely this assumption of OSR that is arbitrary and thus a dubious premise, since it is based on an epistemic optimism, for which OSR requires convincing arguments. According to van Fraassen, it cannot be ruled out and is not unlikely that we are dealing with a bad lot of theories in most cases, to which the argument owes its name.

Peter Lipton sees the assumption, which is necessary for the premise of the bad lot argument, that scientists can indeed rank theories according to their probability of truth, as a good reason to reject the second premise that scientists cannot be sure that they are dealing with a bad lot (cf. Lipton 2004, pp. 151–163). In contrast, Wray argues convincingly that it is sufficient for scientists to be able to assess predictive accuracy (Wray's own criterion) or empirical adequacy and strength (van Fraassen's criteria) in order to weigh theories against each other, but that this does not amount to an assessment of probability of truth (cf. Wray 2008, 2018, pp. 49–50). The crucial point is that Lipton already implies a realistic theory in his counter. The idea that scientists are reliable *in terms of determining the probability of truth* assumes OSR. An empirically barely adequate theory is most likely false. According to Wray, however, anti-realists merely hold that from T_1 *is more accurate/empirically adequate than* T_2 it does not follow that T_1 *is probably true*.

Stanford's modal version of the bad-lot argument differs in that it does not contrast the set of historically proposed theories with the set of all meaningful or reasonable theories. It is more radical, as the set of historically proposed theories is limited, but growing. Instead, Stanford contrasts the set of all conceivable theories with the set of all possible theories. Here he surmises that in some (or even all) areas of phenomena it may be the case that the true theory T_x is not even included in the first set, which means that any hope of ever finding it is in vain. Although Stanford presents a harder conclusion, even this minimal reconstruction shows that many strong premises are needed for the argument. Thus, in my view, this argument rests on feet of clay with the weighty assumption that both sets can be compared (in the way that the argument does) and, furthermore, that it can be ensured that both sets are not identical for a given range of phenomena. Stanford offers a different, and in my opinion stronger, historical argument that builds on contrastive UTE without needing the problematic premises of the modal bad-lot argument. I will reconstruct that argument in my next subchapter.

Similar thoughts to those on contrastive UTE motivate van Fraassen to introduce a further argument, the so-called model argument. He defines a model of a theory as an abstract structure permitted by the theory, which can be embedded in the theory and in which all relevant parameters of the theory have specific values. Van Fraassen's use of the term model, inspired by logics and metamathematics, thus differs from typical scientific definitions. There, MODELS are usually understood as abstract or concrete structures that represent a target system and which can be constructed from a theory and, if necessary, make the theory applicable to a context (Frigg 2010b; Giere 2004; Thomasson 2020a). However, these are rarely structures in which all relevant parameters have specific values; in fact, this is an absolute exception rather than the rule. For the following reconstruction of his argument, I use van Fraassen's definition, even if I use the general and less demanding concept of a model in terms of Giere and Frigg in the rest of this book.

Van Fraassen constructs the model argument from the example of the absolute motion of the solar system in Newton's mechanics: Ptolemy assumed that the earth is a body at rest around which the celestial bodies orbit in spheres. Copernicus postulated that the sun was the body at rest around which the earth and the other planets orbited. "Finally, Newton, in his general mechanics, did not assume that either the earth or the sun is stationary. And he generalised the idea of apparent motion—which is motion relative to the earth—to that of motion of one body relative to another." (van Fraassen 1980, p. 45). In Newton's theory, all bodies have an absolute velocity in relation to space, understood as space in the absolute sense. Yet bodies only ever have velocities relative to other bodies (and therefore also to every observer). Although Newton hypothesised that the sun rests in absolute space, he admitted that this cannot be proved by observation. Since our solar system (or even the entire observable universe with the fixed star sphere still assumed by Newton) could also move as a whole at any speed relative to absolute space, there would be no other observations, provided that all relative movements of the planets, sun and moons remain identical to each other.

Now, van Fraassen imagines that Newton assumes the empirical adequacy and the truth of his mechanics, and that he also considers the thesis (called TN(0)) that our solar system has an absolute velocity of 0 m/sec to be empirically adequate, but not true because there can be no proof. In a next step, van Fraassen introduces the fictitious philosopher Leibniz$*$, the only conviction of whom that differs from Newton's is that no absolute space exists. For this reason, he considers the thesis TN(0), like all related theses TN(v), where v stands for all conceivable velocities, to be nonsensical. Furthermore, Leibniz$*$ does not believe in the truth of Newton's mechanics, as it presupposes absolute space in its axioms. However, Leibniz$*$ considers both Newton's mechanics and TN(0) as well as all TN(v) to be empirically adequate and is therefore, like van Fraassen, a constructive empiricist.

The model argument now states that Leibniz$*$ indeed has a theory about the motion of celestial bodies, while realists would have to claim that it is not the case that he has one. According to van Fraassen, realists must attest to the following situation: Since Leibniz$*$ on the one hand considers Newton's theory to be wrong and must therefore reject it, but on the other hand also proposes no alternative, he has no theory about the motion of celestial bodies. Van Fraassen argues against this:

> The theory which Leibniz$*$ holds of the world, TNE, can nevertheless be stated, and I have already done so. Its single axiom can be the assertion that TN(0) [and Newton's mechanics; J.V.] is empirically adequate: that TN(0) has a model containing motions isomorphic to all the appearances. Since TN(0) can be stated in English, this completes the job. van Fraassen 1980, p. 47.

By *appearances* van Fraassen means all actual observations of the phenomena. It is sufficient that all observed, concrete structures can be embedded in one of the abstract models (e.g. with TN(0)) that the theory allows.

What is crucial about van Fraassen's argument is that, although it does not support the conclusion that OSR is false, it uses contrastive UTE and the possible scientific attitudes towards it to show van Fraassen's constructive empiricism and other anti-realisms to be equally plausible at the expense of OSR.[8] Any OSR must muster a defensive manoeuvre to explain the situation between Newton and Leibniz∗. Because Newton has to admit that several theoretical assumptions about the relative velocity of space could be true and that it seems impossible within the framework of his theory (at least that which is limited to celestial mechanics) to provide proof of this, he opens the door to his scientific critic Leibniz∗. The realist must explain both Newton's openness with regard to all TN(v) and the fact that Leibniz∗ claims to have a theory. According to van Fraassen, this is exactly what he succeeds in doing from his constructive-empirical perspective, but not from that of OSR. Here I follow van Fraassen, although the anti-realist solution does not seem much better to me. The situation seems to be superiour with respect to the model argument, but not for an understanding of knowledge claims, for example in scientific causal explanations. There, OSR seems to be the obvious position. However, neither OSR nor anti-realism seem to be particularly attractive.

5.1.5 *Interim Conclusion on Underdetermination and the Possible Reactions of Objective Realism*

What conclusions can be drawn from the UTE arguments considered? I start with the results of the previous section: Contrastive UTE, unlike experimental and holistic UTE, does not concern *a given* theory or hypothesis, but instead the allegedly best theory *in relation to a set of alternatives*. The relevant problems with contrastive UTE arise from the limitation of historically contingent alternatives and the disproportion of the set of considered alternatives to the set of meaningful but unproposed alternatives.[9]

The counterarguments to the realistic abduction argument derived from contrastive UTE succeed in invalidating it. According to the bad-lot argument and the model argument, there is no reason to believe that our best current theory is true. On the contrary, both arguments compel any OSR to explain how to deal with discursive reticence within science and "negative" theories (model argument) and to justify the epistemic optimism behind OSR (bad-lot argument).

[8] Cf. for this conclusion, see also the critical analysis of the bad-lot argument in the variants of van Fraassen and Wray in Mizrahi 2013a.

[9] For an application of UTE in relation to quantum physics, cf. Egg and Saatsi 2021.

It is therefore hardly surprising that modern, cautious variants of OSR attempt to take the challenges of UTE into account from the outset, without engaging in excessively strong theses, as is the case, for example, with Quine's combination of holism and reductionism. Entity realism is a prime example often cited by Chakravartty and other OSR proponents of a defensible form of realism, in view of the historical arguments and UTE. In order to better understand Chakravartty's position, entity realism, represented in different variants by Ian Hacking, Nancy Cartwright and Ronald Giere (Hacking 1982, 1983; Cartwright 1983, esp. pp. 74–86; Giere 1988, esp. pp. 92–140), is outlined here together with Chakravartty's critique of it.

5.1.5.1 Entity Realism as a Reaction of OSR to UTE

Entity realism provides a template for Chakravartty's semirealism, as it is also a selective scepticism. In this, the existence of theoretical entities is affirmed ontologically in a realist voice precisely when there is causal contact with these entities (cf. Hacking 1983, pp. 26–27). If, for example, electrons can be detected in various experiments and directly manipulated or used to manipulate other entities, there is good reason to commit to the existence of the entities, regardless of changing theories and their threat by UTE. The different variants of entity realism have in common that they assume a realism for entities without committing themselves to the approximate or partial truth of theories or the scientific ontologies that contain them. Depending on the variant of entity realism, they recommend an agnostic or even pessimistic attitude towards them.

Ian Hacking proposes, against naturalists who see ontological commitment arising from the truth of our best theories, recognising successful experimental work as the strongest argument for the thesis of ontological commitment to scientific entities (cf. also for the following Egg 2018, p. 121; Hacking 1983, p. 262). We successfully manipulate unobservable scientific entities to produce new phenomena and investigate other aspects of nature. This gives us a reason to assume their subject-independent existence, even if our theories are not true. According to Hacking, mere theoretical concepts can be conceived anti-realistically, e.g. as heuristic devices. As soon as an intervention is performed on or with an entity, it should be understood realistically.

Since theories involving fundamental laws of nature need not be true in order to be explanatorily successful—think of the ether theory—theories based on the validity of the asserted laws of nature should be viewed antirealistically and even sceptically (cf. also for the following Cartwright 1983, pp. 58–62): This is a central thesis of Nancy Cartwright. According to her, arguing that the inference to the *best theoretical explanation* (cf. Suárez 2008) resolves the contrastive UTE, is an inadmissible manoeuvre on the part of OSR. However, she agrees with Hacking that this does not apply to causal explanations that presuppose certain entities (for a defense of this thought with regard to instrumentally mediated empirical evidence cf. Khalili 2024). With regard to these, she is a realist. So if *X* has causally caused

Y in some cases, i.e. by showing through experiment, observation and statistical analysis that without *X Y* would not have occurred, then realism should be argued for *X*. This is usually a specific entity, such as an object, event or process. It can also be an unobservable entity. However, a parallel argument for fundamental laws with a counterfactual statement "[i]f law *L* did not hold, phenomenon *y* would not occur" (Cartwright 1983, p. 9) cannot be given, since it cannot be proven by experiment, observation and statistical analysis.[10]

A number of authors have voiced several pertinent criticisms of this idea (cf. esp. Hitchcock 1992; Morrison 1994; Pierson and Reiner 2008). With regard to the UTE arguments, I consider the following to be central: Even if it is conceded that entity realists have a good reason to insist on an ontological-realist understanding of unobservable theoretical entities via the causal manipulation criterion, they can neither attribute properties to these entities nor understand their manipulability without the descriptions of scientific theories. Electrons, for example, have different properties in the various theories that have historically been advocated, enter different relationships to other entities and are embedded in different structural contexts. *What* electrons are, the critics argue, changes with the theory. Entity realists would either respond agnostically, like Giere, that nothing certain can be known about entities, or that the best theories are, strictly speaking, wrong.

Experimental and holistic UTE, however, points precisely to the special theory-ladenness of experiments. Chakravartty therefore criticises that entity realism is "too crude" (also for the following Chakravartty 2007, p. 32): "what the realist requires is something more refined: a knowledge of the specific properties and relations on which existential claims are based, and that are likely to survive over time." In the electron example and conceivably many others, however, this would not be the case. Charkavartty therefore believes that entity realism fails due to the impossibility of drawing a line between the entities in question that are to be located in nature and those that must be viewed sceptically. This argument appears to me to be correct. Entity realism leaves the realist with too few resources to say anything substantive about the entities in question.

It seems possible here to argue further that in a consistent reading entity realism is not OSR, but a form of epistemological anti-realism that doubts the knowability of properties of theory entities and basal structures of nature and its laws. This is because the epistemological thesis of scientific-philosophical realism is undermined when the partial or approximate truth of the best current theories is *not* assumed.

5.1.5.2 Semirealism as a Reaction of OSR to UTE

In line with this criticism of entity realism and the immense problems of standard formulations of OSR, such as reductive naturalism in dealing with UTE,

[10] I analyse this argument in detail as an objection to the structural thesis of OSR in my Sect. 5.3.4, pp. 151–156.

Chakravartty's special form of selective scepticism is not surprising. As explained in Sect. 4.2.2, Chakravartty does not respond directly to the individual UTE arguments, but instead outlines a position that is so subtle and cautious that it does not come into conflict with these arguments:

With regard to the truth of theories and ontologies of science, Chakravartty recommends realism only for particularly hedged parts. Therefore, it is not necessary for semirealism that the best current theories are completely true and our science ontologies are fully correct. Whether we are dealing with a bad lot can be decided by Chakravartty on a case-by-case basis. If there are no detection properties, we cannot conclude that a concrete entity exists. In that case, his meta-theory does not recommend realism. Experimental and holistic UTE must be treated within the methodology of a particular discipline. However, they can be based in the background on the same, well-established hypotheses that the detection properties and proven entities offer us. Introducing additional auxiliary properties seems unproblematic as long as their status remains clearly recognisable. Alternative theories, on the other hand, have the same detection properties, and in their ontologies they are related to the same natural entities. If additional entities with only auxiliary properties, i.e. merely postulated entities, appear, Chakravartty's usual caveat applies: this in no way restricts the legitimacy of resorting to postulates in theory construction. After all, it is possible that an auxiliary property becomes a detection property in the course of research and a postulated entity becomes a natural entity, which must then be included in the realistically interpreted scientific ontologies.

Even the possibility that auxiliary properties and postulated entities could be discarded in a later, better theory does not, according to Chakravartty's conception, provide an argument against semirealism: This is due to the discontinuities of scientific progress. Chakravartty considers his position subtle enough to evade the UTE arguments. The price for this is selective scepticism and an abandonment of strong naturalist theses such as RP or holism. In addition, part of the resolution of UTE-challenges is relegated to the disciplines that explore a section of the same, structured nature. In the next two subchapters, I will show how this constellation helps Chakravartty's cautious OSR in relation to historical arguments, but makes it particularly vulnerable to antitotality arguments.

5.1.6 Consequences of Underdetermination

Before doing so, however, I would like to summarise my consideration of the various arguments for underdetermination. These arguments sow doubts about OSR and make some anti-realisms plausible. The contrastive UTE does not directly refute OSR in any of its variants. However, as I will show, it does play a role in enabling certain historical relativist arguments that aim to do so. More importantly, contrastive, experimental and holistic UTE impose defensive manoeuvres on OSR, as I have shown using the examples of entity realism and Chakravartty. This shifts the burden of the argumentation onto the side of OSR.

In order to justify confidence in the theories and their ontological implications despite experimental, holistic and contrastive UTE, an OSR has the option of employing a form of selective scepticism in line with Chakravartty and of conceding that parts of our best theories are wrong. This option must prove capable of also fending off historical counterarguments, and it is then particularly threatened by antitotality arguments, as I will show in the next two subchapters.

For OSR, an alternative solution to selective scepticism is to rely on holism. As I have already demonstrated in detail using the example of Quine's naturalism, this amounts to contradictions with the basic assumptions of OSR. It turns out that, regardless of the path it chooses, OSR must undergo a cut or make a concession in order to cushion the implications of UTE.

A final strategy in OSR is to invoke the idea of scientific progress (exemplary Popper 1984, 2002, when interpreted and developed as a realist, as in: Niiniluoto 2014). The objective realist seeks to make plausible that the best theories are both approximately true and sufficiently true to give us the correct ontology of the single real external world, if only the sciences have matured far enough. I will call this strategy the *easy way out (of UTE)*, since neither heavy epistemological and ontological theses have to be defended (as with the combination of naturalism and holism), nor must a yielding, subtle stance has to be explored (as in selective scepticism). The easy way out of UTE is therefore traditionally a prime target of the anti-realist counterarguments that I discuss in the next subchapter: historical arguments. One of my concerns is to show that the idea of scientific progress or a future science alone is not sufficient against UTE arguments.

5.2 Historical Arguments

In this subchapter, I pursue several goals: In keeping with the overarching task of the chapter, the primary aim is to analyse some of the historical arguments for anti-realism. These are arguments that draw on the history of the sciences to attack OSR on this empirical basis. Many variants of OSR are incapable of dealing with historical arguments. But I will argue that they can be dealt with in Chakravartty's semirealism in a similar way to UTE arguments. Yet again, I will clearly mark that Chakravartty succeeds in doing so only by making certain concessions and only with certain presuppositions that—like the UTE arguments—make his position vulnerable to antitotality arguments, as I will show in the next subchapter.

A secondary aim here is to reject the *easy way out (of UTE)* mentioned at the end of the previous subchapter on the basis of historical arguments. The idea of scientific progress or the reference to future sciences alone is not sufficient for OSR to defend itself against UTE arguments. These ideas must in turn be defended against historical arguments, which is possible, but requires a more differentiated picture of the sciences and their history than is assumed in the easy way out. The combination of experimental-holistic UTE, contrastive UTE, and historical arguments is therefore sufficient to refute many OSR positions or to impose so many defensive manoeuvres against them that they yield unsatisfactory philosophies of science—but not all of OSR is refutable with only these two families of counterarguments.

Instead, the combination of UTE and historical arguments shows that rescuing OSR requires hard work. Chakravartty, with his semirealism, is cited as a positive example of this. Unlike Quine, he does not get carried away with overly strong theses such as RP or holism, which conflict with each other and lead to unsolvable problems, even in isolation. Instead, Chakravartty opts for a moderate and maximally cautious OSR by integrating selective scepticism. His stated aim is to sail between the Scyllian monstrosity of UTE and the Charybdian maelstrom of historical arguments. I will examine whether he succeeds and what concessions, if any, need to be made by OSR following in Chakravartty's wake.

To this end, I will focus on a number of selected and strong historical arguments. In order to reduce the large number of contributions to the debate, I will present more pointed variants of the two most influential arguments in the Anglophone discussion: the incommensurability argument (5.2.1) and the so-called pessimistic meta-induction (5.2.2). Both claim to refute OSR directly by attacking its epistemological thesis. I will argue that they fail, even though both arguments have problematic consequences for OSR. In addition, I will consider an argument that is particularly suitable for entangling OSR in defensive manoeuvres together with UTE considerations: Kyle Stanford's Unconceived-alternatives-argument (5.2.3).

Finally, again using Chakravartty's semirealism as an example, I will assess which argumentative resources are still available to OSR and which positions similar realist strategies must commit to (5.2.4). Here I will also summarise why the *easy way out (of UTE)* fails. Finally, I show that empirical-historical arguments should also be taken seriously by realist positions because they provide an important basis for any critique of science. I argue that there must be room for the critique of science in any good realist philosophy of science.

5.2.1 The Incommensurability Argument

The title of this section is misleading because there are in fact a number of incommensurability arguments. Two different concepts of incommensurability are proposed by Thomas S. Kuhn and Paul Feyerabend (Feyerabend 1958, 1962; Kuhn 2008). Both subsequently use them for several arguments; both have refined these arguments over decades (cf. e.g. Feyerabend 1975; Kuhn 2000). These were not initally designed to address the realism-anti-realism debate in philosophy of science because they preceed it, though that makes them no less influential (cf. Chang 2021; Wray 2021). Feyerabend's and Kuhn's ideas were taken up by many other philosophers of science (particularly noteworthy is the influential synthesis of Kuhn's ideas with those of Popper by Lakatos 1970), and the interpretation and validity of these arguments is the subject of a debate of its own (cf. e.g. Brown 2001; Hoyningen-Huene and Oberheim 2009; Sankey 2016). It is therefore misleading to speak of a single incommensurability argument. However, the title is justified. In what follows, I will concentrate on reproducing a certain reading by the debaters of Kuhn's concept of incommensurability as the strongest argument against OSR.[11]

To do so, it will not be necessary to take Kuhn's entire philosophy and history of science as a basis. Kuhn's incommensurability thesis is based on his concept of a scientific paradigm, which therefore requires at least a rough characterisation. A paradigm can be understood as a fundamental theory, generally recognised as true,[12] which forms the basis for practical work in an empirical scientific discipline or sub-discipline. An example from physics is Newton's mechanics. If this is accepted as a paradigm, explanatory problems arise, such as for planetary motion or other motion phenomena. A methodology follows from the paradigm. This in turn determines the type of experiments that can reasonably be carried out and observational set-ups that are valid. Results are accepted if they are based on the fundamental, paradigmatic theory. A paradigm thus organises the entire discipline or sub-discipline.

Kuhn assumes that paradigms only ever fulfil their role successfully for a certain stretch of time—his analysis of the history of science in *The Structure of Scientific Revolutions* is designed to justify this empirical premise—and then run into crises. When this happens, and several historical conditions are met, the most important of which is the availability of a new paradigm, a scientific revolution occurs. The old paradigm is replaced by the new paradigm.

According to Kuhn, paradigms are always incommensurable (cf. Kuhn 2008, pp. 155–168 and pp. 209–211). This essentially means that two theories describe points of view (e.g. about the legitimate methodology, the relevant problems, the ontology of science to be recognised, and the vocabulary to be used, which Kuhn calls a taxonomy), whose communication with each other is disturbed. In contrast to two competing theories working under the same paradigm, proponents of paradigm A not only claim the falsity of paradigm B. The research questions, methodological approaches and solutions in paradigm B are no longer fully *understandable* for proponents of paradigm A if Kuhn's historical analyses are correct.

Kuhn goes even further. In the postscript to his famous book, he writes that scientists who adhere to different paradigms perceive the world systematically differently, as their perception is already predetermined by the paradigm they hold. These groups would therefore "*in a sense* actually live in different worlds" (Kuhn 2008, p. 204). Each group talks about different things, the meaning of terms shifts, they have different taxonomies. According to Kuhn, there is neither complete translatability from one paradigm to another nor sufficient agreement on methods

---

[11] In Feyerabend's case, it is unclear if his incommensurability argument is directed against realism, as he is sometime interpreted as a realist. Cf. Crull 2024.

[12] That contemporary physicist tend towards scientific realism with regard to their current foundational theories is empirically suggested in Henne et al. 2024.

and practices. This does not mean that two paradigms cannot be compared on the basis of empirical predictive power or historically. In fact, Kuhn presupposes both for his own analysis. It is also possible to reintroduce some of the old problems or old concepts in the new paradigm as special cases. However, incommensurability then means that not all aspects of the other paradigm are comprehensible from the point of view of the truth of one paradigm and therefore communication difficulties necessarily arise between proponents of two paradigms. Kuhn justifies this in *The Structure of Scientific Revolutions* mainly using empirical studies of different historical paradigms.

If Kuhn's empirical findings are granted for the argument, this means that the two representatives of the old and the new paradigm are in at least some cases talking beyond each other, as they do not recognise the same concepts, the same scientific ontology and/or the same methodology. Although the new paradigm is seen as more accurate than the old paradigm after the scientific revolution, it will eventually reach a crisis and (in the appropriate circumstances) be replaced by another, third paradigm. From this, the following argument against OSR can be formulated:

(1) A paradigm is successful exactly when it explains the phenomena in its intended target context, allows correct predictions within it and enough further research (e.g. on special cases).
(2) If a paradigm is replaced by a new one, they are incommensurable with each other.
(3) If a paradigm is only recognised as true or approximately true as long as it is successful, and
(4) if a paradigm can no longer be considered successful from the point of view of another paradigm due to their incommensurability, and
(5) if the history of science shows that even the most successful paradigms of the past have entered into crisis and thus fallen victim to a scientific revolution, and
(6) if nothing fundamental has changed in the historical process of science,
(K) then our current paradigms will no longer be recognised as true or approximately true in the future.

This argument is directed against the ontological realism of OSR in the sense of what realistic truth guarantees: the (approximate) truth of the best theories and their ability to provide a partially correct ontology of the single real external world is called into question if they are based on paradigms. In that case, current paradigms would be replaced in the future, and the best theories would be incommensurable with future paradigms and thus no longer recognisable as true. Realistic concepts of truth (such as the deflational or that of correspondence theory) require that the truth of theories about general relationships in nature does not depend on the time of observation. According to Kuhn, however, scientific ontologies and theories are historically fleeting. If we take Kuhn's perspective on the process of the history of science and the role that paradigms play for today's disunified sciences, then there are good reasons to assume that current theories and scientific ontologies will be discarded in the future.

Against this idea of paradigms, a realist can try to show that the sciences are historically structured differently, i.e. attack premises (4) or (5). A realist can also formulate the following objection to premise (6): In physics (or any other science in relation to which she is a realist), the step from Newton's mechanics to relativity physics on the one hand and quantum physics on the other has changed the historical process of science itself and a refutation of the new paradigm is therefore unlikely. This idea has the same meaning as the thesis that such a science is a "mature science" (Metaxopoulos 1986).

With all the above-mentioned answers from the realist, the burden of argumentation lies on the side of OSR. They need a strategy against the incommensurability argument and its variants. However, many realists, including Quine and Chakravartty, agree with Kuhn that there will continue to be significant disruptions in the sciences and changes of our theories. As I will show below, Chakravartty offers an alternative to the picture that paradigms are necessarily incommensurable, in which true insights and subordinate theories (or parts of theories) are transferable from one paradigm to the next.

Another possibility of OSR is to reject Kuhn's suggestion of a historical necessity for paradigm shifts. It is then pointed out that it is not supported by his sociological and historical considerations of different sciences.[13] For example, OSR could insist that paradigm shifts are rational and that the thesis of genuine incommensurability between two paradigms, such as Newtonian and contemporary physics, is untenable. OSR is not on the defensive here, but even in this case it seems appropriate to demand an explanation of how theory and paradigm change can be integrated into the respective OSR.

5.2.2 *The Pessimistic Meta-Inductions*

The difficulty of reconciling OSR and theory change is also the basis of a strand of arguments often referred to as PESSIMISTIC META-INDUCTION (PMI). According to this, it must be empirically-inductively inferred from the history of science that pessimism about scientific progress and/or our best current theories is appropriate. The IBE is thus used here for the anti-realist counterargument. A standard formulation of the PMI is: The best explanation for the falsity and simultaneous acceptance of scientific theories in the course of the history of science is that there is no scientific progress in the sense of a gradual approximation to the truth about the external world, which is why our best current theories are probably also false.

The first point to make is that certain forms of the PMI are problematic for a number of reasons. For example, van Fraassen considers the PMI in its inductive

[13] I would like to thank Jens Rometsch for this further objection and the invitation to take a harder look at Kuhn here.

formulations to be a simple fallacy, e.g. in Larry Laudan's classic variant, and rejects the use of the IBE here (cf. van Fraassen 1980, p. 71). Since van Fraassen criticises the IBE more generally, as I have already shown in my treatment of UTE, in his view applying it in anti-realist arguments is a mistake. Others, however, including Chakravartty, consider the PMI to be the most relevant attack on OSR (cf. Chakravartty 2007, p. 28). This confusing picture can only be resolved by looking at the various arguments that operate under the heading of PMI.

5.2.2.1 Real Pessimistic Abductions and Inductions from the Historical Record of Science

Chakravartty summarises the basic idea as follows: "First, there is an assertion to the effect that the history of science contains an impressive graveyard of theories that were previously believed [to be true; J. V.], but subsequently judged to be false [. . .]. Second, there is an induction on the basis of this assertion, whose conclusion is that current theories are likely future occupants of the same graveyard." (Chakravartty 2008, p. 152). Wray characterises at least four different arguments that build on this basic idea. The first is formulated by Hilary Putnam in *Meaning and the Moral Sciences* (also for the following Putnam 1978, pp. 24–25; Wray 2018, pp. 68–86):

The argument starts with the premise that some entities have different properties than was assumed in past theories (e.g. Bohr's electron), while others did not exist at all (e.g. phlogiston). The second premise is that these past theories were also believed to be true and also had a justification based on scientific methodology. Putnam concludes that neither the correct application of scientific methods nor the agreement of the scientific community guarantee that a theory held to be true is also true. He emphatically defends this conclusion. He goes on to claim that all or at least the vast majority of theories of the past did not refer to entities in the extarnal world, but that the theories of that time were superseded by the next generation. Since nothing fundamentally changes in the scientific research process, it is likely that today's theories will also be superseded, that their theoretical terms do not refer to anything in the external world and that the assumed entities do not exist, or exist and possess wildly different properties.

Putnam does not wholeheartedly endorse this second conclusion in the further course of *Meaning and the Moral Sciences*. In my view, there is a good reason for this: The premise that all or the vast majority of past theoretical terms do not refer to any external world entity is too strong. Anti-realists usually argue that this does not have to be assumed for the argument:

In the well-known *Confutation of Convergent Realism*, Laudan advocates a weaker premise in his variant of the argument (cf. also for the following Laudan 1981, esp. p. 26): *In the past, there have been more examples of successful theories that turned out to be false than of those that are still considered true.* Lauden gives a long list of cases.[14] Laudan recognises, however, that this weaker empirical hypothesis is also not covered by his work—for this to be the case, *all* theories of the past would have to be historically researched and evaluated. Laudan believes, however, that there are so many individual examples of once successful but false theories that the empirical claim to the contrary is highly implausible.

The first powerful counterargument to this can already be found in Boyd, but is clearly elaborated by Sherrilyn Roush (cf. also for the following Boyd 1983; Roush 2010, esp. p. 55). The methodology of almost all sciences has changed considerably, and theories from two hundred years ago cannot be subsumed under today's theories. In short, the empirical data on which Laudan's induction or abduction conclusion is based is inconsistent and the argument leads us to a false conclusion.

Moreover, even if more than half of all the proposed theories are wrong, it should not be concluded that our current theories are probably wrong. On the contrary, it seems clear that if, for example, there are five theories about a phenomenon that have been successful at different times and are mutually exclusive, at least four of them must be wrong, even from the perspective of realism. The connection suggested by history, that if X is a successful theory, then at some point X will prove to be false, seems incomprehensible from the mere fact that more than two mutually exclusive theories of a phenomenon have been proposed in the history of science.

However, if this second PMI argument were grounded in the suggestion by Peter Vickers (2013), in the course of expanding Laudan's list, that the majority of all theories put forward exactly 200 years ago, for example, were wrong, then this would have to be proved either by an evaluation of all theories put forward at that time or by a random sample. However, Moti Mizrahi, one of the fiercest critics of the PMI, raises a powerful objection (cf. also for the following Mizrahi 2013b, 2015): The arguments of Laudan and Vickers would not acchieve a full analysis or take a random sample. They only collected relevant negative examples, for which it is also questionable whether they are fully-fledged theories, individual hypotheses or paradigms. Particularly in the case of more complex theories, it is often not possible to say that they are completely false (in the sense that the entities they postulate do not exist for the most part, have completely different properties or are related to completely different natural laws). Laudan and Vicker's "cherry-picking" (Mizrahi 2015, pp. 129–130) prompts Mizrahi to make the eponymous accusation: "The pessimistic induction: a bad argument gone too far" (Mizrahi 2013b, p. 3209).

All these objections clearly show that critics of OSR would have to initiate a broad induction to refute the epistemological thesis, but are unable to do so. They would have to avoid talking about all theories of the past in an undifferentiated way, as otherwise the argument would lose its impact. The empirical premises, the

[14] For forty years now, there has been an ongoing debate as to whether all entries on Laudan's list are justified. A relevant extension and improvement of the list was proposed in Vickers 2013, esp. pp. 191–194 criticism of this is expressed, for example, by Mizrahi 2016; for a current overview of the debate, see Lyons and Vickers 2021.

inductive data basis of these PMI variants in maximum generality, cannot be justified by the historical evidence actually cited. The premise *there is theory change and there have been major errors in the course of the history of science* does not fulfil the purpose of Putnam's, Laudan's and Vicker's arguments.

5.2.2.2 Deductive Reformulations of the PMI

Given the failure of the abductive-inductive variants of PMI, it is not surprising that alternative variants of the argument have been proposed. A better PMI variant is that of a *reductio ad absurdum*, which I reconstruct here on the basis of Stathis Psillos' and Wrays' explanations (Psillos 1999, p. 102; Wray 2018, pp. 74–75).

The following applies to any empirical scientific theory about any domain *B* of the external world:

P1 Assumption for the sake of argument: Our current successful theories about the domain *B* are approximately true (partial thesis of OSR).
P2 If our current successful theories about *B* are approximately true, then discarded theories about *B* from the past cannot have been true.
P3 False, discarded theories about *B* were nevertheless successful in the past.
K1 Therefore, empirical success of a theory is not necessarily linked to the approximate truth of a theory.[15]
P4 In order for *X* to serve as an explanation for a phenomenon *Y*, there must be a necessary link between *Y* and *X*.
K2 Approximate truth of a theory about *B* cannot explain its theoretical success.
K3 Therefore P1 is unfounded.

Clearly, K3 follows from K2 only conditionally, not by the premises alone. I consider the Psillos-Wray interpretation of the argument as a *reductio ad absurdum* to be wrong-headed, since K2 does not lead to any logically contradictory expression with P1 and does not contain any obvious (e.g. conceptual or logical) falsehood. A problematic consequence for OSR that Psillos and Wray see here nevertheless follows: superficially, since K2 firstly contradicts the no-miracle-argument for realism, secondly because the approximate truth of a theory intuitively seems like an effective explanation for the success of a theory. Even many anti-realists would reject K2, since they usually only doubt that there are true theories in the sense of realism or that it is possible to be certain of their truth. Not all anti-realists, on the other hand, want to argue that if a theory were true or approximately true, this would not be a good explanation for its success. However, this is what K2 demands. In this respect, it could be argued that K2 contains an absurd consequence, that it is a thesis that should be rejected, and thus provides another ground for rejecting one of its premises.

[15] The word "necessarily" can be read here in the sense of causal, natural law, logical or ontological necessity, the negative conclusion should apply in all cases according to the argument.

This variant of PMI is not concerned with the reference of theoretical terms or the existence of theoretical entities, but with the truth and falsity of theories. But what is more significant in this context is that we deal with a deductive argument with an empirical premise (P3), not an abductive or inductive reasoning pattern, which the name PMI would suggest.

Surprisingly, the argument is often reproduced in this or similar terms. Psillos criticises premise P2, claiming, like Chakravartty, that formerly successful but rejected theories often referenced in part and contained some true hypotheses that are preserved in contemporary theories. Psillos does not want to concede the consequence of P2. However, this makes it necessary for OSR to explain why theories of the past, such as Newtonian mechanics, which are partly false and are no longer defended, should nevertheless be considered *possibly true*, *approximately true* or *partially true*. Here OSR owes the counterargument a defensive manoeuvre.

The last variant of PMI arguments that I would like to consider is the reconstruction of Laudan's argument by Timothy Lyons (cf. also for the following Lyons 2002). This endeavours to remedy the apparent fallacy in Laudan's variant and to present the argument as a double *modus tollens*: According to Lyons, Laudan wants to show that (1) a theory need not include exclusively referring theoretical terms in order to be historically successful, nor (2) must it be true. A *modus tollens* argument is about showing that q is always the case if p is the case and non-q holds in at least one case, and that thus non-p must be true.

Lyons' Laudan claims that the realist relies on the premise *if a theory is successful, it must also be approximately true* or …, *all its theoretical terms must refer*. According to Lyons' Laudan, this is the case because the realist justifies her view of our best current theories by arguing that their success can only be explained by these two premises being true. Lyons' Laudan thus assumes the no-miracle argument in a rudimentary form for every OSR, according to which the only explanation for theory-success is the truth of the theory. Therefore, a single case of a successful and not nearly true theory is sufficient to undermine the justification of OSR. Historically, this can be proven: Theoretical success is not a sufficient condition for (approximate) truth. This also applies to the connection between reference and theoretical success. According to Lyons' Laudan, a counterexample can be cited for both cases. In this respect, he aims to state that Laudan's much-criticised list of historical cases is even "overkill" (Lyons 2002, p. 76), since for each premise a single case of falsity with success and non-reference with success would have sufficed. Here, too, a deductive argument with two empirical premises is used.

Firstly, it can be objected to this that not every OSR is forced to defend the no-miracle argument. Second, according to Mizrahi, it is possible to formulate the no-miracle argument in such a way that it does not claim that the *only* reason for the success of a theory—explaining phenomena, making correct predictions, acceptance in the scientific community, compatibility with other theories—is its (approximate) truth.[16] This premise, which is assumed by OSR, seems to be too strong and to jeopardise the argument in the first place. But here, as in the Psillos-Wray variant, there is pressure on the realist. She must explain why partially false theories can be successful and how partial truths formulated using them persist in subsequent theories.

5.2.2.3 The Challenge to Realism and the No-Miracle Argument

Despite the many variants of the argument and the efforts of realists like Chakravartty to respond to the PMI, some anti-realists, such as Wray and van Fraassen, do not see a challenge from the PMI to OSR. They also do not think that forms of PMI motivate philosophers or scientists to take an anti-realist position in the debate. This is not because they consider all formulations of PMI and other historical arguments to be fallacies (although van Fraassen is, as I noted, a great critic of the IBE and therefore rejects the abductive and inductive variants of the PMI). Rather, van Fraassen and Wray believe that a moderate OSR is able to reconcile such insights into the history of science with his attitude towards contemporary theories, which is in line with Chakravartty's stance. After all, no one denies theory change and assumes that all the claims of our current sciences are true.

I agree with Mizrahi, after reviewing the PMI variants, that Laudan's classic article cannot be reconstructed favourably as a genuine abduction or induction, because then the argument is flawed and lacks the necessary data base. It should be read as a criticism of OSR's conclusions from the success of theories rather than as an attempt to refute the no-miracle argument outright. In particular, it shows two things in the variants of Psillos-Wray and Lyons: First, approximate truth cannot be the only reason for the empirical success of theories. Second, successful theories do not need all their theoretical terms to refer and all of their postulated entities need not exist in the external world. Therefore, a premature IBE of OSR from the empirical success of theories to their approximate truth and a correct scientific ontology associated with them is not easily justifiable and instead faces scrutiny.

The consideration of the deductive forms of PMI produces a better understanding of how the no-miracle argument for realism in philosophy of science should be framed. In the previous subchapter on UTE, I have already shown that Smart's abductive variant of the no-miracle argument cannot stand up to criticism. The arguments considered here make it clear that such an argument must not be based on the premise that truth is the only reason for theoretical success. Alternative, anti-

[16] Mizrahi even wants to show that all deductive formulations of PMI fail completely and that they too must be abandoned. But Mizrahi's counterarguments, e.g. in Mizrahi 2013b, pp. 3211–3216, show this only for strong variants of these PMI arguments. I therefore consider the more limited successes of PMI, which I emphasise in the next section, to be justified by the two variants reconstructed in this section. Cf. also Lyons 2013, 2017; Cf. for Mizrahi's own view Mizrahi 2020, 2021.

realist explanations for the success of the sciences should also be considered,[17] when the no-miracle argument is approximated by Putnam's words: "The positive argument for realism is that it is the only philosophy that doesn't make the success of science a miracle." (Putnam 1979, p. 73). My formulation of the argument, which follows Putnam and includes the preceding analytical results, reads:

P1 If all parts of a scientific theory that enable it to make correct observational and experimental predictions, successfully explain phenomena, make phenomena (partially) controllable and/or develop a functioning technology from them were wrong, then this would be tantamount to a miracle, i.e. a particularly improbable coincidence.
P2 There have been such successes of scientific theories in the past and present.
P3 Either all such successes in the past and present are particularly improbable coincidences or the parts of the scientific theories responsible for the success are not all completely false (i.e. some are partially true).
P4 We should not explain all such successes by particularly improbable coincidences.
K Parts of our scientific theories are true.

Anti-realists will try to argue against P1 by explaining the success of a given, completely false theory by something other than a true part. Whether such an argument would be convincing is another matter. I remain neutral here on the question of whether the argument in my version can count as a motivation for OSR. Neither Quine's position, nor Chakravartty's, nor my own proposal for a new scientific realism require the argument.

It should merely be noted here that in my version the no-miracle argument does not run into difficulties with the analysed UTE and valid PMI arguments. However, it only motivates an OSR that advocates selective scepticism. It is compatible with anti-realisms, such as van Fraassen's constructive empiricism, that are agnostic about the existence of unobservable theoretical entities. This is the case regarding the motivation for selective scepticism and the compatibility with some anti-realism, since the argument does not support the conclusion that theories as a whole are true.[18] In the second part of the present book, I will introduce a new interpretation

[17] Cf. e.g. the evolutionary explanation for success in van Fraassen 1980, pp. 39–41, esp. p. 40: "I claim that the success of current scientific theories is no miracle. It is not even surprising to the scientific (Darwinist) mind. For any scientific theory is born into a life of fierce competition, a jungle red in tooth and claw. Only the successful theories survive—the ones which *in fact* latched on to actual regularities in nature." For an overview of the convincing arguments that the evolutionary explanation and realism are not alternative in any meaning full way for the debate cf. Park 2022b.

[18] I will not reproduce the long and cumbersome debate on the many no-miracle arguments here, since my variant of the argument fulfils the requirement of a motivation for scientific realism without wanting to justify it strictly. My form of the argument follows directly from the critique of the famous versions by Smart (and Boyd), which I discussed on p. 112 of this book, and that of Putnam, on which my variant is based.

of Meillassoux's ancestrality argument, which, among other things, provides a reason to reject certain forms of anti-realism—fully independent of the no-miracle argument. Nor do I claim here to justify why we *must* be realists in philosophy of science, although I believe that realism is the more appropriate position.

It should be noted that certain forms of OSR and the no-miracle argument cannot defend themselves against the better variants of PMI, while Chakravartty does not yet seem to get into major difficulties at this point. My version of the no-miracle argument supports his stance. His cautious semirealism has no problems with historically given, false parts in theories, but he owes it to the historical arguments to say more about the idea of his partly optimistic, partly pessimistic picture of the history of science. Before turning to this consideration, I would like to look at a final historical argument that makes this challenge to Chakravartty even clearer: Stanford's argument of unconceived alternatives.

5.2.3 Stanford's Unconceived-Alternatives Argument

Stanford motivates and justifies his anti-realist position, epistemic instrumentalism, with a historical argument.[19] He argues as follows:

> [W]e have, throughout the history of scientific inquiry and in virtually every field, repeatedly occupied an epistemic position in which we could conceive of only one or a few theories that were well-confirmed by the available evidence, while subsequent history of inquiry has routinely (if not invariably) revealed further, radically distinct alternatives as well-confirmed by the previously available evidence as those we were inclined to accept on the strength of that evidence. [...] [W]e have repeatedly found ourselves encouraged or even forced under the impetus provided by recalcitrant phenomena, unexpected anomalies, and other theoretical pressures to discover new theories that had remained previously unconceived despite being well confirmed by the evidence available to us[.] Stanford 2006, p. 19.

According to him, there are good reasons to believe that today's best theories will be replaced by *unconceived alternatives* in the future. Therefore, the epistemological claims of OSR, especially if they are accompanied by the thesis of cumulative progress in scientific knowledge, are not tenable in Stanford's eyes. In contrast to PMI, it is neither claimed that future observations would take care of the problems of our best current theories, nor that these theories themselves are the actual problem. The latter is due to the epistemic situation of the empirical sciences. Due to the contrastive UTE, we are not in a position to weigh up all relevant, empirically different theories against each other if we have not (yet) generated them, as the historical record of the sciences shows:

> Eliminative inferences are only reliable when we can be reasonably sure that we have considered all of the most likely, plausible, or reasonable alternatives before we proceed to eliminate all but one of them (or, in the limiting case, simply rest content with the lone

[19] Cf. Stanford 2006, pp. 188–214; and cf. for a critique of it Fine 2008, whereas some of this was adopted in: Stanford 2021.

> contender). But the history of science shows that we have repeatedly failed to conceive of (and therefore consider) alternatives to our best theories that were both well confirmed by the evidence available at the time and sufficiently plausible as to be later accepted by actual scientific communities. Even more briefly, the historical record suggests that in science we are typically unable to exhaust the space of likely, plausible, or reasonable candidate theoretical explanations for a given set of phenomena before proceeding to eliminate all but a single contender, but this is just what would be required for such eliminative inferences to be reliable. Stanford 2006, p. 29.

The historical archive that Stanford defends in the subsequent course of his book consists of case studies that are supposed to prove that such cases existed. He mainly cites examples from biology, but does not overplay his hand. There is no need to show that this is always or even usually the case in order to teach caution. With his argument, Stanford does not want to establish either a global scepticism or a metalocal anti-realism in philosophy of science (cf. also for the following Stanford 2006, pp. 35–37). Instead, he recommends a local scepticism with regard to those theories (and the associated ontologies of science) for which there are good reasons to assume that not all reasonable alternative theories have been formulated. For him, however, these are *de facto* all empirical natural sciences that work with complex theories and models.

Firstly, it should be noted that the argument is an inductive conclusion: It infers the future from the majority of past cases.[20] As Wray correctly points out, it is remarkable that Stanford can admit the realist objection that the methodology and instrumental as well as the theoretical work of sciences today differs greatly from that of other epochs (Wray 2018, p. 83). However, instead of concluding from OSR that today's theories are therefore much closer to the truth, his argument undermines the plausibility of OSR's epistemological thesis.

The main difference that Stanford sees in *Exceeding Our Grasp* between his argument and the classical formulation is that he is not concerned with a direct comparison of contemporary with historical examples or with a fundamental pessimism about all theories and hypotheses that would be hard to redeem. For him, the problem lies elsewhere: "[P]resent theorists are no better able to exhaust the space of serious, well-confirmed possible theoretical explanations of the phenomena than past theorists have turned out to be." (Stanford 2006, p. 44). Stanford is thus concerned with showing that our epistemic ability to evaluate the contrastive UTE of theories is not fundamentally different from the situation of past generations. It is therefore reasonable in his view to assume that further changes in theory will prove most current theories wrong. OSR is therefore not justified in assuming that we are dealing with a theory change towards a more correct or even ultimate description of the external world. Stanford's unconceived-alternatives argument is thus a combination of a historical abduction and the idea of contrastive UTE.

[20] Timothy Lyons offers a reconstruction of this argument in which it is a deductively valid *modus ponens* argument with some empirical premises that are themselves only inductively justified Lyons 2013, p. 372.

Mizrahi argues against this argument that the cases in which scientists were unable to formulate alternative theories in the past, even though they could have been confirmed by the empirical evidence, do not provide a clear picture (cf. also for the following Mizrahi 2015). Mizrahi admits that there are Stanford-cases—that is, alternative theories that are just as well (or better) confirmed by the evidence already available at the given time than the historically successful theories. But he thinks there are also many divergent cases, meaning that the historical basis of the inductive premise is not clear.

Firstly, I think it is questionable whether Mizrahi's argument, which relates purely to the historical situation, can be used to counter Stanford's argument. After all, it concerns the problematic epistemic situation of any inquiring community, a form of contrastive UTE. Moreover, the question of the historical situation can only be resolved by a detailed, historical examination of a particular discipline in a particular period of the past. For the argument to be valid, there should be more Stanford-cases than Mizrahi-cases (cf. McAllister 2023). This is an interesting historical question, but there is a better way to deal with the problem raised by Stanford.

It is taken by Chakravartty, who claims that a real solution to PMI for OSR is also a solution to Stanford's argument (cf. also for the following: Chakravartty 2008, esp. p. 149). We can agree with this if we share Chakravartty's conviction that a refutation of these anti-realist arguments from the history of science must essentially establish that certain structural parts of theories are true and are preserved despite the change of theory. In this picture, Stanford's pessimistic conclusion applies to many of the best current theories as a whole, but not to the well-supported parts of these theories and ontologies of science (detection properties, natural entities, and their mathematically expressible relationships). Below, I explain how Chakravartty deals with historical arguments.

5.2.4 *Chakravartty's Semirealism, Historical Arguments and the Critique of Science*

Chakravartty responds to the challenges posed by anti-realist arguments from the history of science by defending his OSR in a way that echoes his defence against UTE arguments. He believes that a response to the valid variants of PMI and the unconceived-alternatives argument must be found together. With his argument, Stanford also explicitly opposes selective realism, especially in the variants of Philip Kitcher and Psillos (cf. also for the following Stanford 2006, pp. 164–167; for criticism: Saatsi 2009, pp. 361–362). According to Stanford, the realist faces two problems: It is inadmissible to analyse formerly successful theories by attributing their success to true parts that can only be shown to be true because they occur in the best current theories. Moreover, the realist must be able to specify which current local theories (or parts of theories) she invests in ontologically and why (cf. Stanford

2006, pp. 181–185). Chakravartty essentially takes up these challenges in defence of his semirealism.

I will first show Chakravartty's way out and briefly discuss why he relies on considerations of entity realism and structure realism. Then I will explain why the *easy way out* of UTE arguments is rendered unworkable by the historical arguments. Finally, I will briefly show why the presented and related historical arguments give rise to various forms of criticism of science.

5.2.4.1 Chakravartty's Synthesis of Historical Optimism and Pessimism

Chakravartty distinguishes several ways in which realism can counter historical arguments. Some rely solely on the fact that theories are increasingly improving, and are thus closer to the truth than theories of the past. Those replies are searching the easy way out, as I have called the strategy.

> "Others contend that the pool of data serving as the ground of the induction is too small to permit a credible conclusion. Many responses to Larry Laudan's (1981) version of PI [=pessimistic induction; J. V.], for example, take issue with the list of theories he cites as evidence, arguing that if one factors in the further qualifications assumed by realists – the notion of maturity, the exclusion of ad hoc theories, the importance of novel predictions, and so on – the data for pessimism are greatly reduced." Chakravartty 2007, p. 29.

But to provide this proof, realists would have to present immense historical studies. Chakravartty believes another option is better:

> "Not all aspects of scientific theories are to be believed. Theories can be interpreted as making many claims about the nature of reality, but at best one has good grounds, or epistemic warrant, for believing some of these claims. Only some aspects of theories are likely to be retained as the sciences march on. I will refer to any approach that takes this advice seriously as a form of *selective scepticism*. The primary motivation for this modification to realism *simpliciter* is to pick out, from among the numerous claims embedded in theories, the ones that are most epistemically secure and thus likely to survive over time." Chakravartty 2007, p. 29.

I have already sufficiently outlined Chakravartty's proposal to divide scientific ontologies and theories into secure and insecure parts as well as entities with detection and auxiliary properties. However, the problem of selective realism addressed by Stanford with regard to current theories remains: It is difficult to say in individual cases—e.g., in the theory of continental drift in geology—which of the implied empirical assertions are epistemically most certain and protected by detection properties and therefore stand the test of time. It *seems* quite certain that the Himalayas were piled up when the Indian continental plate slid under the Eurasian continental plate.[21] It also seems certain that the Indian continental plate in the Triassic was connected to Antarctica and Africa in the supercontinent

[21] Also for the following in general on plate tectonics: Anderson 2007; Frisch and Meschede 2013; Windley 1995, and on the case of the Himalayas: Decelles et al. 2002; Kind and Yuan 2003.

Gondwana. However, what exactly is understood to be contained by the theory of continental drift is essential in order to decide whether it must be adapted as a whole, e.g. by the possible discovery that Gondwana was never a continent of our planet, or whether only a conclusion from the completely correct theory to the theory-external and merely implied hypothesis that there was a continent of Gondwana has been refuted. Chakravartty's solution to PMI and Stanford's argument heightens the urgency of stating clearly how theories are to be understood, and what the body of a theory really contains.

> "A knowledge of unobservable properties and relations, I submit, is no pyrrhic victory for the realist; it is substantial. Stanford often writes as though theories and entities are the units of scientific knowledge to which realists must commit, and thus fails to note how well-confirmed properties persist through changes in theories and the entities they describe. A sophisticated realism is finer grained than he allows." Chakravartty 2008, p. 155.

Chakravartty takes this point as a lesson from entity realism, which explicates that theories need not be fully true in order to ascribe existence in the sense of OSR to certain causal properties and the entities that support them with good reason.[22] Against entity realism, he maintains that the realist must invest ontologically in the parts of the theories that determine concrete structures and detection properties. In doing so, he emphasises that she is committed to the properties about which causal knowledge exists, and also to the fact that these properties must be attributed to certain natural entities. In the geological example, this means that the hypothesis of the continent of Gondwana is not part of the core of the theory of plate tectonics and continental drift. However, since we are in causal contact with properties of continental plates (e.g. their relative motion), selective realism is appropriate for these parts of the theory.

> "Once these key points regarding causal knowledge and causally efficacious properties, suitably elaborated, are taken more seriously, it would seem that several other realist insights can be rescued from Stanford's scepticism. Structural realists, for instance, point out that in some fields of scientific research (including many branches of physics), there appears to be a great deal of preservation of mathematical structure across theories over time. Stanford complains that the definition of structure here is vague, and no doubt, in the earliest literature that gave rise to the recent revival of structural realism, that was so. But it is possible to define the term "structure" very precisely—in terms of relations between properties of the sort I have indicated, for example." Chakravartty 2008, p. 155.

Here, Chakravartty refers to his definition of concrete structure.[23] It was defined by Chakravartty as the relations between the first-level causal properties of natural entities. If it is assumed that all concrete structures form a whole, then that is essentially the structural thesis of the single real external world in Chakravartty's OSR. As mentioned in the quotation, he makes use of some of the ideas of structural realism. With these two manoeuvres, Chakravartty secures his understanding of theories as descriptions of concrete structures and explains the extent to which he

[22] Cf. my presentation of Chakravartty's critique of entity realism in Sect. 5.1.5, p. 120.

[23] Cf. Chakravartty 2007, pp. 39–41 as well as my presentation in Sect. 4.2.2, p. 100.

wants to be a selective sceptic *and* a selective realist about theories, properties and natural entities.

> "[R]ealists can admit a pessimistic induction on the history of past science *simpliciter*, while simultaneously asserting an optimistic induction on the parts of theories to which they commit, to the extent that these parts tend to survive over time. The structures to which realists should commit [...] involve properties and relations that are essential to describing our causal connections to the world. Realists should commit to concrete structures that are detected and described in scientific theories." Chakravartty 2007, p. 47.

With regard to historical arguments, this means that Chakravartty recognises the discontinuity of the sciences on the one hand, while also accepting the limited arguments of the deductively valid PMI and Stanford's argument. On the otherhand, he argues in favour of the validity of a second, optimistic induction with regard to specific parts of theories. This means, contrary to Stanford, that we are able to take a realistic position for the most secure parts of the best current theories, despite the contrastive UTE, the historical archive of the sciences, and the inferable theory change in the future.

It is advisable for OSR (and any realism that wishes to counter the historical arguments for that matter) to adopt a form of realism that is cautious enough to deal with theory change and flawed parts of theories. I show in the next subchapter that, in addition to Chakravartty's OSR, different versions of structural realism have been proposed that share this programme. One of those attempts to argue for a better version of selective scepticism than OSR, another to take the route of metalocal realism about theoretical entities and also about theories themselves, but by limiting the permissible kinds of entities in relation to which realism is appropriate to structures. Before doing so, however, the question arises as to why the easy way out (of UTE) fails due to historical arguments and why realists should take historical arguments seriously in order to maintain the possibility of criticising science.

5.2.4.2 The End of the *Easy Way Out*

I will now show why the *easy way out* (of UTE arguments) is not viable. With this strategy, some representatives of OSR[24] try to escape UTE arguments. The easy way out is to assume that, while the best current theories are threatened by underdetermination, future theories will provide us with complete and correct domain ontologies when the exploration of the external world is sufficiently advanced.

But the historical arguments presented in this subchapter clearly show that such optimism is empirically inappropriate. Moreover, if the history of science is discontinuous and can never dispense with auxiliary properties, postulated entities and idealised formulations of natural laws, the point envisaged by the easy way out will never be reached: the end of research, the moment in which the (humanly

[24] Cf. the last Sect. 5.1.5 of the previous subchapter, p. 122.

accessible) external world is fully understood and grasped by our theories and we have the (cumulative) ontology of the single real external world domains epistemically accessible to us. With the discontinuity of the historical process of science and the indispensibility of postulations and idealisations, we should be pessimistic about such an end to all research. There is no easy way out of UTE arguments. This is particularly evident in Stanford's argument, which combines considerations of contrastive UTE with the historical record.

What a proponent of OSR can hope, with Chakravartty in mind, is that increasingly more parts of science will gradually improve. This means more possibilities for causal contact with natural entities and greater knowledge of detection properties (and consequently more of the over-all concrete structure of the single real external world). This optimism balances out the pessimism about the envisaged end of science. It is to be clearly distinguished from Popper's falsificationism, which takes up the idea of progress, but is not selective realism. It does not claim that parts of the best current theories are true, but merely that they can *count as* provisionally true (cf. Popper 1984, pp. 62–67, 2002).

One way of rehabilitating the easy way out could be seen—following Popper—in the appeal to the completed sciences as a regulative ideal. This might help to specify the theoretical assumptions to which realism is committed. They could be those that are consistent with the theoretical assumptions of the (never fully attainable) regulative ideal. Even though I will return to this idea in the last chapter of this book, I would like to argue here that realism is not served well by this argument. This manoeuvre makes it impossible to say which current beliefs about (partial) theories and entities are true or false, without it being obvious why a commitment to a regulative ideal for the present day would provide a better philosophical position than, say, a sophisticated anti-realism about theories and unobservable entities. The fallibility of current theories is shifted to an epistemic ignorance hidden in the ideal sciences.

5.2.4.3 Critique of Science and the Question of Realism

Finally, I want to note that a complete rejection of historical arguments on the part of OSR or new scientific realism would be inappropriate. The argument, apart from philosophical honesty, for this is as follows:

My first premise is that no critique of science can be convincingly conducted without recognising historical arguments. By this I mean any form of scientific and/or philosophical analysis of science as a social system, historical factor and worldview that aims to discover and correct unwarranted assumptions, errors, problematic effects, ethical conflicts and dogmatic scientistic ideologies. I will argue in favour of this premise with due brevity in a moment.

The second premise is that no philosophical position (and certainly no realist position) should dispense with the critique of science. I will not argue in favour of this premise because, firstly, I think that it is largely uncontroversial and, secondly, there are already many sound arguments for it.[25]

Why are historical arguments needed to criticise science? One reason is that only from a historical perspective and with evidence of historical errors can it be made clear why science is able to explore reality (at least from a realistic perspective) and deliver true results, yet at the same time explain why the fallibility of scientists and scientific communities is not cancelled out. Without historical examples and their argumentative systematisation, it is impossible to understand how scientific knowledge claims and the mistakes of scientists and their communities fit together. This is decidedly *not* to assert that the fundamental epistemological fallibility of people can only be established by means of historical arguments in philosophy of science, but that only they explain how the dynamics of scientific objectivity claims, concrete errors and mistakes in theories and theory change are organised.[26] Without historical arguments, it is also much more difficult to argue against the deconstructivist consequences of historical criticism of science in certain areas. For example, when Hans-Jörg Rheinberger concludes from his critique of the science of microbiology that the experimental systems and partly contingent contextual factors in protein synthesis in the test tube determine the allegedly objective objects of knowledge in a strong sense, realists should be able to give an alternative interpretation of protein synthesis (cf. Rheinberger 2001, p. 26).

Another reason is that historical arguments play an important role in drawing the line between a legitimate scientifically informed view of reality and a problematic scientistic ideology (cf. e.g. the classic attack on scientism in Midgley 1992). By a problematic scientistic ideology I mean any system of beliefs that grants unrestricted authority to the natural sciences and thus justifies power structures, resource distribution and social relations. Examples include sexism and racism that are allegedly based on biology.[27] In order to differentiate between problematic scientistic ideology and legitimate scientific conceptions of reality, it must be understood that the sciences themselves are a complex social system that serves more purposes than simply the generation of true theories and models (cf. Srinivasan 2016). In order to understand the sciences as a social system, the specific and sometimes contingent historical development of the sciences must be traceable. The history and sociology of science therefore require premises from philosophy of science (such as the discontinuity of scientific theory development). To ignore the results of these empirical subdisciplines outright would be unscientific, as they have direct implications for philosophically relevant questions, e.g. for the (dis)unity of the sciences.

[25] Here, reference could also be made to fundamental works in the sociology of science cf. e.g. Longino 2019, as well as to the debate under the heading *Scientism* as ideology and philosophical paradigm cf. e.g. de Ridder et al. 2018.

[26] I would like to thank Jens Rometsch for criticising this passage, which has given rise to clarification.

[27] For an overview, see Appiah 2006; Hankinson Nelson 2017; Jackson and Weidman 2004.

Finally, historical arguments are an integral part of many individual debates in the criticism of science in certain disciplines. Without the historical evidence of, for example, sexist theory generation and correspondingly deficient experimental designs, which are subject to bias, feminist criticism of some explanations of primatology would not be possible. What is needed in such cases is both a detailed critique of individual studies and experiments as well as evidence from the social sciences for historical constellations that support certain research programmes for non-scientific, ideological reasons.[28]

* * *

My analysis of the historical arguments has yielded several results. The easy way out of UTE does not help OSR. As Chakravartty rightly argues, the best prospect for OSR is to advocate a selective scepticism and realism that can navigate between UTE and historical arguments. To do this, Chakravartty proposes a realism about theories and natural entities that relies heavily on causal properties and concrete structures. To decimate or completely ignore historical arguments in their relevance to the debate is not only philosophically dishonest, but would also produce the consequence that critique of science is no longer adequately possible. Thus, a subtle and moderate OSR is the only viable option in the face of UTE and historical arguments. In the next subchapter, I show that the thesis in Chakravartty's semirealism about a fixed structure of the single real external world, already developed in Sect. 4.2.2, is problematic. However, this only allows Chakravartty to make the necessary assumptions of concrete structure and causal properties in his variant of such an OSR. A third type of argument, antitotality arguments, refute the structural thesis of OSR, which, in combination with the UTE and the historical arguments, proves the first stalemate of OSR against the neo- and anti-realist counterarguments.

5.3 Antitotality Arguments

The UTE and the historical arguments are classics in the discussion about scientific realism and motivate, for example, Chakravartty to strive for a course between the two. In contrast, antitotality arguments seem to be at home in the global ontological and metaphysical debate or are associated with special debates in philosophy of science, such as those regarding the laws of nature. In my view, however, they represent an underestimated challenge for OSR. The idea of a single real external world with a fixed structure, e.g. of natural laws, lies at the heart of OSR, as I demonstrated in the previous Chaps. 3 and 4. This picture comes under attacked by antitotality arguments in two ways. On the ontological level, these arguments object that there can be a single real external world as a totality (sum, set, mereological

[28] Cf. for a specific critique of primatological experiments Lloyd 1993 and the general research programme Haraway 1991, p. 21–42.

whole, ...) of subject-independent entities (type 1). On the level of philosophy of science or metaphysics, they argue that this external world has necessary, natural-law-like linkages that need to be mapped by theories (type 2). Type 1 arguments are put forward by Markus Gabriel for his new realism and by Carol Rovane for her metaphysical relativism, among others, in order to criticise the image of the single real external world. Nelson Goodman and Alain Badiou also propose such arguments. A type 2 argument is put forward by Nancy Cartwright, among others, to reject the concept of fundamental laws of nature, or by Quentin Meillassoux to insist on the necessary contingency of all concrete objects, structures and laws of nature. In what follows, I will reconstruct selected arguments from these and other authors to show why the assumption of a single real external world that can be depicted in its ontological structure as fixed by natural laws is problematic, without going any further into the positive theses of the New Realisms or anti-realisms.

5.3.1 The World Problem and Set Paradoxes

Gabriel argues against the definition of the "world, in the sense of a totality of all modally robust facts". (Gabriel 2020, p. 43). Using Bernard Williams' formulation of the "absolute conception of reality" (Williams 1978, p. 64) as an example, he argues that modally robust facts should be those that are subject-independent. This conception of the world essentially corresponds to the single real external world of the subject-independent entities of OSR. Gabriel puts forward several arguments against this idea:

The first argument, in *Limits of Epistemology*, is a variant of the argument of pseudo entities. Gabriel specifies this with regard to the statement that this conception of the world would fail because "there are facts that are not modally robust." (also for the following Gabriel 2020, p. 43). Facts about "ourselves as minded beings, the state, the future" and the like are not then components of the world. This leads to a problem of self-application: the supposed fact itself "*that* the content of knowledge is exhausted by the totality of all modally robust facts" would not be part of the objective (external) world. However, this is contrary to the claims of OSR, which requires second-order objective knowledge in order to be formulated without contradiction. In addition to the pseudo entity problem I outlined above in Sect. 4.1.2, this first argument attacks OSR more deeply.

With G.E. Moore, an attempt could be made to invalidate this argument by presenting the external world as the set of (mesoscopic, "common sense") individual things that we encounter in everyday life (Moore 1939). According to Gabriel, however, Moore is already making a mistake because this leads to extreme metaphysical realism. This not only assumes a single true description of the external world, but also requires that this can be realised through unproblematic references in our knowledge practices: If Moore holds up his hands, then his "common-sense" assertion that there are two hands would be the only true description. This would justify a "*naive ontology of individuals*" (Gabriel 2020, p. 50, 2015a, p. 92), because even the perspective of a particle physicist, according to which there are no hands but a configuration of particles, would have to be false, since it deviates from the true description. However, this leads to problems that hardly any representative of OSR would accept.

Goodman proposes a similar argument for irrealism in *Ways of Worldmaking*: In order to defend OSR and other strong versions of OR, such as metaphysical realism, it would be necessary to refer to the world as independent of subjects. Goodman considers this premise to be fulfilled by the definition of the external world as SI. However, any further statement that, as in OSR, is intended to specify the external world and our relationship to it, fails: "We are confined to ways of describing whatever is described. Our universe, so to speak, consists of these ways rather than of a world or of worlds." (Goodman 1978, p. 3). Goodman's external world—*our universe*—thus disintegrates into our modes of access, whereby an anti-realist position against any OR is to be justified. Goodman's argument is based on the indescribability of the totalised external world, when every epistemic mode of access is shown to be arbitrary because it belongs to the subjects. One could never say anything about the world without a frame of reference, so the question of whether the external world is independent of subjective frames of reference "is then either senseless or itself defines a new frame of reference" (Gabriel 2020, p. 269, en. 8), as Gabriel writes.

Finally, Gabriel offers, especially in *sense and existence*, some related arguments against the usual ways of totalising the (external) world. Arguing against the idea that the world can be totalised as a maximally comprehensive set *of all other entities*, Gabriel recalls variants of Russell's set paradox (Russell 1938, chap. X and §106, according to which, the set of all sets that do not contain themselves may not contain themselves on the one hand (since, contrary to its definition, it would otherwise contain a set that contains itself), but must on the other hand contain itself (since otherwise it would not be complete, but would exclude a set that does not contain itself), cf. Gabriel 2016, p. 142. Russell or Ernst Zermelo would have spoken of classes in today's terminology, but that can be postponed in this context and transferred to ontology. Representatives of a set ontology such as Alain Badiou shy away from forming something like the absolute external world that contains everything in order to protect themselves from the resulting objections (cf. Badiou 2005, 2009).

Gabriel formulates this consideration in a more accessible way than his "*argument from the list*" (Gabriel 2015a, p. 17): If a real number is assigned to each element of the all-encompassing, absolute set of the external world, this can be written down on a potentially infinitely long list. Since some numbers are assigned to concrete entities, further numbers must be introduced to represent the numbers already assigned. This results in an initial extended list, on which the original list itself, which is an entity, is also noted. Now, however, the first extended list would have to be noted, resulting in the second extended list, and so on. These considerations, orientated on Cantor and Russell, which Gabriel elaborates with a crical and distanced view on Badiou, boil down to the fact that a larger set can always be formed and that any concept of an absolute, maximally comprehensive set is misleading. In the end, Gabriel does not defend the argument from the list, as he does not share Badiou's assumptions about the coextensiveness of set theory and ontology, but instead presents two main arguments of his own in favour of the thesis.

5.3.2 Gabriel's Main Ontological Arguments

Gabriel's first main argument presupposes the concept of a field of sense. "What individuates an ontological field of sense in contrast to other fields are the rules of arrangement to which objects are subject, insofar as they belong to it. I call the rules of arrangement "sense" and the relations that exist between a field and the objects occurring in it "appearance". Objects appear in fields of sense, that is, in domains that are determined by different senses in relation to one another." (Gabriel 2016, p. 163; my translation.) The first main argument states that there is no description of the world as a whole—as a field of sense of all fields of sense. Otherwise, every ontological domain (every field of sense in Gabriel's terminology) for which a description of assignment exists, according to which entities are functionally assigned to it and a rule how they are arranged in it, would have to be included in the description of the world.[29] However, because, according to Gabriel, we need to admit that there are an incalculable number of such domains, some of which are incomplete (e.g. the future), this full integration is impossible. Since a detailed discussion of the concept of a field of sense and Gabriel's underlying arguments would be too extensive here, I will concentrate on the second main ontological argument, which can be reconstructed without appealing to fields of sense:

> The crucial difference between the world and any other field of sense that could be said to appear in itself is that the world is said to be all-encompassing *ex hypothesi*. If it appears in itself, everything that appears in the world appears not only in the world, but also in the world appearing in itself. But this generates a problematic infinite regress: if everything that appears in the world (i.e.: everything that exists at all) appears not only in the world but also in the world appearing in the world, this is also repeated for the world appearing in the world, etc. *ad infinitum*. Gabriel 2016, p. 230; my translation.

Gabriel constructs this argument as a *reductio ad absurdum*. The prerequisite for the argument is that the world (or external world), in order to exist and be recognised by subjects, must appear as an entity within itself. This "appearance" refers to Gabriel's ontology of fields of sense. However, it can also be broken down here in such a way that this concept need not be presupposed for the second main argument: If the world or external world is to be an intelligible entity that we can recognise and know, and to which we can refer linguistically without contradiction, such that the assumptions

[29] Cf. also for the following Gabriel 2015a, 2016, esp. pp. 224–229. Gabriel thus immediately criticises the concept of a worldview in the original sense, which is introduced in Kant 2009, AA 5, p. 255.

of OR (and thus also of OSR) can be formulated, then it must necessarily be situated as an entity in a certain context.[30] The only context and ontological domain that is conceivable for such a totality is the totality itself. However, this leads to the infinite regress that Gabriel speaks of in the quotation. Here, one sees a parallel to the argument from the list, with the difference that set-theoretical considerations do not have to be introduced into the ontology for it to be successful. At this point, a defender of totality could object that the regress is unproblematic or can be averted:

> There are at least two ways of conceptually approaching such a totality. Firstly, the totality could be conceptualised additively. Let us call this *additive totality*. The second insists on the unification of totality [. . .]. There must be some principle or set of principles that assign a status to everything in the ordered structure of the world. Let us call this *qualitative totality*. Gabriel 2016, p. 230, my translation.

In Gabriel's view, what speaks against the first option is that it is unclear how it can be ensured that all domains are taken into account in the addition if there is neither an overall context nor a principle that binds everything together. Here Gabriel explicitly asks about the idea of the identity of world and universe, which comes close to OSR's thesis of the external world and is, for instance, advocated by Jonathan Schaffer (cf. 2010a,b, 2013): Isn't the addition of everything physical a hidden metaphysical principle? In order to say that the external world is e.g. the sum of all concrete entities or all entities situated in space-time, "a somehow manageable structure of natural laws" (Gabriel 2016, p. 232, my translation) must be presupposed. Without this structure, such a sum simply cannot be formed, Gabriel argues.

This makes the second option much more attractive, according to which the world or external world is the ontological domain of all other ontological domains. For this to be the case, it must either be situated in another domain or emerge in itself. The argument against situatedness in the other domain D^w (with which the regress is to be averted) is that this would then contain at least the world and perhaps other ontological domains. D^w would be a superworld and an equally large or larger totality, so the world would also have to contain this superworld. This vicious circle is not resolvable.

The better option seems to be to let the world emerge in itself. This means assuming a qualitative totality in such a way that the defender of totality can show that the regress is not problematic.

What options are available in the case of OSR? Gabriel argues that if the SI-entities in an OSR are identified as those that occur in spacetime or that enter into causal interactions, then the external world does not belong to these SI-entities: "The world itself cannot be encountered in space." (Gabriel 2008b, pp. 81–82, my translation). Causal interactions occur in it, not with it as an interaction partner. It

[30] Gabriel's ontology of fields of sense argues that this is not only an epistemological necessity, but an ontological necessity without which no entity (no object in Gabriel's vocabulary) can exist. For him, *to exist* means *to appear in a field of sense*.

is either identical with spacetime or in turn contains spacetime. What are the deeper problems here?

Causal interaction only offers an unfavourable theory option. If the world is in causal interaction with other objects, i.e. with its parts or elements, there would have to be some form of *top-down* causality.[31] Even if one is prepared to take this step, the difficulties do not end, as I would like to argue, as there would then have to be at least one causal interaction in which all elements of the external world or all parts of the single real external world interact with a particular element or part x at once. Not only is this example difficult or even impossible to imagine (especially if, for example, abstract entities are to belong to the external world or laws of nature governing it), but it also requires that x interacts causally with x. In my view, however, this is contrary to all the usual notions of causality. In any case, the burden of argumentation lies with the one who holds such a position.

The second option—the external world is identical with spacetime or contains spacetime—also leads to major problems. An veritable obstacle to this option is the problem of pseudo entities (Gabriel's keyword here is "*zügellose[r] Nominalismus*" Gabriel 2016, p. 91, translatable as *intemperate nominalism*) and Gabriel's self-application problem mentioned above. A further difficulty lies in the fact that, at least if one were to accept the identity thesis of spacetime and the external world, the physical theories of spacetime and the actually local realism-anti-realism debate in philosophy of time would need to be considered ontologically and metaphysically primary in relation to the formulation of OSR: One would first have to be empirically certain in the formulation of the physics of time, and have defined a secure position in the local realism debate about time, before one could embark on a justification of OSR. If the positions expressed in both respects are not in harmony with each other and with OSR, insoluble problems arise. As was pointed out in Chap. 1 on the realism-anti-realism debate in philosophy of time, this is quite a difficult task, especially if one cannot already build on an OSR or an OR to position oneself in the debates. Ultimately, it seems quite unclear how one would try to resolve this problem and the problem of self-application in a position that increasingly resembles eliminativist materialism.

Two things need to be noted here: Firstly, the burden of proof that the regress of totalities is infinite but unproblematic lies with OSR. Gabriel argues from the position of the field of sense-ontology that, so far, no argument is in sight that could resolve this problem (cf. Gabriel 2024a, pp. 66–73). Secondly, all totalisation principles described above fail to do justice to reality. According to Gabriel, they all make the mistake at some point of reducing the "colourful bouquet of the existing to some puny branch" (Gabriel 2016, p. 245). Gabriel considers, among other things, network ontologies by Bruno Latour and Eduardo Luft as well as naturalisms that equate existence with a domain-specific property (such as the aforementioned

[31] Cf. for a proposal of this theory option without the largest element: Ellis 2016; Ellis and Gabriel 2021.

properties of being spatiotemporally and/or causally embedded). Here as there, the attempt fails (cf. also the discussion in Gabriel 2024c; Priest and Gabriel 2022).

If Gabriel is correct in this, all principles of totalisation suffer from the claim to be an alternative-free, yet informative description of reality as a whole, which is capable of seamlessly summarising this whole of reality into a totalised world or even a totalised, single real external world. Despite some criticism of the notions of a sense (cf. Benoist 2024) and regardless of the question whether the impossibility of any metaphysics whatsoever is truely established by Gabriel (cf. Stang 2024), and even if the idea of the world can be rescued in a non-totalised way (cf. Travis 2024), his arguments against such totalities are strong and need to be answered by OSR. His conclusion is that the seamless and informative description of the whole of reality does not exist and therefore the concept of the world, let alone the single real external world, cannot fulfil what OSR demands of it.

5.3.3 Rovane's Arguments Against Unimundialism

Carol Rovane also argues against the single real external world on which OSR is based. She ventures to show that there are truths that are mutually exclusive. She calls these *alternatives*: "alternatives [are] truths that cannot be embraced together." (Rovane 2013, p. 58, emphasis deleted). *Truth* here does not primarily mean a true statement, but a *true fact*. From this, Rovane argues, it can be concluded that two mutually exclusive statements can be true, but are not simultaneously justifiable. According to Rovane, this does not mean that the statements or facts have to be inconsistent or consistent. Rather, there is *no* logical relationship between them at all, but "normative insularity" (Rovane 2013, p. 79). The two alternatives are isolated in the sense that if one is accessible, the other must be inaccessible. They are normatively isolated because they are not logically related, so that truth and falsity being in the domain of one alternative does not justify drawing conclusions about the truth and falsity of facts concerning the other alternative. She considers the resulting form of metaphysical alternative relativism represented by philosophers as diverse as Kuhn, Goodman, Feyerabend and Rorty (cf. Rovane 2013, pp. 72–73). Beyond the internal differentiations between them, according to Rovane they all doubt that logical relationships exist between all facts in her sense. She thus argues against the following thesis:

> "If logical relations do run everywhere among all truth-value-bearers, then all of the true ones must be consistent and conjoinable. It follows that there is a *single, consistent, and comprehensive body of truths*, and this amounts to a metaphysical commitment to the oneness of the world, or *Unimundialism*." Rovane 2013, p. 79.

This is a clear component of OSR: OSR cannot do without the thesis of the *single* real external world.[32] Rovane considers an alternative way to gain the external world, namely to conceive of it as a mere set of all subject-independent facts. But she rejects this idea for reasons that are similar to Gabriel's arguments against additive totality: there is no criterion by which one can determine that this set formation is feasible, since it presupposes the connectivity of all facts. It is debatable whether Rovane attacks unimundialism on metaphysical, ontological or logical grounds, but this question can be postponed here. What matters is that she has a specific purpose, which I will briefly outline before reiterating her central argument in favour of alternatives in the domain of nature:

Rovane's aim in *The Metaphysics and Ethics of Relativism* is to defend alternative relativism against the frequently advanced attack of incoherence. To do this, she must either refute unimundialism or at least encounter difficulties in order subsequently to support the so-called multimundialism on which her alternative relativism is based. This stance falls into the category of global aletic relativism in Dorothee Schmitt's helpful typology of relativisms, the core thesis which is that "two statements that are translations of each other can be given different truth values within different frames of reference" (Schmitt 2018, p. 65; my translation). This is subject to Rovane's metaphysical relativism and is compatible with the position of the same name in Schmitt's systematics (cf. Schmitt 2018, pp. 95–97). Schmitt's argumentative goal of showing that global relativisms cannot be countered by self-refutation arguments, if they keep a few key points in mind, thus complements Rovane's goal. All of the following three central challenges are considered by Rovane: "Does the theory contain absolutist residues that conflict with the other components of the theory? Is normative distance ensured? Is there the necessary semantic and epistemic access to other frameworks ensured to formulate and justify the theory?" (Schmitt 2018, p. 375; my translation).

Rovane argues against the single real external world by attacking unimundialism, which eradicates absolutist residues from her theory. This rids the world of the predicate to be the *singularly real* and creates space for the assumption of her metaphysical relativism that there are several, equally real worlds. Logical relations between truth-bearers therefore do not exist everywhere, but are a regional matter, as multimundialism suggests: "Relativists affirm instead that some truthvalue-bearers do not stand in logical relations to one another, that there are many noncomprehensive bodies of truths that cannot be conjoined, that there are many worlds rather than one." (Rovane 2013, p. 91).

According to Rovane, multimundialism and thus also alternative relativism are neutral with respect to realism and anti-realism in metaphysics, since she classifies this discussion as the one that asks whether reality is subject-dependent or not (cf. Rovane 2013, pp. 92–93). The neutrality thesis may be correct in Rovane's

[32] This does not apply the other way round: Rovane shows in Rovane 2013, pp. 85–88 and elsewhere that unimundialism does not imply realism, if only because a single world is conceivable that is dependent on subjects as in subjective idealism.

understanding of the debate—she does not elaborate on her view—but I think it is premature for two reasons. First, this characterisation based solely on the criterion of SI is not compelling. In Chap. 2 I argued in favour of qualifying anti-realism differently, namely as ontological or epistemological anti-realism. Rovane's multimundialism may allow for ontological realism—and even presupposes it for her argument in the domain of nature, as I will show shortly—but it is a variant of epistemological anti-realism with respect to some alternatives: Even if all alternatives exist, they can never all be known by one subject, since some of them do not stand in logical relations and thus admit only fragmented and isolated bodies of knowledge that can never be unified. Secondly, the totality assumption plays a central role in OR and is closely linked to the SI assumption, as I have already shown, which is why her arguments are of interest here.

Rovane shows that it is not easy to defend unimundialism. This is all the more true since, according to her, there are coherent alternatives, among them namely her alternative relativism. It should be emphasised that Rovane meets Schmitt's second and third demands: there is normative distance and Rovane's relativism of alternatives ensures that semantic and epistemic access to other frameworks, to other worlds, exists at all times. Normative isolation ensures that not all alternatives can be known from the various frames at the same time. It does not mean that other frames cannot be known and named. Together with Gabriel's antitotality arguments, this puts OSR on the defensive, as I will shortly explain in greater detail.

Rovane builds her central argument for multimundialism in nature on a debate reconstruction. Here Carnap provides her with the model of a multimundialist (cf. also for the following Rovane 2013, pp. 133–135). According to Rovane's Carnap, we need a scientific language to grasp the external world, which we define ourselves through conventions. Various such scientific languages can be defined (the meanings of the sentences of which depend on verification criteria), which are normatively isolated from one another. The truth-makers in different, normatively isolated scientific languages are alternatives in Rovane's sense, which Rovane's Carnap justifies by saying that the choice between different scientific languages is open and depends only on practical considerations. Rovane's Carnap therefore draws the conclusion that every scientific language forms an alternative, describes a different world and enables a separate corpus of knowledge.

It is sufficient to outline here the next steps leading to Rovane's final argument in favour of a multimundial position regarding natural objects (i.e. an alternative relativism about the external world). First, Rovane reconstructs Davidson's argument against relativism from *On the Very Idea of a Conceptual Scheme* (Davidson 1973) as one that speaks against alternatives. The interesting core of the argument for Rovane is that if we grant others a belief that is true for them, the *principle of charity*, which we need as a hermeneutical means of our interpretation, forces us to recognise this belief as by and large also true from our point of view. From this, Rovane develops her own holistic argument against multimundialism as an attitude in the sciences:

> "In all cases where we can make sense of what others believe, we will find not only that they by and large agree with us about many matters in just the way that Davidson contended, but also that, at the points at which others diverge from us, their beliefs still stand in myriad logical relations to what we already believe, and as a result we shall not find any occasion for the thoroughgoing epistemic indifference of the Multimundialist that would follow upon an encounter with normative insularity; on the contrary, we shall have no choice but to be epistemically engaged in the ways that go together with the Unimundial stance." Rovane 2013, p. 146.

Contrary to Rovane's Carnap, this argument shows that if we assume that we can comprehend the (scientific) beliefs of others, we must not assume a normative isolation of the underlying scientific languages or epistemic attitudes.

The next step in Rovane's derivation is a consideration of Kuhn's arguments concerning the incommensurability of scientific paradigms and Feyerabend's considerations based on them. Here Rovane argues that both are perfectly compatible with a unimundial attitude, even if they leave open the option of multimundialism in the domain of natural facts. For two reasons, everything initially seems to amount to a unimundial *stance* in the sciences:

> First, insofar as it is a fundamental aim of science to offer systematic explanations, and to pursue the possibility of arriving at maximally comprehensive explanations, scientific investigations must proceed on the assumption that logical relations really do run everywhere in the domain of natural facts open to such investigation, even if they are not yet all self-identical in the way that they would be if we already possessed explanatory frameworks that made them perspicuous. Second, if some variety of empiricism is fundamental to the scientific endeavour, which would seek to find ways in which the world itself might correct our errors, and if the unit of empirical confirmation is—as Quine, Kuhn, and Feyerabend all held—a whole theory, then we must find ways to pit whole theories against one another through decisive experiments as Feyerabend recommended, and the very effort to design such experiments presupposes that there are points of direct logical contact among all competing theories waiting to be discovered, even in cases where they are not yet evident because the theories are incommensurable and hence fail to share meanings at the theoretical level. Rovane 2013, pp. 177–178.

At this point, one could assume that this represents the end of the relativistic challenge of OSR. However, Rovane sees the partial thesis of ontological realism with respect to the facts of nature, to which OSR is committed, as the starting point for a new argument for her multimundialism. For Rovane, it is the central argument in this debate for taking the option of alternatives in nature seriously, even on the condition that they are not known: "[W]e are confining our attention to the domain of natural facts open to scientific investigation, and we are accepting that the Argument from Holism Against Normative Insularity rules out the possibility of our ever coming to encounter a multiplicity of worlds within that domain." (Rovane 2013, p. 186). Ontological realism, however, implies that the facts of nature exist regardless of whether we can ever come to know them or not. The point is that perhaps a unimundial attitude is advisable in the *practice of science*, but a multimundial attitude to *nature* as an ontological domain is advisable in our philosophical reflection. It can be put this way: if we want to think of nature as radically independent by discovering it, we should affirm the possibility that there are Rovainian truths, i.e. facts that we cannot know not only accidentally but in principle. But why should we accept such alternatives that we cannot know in principle?

There are several reasons for this. According to Rovane, the results of cognitive science speak in favour of taking alternatives that are not knowable seriously nonetheless (cf. for the following Rovane 2013, pp. 187–190). To this end, Rovane relies on Noam Chomsky's concept of cognitive abilities as specialised and limited (cf. Chomsky 2000). Since not all knowledge is propositional, scientific language cannot contain all knowledge. If scientific language is limited and specialised, then so is the knowledge that can be expressed through it. Rovane sees this as an argument against unimundialism in the domain of nature, since she believes a *View from Everywhere* is also not possible for the natural sciences. She turns this into a positive argument in favour of multimundialism:

> We cannot know other minds without knowing what they know; but if Chomsky is right, what we know about our own minds gives us reason to accept the possibility of other kinds of minds that can know things we cannot know, because their cognitive abilities are different from ours; when we take this possibility seriously, we must accept that there may be alternatives-unknowable-by-us—and hence many worlds rather than just one. Rovane 2013, p. 190.

Rovane presupposes here an ontological realism about nature and that some facts of nature are knowable by scientific methods (cf. also for the following Rovane 2013, p. 190–192). Rovane's epistemological anti-realism, on the other hand, concerns the idea that the facts of nature form a unity and are all knowable, at least in principle, from a scientific point of view (e.g. our point of view). Instead, Rovane claims that the facts of nature are partially isolated; it is conceivable that nature contains several worlds (in her sense), each of which can be understood ontologically realistically, but only one of which is accessible from our point of view. There are therefore alternatives in nature, facts (truths in Rovane's vocabulary) about other worlds that we cannot know together with the facts accessible to us. At the same time, Rovane argues that if we share a scientific language, we should and can follow an operational unimundialism in our common research practice. This is true because we are then exploring and gaining knowledge about the same world within the larger domain of nature. This also follows from the underlying ontological realism and its epistemology, which allows direct knowledge of natural facts. Operational unimundialism anchored in the common practice of science and the metaphysical recognition of an ontological multimundialism with regard to nature are not mutually exclusive, but are both required from Rovane's perspective.

5.3.4 Problems of Laws of Nature as Fixed Structure of the External World

Another argument, this time from a related debate in philosophy of science, problematises a different aspect of the single real external world. As I showed in

Sect. 3.2.3, the *structure thesis* is derived from OR's conception of the external world. This is the thesis that the external world has a fixed structure, which is usually realised via natural causality and laws of nature. The consistency of the external world in OSR is based on the structure thesis. The most obvious, fixed structure that naturalists can point to through their theses N2 and N3 are the most fundamental laws of nature, which are described by scientific theories and thus belong to scientific ontologies. Chakravartty also relies on the structure thesis in the form of his concrete structure of the external world, which is formed by the relations between first-level causal properties of natural entities, as I have shown in Sect. 4.2.2. In her book *How the laws of physics lie*, Nancy Cartwright offers a classic argument that fundamentally argues against this move.

She distinguishes between *fundamental laws of nature*, which have explanatory power and are generally applicable, and *phenomenological laws*, which depict local natural relationships by means of scientific models in equations that each describe only a specific physical interaction. According to Cartwright's argument, fundamental laws of nature are not true, but explanatory—which is why she proposes a sceptical anti-realism with regard to them. Phenomenal laws of nature, on the other hand, are true but only explanatorily limited. Her central argument in favour of this idea consists of three steps:

> (1) The manifest explanatory power of fundamental laws does not argue for their truth.
> (2) In fact the way they are used in explanation argues for their falsehood. We explain by *ceteris paribus* laws, by composition of causes, and by approximations that improve on what the fundamental laws dictate. In all of these cases the fundamental laws patently do not get the facts right.
> (3) The appearance of truth comes from a bad model of explanation, a model that ties laws directly to reality. As an alternative to the conventional picture I propose a *simulacrum* account of explanation. The route from theory to reality is from theory to model, and then from model to phenomenological law. The phenomenological laws are indeed true of the objects in realty—or might be; but the fundamental laws are true only of objects in the model. Cartwright 1983, pp. 3–4.

It should be emphasised here that this means that Cartwright does not need to present a comprehensive scepticism in philosophy of science,[33] but should be better understood as an epistemological anti-realist with regard to fundamental laws of nature, which are particularly to be found in physics. I will shortly examine the central first two steps of her main argument in more detail, but first I want to show more clearly why it challenges OSR and why I consider her argument to be of the variety of antitotality arguments that have received little attention in the metalocal realism-anti-realism debate of philosophy of science.

One might think that the ignorance that this argument often meets with in the realism-anti-realism debate is mainly based on a conviction of many OSR proponents about the philosophy of physics. After all, it is the standard view that any anti-realism about fundamental laws is wrong, and that they should instead be

[33] Although she did so at the time of the original development of the argument according to her own statement; however, later, without changing the argument, Cartwright co-developed entity realism.

an integral part of what one wants to be a realist about.[34] Cartwright's thesis is thus seen as a special case in the debate about laws of nature, the falsity of which would be proved *a limine* if the debate about theoretical entities were to be decided in favour of realism. Even if I do not deny Cartwright's importance in the local debate about laws of nature, —after all, her contribution is often considered central—, for me this is not where the actually explosive force of Cartwright's argument lies. My map of the debates in Chap. 2 shows why: the local debate about laws of nature is an important part of the realism-anti-realism debate in philosophy of science, which is metalocal. The fact that this is often overlooked means that protagonists believe that Cartwright's arguments do not concern the bigger issue.

The crucial point is that OSR *presupposes* a single real external world and thus the structure thesis. Naturalism in particular will always assume that the dynamic external world, nature, has a fixed structure, since all its entities are in a natural law-like context. Parts of this connection are investigated by the individual sciences and revealed by the best scientific theories. The external world is conceived by OSR as dynamic in nature, but is nevertheless subject to unchanging structural principles, the laws of nature, which apply always and everywhere, or at least (in certain interpretations of quantum physics) are generalisable and probabilistic (cf. Healey 2017, pp. 238–245). The cohesion of the external world is established by OSR directly or indirectly by referring to the most fundamental laws of nature. But like the arguments of Gabriel, Goodman, and Rovane, Cartwright attacks this strategy. If her argument is correct, OSR must be able to provide an answer as to how the connection that is necessary for for making plausible the thesis of the single real external world can be established *without fundamental laws of nature*. The structure thesis cannot be substantiated in this way, if Cartwright is right.

This poses a problem for semirealism. For Chakravartty's concrete structure, the challenge that is presented means the following: He must show that the causal properties that have to stand in fixed relations do not do so only because fundamental laws of nature apply that determine these fixed relations with natural necessity, always and everywhere. Chakravartty rejects Cartwright's entity realism, as I have already shown at the beginning of this chapter.[35] One reason for this, which is in the background and is only now becoming clear, is that only *local* causal connections are justified in entity realism. This cannot be acceptable to Chakravartty. But his reference to Cartwright is not purely negative. The fact that Chakravartty chooses first-level causal properties of natural entities to play a role in his relational concrete structure at all is drawn from Hacking's and Cartwright's proposal of entity realism. The problem is that he now has to move from local causal structures to the concrete structure of the entire external world.

---

[34] Cf. e.g. Carroll 2020, emphasis as in the original: "The majority of contemporary philosophers are *realists* about laws". Among empiricists, however, there are also other anti-realists concerning laws of nature, such as van Fraassen 1989 and Tetens 1982.

[35] Cf. Sect. 5.1.5 of the present book, p. 120.

> They [causal properties; J. V.] regularly cohere to form interesting units, and these groupings make up the particulars investigated by the sciences and described by scientific theories. Chakravartty 2007, p. 119.
>
> [T]he necessary character of causal phenomena in the context of singular causation immediately yields an account of general causation – an account in terms of laws of nature. Chakravartty 2007, p. 121.
>
> Every case of singular causation is thus an instance of general causation, because singular causal processes incorporate instances of properties that are subsumable under general laws. Chakravartty 2007, p. 122.

The first quotation still reads close to entity realism—with the exception of the word *regularly*. The other two quotations reveal that the reason for the occurrence of individual instances of causal properties lies on a general level. Chakravartty only succeeds in securing the regularity of causal properties because he assumes that fundamental laws of nature—*general laws*, to use his term—are relations between causal properties of natural entities that actually apply always and everywhere. Thus, Chakravartty can only connect all individual structures to the overall structure of the external world by way of fundamental laws of nature, for which (partly true) theories can then be formulated. If Cartwright's argument is valid, Chakravartty's metaphysics of realist philosophy of science lacks a necessary ingredient to justify the structure thesis on which he bases his take on the single real external world.

Cartwright's main argument[36] for the thesis that fundamental laws of nature do not apply in the single real external world—that is, for the above-mentioned first two steps of her overall argument, with which I am concerned—is a *modus tollens* argument: If the fundamental laws of nature apply in the single real external world, then all external world entities for which a certain fundamental law of nature applies must always exhibit the behaviour that can be derived solely from the laws of nature. In most cases, however, external world entities do not exhibit such easily deducible behaviour, as Cartwright argues in the second step. Therefore, fundamental laws of nature are not true.

Instead, Cartwright proposes that fundamental laws of nature only hold in highly idealised models, not in or for the external world. If one nevertheless sticks to basic realist convictions, this amounts to her form of entity realism. Without going into entity realism in greater detail at this point, it can be stated that the fact that the fundamental laws of nature apply in highly idealised models explains why we attribute such high explanatory power to them: they obtain this through the explanatory power and generality of the idealised models (cf. Cartwright 1983, pp. 44–53). According to Cartwright, explanatory power and generality on the one

[36] Cartwright supplements this argument in several works with further arguments, which I will not go into here for reasons of brevity. They include the argument that in practice phenomenal laws are not derived from fundamental laws of nature, and in this sense the former are not special cases of the latter. Also, Cartwright's alternative understanding of models and the insight that fundamental laws of nature are a heuristic in the process of searching for suitable models rather than anything else can also be understood as an argument in favour of an epistemological anti-realism of an instrumentalist kind with regard to fundamental laws of nature. Cf. on the main argument reproduced here in particular Cartwright 1983, Essay 2 & 3, pp. 44–73.

hand and truth on the other are antiproportional in the case of laws of nature, which orders more fundamental and more phenomenological laws. More important for my argument than this is the second step in her argument. It explains why fundamental laws of nature do not apply in the single real external world. Consequently, Chakravartty should not refer to them in order to justify his variant of the structure thesis, but he does so anyway—and has to, given his way of getting out of UTE and historical arguments.

Carwright's most simple example of the falsity of fundamental laws of nature is Newton's law of gravitation ($f = Gm_1m_2/r^2$). This law expresses the force of attraction f between two objects with the masses m_1 and m_2.[37] However, this predicted force of attraction can almost never be measured exactly (cf. Cartwright 1983, esp. p. 58, also pp. 62–67). According to Cartwright, the constant deviation of the measurement can only be explained if one of two strategies is chosen: Either a *ceteris paribus* clause must be inserted, according to which f is measured exactly if only and exclusively (Newtonian) gravitation would apply in nature and, moreover, the masses m_1 and m_2 would be the only objects in nature; or the force must be treated as a mere component (f_g) of another force f_{total} according to the law of gravitation, which could then be measured. Cartwright excludes the last option, since in that case f_g no longer refers to real things in the external world, but only clarifies a certain aspect, whereas the composite force f_{total} always depends on the concrete, local situation and would therefore no longer be general or fundamental. Conversely, a resultant force f_{total} cannot be composed of two real forces f_g and f_n, because these never occur and are mere epistemic tools in explaining f_{total}.[38] From this, Cartwright can draw the following conclusion:

> *Ceteris paribus* generalisations, read literally without the "ceteris paribus" modifier, are false. They are not only false, but held by us to be false; and there is no ground in the covering-law picture for false laws to explain anything. On the other hand, with the modifier the *ceteris paribus* generalizations may be true, but they cover only those few cases where the conditions are right. For most cases, either we have a law that purports to cover, but cannot explain because it is acknowledged to be false, or we have a law that does not cover. Either way, it is bad for the covering-law picture. Cartwright 1983, pp. 45–46.

For Cartwright, the *ceteris paribus* condition does not mean that all other external factors remain the same, but that no other factors have an effect. Fundamental laws of nature only apply if corresponding *ceteris paribus* conditions are fulfilled. However, these conditions are only fulfilled in rare cases, namely when no other laws of nature or other local factors exert an influence. *Fundamental* laws of nature

[37] Cf. also for the following Cartwright 1983, pp. 56–58. Cartwright also discusses several examples from quantum physics. Those theories can be considered more up to date (though her original contribution is from 1983), but the mathematics of which should not be assumed here for reasons of broad accessibility. I thus remain in the simpler Newtonian mechanics.

[38] Cf. to this Cartwright 1983, pp. 58–62. This part of the argument has been widely criticised. For a defence and extension, see Giere 1999, who does not believe in implicit *ceteris paribus* conditions of laws of nature, but follows Cartwright's conclusion that there are no fundamental laws of nature.

should apply always and everywhere. Cartwright's conclusion that they do not apply to all external objects in a single real external world is justified. Therefore, no strictly fundamental laws of nature exist outside of our models.

An important criticism of the argument comes from Mehmet Elgin and Elliott Sober (Elgin and Sober 2002): They argue, in a nutshell, that Cartwright overlooks the fact that from the conditional *only if the* ceteris paribus *condition holds, then the fundamental law of nature also holds* and the fact that the antecedent is false, the falsity of the consequent does not follow. However, they present an alternative partial argument that seeks to secure Cartwright's essential conclusions. Moreover, it is not a foregone conclusion that their reconstruction matches what Cartwright had in mind: In my reconstruction of the argument, the logical flaw in her argument that Elgin and Sober criticise does not appear.

Cartwright's argument shows that the structure thesis of OSR, i.e. the assumption that the single real external world has a fixed structure, cannot be justified by fundamental laws of nature. Phenomenal laws of nature do not establish this structure, since they can only be linked to each other via fundamental laws of nature. However, these fail as a fixed structure themselves because, as the argument shows, they are not true of nature and do not apply always and everywhere. Thus, Cartwright's arguments hit the same note in philosophy of science as Gabriel, Goodman and Rovane strike on mainly ontological grounds. However, in the face of inconsistency, even in the natural sciences, it seems difficult for OSR to find a justification for the structure thesis that is completely independent of fundamental laws of nature.

I do not want to leave unmentioned here that at least one new realist, using completely different arguments, puts forward a very similar thesis to Cartwright's: Meillassoux argues—not from the perspective of philosophy of science but ontology—that it is better to assume that the laws of nature (whereby he does not distinguish between fundamental and phenomenal) are stable, but contingent and only contingently stable. They can therefore change at any time and thus never serve as the basis of a single real external world. While he only speculates about their stability and the reason for this, he believes that he has proved a philosophical concept of time, which justifies the possibility of the laws of nature being able to change, on the one hand, and the contingency of all existing entities, including the laws of nature themselves, on the other.[39] Based on this, he advocates a fragmented picture of reality, consisting of several ontological domains, which he calls worlds. However, all of these ideas are based on Meillassoux's positively developed ontology, which I will not present in full for the methodological reasons mentioned at the beginning of this chapter.

[39] Cf. on the former Meillassoux 2014 and on the latter Meillassoux 2008, p. 53.

5.3.5 Consequences from the Antitotality Arguments

The reconstructed antitotality arguments of Gabriel, Goodman, Badiou, Rovane and Cartwright are based on different premises and aim at different conclusions. Nevertheless, they can be grouped together in the context of my argument: All of them criticise the idea of a single real external world essential to OSR. I have already shown that, as in Chakravartty's case, this must be framed against the underdetermination and historical arguments as a subtle realism that involves selective scepticism. These subtle forms of OSR are all the more dependent on the thesis of the single real external world and, consequently, on the structure thesis. The subtlety, and reticence to adopt heavyweight premises, such as those necessary in Quine's RP to counter the UTE and historical arguments, make it necessary to accept this core thesis of OSR. I have paradigmatically demonstrated this with Chakravartty's semirealism. However, the analysed antitotality arguments speak against the tenability of this core thesis.

According to my reconstruction of the antitotality arguments, I consider the prospects of defence of OSR, especially Chakravartty's, to be minimal. Any OSR would have to explain why Goodman's and Gabriel's arguments fail. A proponent of OSR would have to be able not only to *reject* alternatives in Rovane's sense, but to rule them out by showing the impossibility of their existence. It would have to be possible to justify the structure thesis independently of fundamental laws of nature or general natural causality, without resorting to other strong premises such as holism or reductionism. None of this seems promising to me:

The state of the debate forces OSR either to immunise the thesis about the single real external world or to defend a metaphysics of philosophy of science against all antitotality arguments. The immunisation strategy seems to be fraught with a number of difficulties. One option might be to claim that every opponent of OSR shares the thesis of the single real external world and the structure thesis. I take this to be an obvious fallacy: One only has to refer to the authors analysed in this subchapter. Another immunisation strategy consists in the argument that OSR merely provides a meta-theory of the sciences and only needs to presuppose what the OR generally justifies for ontology. However, this tactic amounts to shifting the discussion to ontology and epistemology, to merely postponing the problem as the arguments of Gabriel, Goodman and Rovane are at home in those debates. Such explicit immunisation strategies do not seem viable to me.

Are there implicit immunisations? My hypothesis, which cannot be realised by an investigation here, is that many proponents of OSR do not explicitly commit themselves to the thesis of the single real external world or the structure thesis, but imply it in order to evaluate OSR as the best current meta-theory of the sciences. Even then, my criticism would be appropriate. This option would result in an implicit acceptance of the single real external world thesis, with the same problematic consequences.

I have examined the option of a defence of OSR on the basis of a metaphysics for scientific realism using Chakravartty's proposal. The OSR, even in the subtle variant of metaphysics formulated in Chakravartty's semirealism, does not escape the combination of the three groups of arguments. However, the analysed arguments also do not provide any justification for advocating a particular anti-realism, for seeing anti-realism as an adequate solution in the field of philosophy of science. For although with many of the arguments anti-realists aim to present their form of anti-realism as the best or only possible variant of philosophy of science, these *further-reaching* conclusions of the anti-realist arguments in all the cases analysed—from van Fraassen's bad lots, to Stanford's unconceived alternatives, to Wray's combination of historical and UTE arguments—are ultimately never justified. This leaves OSR and the various anti-realist arguments in a stalemate: OSR seems indefensible insofar as no way out of the threefold attack by UTE, historical and antitotality arguments is foreseeable. However, no variant of anti-realism can itself be justified with the arguments analysed. A solution to this realism-anti-realism debate does not appear to be possible under the alternatives considered thus far.

This first conclusion justifies half of the main thesis of the first part of this book: *OSR and anti-realism in philosophy of science are in an argumentative stalemate.* I consider this an excellent motivation to search for a new realism in philosophy of science. This new scientific realism should not be based on fundamental laws of nature and the structure thesis, nor on the single real external world, but should nonetheless be an ontological and epistemological realism in the metalocal debate. Before I turn to the second task of this book—the search for the outlines of such a new scientific realism—it is necessary to justify the second half of the first main thesis: that of the stalemate between different variants of OSR. This is of interest because the frustrating state of the debate marked by the first stalemate has not gone unnoticed, even though my argument for the stalemate as such is new. An explicit and detailed justification of the stalemate was not yet available—it is explicitly one of the main tasks of my work. However, it is no secret that realists, although they represent a majority in the debate, only provide divergent and ultimately not fully satisfactory solutions to the anti-realist counterarguments (even from the point of view of most realists, as the range of different realisms and naturalisms emphasises).

The debate between classical variants of OSR, especially (reductive and non-reductive) naturalisms and selective realisms, and the various anti-realisms is accordingly perceived to be inadequate. One attempt to formulate realism differently within the debate and thus escape the long-standing problems is *structural realism*. This raises the question of whether structural realism is already a form of new scientific realism in my sense and thus able to escape the problems of OSR. In the conclusion to this subchapter, therefore, I analyse structural realism in comparison to OSR. I aim to show that structural realism is not a new scientific realism, but a special and sophisticated variant of OSR. Although the best form of structural realism has a number of advantages in comparison with semi-realism, it also has important disadvantages. Therefore, I argue for the persistence of a second stalemate, the argumentative stalemate of the best variants of OSR: Again, both sides raise effective objections to each other, but the solutions are less convincing on both sides.

5.4 Objective Realism III: Structural Realism

Structural realism can be roughly characterised as the form of scientific realism that attempts to introduce structuralist ideas into the discussion of the realist framework. The most important goal of structural realism in the context of the present analysis is to develop a position that is realistic, but at the same time able to escape the dilemma of historical and UTE arguments. The force of the counterarguments is to be mitigated by replacing the threatened theoretical objects of different naturalisms with structures as central elements of our theories. There are different variants of this position. In the research debate, these are divided into epistemic structural realism (EpiSR) and ontic structural realism (OntSR). Steven French, representative of OntSR, has provided his working definitions in "slogan form" (French 2014, p. 2):

> ESR [=EpiSR; J.V.]: all that we *know* is structure
> OSR [=OntSR; J.V.]: all that there *is*, is structure French 2014, p. 2.

I will present these positions on the basis of their most important proponents and differentiate each of them from Chakravartty's semirealism. Finally, I will examine which structural realism should be categorised as OSR and evaluate whether EpiSR or OntSR have better prospects for escaping the anti-realist objections and criticism by Chakravartty and other structural realists.

5.4.1 Epistemic Structural Realism

In his seminal article *Structural Realism*, John Worrall aims "to give the argument from scientific revolutions its full weight and yet still adopt some sort of realist attitude towards presently accepted theories in physics and elsewhere." (Worrall 1989, p. 99). Worrall thus starts from the anti-realist conclusion of the incommensurability argument and PMI and wants to reconcile realism about scientific theories with it. To this end, he draws on Henri Poincaré's preliminary work and thus makes use of the early structuralist tradition in philosophy of science.

Worrall first shows that, in view of the historical arguments, all classical forms of OSR retreat to minimal positions. He refers to his own earlier position (cf. Worrall 1989, p. 111). Alternatively, a classical OSR is only able to evade the historical arguments rather poorly, according to him. Worrall cites Boyd as an example of the latter. To solve this problem, Worrall attempts an alternative that is in a sense the complementary manoeuvre to entity realism: He assumes that while neither the theoretical objects, nor their direct properties—that is, the explicit ontologies of science—outlast scientific revolutions, their structural properties do. In Worrall's view, this always applies if the earlier theory was successful in the sense that it also enabled predictions to be made about new phenomena. Even discarded theories often provide mathematical descriptions of a phenomenon that later appear as borderline cases in the new theories: "[T]he theory must somehow or other have latched onto the "universal blueprint"." (Worrall 1989, p. 101).

5.4.1.1 Ether and Electromagnetic Field Theories as the Starting Point of Epistemic Structural Realism

Worrall's example is Augustin Fresnel. Fresnel attempts to describe mathematically the propagation of light at various interface surfaces as oscillations in an elastic medium, the ether. To do this, he sets up systems of equations for various optical phenomena. Consider an unpolarised ray of light that strikes a transparent medium and is divided into one polarised ray of light as a reflection and one that is refracted by the medium. Let I^2 be the intensity (with I as the wave amplitude) of the incident ray, R^2 that of the reflected ray and X^2 that of the refracted ray, which is polarised towards the plane of incidence.[40] Let $I^{2'}$ be the intensity (with I' as wave amplitude) of the incident ray, $R^{2'}$ that of the reflected ray and $X^{2'}$ that of the refracted ray, which is polarised parallel to the plane of incidence. If α is the angle of incidence to the interface surface and β is the angle of the refracted light ray to the interface surface, the Fresnel formulae result:

$$\frac{R}{I} = \frac{\tan(\alpha - \beta)}{\tan(\alpha + \beta)}$$

$$\frac{X}{I} = \frac{2 \sin\beta \cdot \cos\alpha}{\sin(\alpha + \beta) \cdot \cos(\alpha - \beta)}$$

$$\frac{R'}{I'} = \frac{\sin(\alpha - \beta)}{\sin(\alpha + \beta)}$$

$$\frac{X'}{I'} = \frac{2 \sin\beta \cdot \cos\alpha}{\sin(\alpha + \beta)}$$

James Clerk Maxwell's theory, which is no longer based on the ether but on electromagnetic fields, rejects the idea that I, R, X, etc. are measures of the amplitudes of the vibrations in the ether, i.e. the magnitude of the deflection of the ether particles by the light wave. The angles cannot be interpreted as the polarisation directions of the ether light wave either. It is therefore surprising that Maxwell's field theory adopts Fresnel's equations with newly interpreted variables. In Maxwell's field theory, the amplitude vectors A of all three electromagnetic waves (incident, reflected and refracted) are each divided into two vector components A_p parallel and A_s perpendicular to the plane of incidence. The parallel vector components can then be labelled I for the incident electromagnetic wave, R for the reflected wave and X for the refracted wave, while the perpendicular components can be labelled I', R' and X' accordingly. The angles result from the propagation vectors in the field and the formulae are still valid. Of this, Worrall writes:

[40] In the naming I follow Worrall 1989, p. 119; whereas Demtröder 2017, pp. 222–224, notes that the formulae read as modern notations are also comprehensible.

> There was continuity or accumulation in the shift, but the continuity is one of *form* or *structure*, not of content. In fact this claim was already made and defended by Poincaré. And Poincaré used the example of the switch from Fresnel to Maxwell to argue for a general sort of syntactic or structural realism quite different from the anti-realist instrumentalism which is often attributed to him. This largely forgotten thesis of Poincaré's seems to me to offer the only hopeful way of both underwriting the 'no miracles' argument *and* accepting an accurate account of the extent of theory change in science. Roughly speaking, it seems right to say that Fresnel completely misidentified the *nature* of light, but nonetheless it is no miracle that his theory enjoyed the empirical predictive success that it did; it is no miracle because Fresnel's theory, as science later saw it, attributed to light the right *structure*. Worrall 1989, p. 117.

The form of structural realism that Worrall discusses here assumes that the success of Fresnel's equations in the old paradigm can be explained if it is conceded that they correctly represent the structure of the phenomenon, albeit through ether theory in an incorrect scientific domain ontology. He attributes this thesis to Poincaré, but introduces it into the debate as a new idea by characterising it independently of which reading of Poincaré is more appropriate[41]—as a form of realism that considers only the *knowledge of structural relations* in nature to be important and empirically testable, not assertions about deep ontological or metaphysical inner natures of entities. Inner natures in this sense denote the actual ontological constitution of the concrete entities of the external world, as they are independent of our scientific access to them.

Formally and structurally, there is a cumulative growth of scientific knowledge in Worrall's structural realism, even if the incommensurability argument and the other historical arguments prove that there is a paradigm shift at the level of the description of reality by theory and that one scientific ontology is replaced by another (cf. Worrall 1989, p. 120). But the structural knowledge about the relations of these entities remains and keeps growing even beyond the paradigm shift. According to Worrall, Fresnel's equations are not representative insofar as they (reinterpreted with Maxwell's scientific ontology in his field theory) are completely preserved in the new paradigm. However, Worrall believes that it is more often the case that old equations appear as borderline cases in the new theory. He considers this to be the rule for successful theories that have been superseded by new theories.

What follows from this? With reference to the structural realist, Worrall writes: "He insists that it is a mistake to think that we can ever "understand" the *nature* of the basic furniture of the universe. [...] On the structural realist view what

[41] Worrall refers to passages such as this in Poincaré 2018, p. 115: "The equations express relations and if these equations remain true, it is because the relations preserve their reality. Now as before, they teach us that there is a particular relationship between something and something else. We formerly called this something *motion* and we now call it *electric current*, but these labels were only images standing in for the real objects that nature forever hides from us. The true relations between these real objects are the only reality we can reach, the only condition being that the relations between the objects are the same as those between the images standing in for the objects. It does not matter whether we find it useful to replace one image by another, as long as these relations are known to us."

Newton really discovered are the relationships between phenomena expressed in the mathematical equations of his theory, the theoretical terms of which should be understood as genuine primitives." (Worrall 1989, p. 122). With scientific theories we therefore only ever come to know the structures, i.e. relations that natural properties, objects, processes and other entities have to each other, not the entities themselves, whatever their inner nature may be. Worrall does not deny their existence, nor does he deny that the structure is conditioned by these elements. He merely rejects the knowability of these objects, properties, etc.

The agnostic position towards natural properties, objects, processes and events in combination with realism regarding their structures has earned Worrall's position the name EPISTEMIC STRUCTURAL REALISM (EpiSR). The position is thus a selective scepticism in Chakravartty's sense. Worrall presents it as a realistic alternative that can take account of the no-miracle argument and the historical arguments, while at the same time doing justice to current theories of gravitation and quantum mechanics in physics. Worrall's proposal does this, but criticisms were subsequently levelled that motivated a new variant of structural realism, which I will present in the next section.

5.4.1.2 The Problem of Abstract Structures and Unknowable Objects

The inner-theoretical necessity of structure, which bears the burden of argument in Worrall's EpiSR, is presented by James Ladyman in his article *What is structural realism?*, which made a central contribution to the development of OntSR:

> Since there is (says Worrall) retention of structure across theory change, structural realism both (a) avoids the force of the pessimistic meta-induction (by not committing us to belief in the theory's description of the furniture of the world), and (b) does not make the success of science (especially the novel predictions of mature physical theories) seem miraculous (by committing us to the claim that the theory's structure, over and above its empirical content, describes the world). Ladyman 1998, pp. 409–410.

Ladyman goes on to ask how we know the unobserved entities that ultimately remain unknowable in Worrall's proposal, but which form this structure. His answer is that we only know them through a Russellian description, i.e. only through their structural properties (cf. also for the following Ladyman 1998, p. 411). As Ladyman shows, Grover Maxwell tried to capture these structural properties of unknown entities via Ramsey sentences even before Worrall: Where a theory asserts the structure S is formed by the related theoretical objects T_1 to T_m and makes certain observational predictions O_1 to O_n, this theory can be formalised as a Ramsey sentence. It assumes only the structure, but eliminates the theoretical entities by indirectly requiring only the existence of a set of structural elements:

> If we formalise a theory in a first order language: $\varnothing(O_1, \dots, O_n; T_1, \dots, T_m)$, then the corresponding Ramsey sentence is $\exists t_1, \dots, \exists t_m \varnothing(O_1, \dots, O_n; t_1, \dots, t_m)$. Thus the Ramsey sentence only asserts that there are some objects, properties and relations that have certain logical features, satisfying certain implicit definitions. It is a higher-order description, but ultimately connects the theoretical content of the theory with observable

> behaviour. However, it is a mistake to think that the Ramsey sentence allows us to eliminate theoretical entities, for it still states that these exist. It is just that they are referred to not directly, with theoretical terms, but by description or by logical form, that is via variables, connectives, quantifiers and predicate terms whose direct referents are (allegedly) known by acquaintance. Thus Maxwell (and Russell) claimed that knowledge of the unobservable realm is of its structural rather than intrinsic properties, or, as is sometimes said, is limited to knowledge of its second order properties. It is arguable that this is the purest structuralism possible, for the notion of structure employed refers to the higher-order properties of a theory, those that are only expressible in purely formal terms. Ladyman 1998, pp. 411–412.

This approach is *intended to defend OSR* by advocating an ontological and epistemological realism only regarding structures of nature, and an epistemological anti-realism about other entities (especially unobservable objects, processes and events). In this respect, one can see why EpiSR is a selective scepticism. According to Psillos, this applies not only to Maxwell, but also to Worrall, since the notion of abstract structure, which both presuppose, has the same problematic consequence.[42]

Here, however, the Newman problem arises, which has already decisively damaged Russell's version of structuralism. Max Newman proves that an abstract structure understood in this way, which is asserted by a theory, can trivially be found in any set with enough elements, i.e. also in nature. Ladyman describes this problem as follows:

> Suppose that the world consists of a set of objects whose structure is W with respect to some relation R, about which nothing else is known. *Any* collection of things can be organised to have structure W provided there is the right number of them. As Demopoulos and Friedman point out, this is because if ∅ is consistent, and if all its purely observational consequences are true, then the truth of the corresponding Ramsey sentence follows as a theorem of second-order logic or set theory (provided the initial domain has the right cardinality, and if it does not then consistency implies that there exists one that does). The *formal* structure of a relation can easily be obtained with any collection of objects provided there are enough of them, so having the formal structure cannot single out a unique referent for this relation—in order to do so we must stipulate that we are talking about the *intended* relation, which is to go beyond the structural description. Ladyman 1998, p. 412; original emphasis.

Moreover, Jane English has shown that two Ramsey sentences that are incompatible cannot have the same observable consequences (cf. English 1973). This means that even Worrall's EpiSR cannot make more realistic claims than van Fraassen's constructive empiricism, which is an explicit anti-realism. If Worrall's structure of nature, which we express in a theory, were an abstract structure that we can formulate as a Ramsey sentence, according to English's proof the realist assertion of the reality of abstract structures of a theory in EpiSR is equivalent to the (anti-realist, constructive-empiricist, instrumentalist) assertion of the empirical adequacy of all empirically equivalent theories.

EpiSR falls back into a form of epistemological scientific anti-realism because of its reticence with regard to the existence or non-existence of objects (or processes,

[42] Cf. Psillos 1999, p. 151, whereby the defence in Worrall 2007 is not able to resolve the dilemma I present below, but only successfully defends itself against the direct accusation of the Newman problem.

events, . . .). In EpiSR, it can only be said that theories that describe a certain abstract structure are empirically adequate. Alternatively, contrary to its motivation, EpiSR would have to assume the knowability and existence of theoretical objects, processes and events and thus fall back into a classical OSR. In no case does the EpiSR really satisfy the counterarguments.

5.4.2 *Ontical Structure Realism*

This is where OntSR comes in. It is conceived as a selective scepticism that, like semirealism, can specify what it is realist or anti-realist about. It should, as Chakravartty demands and realises, find a way to make meaningful decisions about its own scope. To do this, OntSR must confront the classical counterarguments, and the historical and UTE arguments, as well as the problems of EpiSR with abstract structures.

5.4.2.1 Steven French's and James Ladyman's Variants

With regard to the Newman problem, Ladyman rejects the idea of Ramsey sentences (cf. also for the following: Ladyman 1998, p. 416). This idea is based on the syntactic understanding of theories, in which these are interpreted axiomatic systems. Ladyman aims to build his variant of OntSR on a semantic understanding of theory, according to which theories are structures or (collections of) models that represent their target systems and interactions. They are therefore themselves extralinguistic entities that can be described and characterised in different ways. A theory or model and a target system are thus partially isomorphic. French, on the other hand, wishes to remain neutral and use semantic and syntactic understanding situationally without committing to either.

Ladyman regards conventional OSR proposals as outdated not only because of the UTE and the historical arguments, but also because they have nothing to counter Cartwright's arguments against fundamental laws of nature (cf. Ladyman 1998, p. 417). However, he believes that even highly theoretical models are successful and that abstract mathematical structures in theories must therefore have a "grip on reality". "This means taking structure to be primitive and ontologically subsistent." (Ladyman 1998, p. 420). Ladyman describes the programme that he pursues, and in which French can also be classified, as follows:

> The demand for an individuals-based ontology may be criticised on the grounds that it is the demand that the structure of the mind-independent world be imaginable in terms of the categories of the world of experience. Traditional realism should be replaced by an account that allows for a global relation between models and the world, which can support the predictive success of theories, but which does not supervene on the successful reference of theoretical terms to individual entities, or the truth of sentences involving them. Ladyman 1998, p. 422.

In a series of partly joint articles and monographs, in which French's *The Structure of the World* and Ladyman's and Ross' *Every Thing Must Go* in particular stand out, closely related variants of OntSR are developed. The positions are so similar in their basic features that I will particularly outline them together, but in certain details I will reconstruct French's variant.[43] First, I clarify what is meant by structure in OntSR and why OntSR does not run into EpiSR's problem of abstract structures, which is linked to the relapse into object-oriented naturalism or anti-realism explained above. I then explain how OntSR can deal with the UTE and historical arguments. Finally, I argue that the antitotality arguments also pose an unresolved problem for OntSR, as it follows from further criticisms that French must hold a particularly strong object-realist structure thesis.

5.4.2.2 Structure as Ontologically Primitive

What does the proposal that in OntSR structures are ontologically primitive and objects are eliminable mean? And how does this view of structure solve the problems that EpiSR encounters? First, OntSR denies the dichotomy of the structure formed by the natural entities on the one hand and the inner nature of the natural entities on the other. It does not assume that we have knowledge of the structure formed by natural entities as we represent them as theoretical entities, nor of their inner natures. Instead, OntSR questions what conception of natural versus theoretical entity EpiSR implies. For regardless of whether we are dealing with objects, processes or events, EpiSR treats natural and theoretical entities as single things. French and Ladyman consider this to be a mistake.

In addition to the Newman problem described above, which arises from this dichotomy and the view of structures as abstract entities, OntSR can also be motivated by "metaphysical underdetermination" (French 2014, p. 33–38). This concern arises from the following problem: In metaphysics and the (ontologically orientated) realism debate, it is often overlooked that in philosophy of physics it is openly discussed whether quantum objects can be understood as individuals or better as "non-individual[s]" (French and Ladyman 2003b, p. 37; for an overview of the debate on wave function realism cf. Ney 2023; Schroeren 2022). For the realist, this is an important point, because for her, conditions of existence apply to objects, ontologically at least the condition of (self-)identity. However, quantum objects (if they exist in the realist sense) violate this condition according to their interpretation as non-individuals.

From the perspective of philosophy of physics, French and Ladyman argue that both metaphysical interpretations (quantum objects as individuals and as non-individuals) are justified and that quantum objects are therefore metaphysically underdetermined (cf. French and Ladyman 2003a, p. 74). This means that the scientific ontology of physics can never be read off *directly* from physical theories.

---

[43] All such passages in which I go into specifics are explicitly labelled.

OntSR considers the dispute about the status of quantum objects and the idea of natural individuals of EpiSR to be misguided. French's and Ladyman's solution consists in rejecting an account according to which individuals and non-individuals are fundamental ontological entities in order to interpret them instead as representations of real, fundamental structures: "There are no epistemically inaccessible objects lying behind the structures which we can know." (French and Ladyman 2003b, p. 39). Structures are instead ontologically primary in OntSR, the variously conceived quantum objects and all other objects, events and processes are one-way dependent on them. What the best formal framework for this relation is remains until now open at least for French's OntSR (cf. French 2023).

French's and Ladyman's example is an electron: the relational, structural property of electric charge is real, it occurs in laws, formulae and our experience. However, the term *electron* is a term that corresponds to our linguistic or metaphysical preferences and is merely an idealised model. Accordingly, electric charge has no object as a carrier that is ontologically on a par with it.

5.4.2.3 Elimination of Objects and the Problem of Relations without Relata

How can there be structure without relations and how can there be relations without relata? It could be argued that objects, processes and events creep back in through the concept of structure. Would OntSR then be as confronted with the Newman problem as EpiSR? French's and Ladyman's idea is to understand objects, events and processes as heuristics. From their point of view, the counterarguments raised result from the prejudice of our own mesoscopic nature, according to which structures only arise from the relations of objects as relata (cf. French and Ladyman 2003b, pp. 42–43). The structures that French and Ladyman consider to be real are physical, not abstract and mathematical. They admit that the boundaries between the two are "blurred" (French and Ladyman 2003a, p. 75). However, there is a partial isomorphism between physical structures and observable phenomena, as well as between these and our (partly mathematical) theories (cf. French 2014, pp. 103–104, According to French, further partial isomorphisms exist between different theories about a certain physical structure and between the theory and its scientific models.). A concrete, spatiotemporal, i.e. physical, structure is therefore always partially isomorphic with the structure of a partially true theory about it (cf. also for the following French 2014, p. 104; French and Ladyman 2003a, p. 75; Landry 2007, esp. p. 2). The physical structures are the causes of the phenomena and in this sense are strictly separated from abstract mathematical structures. Partial isomorphism can be used to represent the physical structure in a model or theory. Scientists therefore describe the physical structures with the help of mathematical structures, but there is no identity relationship here. It is worth taking a look at Ladyman's version of OntSR:

> The idea then is that we have various representations of some physical structure which may be transformed or translated into one another, and then we have an invariant state under such transformations which represents the objective state of affairs. Ladyman 2020, p. 421.
>
> [R]elational structure is ontologically more fundamental than individual objects. [...] [O]ntic structural realism is a form of realism about the modal (causal or nomological) structure of the world [...]. Ladyman 2007, p. 24.
>
> Individuals need not have self-individuating elements, nor be individuated independently of the other individuals in the relational structures of which they are parts. Ladyman 2007, p. 25.
>
> The argument is that without distinct individuals that are metaphysically prior to the relations, there is nothing to stand in the asymmetric relations that are supposed to confer individuality on the relata. Ladyman 2007, 31.

Following this, Ladyman cites Charles Parsons (reference is made to 2004, p. 75), to show that the concept of object in the structuralist sense is a formal concept. He discusses mathematical graph theory, in which the abstract structures of graphs consisting only of nodes and (directed or undirected) edges are analysed. In this picture, the graphs correspond to the structure of Ladyman's OntSR, while the nodes correspond to objects, processes and events. Individuation and the diversity of the latter are determined purely structurally, via the overall structure of the respective graph. Nothing sets aside one node from the other, except their respective position within the structure.

Ladyman applies these considerations in favour of structuralism in mathematics to physical objects. He argues for contextual individuation, which (at least for fermions and spacetime points) assumes a weak relational distinctness (cf. Ladyman 2007, p. 39). This means that they have no intrinsic natures or identities, but are only different insofar as they exhibit a weak distinctness within a certain physical structure. Weak distinctness exists between two entities iff both are in a dual irreflexive relation. A example helps here (cf. Black 1952): Max Black's thought experiment of a possible world that contains only two identical spheres of iron that are one mile apart exemplifies this macroscopically: The two spheres are in a dual relation to each other that is irreflexive (sphere A is one mile away from sphere B, but not one mile away from itself), albeit symmetrical. Therefore, the two are only weakly distinguishable from each other. With Ladyman, physical objects (at least fermions and spacetime points) would now be weakly relationally distinguishable in precisely this way via their respective structural context.

Does this already solve the problem mentioned? Hardly. All entities in Ladyman's OntSR are structuralist in nature. Objects, processes or events exist independently of the subject, but are constantly dependent on the structures that underlie them. These structures are Ladyman's ontologically fundamental entities. In order to avoid the problems of EpiSR, they are not conceived as abstract—especially not mathematical—structures. This raises the question of what physical structures are without ontologically equivalent objects, processes or events that stand in relations and form these structures as the relata. Critically addressed by van Fraassen (cf. van Fraassen 2006, pp. 292–293), Ladyman explains:

> Physical structure exists, but what is it? What makes the worldstructure physical and not mathematical? Ladyman and Ross advocate a kind of neo-positivism according to which when questions like this arise it is time to stop; however, perhaps van Fraassen is right that I should have stopped earlier. Ladyman 2007, p. 40.

I think that Ladyman should not have stopped earlier—after all, he openly admits to proposing a metaphysics of nature—but that he simply owes the answer to van Fraassen. This means that, in the end, Ladyman's OntSR cannot say anything about what physical structures are ontologically, but aims to advocate a form of neo-positivism. This means, however, that what we have here is a metaphysics of nature that has no answer as to what its most fundamental entities, namely physical structures, are; and what makes them different from abstract structures. I take this to be a grave mistake, as one cannot evaluate this proposal further at this point. For instance, if the concept of emergence of causation from a more fundamental pattern dynamism is successful depents on this (cf. Dvořák 2023).

It is therefore worth taking a closer look at French's alternative solution.[44] This can also be motivated by a point of criticism by Psillos. He asks how OntSR can separate the structure in which entities are located from their inner nature. Psillos' conviction is that this is not possible, since the structures of scientific entities form a continuum with their inner natures (cf. Psillos 1999, pp. 155–156). His argument is that structural descriptions of an entity or property (his example is mass) always presuppose a causal agent and an interpretation of the mathematical description or definition is necessary in order to be able to give a structural description that explains something scientifically. Psillos reasons that there is never a separation into mere structure or the inner nature of entities. This idea belongs firmly to medieval scholasticism, which chooses form and substance as the basic distinction, not to modern natural sciences.

In response to these concerns, French admits that structures consist of relations and that these relations require relata. What is the argument against these relations being structures or relations themselves?

> The structuralist offers [a] "top-down" picture in which we start with the laws and principles "presented" (on the surface as it were) by the theory, interpret these, at least minimally, in terms of relations and properties, but then resist the temptation to take that further metaphysical step and regard these last as possessed by [...] objects. In particular [...] there is nothing in the theory itself, or in the laws and principles as they are presented, that *requires* us to posit objects qua property possessors. On this view, these relations and properties are features of the fundamental structure of the world [...] and what we standardly designate as "objects" are indeed mere nodes in this structure. French 2014, p. 114.

[44] Other alternatives as in Esfeld 2004, 2009, 2017; Esfeld and Lam 2008, 2011; Lam 2017; Lam and Esfeld 2012 attempt a more moderate form of OntSR, but fall back into a classic OSR with objects, which in my view has no obvious advantage over Chakravartty's semirealism as far as the debate analysed here is concerned. The same applies to Floridi 2008. For a critique of Esfeld and Lam's moderate OntSR cf. Busch 2003; French 2014, pp. 178–181 and pp. 231–254 as well as Ladyman 2020. I therefore postpone consideration of the moderate OntSR in this book.

The nodes in French's explanations are strongly reminiscent of those that Ladyman introduces into his OntSR via the analogy of graph theory. However, French must also explain the existence of relations and properties. He wishes to deny that he only has a mathematical, e.g. group- or set-theoretical structure, which could then always be completely reformulated into Ramsey sentences. In this way, he avoids the dilemma that follows the Newman problem (cf. also for the following French 2014, pp. 116–121). Instead, not all parts of theories can be "ramseyfied" according to French, since, for example, dependency relationships or part-whole relationships cannot be replaced by variables. For French, theories and models are representations of physical structures of the world, that is, natural systems (French 2014, p. 135. Here he quotes; Bueno and French 2011; French 2003). A theory *presents* with named objects as a heuristic a structural interrelation that exists in a part of the natural world. He tries to go beyond Ladyman to provide an answer to the questions raised, especially about the non-existence of objects, when he writes:

> So, we can accept that "Tables exist" is true but refrain from any ontological commitment to tables, because "Tables exist" is made true by the relevant "simples" [...]. French 2014, p. 176.
>
> [O]ne can take the relevant "features" of structures as acting as the appropriate "simples" or truthmakers. [...] Here they will include symmetry principles and fundamental laws and the truthmaking relation will be reversed of course, in so far as it is not objects and properties that make true law statements and the like, on this view, but rather laws, and symmetries, that ground the properties and behaviour of the putative objects. [...] But nor will these simples be spatio-temporal [...]. It has long been part of the structuralist programme to incorporate space-time within this ontology, and the structure of the world has been taken to include space-time structure [...]. French 2014, p. 183.
>
> But in this context at least I see no reason why we cannot release simples from such a (spatio-temporal) constraint and allow them to be the kind of "building block" from which one constructs space-time, elementary particles, and so on. French 2014, p. 177.
>
> Each fundamental physical object depends on the structure to which it belongs. French 2014, p. 178.
>
> At the fundamental level there are no objects, only structures, and ultimately it is in these terms that particles should be understood [...]. [T]he very constitution (or "essence") of the putative objects is dependent on the relations of the structure. French 2014, p. 181.
>
> Our putative objects only exist if the relevant structure exists and the dependence is such that there is nothing to them—intrinsic properties, identity, constitution, whatever—that is not cashed out, metaphysically speaking, in terms of this structure. [...] [W]e could take "Particles exist" to be (literally) true but maintain that what makes the sentences true are not particles as objects; that is, the truthmakers are structures or aspects thereof [...]. French 2014, p. 182.

In essence, this means that in French's scientific ontology only fundamental laws of nature and principles of symmetry exist as ontologically primary. They constantly and causally determine an overall structure of the world, an all-encompassing nature. This world structure unfolds itself in a differentiated way and contains, among other things, our spacetime and within it local interconnections of relational substructures that form complex patterns. Thus, relational properties emerge whose relata are each different local aspects of structure. French then goes through several options regarding how the alleged objects, such as tables, can be reconstructed from this purely structuralist ontology. He favours a ""bundle" view of objects" (French 2014, p. 184):

> I shall suggest that we should reverse the current understanding of the relationship between (instrinsic) properties and laws by taking the latter to have ontological priority as features of the structure of the world, with the former as derivative, or dependent. [...] A "global" bundling of the relevant polyadic properties understood in group-theoretic terms will yeald the blob as structure of the world, with a "local" bundling of the relevant properties giving us the putative "objects". French 2014, p. 189.

This means that all ordinary and scientific objects are one-way dependent on more fundamental physical structures. Physical particles must be understood as local bundles of properties that are one-way dependent on the fundamental laws of nature and on the symmetries that make up the structure of the world. Therefore, the metaphysical underdetermination of quantum objects is an unresolvable fact: they can be understood as individuals or non-individuals. Both epistemic options are legitimate in describing the structural aspects that are only minimal bundles of properties.

French's OntSR thus contains a comprehensive ontology of science as well as a structuralist metaphysics of nature that defends an object-eliminativism.[45] French thus takes on a heavy burden in order to avoid the problem of the metaphysical underdetermination of quantum objects and the difficulties of Worrall and Ladyman. In what follows, I must clarify two questions in order to assess the potential success of this proposal: Does French's OntSR solve the UTE and historical arguments as effectively as Chakravartty's semirealism? I explore this question in the next subsection. The subsequent subsection examines whether French's metaphysics, based on fundamental laws of nature, symmetries and causality, can convincingly hold up against antitotality arguments and targeted counterarguments.

5.4.2.4 Underdetermination and Historical Arguments

The strategy that OntSR employs against the usual UTE and historical arguments is identical to that of semirealism: selective scepticism. In the case of OntSR, the objects entailed by the theories (as well as processes, events, etc.) are underdetermined and change with the theories. However, descriptions of structures and relations by the theories are (largely) true. Firstly, this explains the success of false theories such as the ether theory used by Fresnel or the calorific theory of heat. Science is also cumulative, as structural insights are gained and preserved (cf. French and Ladyman 2003b, p. 45).

---

[45] I thank Alex Englander for pointing out that there is a proximity to the approaches of German idealism here, which also explains why French explicitly distinguishes himself from information idealisms such as that of Tegmark 2007 or Floridi 2008, which also—consciously or unconsciously—choose similar philosophical approaches to German idealism. For criticism of Floridi's position as incompatible with scientific realism cf. Wheeler 2022.

Furthermore, contrastive UTE as a counter can be dispelled. French explains that if it is not possible to decide between T_1 and T_2 on an empirical basis, then theories presuppose different objects and processes, which argues against object-oriented OSR, but that the same structure would be described, which would result in an argument in favour of OntSR from the contrastive UTE (cf. French 2014, pp. 22–24).

Thirdly, upheavals of theories are comprehensible. Ladyman argues that the success of phlogiston theory cannot be explained adequately by any OSR that is not a selective scepticism. Classical forms of OSR fail because of the combination of UTE and historical arguments in this case:

> The structural realist solution to this problem is to reject the claim that the nature of unobservable entities is successfully described by science, and to argue instead that successful scientific theories give increasingly accurate descriptions of the structure of the world. Theories can be very different and yet share all kinds of structure. The task of providing an adequate theory of approximate truth that fits the history of science and directly addresses the problem of ontological continuity has hitherto defeated realists, but it is easily possible to display the structural commonalities between different theories. Hence, a form of realism that is committed only to the structure of theories might not be undermined by theory change. Ladyman 2011, p. 97.
>
> [W]e can say that phlogiston theory identified a number of real patterns in nature and that it correctly described aspects of the causal/nomological structure of the world as expressed in the unification of reactions into phlogistication and dephlogistication. Ladyman 2011, p. 100.

Holger Lyre argues that the essential argument in favour of OntSR is its ability to retain scientific realism and to follow insights such as the no-miracle argument, but at the same time to meet the anti-realist challenges more successfully than competing variants of OntSR:

> The structural realist claim therefore consists in being able to escape both Kuhnian incommensurability concerns and the pessimistic meta-induction. Furthermore, in the cases usually discussed, underdetermination does not occur with respect to the mathematical structure, but only with respect to object-like entities. Conversely, according to the structural realist view, it is precisely due to the invariance of the structure of possible rival theories in the context of underdetermination scenarios that they are empirically equivalent-adequate theories. Seen in this light, [...] structural realism does indeed offer an optimistic option in the context of today's debate about scientific realism. Lyre 2004, p. 98; my translation.

In addition, there are weighty reasons from the current philosophy of physics in favour of OntSR. Without going into further detail here, it should be mentioned that there is no agreement on the various options for quantum ontology and spacetime ontology, nor is an integration of the two physical theories in sight. For this reason, the interpretation debates seemingly cannot be decided under the assumption of object-centred realism. This leads to a metaphysical underdetermination of OSR in philosophy of physics, which OntSR cancels out (cf. French 2014; Lyre 2004, pp. 33–42). Since the mathematically describable structure of the theories is not in question, the OntSR offers itself as a theory option in philosophy of physics.

In conclusion, this means that OntSR, like semirealism, is able to meet the two classical challenges of realism. On this positive side, these philosophical options are in a counterbalanced position, in a stalemate. However, as I will show in the next section, this is also true in at least one other sense: OntSR is challenged by antitotality arguments, whereby I will argue that the elaborated variant of OntSR according to French is a special form of OSR and is thus unable to deal with all three families of counterarguments.

5.4.3 Traps for French's Ontic Structural Realism

In the last section of this subchapter on structural realism, I will criticise the most successful variant of structural realism analysed here, French's OntSR, and show why—like Chakravartty's semirealism—it cannot deal with antitotality arguments in addition to UTE and the historical challenges. In doing so, I will show that French's OntSR and semirealism are in an argumentative stalemate not only in a positive sense—in overcoming the classical challenges of OSR—but also in a negative sense. To prove this, I take two steps: First, I address the problem of object-eliminativism and the reduction of objects to nodes in structures necessary for it, as well as their indispensability of things other than structures for any theory of causation, which Chakravartty and J. E. Wolff, among others, raised against French. French's partly still open answer shows that this position cannot conclusively justify itself against the criticism.[46] More importantly, French's OntSR is also an OSR. This prepares the ground for the second important objection, the antitotality arguments, which hit French's OntSR as an OSR. This second step reveals the negative stalemate between the OntSR and semirealism.

5.4.3.1 The Problem of Object Elimination, Irreducibility and Causation

In the criticism that she formulates in *Do Objects Depend on Structures*, Wolff returns to the fact that structure is defined as relational. If it is to get by without relations that are *not* structures, French (and ontic structure realists of the object-eliminativist persuasion) find themselves in a regressive movement (cf. also for the following Wolff 2012, pp. 607–608). If, like French, one proposes to eliminate objects completely from ontology, then an object must be reducible to its structural location, i.e. to the node with its relations, which is determined by the over-all structure and is part of it. A node, a mere position in the network of relations, is ontologically dependent on the overall structure. If, like French, you wish to eliminate objects (especially physical objects such as quantum particles and

[46] For the sake of brevity, I will leave out the subsequent critique of French on the notion of "theory" in philosophy of science from his standpoint of OntSR, as OntSR must first be defended in order to arrive at his genuine proposal for theories, even though some of the criticism of the notion remains valid regardless of one's position on OntSR; cf. French 2020, 2024a.

atoms), you must show that they are reducible: "If objects are reduced to positions, and positions depend on the structure they belong to, (re-conceptualised) objects depend on particular structures." (Wolff 2012, p. 611). Here, one could also ask whether there is an interdependency relationship between structures and their nodes. However, Wolff's subsequent point of criticism arises even without this additional complication for French and without the issues raised recently by Federico Benitez (cf. 2022).

Wolff then identifies the options that remain for OntSR: In general, objects are not identical to their positions in structures. Wolff's example is Barack Obama, who is clearly identifiable in his social positions as Michelle Obama's husband and as President of the USA in 2011, but who is not identical to these social positions. For Barack Obama to be reducible, it would have to be (logically, nomologically or ontologically) impossible that he was *not* president of the USA in 2011. This does not seem to be the case.

For Wolff, this argument is not yet decisive for French's OntSR, as that position does not treat social objects such as people with object eliminativism, but mainly fundamental scientific objects. However, the requirements are clear: at least for physical objects, it must be the case that their positions in the causal-nomological-modal structure of the world cannot change, so that we can speak of a reduction of objects to the structure.

> Is such a reduction claim plausible for the particles of fundamental physics? Structural realists like to appeal to familiar phenomena from quantum mechanics to support their position. Two electrons in a singlet state have opposite spin in any spin direction, but no definite spin in any direction is assigned to either one of them. What tempts structural realists to suggest that the electrons are nothing but positions in a structure seems to be that, within a singlet state, electrons are not distinguishable from each other. [...] If a singlet state of two electrons is the relevant kind of structure in this context, then for particles to reduce to structure, electrons must not have properties other than the ones described by the state. But particles have state-independent properties in addition to state-dependent properties. True, these don't help to distinguish the two electrons from one another, but they do distinguish singlet states of electrons from singlet states of other kinds of particles. How do we make sense of the difference between muons and electrons if not through the difference in mass between muons and electrons? Mass is a state-independent property, not a state-dependent property. Some properties of fundamental particles, then, do not depend on the state the particles are in. That means that the occupants of these states have properties that aren't reducible to the state. And that in turn means that the occupants themselves cannot be reduced to a particular role within a state, since the roles in the state are described only with respect to the state-dependent properties.
>
> Different kinds of systems can be in a singlet state, and while it is true that neither particle can be said to be the spin-up or the spin-down particle, it makes a difference for physics what kind of particles are in a particular singlet state, e.g. whether they are electrons or muons. Wolff 2012, pp. 614–615.

The OntSR cannot simply reply that the structure to which e.g. electrons are reducible is greater than their singlet state, as Wolff shows in the next step of her argument. However, the following argument is more important: French claims in his OntSR that all other properties such as mass can also be interpreted as structural properties and that they therefore do not have to be attributed to ontologically independent objects. But Wolff refutes this:

> The starting point for a structural account of properties like mass and spin is the classification by Wigner (1939) of particles as irreducible representations of the Poincaré group. Wigner classifies different solutions to free relativistic field equations in terms of different values for mass and spin as different irreducible representations of the Poincaré group. [...] Wigner's classification reveals how the different kinds of particles belong to a common structure, the structure of the Poincaré group, by showing how different kinds of (elementary) particles can be thought of as different irreducible representations of that group.
>
> There are some general worries about this structuralist strategy. For example, it is unclear how the Poincaré group can be understood as a concrete structure, unless we take the particles to instantiate this structure, in which case we might worry that the particles are far from being eliminable. More importantly for the present context, though, even such a structural account of state-independent properties does not yield a reduction of state-independent properties to structure. For such a reduction to be the case, we would once more need supervenience, in this case the supervenience of particular representations of the Poincaré group on the group structure of Poincaré group. But once again, no such supervenience relationship holds. In this case, this just follows from the relationship between a group and its representations. The different kinds of particles have different values for mass, and accordingly they have to be thought of as different irreducible representations of the same group. So a change in representation is precisely not a change of the group, or group structure. That means representations do not supervene on group structure, and are hence not reducible to the group. Since the representations differ by the exact values for mass and spin, this structural account of state-independent properties does not help the reductionist structuralist. Wolff 2012, p. 617; she refers to: Wigner 1939.

This means that the elimination of objects in French's OntSR cannot succeed by reducing the alleged objects to nodes in the structure.[47] This counterargument is particularly problematic for French for two reasons. Firstly, it attacks the object eliminativism with which French's variant of the OntSR responds to the metaphysical question to which Ladyman gives no answer. Secondly, as shown at the end of Wolff's quote, the non-relational properties of objects are French's undoing if they cannot be captures by the account. They are not dependent on the concrete physical system in which the objects are located. This in turn raises the question of how such elements, which French cannot explain by the immediate physical structure but which play an important role in our theories, can be integrated into his OntSR at all. Wolff's analysis therefore poses a problem for French.

> More acutely perhaps, Wolff questions the metaphysical grounds for adopting an eliminativist stance in this context to begin with. Let's recall that in the face of the underdetermination between quantum particles as individual objects and quantum particles as non-individual objects, the structural realist suggests that you pull in your metaphysical horns, as it were, and insist that quantum "particles" should not be regarded as objects in the first place. What are they then? Well, first of all, the "they" here refers to whatever it is that lies behind/gives rise to/however you want to put it, those features that physicists attribute to the behaviour of "particles". Which features? Consider the beautiful curves carved out

[47] This arguably also does not change with assuming a "math-first" structural realism as proposed in Wallace 2022.

> in a bubble chamber, for example. These might be attributed to something of a certain mass and a certain charge passing through the imposed magnetic field. What is it that binds together these properties, such that we get this behaviour? Standardly, the answer would be a "particle", understood, within the framework of an objectoriented metaphysics, as an object, whether individual or non-individual. So the structural realist needs to be able to account for that "binding" together of properties – what Chakravartty calls the "sociability" of such properties – and she can, by noting that they "drop out" of the symmetries, described by group theory – such as the Poincaré group for example. It is in this sense that "particles-as-objects" come to be reconceptualised as features of structure; that is, those features that lead physicists to invoke "particles", which most metaphysicians and realists then understand as objects, are "re-situated", if you like, as features of the more fundamental structure, as captured by the relevant group. French 2019, p. 27.
> What is needed, it might be said, is some justification for the claim that the identity of the object depends on the structure, *and nothing else* (Wolff 2011). Here the onus issue arises again: the non-structuralist asks for just such a justification; the structuralist asks what else *could* fix this identity? Indeed, as far as the structuralist is concerned, this demand assumes precisely that which she denies, namely that there is anything "beyond" the structure that pins down the identity of objects. French 2014, p. 179.

French starts from his metaphysical picture of OntSR and claims that this fits the relevant, fundamental physics, whereby—assuming the metaphysics of OntSR—mass and spin can also be formulated in structuralist terms: "Furthermore, the notion of structure [...] is broader and certainly encompasses kinds, since these are given, or better perhaps, *presented*, group-theoretically and hence structurally." (French 2014, p. 180, fn. 25) French does not refute the objection, but attempts at this point to redistribute the burden of argumentation. Wolff would have to show that the idea of "it's relations all the way down" (Stachel 2006, p. 54) is not tenable for metaphysical or physical reasons. Since, according to French, everything is dependent on structures, no objects are needed and Wolff's object assumption can then be justified against Occam's razor.

Wolff makes use of Kit Fine's and E. J. Lowe's arguments that *A*'s ontological dependence on *B* does not allow for the elimination of *A* without an additional reduction of *A* to *B* being possible (cf. Fine 1995; Lowe 2005).[48] Thus Wolff aims to argue against French that even assuming a dependency, a weak existence of objects in ontology is still necessary (Wolff 2012, p. 622, which is directed in particular against French 2010). This would be an OntSR position closer to Ladyman's OntSR. According to Wolff, French fails to show how objects can be reduced. She writes: "To show that the objects are "thin" enough for structural realism, then, [OntSR] needs to claim more than just dependence. An ontological dependence claim by itself yields a weak form of structural realism, one according to which objects depend on structure, but not vice versa." (Wolff 2012, p. 622). However, this form of OntSR, unlike French's, retains objects in its ontology.

Wolff's first question is whether these thin objects have any applications at all (cf. also for the following Wolff 2012, pp. 623–624). She argues here in favour of OntSR that such an interpretation is appropriate in the case of electrons and

[48] Whereby the last contribution does not correspond to the current one in Tahko and Lowe 2020.

other quantum particles. However, this could not satisfy French: Firstly, his solution to the problems of Ladyman's OntSR approach with thin objects consisted in object eliminativism. Secondly, this only works for quantum particles. The general reductionism of all objects to structures is refuted not only by her but also by the arguments of Chakravartty, Jacob Busch and Psillos, as I will show shortly.

French counters this argument in a footnote: "I would insist that if *x* is dependent on *y in the right sort of way*, then *x* can be eliminated in favour of *y* and what we have in the case of physical particles is just the right sort of dependence that can sustain eliminativism." (French 2014, p. 182, fn. 31). This means that French is left here with nothing except repeating that objects do not exist, but that propositions about them can be true, since they refer to bundles of *simples*. For French, these are metaphysical parts of a structure that is built on the laws and symmetries of nature as a whole. He must hope that this metaphysical world view is convincing enough despite the doubts sown by Wolff, e.g. through his solution to the problem of metaphysical underdetermination of quantum objects in philosophy of physics.[49]

I have already shown in Sect. 4.1 on reductive naturalism that reductive physicalism has to solve the problems of the disunity of science and of pseudo entities. The eliminativist solution, i.e. the assertion that all objects can be eliminated and reduced to quantum objects, was not successful. Whether French would go down this path and whether the problems are meaningfully applied to OntSR is not a question I am pursuing here. However, it is clear that Wolff is cornering French. This is closely related to the accusation that his OntSR is interesting for some areas of physics, since it is clearly orientated towards their problems, but cannot deal with other natural sciences or non-mathematical sciences (cf. Newman 2005, pp. 1377–1378). I put this domain of problems aside, since the challenge presented below is sufficient to show that antitotality arguments are not solvable even from French's structuralist perspective. Crucially for me, French insists, in response to Wolff's argument, which would pose serious difficulties for his OntSR if successful, that all conceivable objects must be explainable by bundled structural aspects. This, in turn, is only guaranteed because everything (all local physical structures in French's picture) results from the fundamental symmetries and laws of nature that apply supratemporally and spatially (they also contain our spacetime). This sounds suspiciously like a structuralist variant of the single real, subject-independent external world of OSR, an assumption successfully called into question by antitotality arguments. To substantiate this suspicion, a further look at the criticism of causality in OntSR without objects suffices.

The aforementioned criticism by Busch, Chakravartty, Psillos and others argues against French's OntSR with the necessity of objects in theories of causation and causality (cf. Busch 2003; Chakravartty 2003; Psillos 2006). The main problem is that without objects, processes and events as fundamental entities of the natural sciences, causal explanations can no longer be made. However, these explanations

[49] Though this argument is itself target of different criticism, cf. Luty 2024, which I take to be only partially successfull and therefore set aside here.

are ubiquitous in the natural sciences, as theories of causation cannot do without them. The critique sets French the task of enabling accounts of causation without objects within his metaphysical conception of nature or the single real external world of structure. In addition, there are direct counterarguments that build on the criticism, such as this, proposed by Chakravartty:

> We might call this a *causal dependence*. One of the most important explanatory roles served by objects is to provide a means of change. Objects have properties, and it is because they have these properties that things happen to them. Indeed, the mathematical equations that typify law statements in physics are generally equations that are interpreted as representing relations between properties of objects. If we increase the pressure, says Boyle's law, the ratio of the temperature to the volume must also increase. If we apply a force, says Newton's second law, the body must experience an acceleration that is proportional to the force and inversely proportional to the mass. Objects with properties are thus explanatorily central in the entrenched ontological framework. How does an objectless ontology account for change?
>
> [...] It is reasonable to expect any account of scientific realism to possess the ontological resources with which to explain how we get from one state of affairs to another. If one accepts the ontological framework of OSR, one appears to be left with explanatory gaps—missing links—between subsequent states of affairs. Given a concrete instance of some set of relations, we have no explanation for what constitutes the active principle that transforms this set of relations into another. In the entrenched ontological framework, the "nodes" of structures are no mere phantoms, posited for heuristic reasons but then relinquished by careful metaphysicians. They have ontological clout. The natures of properties give us something on which to hang explanations of change. [...] If we take the slogan "relations without relata" seriously, however, the replacement framework seems to have insufficient resources with which to provide the desired explanations. Chakravartty 2003, p. 872.[50]

According to Chakravartty, we need objects that are carriers of properties, ontologically, for *processes* that ontologically enable changes in properties and structures. Any causal explanation of a phenomenon therefore depends on objects as carriers of properties that are involved in processes. This is the nut that French has to crack.

Psillos goes even further than this idea. He draws on Stewart Shapiro's distinction between *ante-rem* and *in-re* structures from philosophy of mathematics (cf. Shapiro 1997, pp. 85–89). Put simply, an *in-re* structure is a system that occurs somewhere and is formed by different objects with relations to each other. An *ante-rem* structure is defined abstractly, i.e. logically or mathematically, via relations between empty positions or nodes. It can be instantiated by a specific concrete or other abstract structure as required. The argument that Psillos introduces also refers to the fact that OntSR must present a theory of causation, but assumes the wrong kind of structure:

> Take, then, an *in re* structure W and an *ante rem* one W^I that are isomorphic. Let us add that W is the structure of a concrete physical system and that W has a certain causal unity and role. W^I instantiates itself, but since it is *ante rem*, W^I has no causal unity and plays no causal role. Yet, W and W^I are isomorphic. It follows that the causal unity and the causal role of W are not determined by its structural properties, that is, the properties it shares with

[50] He also argues that the motivation (metaphysical underdetermination of quantum objects as well as a response to UTE and historical arguments) is not sufficient to abandon *all* objects in ontology cf. Chakravartty 2003, p. 871.

W^I and in virtue of which it is isomorphic to W^I. If it were so determined, W^I would have exactly the same causal unity and causal role as W. But it has not. So, if we take structures to be causal, we should not look to their structural properties for an underpinning of this causal unity and activity. This is not surprising. Places in structures and formal relations do not cause anything at all. It's the "fillers" of the places and concrete (*in re*) relations that do. Psillos 2006, p. 565.

From this, Psillos concludes that OntSR needs objects if it wants to give a theory of causation: If OntSR assumes *ante-rem* structures, then the structure is abstract after all and subject to the difficulties that are partly responsible for the failure of EpiSR. If OntSR assumes *in-re* structures, then there is more to the physical structure than relations and substructures. As noted several times, French wants to claim that its physical structure includes causal relations. Psillos rejects this:

Suppose that a causal chain consists in some kind of structural similarity between events *c* and *e*. Suppose also that we accept the strong view that this structural similarity is all there is to causation. Which then is the cause and which the effect? Structural considerations alone cannot afford us a distinction: cause and effect are isomorphic. A corollary of this is that structural considerations alone cannot distinguish between a case of persistence (where an event persists over time) and a case of change (where an event causes another event to happen). Both cases are structurally identical if causation consists in structural continuity. [...] One might try to avoid these by taking structures to be in space and time and by arguing that the cause is the structure that precedes in time the other structures. This would take structures to be *in re*. Only concrete systems can be in space and time. Now, being in space and being in time are not structural properties. So nonstructural properties are necessary for causal relations. Psillos 2006, p. 568.

However, as shown above, this is not French's attempted solution. The structure of the world is not in spacetime, rather it contains (at least our) spacetime. Here, structure is physical without necessarily being concrete. According to this model, the fundamental laws of nature and symmetries are the basis of everything. They contain and enable, among other, our spacetime, which contains different, connected concrete systems, which in turn consist of different substructures as relata and their relations. Then Psillos's problem comes into play: The structural aspects, which can be bundled as events *c* and *e*, are apparently only connected to each other via partial isomorphisms. It is not easy to say why *c* should be the cause of *e*. Psillos continues:

But even if we leave all this behind, we can still question the *rationale* for taking causation to be exhausted by a chain of isomorphic structures. If causation is a relation of dependence between events (where dependence can be understood in any of the standard ways: nomological, counterfactual, probabilistic), then it should be clear that the idea of isomorphism between cause and effect is undermined. There is nothing in causation-as-dependence that dictates that cause and effect should share structure. If '*c* causes *e*' is understood in any of the above dependence senses (e.g., *e* counterfactually depends on *c* and the like), *c* and *e* have any structure whatever. [...]
Another way to highlight the problems that OS has with accommodating causation concerns the causal relata. Standardly, the relata are taken to be either events or facts. On either view, causal relations depend on objects having properties and standing in relation to each other. [...] Most objects (with the exception of fundamental particles) are structured complexes. Events or facts involving these objects will involve their structure. Consequently, their structure (e.g, the structure of a molecule) is causally relevant to what these objects can do. But these are *in re* structures: they depend on the properties and relations these objects

> have. There is no entry point for OS here. To put the point bluntly: the truth-makers of causal claims require objects and properties. [...]
> By going modal, OS promises to close the gap between abstract *ante rem* structures and concrete *in re* ones. But the modal features of the world are not purely structural. Nor can causation be anything like "the cement of the universe" if structure is all there is. Worse, we cannot make sense of causation if structure is all there is. Psillos 2006, pp. 568–570.

Although the points of criticism are not identical and there is a whole series of similar counterarguments in Busch and other authors, a core problem of French's OntSR becomes apparent here (leading some to advocate for EpiSR instead, cf. Hanson-Park 2023): The usual theories of causation presuppose objects with changeable properties. The obvious solutions cannot be realised by the reconstructed arguments of Chakravartty and Psillos. The picture of French's metaphysics of nature described so far is not sufficient to address this concern. It has become apparent, especially in light of the analysis of Wolff's arguments, that French must respond that causal relations are a peculiarity of local systems in our spacetime and that these must be structurally constituted at the ontologically fundamental level. In the next, final section, I explain why this answer means that French's OntSR is also an OSR and what distinguishes it from other variants of OntSR, apart from object eliminativism. In doing so, I also show that French is no better armed against the antitotality arguments than Chakravartty. The double stalemate between the two positions therefore also exists in the negative.

First, I allow myself a remark on Ladyman's (and Ross's) proposal for a different OntSR. Ladyman addresses the problem of irreducibility of objects more pluralistically than French: "Scientific realism as such has no account of the relationship between ontologies, nor of the relationship between causation and law, at different levels, and in different sciences, and this is problematic for its formulation and defence." (Ladyman 2018, p. 102). Based on the map of debates in Chap. 2 and the overview of OSR in Chap. 4, one reason for this can be identified: In OSR, the background assumption of the single real external world is often made, which necessarily contains all phenomena and the entities that determine them, which is why the scientific ontologies should fall into place as if by themselves. Ladyman sees a problem in this assumption. However, he has to show that his OntSR can organise the order of the scientific ontologies (as representations of the physical structures) in terms of the overall structure.

Ladyman strives to solve this by referring to his "rainforest realism" (Ladyman and Ross 2007, esp. p. 233). "In general, Ladyman and Ross argue that ontology is scale-relative, in the sense that different energy levels and regimes, as well as different length and time scales, feature different emergent structures of causation and law. Real patterns are entities of whatever ontological category that feature (non-redundantly) in projectible regularities." (Ladyman 2018, pp. 103–104; for a more in-depth consideration cf. Ladyman and Ross 2007, pp. 226–229.) For Ladyman there is neither a most fundamental level (quanta, strings etc.) nor individual things from which everything else is composed, therefore all real patterns are equal. Applied to Eddington's famous table, this means that it exists as a real pattern on the mesoscopic scale, but dissolves on the microscopic scale.

We can now ask whether the talk of real patterns and rainforest realism (Ladyman and Ross 2007, pp. 233–238) is helpful in solving the problem. In this picture, there is only the fundamental modal or nomological structure of the world (Ladyman and Ross 2007, p. 130). If we observe this from our perspective, which is part of this structure, on a certain size or energy scale, then we find real patterns that are formed by the fundamental structure there and under these conditions (Ladyman and Ross 2007, pp. 145–154 and 200–210). These can then be described in our theories as objects, processes, events, etc. Ladyman thus not only explicitly speaks out against reductionism and eliminativism, but also against (ontological) pluralism. Instead, he assumes a single real, physical structure that is modal in the sense that, under the right conditions, it produces different real patterns on different scales that are determined by fundamental laws of nature (cf. Berenstain and Ladyman 2012).

With the introduction of real patterns that arise locally due to the single fundamental nomological structure of the world, Ladyman's OntSR is less (micro-)physicalist than French's variant. Since in French's metaphysics of science a physically describable microstructure is ontologically fundamental for all observable structural aspects and everything else, especially common objects, is completely eliminated, Ladyman allows for more diversity in comparison. But here too there is an insistence on a single real structure and the objects in Ladyman's and Ross's metaphysics of science are non-fundamental, possessing no intrinsic nature, identity or individuality as patterns (Ladyman and Ross 2007, p. 131). This supports my suspicion that we are also dealing with a form of OSR in his case.

That said, in a discussion of the arguments against French's variant, Holger Lyre pointed out to me that there might be a way to find an interpretation of Ladyman's structural realism as a new scientific realism with a pluralist outline. That might be so. I will put forth key requirements for any new scientific realism in the final Chap. 8 of this book, and will have to leave it open for future research if a novel interpretation of Ladyman's proposal can be developed, that both adherse to the pluralistic picture of nature elaborated later on in this book and solves the problem at hand. Meanwhile, I will concentrate on French's OntSR, as elaborately worked out in this regard.

5.4.3.2 French's Ontical Structural Realism as an Objective Scientific Realism

French's reaction to the criticism of Busch, Chakravartty, Psillos and Wolff is twofold. First, he explains more clearly how his concrete structure relates to exemplary concrete systems and to abstract structures:

> Putting things in broad terms, the "quantum structure", say, does not exist independently of any exemplifying concrete system, it *is* the concrete system. But this is not to accept that the *system*, as such, and as typically conceived of as composed of objects and relations, is ontically prior to the structure. Indeed, the central claim of OSR is that what appears to be a system of objects and relations should be reconceptualized as a relational structure; that is, it is the structure that is (ultimately) ontically prior and also concrete. Hence, the conception

> of structure as abstract is also rejected. What might be seen as "havering" between the two alternatives here is actually an attempt to articulate a notion of (physical) structure that does not fit into the categories imported from the philosophy of mathematics. French 2014, p. 209.

French responds to Psillos' dilemma by clarifying that it is not a problem to allow causal relations between bundles of relational properties. "One could surely allow that [...] the kinds of relations [...] that we are concerned with have—in some sense—a causal aspect and although that causal aspect may not itself be structural this does not introduce the kind of non-structural element that is problematic." (French 2014, p. 211) The bundle theory of structural aspects as metaphysical *simples* was introduced by French for the quantum level of nature. Here, as stated in the quotation, there are causal relations: Think of two mutually repulsive electrons, which are to be understood as structural aspect bundles.

French continues: "OSR does not advocate analysing all macroscopic causal processes in a structuralist fashion [...]. In essence, OSR piggy-backs on the reduction of such chains [...] in terms of ultimately quantum processes and then insists that those have to be understood in structuralist terms." (French 2014, p. 213). French consequently claims that a reducibility of causal connections to quantum physics is always given in the most diverse sciences. Furthermore, causal relationships at the quantum level are unproblematic in his structuralist picture. This is possible because these processes always necessarily occur in the single real external world. Their metaphysically fundamental structure can be described in OntSR with the help of physics from natural laws and symmetries: "rather what we have is the world-structure with a particular configuration, if you like, or family of relations at one time and a different configuration, or a different arrangement or family of relations, at either earlier or later times." (French 2014, p. 214). Since the world-structure also includes spacetime, this passage can be read in the following way with the *principle of charity*: The world structure, although determined by fundamental laws of nature and fundamental symmetries, is dynamically constituted. This is why our spacetime is inextricably linked to the changing relations of local structures. Like all observable local structures, spacetime is a substructure of the world structure. The decisive factor is that the world structure, independent of our observation, brings about change through its dynamic, modal and nomological nature.

The structure of the world according to French

1. exists independently of our scientific access (we merely represent it, its fundamental laws of nature and symmetries, and its local substructures in our various theories);
2. exists with its fundamental laws of nature and symmetries as the one overall structure of the single real external world (and even contains our spacetime and all local structures as parts);
3. can be known in some of its parts, local substructures, and in some of its more fundamental aspects by (us) researching subjects, whereby we possess positive knowledge about it (which is historically even cumulative in the picture of French).

French thus represents a structuralist, metaphysically monistic and object-eliminativist variant of OSR.[51] Furthermore, his philosophy of science, like every other variant of EpiSR and OntSR considered here, is a selective scepticism that does not consider all statements of scientific theories to be true and ontologically binding. In French's OntSR, the structure thesis that the dynamic, single real external world has a structure fixed by fundamental laws of nature and symmetries becomes the explicit, central dictum of a metaphysics of nature that sees itself justified by quantum physics.

At this point, all the antitotality arguments that have already been applied to Chakravartty's semirealism above can be introduced against French's OntSR. In addition, the question arises whether French's answer to the objections of the irreducibility of objects and the problematic theories of causation is convincing. This is especially true because his reductionism of all scientific objects to quantum-physical structural aspect bundles is subject to similar problems as the one I described in Sect. 4.1.2. However, a more limited critique based on antitotality arguments suffices to show French's OntSR to be inadequate. All the antitotality arguments I have mentioned exist independently of an object metaphysics: The world problem and the set paradoxes in the versions of Goodman, Badiou and Gabriel depend on a total structure or total set, which is also given in OntSR. The argument I have reconstructed from Gabriel's main ontological arguments presupposes integrated entities, which also exist in French's metaphysics of structures and substructures. Unimundialism in the sense of Rovane, i.e. a *single real* external world, is advocated by French. Fundamental laws of nature are clearly and explicitly declared by French to be the metaphysically primary basis of the world, so that Cartwright's arguments are applicable.

It would be possible, but unsatisfactory, to present all the antitotality arguments mentioned here individually. This is because French's physically orientated position as structuralism can be positioned differently from the usual OSR when it comes to problems of philosophy of physics, UTE or the historical arguments, but not when it comes to ontological-metaphysical antitotality arguments. These do not presuppose object metaphysics in the sense that French believes can be found in the usual scientific realism. Cartwright's argument against fundamental laws of nature suffices here as an example.

It removes the ground from French's metaphysical model.[52] Everything in French's OntSR as OSR is dependent on the fixed structure, the world structure, which consists in fundamental laws of nature and symmetries. If there are good reasons to doubt the meaningfulness of such fundamental settings in the interpretation of physics to begin with, the metaphysical ground of the world

[51] I could further try to show at this point that French represents a reductive naturalism in my definition, which runs into the difficulties already outlined. Since the weaker thesis of OntSR as an OSR is sufficient for my conclusion, I will skip this step.

[52] In this, French is not alone. Alternative (and in my opinion less well justified) versions of structural realism, such as informational, Floridi 2008, or quantum structural realism, Adlam 2022, would face the same issue if they could be developed.

structure is missing. In my reconstruction, I have shown in detail why fundamental laws of nature and fundamental symmetries are cited by French to explain the overall structure of nature and the world. For French, their interplay determines and generates all local physical structures (e.g. also our spacetime). As already discussed, Cartwright doubts that our scientific, especially physical, theories give us any positive knowledge of such fundamental laws.[53] In particular, there are no physical reasons to believe that the laws of nature that we formulate in our physical theories would be fully true if they are *ceteris paribus* laws in Cartwright's sense. As I see it, French gets three things wrong here: First, he believes that current physics provides the fundamental laws of nature that are suitable as the basis for a general metaphysics of the single real external world. According to Cartwright, this cannot be correct. Second, French assumes that all scientific objects can be ontologically reduced to quantum structures, which leads to the problems of reductive physicalism. Third, he thinks that the authority of physics, together with the merits of OntSR as a philosophy of physics and against the classical anti-realist objections, is sufficient to convince his readers to accept a strong, counterintuitive metaphysics of nature without specifically addressing and invalidating the various antitotality arguments.[54]

All the pitfalls of French's OntSR mentioned so far are intertwined: objects and non-relational properties of our physical theories cannot be completely reduced to nodes in a structure. Without them, causation as an enabling condition of the sciences as a whole seems almost impossible to explain. Both objects and causal explanations are needed in our natural sciences. The laws of nature assumed by French to be metaphysically fundamental and the single real overall structure of the world cannot be proven in a straight line from our physical theories, but are exposed to a whole series of counterarguments. Together, these points call into question French's approach to the metaphysics of nature and thus his OntSR—as much as the antitotality arguments have called into question Chakravartty's semirealism. The two positions, as strong variants of OSR, are thus also in a negative argumentative stalemate with each other. The best that OSR has to offer is not fully convincing because of its basic assumptions.

[53] Paul Hoyningen-Huene notes that it is not always possible to determine unambiguously which structure is described by a given physical theory (cf. Hoyningen-Huene 2012, here p. 185–186). The problem that arises here is the impossibility of categorising what the fundamental structure of a particular theory is if only the purely mathematical-formal apparatus is considered, without presupposing an overall metaphysical structure including objects. For example, what is the structure of Maxwell's field theory of electrodynamics? How can a series of linear partial differential equations describe a single physical structure, *without* resorting interpretatively to objects?

[54] Additionally, one can formulate a regress argument different from Wolff's criticism according to Cumpa 2023, but I will leave this aside.

5.5 Conclusions from the Double Stalemate

Finally, I would like to summarise the conclusions that arise from the double stalemate and thus draw a conclusion from the first part of this book. The results give me the motivation to search for a new scientific realism. But before I come to this conclusion at the most general level of this piece of work, it makes sense to return to the critique of French's OntSR.

Having emphasised the problems of French's OntSR at the end of the last subchapter, I may have lost sight of the positive aspects of this theoretical proposal. However, I would be doing French's philosophy of science an injustice if I were to reject his approach completely. For despite the shortcomings of object eliminativism, reductionism and world structure as the single real external world, his approach is innovative and some aspects of structuralism seem promising. Looking at the map of debates I developed in Chap. 2, we can see why French's OntSR provides a decisive renewal: Apart from object eliminativism, which separates French's OntSR from other OSR options, he can discursively define the realism-anti-realism debate as metalocal, even if French does not make this explicit. The debate in his view is not just about the problem of theory change and unobservable theory entities. French sees this more clearly than many other authors.

French's insight leads to a different position for philosophy of time and philosophy of mathematics. It is therefore only logical that they enter into a new relationship with philosophy of science in the course of his OntSR. They are *not*, as in most variants of OSR, separate local debates, which thus do not belong directly to the metalocal debate of philosophy of science. Nor can they, as in the case of Quine, for example, be handed over to physics in the case of the debate about time or—in the case of the debate about mathematical entities—be fully incorporated into philosophy of science with the indispensability argument. In French's case, the realism-anti-realism debate of philosophy of time must be resolved within general philosophy of science, which itself can no longer be clearly separated from philosophy of physics. Spacetime turns out to be a dependent component of the structure of the world, which his OntSR can describe on the basis of physical insights. The realism-anti-realism debate on mathematical entities and mathematics itself, on the other hand, are where scientific-philosophical insights into abstract structures are gained. They thus constitute a metalocal debate that is inextricably linked to that in philosophy of science. In particular, the relationship between abstract and physical structures, the deepest essence of French's world, can only be addressed together. The OntSR is characterised in particular by the attempt (as was historically the case with Quine's naturalism) not only to mark a new position within the discussion, but also to bring about a shift in discourse by linking existing debates differently. The new realisms will also have to attempt not only to sound out new positions in philosophy of science in terms of content, but also to initiate a shift in discourse, a reorganisation of local and metalocal debates. This is my first important take-away.

A further conclusion of this chapter is that selective scepticism, whether Chakravartty's semirealism or French's OntSR, is OSR's most favourable realist way of addressing ontological questions about nature. The combination of a realist attitude in philosophy of science regarding the ontological significance of scientific theories and the sceptical or agnostic attitude regarding the question of whether all aspects of a theory are true and have ontological implications, serves successfully to overcome UTE and historical arguments. A selective scepticism with regard to theories is also recommended for any new scientific realism, i.e. the view that only some, clearly identifiable aspects of scientific theories should be understood ontologically as true, partial descriptions of nature about which we wish to be realists.

Excessive reductionism and, in particular, reductionist physicalism are not helpful, but rather a hindrance, as the examples of Quine and French show. The reason for this is not only the strong counterarguments that arise from the problem of pseudo entities and the problem of the disunity of the sciences. French's OntSR in particular fails, regardless of these arguments, because of the criticism of Wolff, Chakravartty and Psillos, which arises from French's reductionist metaphysics. This may be parsimonious in nature and in the Ockhamian sense, if metaphysical theories are to be evaluated in the same way as scientific theories. However, it is anything but minimal when it comes to metaphysical consequences and implications. This points to a further lesson to be learnt from the analysis presented. Neither a reductionism to a scientific (sub-)discipline such as quantum physics is advisable, nor a metaphysics of nature, if by this is meant an all-encompassing theory of the single real external world. A new scientific realism should avoid both paths, but especially their intersection.

At the general level of observation, a deeper insight can be gained by summarising the three above-mentioned conclusions on the reordering of debates, selective scepticism and the critique of reductionist metaphysics. Chakravartty and French can serve as examples of contributions to the debate that draw attention to the form of the philosophy of science debate itself. This applies independently of individual theses or arguments: No philosophical theory in philosophy of science undertakes a clear cartography of the debate. When asking about Chakravartty's or French's metaphysics of OSR, it must be clear whether a global answer can be expected that takes the following form: *the one* metaphysics of the single real external world is such-and-such, while *any other metaphysics* is false. In this case, we have been catapulted into the global debate of metaphysics and ontology. At least French's contribution reads as if there are other metaphysical questions (about social structures, for example), but that the single real external world, the whole true reality, *is* essentially the world of quantum physics. This is due to quantum physical reductionism. This world structure can supposedly be clarified in French's OntSR. Philosophy of science is thus not only on a par with the metalocal debate in philosophy of mathematics about abstract objects, but also exhausts the global metaphysical-ontological debate with a naturalistic impetus. This also fits in with French's reductionism in the sciences themselves.

In the background seems to be the idea that the philosophy of science debate in our time is *identical* with the global metaphysical debate, insofar as statements are to be made about physical reality. To prove this as a thesis, however, French would need an argument to the effect that physics and its interpretation can answer the question of world structure as reality. Such an argument is missing, just as this thesis is not anywhere explicitly articulated by French. As Chakravartty, French is not prone to overplaying his hand and advises caution on many occasions. I think he strains the insight gained in logical positivism that it is philosophical arrogance to make claims to knowledge about the nature of the physical world based on metaphysical speculation and without consulting the best current physics. Even if one agrees with the idea that metaphysics cannot replace physics, as I do, this does not mean that it is justified to assume the opposite.

My opposite thesis is: *Physics does not clarify metaphysics*. Nor can interpreted physics do so unless metaphysical assumptions are implied that must be justified otherwise, on their own philosophical merrit. My argument in favour of this is that the conclusions of physics and the (non-metaphysical) interpretations of physics can never prove that all existential questions of ontology in its method, or all questions about the necessary structure of the external world, are answerable. Nor can non-metaphysically interpreted physics show that the physically describable external world is the only (independent) domain of reality. This is because the empirical and theoretical description of nature never provides negative information of the kind we would need to prove these theses. We can positively *discover* new structures in nature and the most fundamental contexts of nature, but we can never know that reality *cannot harbour more* than is accessible to this or that science. These negative modal theses are not possible results of—finally—empirically founded and philosophically interpreted investigations in certain natural contexts, they are derived through ontological and epistemological arguments in general philosophy of science.

French could therefore still argue, like a proponent of the Vienna Circle, that physics makes metaphysics superfluous. But firstly, historically this discussion has not gone clearly in favour of the Vienna Circle, and secondly, this includes, for example, a form of Carnapian deflationism in ontology, which French must reject (cf. e.g. Carnap 1956, and the modern reinvention of such a position in: Thomasson 2015). This is because, in such a picture, object-eliminativism is not on the proverbial table, since the sciences as a whole cannot reduce all their theses to quantum physics, nor a metaphysics of nature, as French has in mind.

French lacks an external argument in favour of interpreted physics being able to clarify metaphysics. This is why the claim to give *the one* metaphysics of the world is not explicitly articulated. Consequently, we are still nominally (depending on one's point of view) in the local or metalocal debate in philosophy of science. As is clear from my analysis, this runs counter to the way French argues. The shift in discourse from the metalocal debate in philosophy of science to the global debate fails, even if French's reorientation towards debates about time and abstract mathematical entities seems sensible and even exemplary. The result here is that any new scientific realism is recommended to clearly understand and articulate the debate as metalocal and to assume that the discussions about (space-)time and the laws of nature are local and important subdebates. Every new scientific realist must also explain how their position relates to the metalocal debate about abstractions and especially to the debate about mathematical entities and structures.

Chakravartty, as the title of one of his works reveals, writes only *a* (not *the*) metaphysics for scientific realism. What does he mean by this? "I investigate the core elements of promising versions of contemporary realism, and develop a metaphysics that makes sense of these commitments." (Chakravartty 2007, p. xii.) Chakravartty uses the definition of metaphysics as a theory of the world as a whole (cf. Chakravartty 2007, p. 18). I have shown that he does not discuss the idea of a single real external world. He makes it a prerequisite, a metaphysical commitment that must be made for good sci entific realism. In the second part of the work mentioned above, he explicates further metaphysical or ontological commitments to which, in his view, realism in philosophy of science commits us (cf. Chakravartty 2007, pp. 89–182): To realistically understand causal relations in nature as processes, to adopt a dispositional property theory for external world objects and a resulting understanding of natural laws, and to make a distinction between natural and scientific kinds.

Chakravartty does not provide as good a solution to the metaphysical underdetermination of quantum objects as French does. But my final criticism is also less severe than French's. I am not convinced that Chakravartty intends to settle the global debate. Nor would that fit with the modesty of his semirealism as nonreductive naturalism. Rather, I believe that he thinks that the external world thesis is unsuspicious and unproblematic and that a clarification of the ontological consequences and implications can only be achieved with a metaphysics. I contradict both of these assumptions with the antitotality arguments. Instead, the external world thesis should be dropped. As Gabriel argues, this entails the rejection of metaphysics (as a theory of the world as a whole), or at least it forces us to scale back the claims of metaphysics. Chakravartty's claim to give clarity to realism in philosophy of science could possibly be redeemed by the following understanding of metaphysics: Metaphysics is an unambiguous, *partial* philosophical description of nature (or the social realm or even all of reality) and its interrelations. This description results from the collected ontological commitments of partially true scientific theories from different disciplines. We could name what we are realists about, and which (in this basic sense) ontological and metaphysical presuppositions arise from the sciences. However, I would advise speaking of *scientific ontology* rather than metaphysics here in order to avoid misunderstandings. Any new scientific realism will also have to present such a scientific ontology.

A further conclusion can be drawn here. The (in Chakravartty's sense) metaphysical attempts at a solution for philosophy of science are not sufficient for OSR to escape the attack by UTE, historical arguments and antitotality arguments. Anti-realists conclude that this means that OSR has to abandon one of its central theses, which, however, means a substantial adjustment in the resulting philosophy of science. Their proposal is to adopt a suitable ontological or epistemological anti-realism in philosophy of science, according to the definition of realism and anti-realism given in Chap. 2. As I will show in the next part of this book, this strategy is neither the best, since it faces the problem of ancestrality, nor is it necessary. A new scientific realism offers a different way: a basic assumption of OSR is not correct, as the counterarguments show. However, it is not the ontological or epistemological basic thesis of every realism that is the problem, but the premise of the subject-independent, single real external world with a fixed structure. If this complex of assumptions is abandoned, the result is an essential adaptation of the corresponding philosophy of science, but not necessarily ontological or epistemological anti-realism. Elaborating on this idea is one of the important tasks of the second part of this book.

Also in view of the argumentative stalemate between realists and anti-realists regarding abductive inference and contrastive UTE, Mizrahi has recently pointed out that van Fraassen's main positive argument in favour of his constructive empiricism can only be meaningfully reconstructed as an inference to the best explanation (cf. Mizrahi 2018). Mizrahi's reconstruction seems fair, but will not be pursued further, since the analysed arguments against OSR have given me no reason to advocate anti-realism in philosophy of science. Moreover, in the follwing chapters, the critique of *correlationism* will provide a good reason from the new realisms why anti-realism in philosophy of science is not an appealing position. Mizrahi's argument is worth mentioning because it points to a loose thread: An argument against OSR can be formulated from van Fraassen's position, damaging the IBE. But this seems to make a positive justification of one's own position either impossible or at least devastingly difficult. Why then still push for anti-realism?

In my view, this is because a realist position that is not an OSR, i.e. a genuine new scientific realism, does not seem conceptually conceivable in philosophy of science today. Here, however, we are dealing with a case of UTE, as the anti-realist would interject. Just because no realism has (allegedly) been proposed in philosophy of science that is not an OSR does not mean that it cannot exist. The apparent dilemma between OSR and anti-realisms is resolved when it is clearly recognisable that the new realism is a theoretical option in philosophy of science. I would like to demonstrate this in the second part of this book. Before I sketch a positive vision for it in the last chapter, I will do so by means of negative insights from new realism in ontology.

The conclusion I draw from this is to debunk the apparent, merely discursive dilemma that objective realism and anti-realism are antipodes. The differentiation between objective realism and anti-realism is *not* a complete distinction between cases. None of the positions belonging to these families of theories is completely correct, as there are several good reasons against them. This, too, illustrates why a new scientific realist position is needed to resolve the double stalemate.

In Chap. 5, then, I have argue overall that the discursive situation between the diverse currents of objective realism and anti-realist critique resembles a stalemate: OSR, in its heterogeneity, has strong arguments in favour of its own realist motivation. The anti-realist challenges of UTE, historical arguments and especially the neo-realist and anti-realist antitotality arguments provide good reasons why none of the OSR positions is fully adequate. This also applies to Chakravartty's semirealism and French's OntSR. In addition, there is a second argumentative stalemate between the two OSR proposals of French and Chakravartty. On the positive side, both are able to overcome the classic challenges of UTE and historical arguments. This is achieved by means of an adapted, selective scepticism: not all parts of scientific theories have ontological weight and are true. In this context, French mentions the structural aspects, and Chakravartty discusses detection properties and natural objects secured by them in their relations. In addition to this positive stalemate, there is a negative one, because, in addition to the internal difficulties of their positions, French and Chakravartty ultimatey fail because of the antitotality arguments that turn against the thesis of the single real external world with a structure fixed by laws of nature. While French is unable to convince with his justification of the external world thesis, Chakravartty uses it as an unjustified metaphysical presupposition on which his proposal depends.

The central conclusion of the first part of this book, which arises from the double stalemate, is the motivation for a new scientific realism. In order to resolve both stalemates, the thesis of the single real external world and its structure necessarily fixed by laws of nature as well as the primacy of the subject-independence of the entities of the external world must be dropped without assuming anti-realism in its stead. The subsequent second part outlines how further negative insights of new realism from the global, ontological debate—in particular the concept of correlationism and Quentin Meillassoux's argument of ancestrality—contribute to formulating requirements for any new scientific realism together with the antitotality arguments in the final step and to considering some philosophical theory options for such a new philosophy of nature and science.

Part II
New Realism for the Philosophy of Science

Introduction and Structure of Part II

At the beginning of this second part of the book, I am faced with the shambles left behind by the first part in the form of the double stalemate. I have shown that OSR is not convincing due to the thesis of the subject-independent, single real external world in combination with the challenges of the UTE and historical arguments. The OSR contains two theses: First, that the best scientific theories provide in some part or in approximation the correct ontology of the single real external world understood as subject-independent (OOSR). Second, that the single real external world is constituted by subject-independent entities and can be known by us in some of their properties through scientific methodology and theorising, which means that the best theories regarding it are approximately or partially true and should be understood literally (EOSR). Related to this is the structure thesis, according to which there is a law-of-nature or causally fixed structure that underlies the dynamics of the external world as a whole and every natural entity in it. All three ideas—OOSR, EOSR and the structure thesis—cannot be defended simultaneously against UTE, historical and antitotality arguments, as shown in Part I. Alas, I could not find any motivation in these arguments to defend anti-realism in philosophy of science—therefore the metalocal debate ends in the first stalemate. The second stalemate arises from the fact that French's OSR and Chakravartty's semirealism do not fare either much better or worse in the face of this challenge. At first sight, the situation of the double stalemate seems depressing for the philosopher of science.

I see an obvious philosophical option for escaping this apparent calamity in attempts to advocate a better anti-realism that is not solely motivated by the three families of counterarguments. I explore this path in Chap. 6. The variants of constructive empiricism by van Fraassen and Wray, the instrumentalism of Stanford, the contextual empiricism of Helen Longino, but also the (de)constructivist philosophies of science of Martin Kusch and Hans-Jörg Rheinberger can be highlighted here.[1] Rovane's results can possibly be used to defend a relativism in philosophy of science that, unlike the classical variant of Bruno Latour and Steve Woolgar, does not lead to contradictions with the discourse within science (cf. esp. Latour and Woolgar 1986). I will show that many of these anti-realisms utilise strategies of *correlationism*.

Another option, represented by variants of correlationist realism, is also the subject of my Chap. 6. Here the *subject-independence* of the external world is discarded, contrary to OSR, in order to pursue a new, weaker variant of the independence criterion. One could call it the natural external world, which correlationism puts in the place of the single real external world of OSR, and which is reduced in its status compared to the latter.[2] The natural external world of correlationism is therefore not considered real because it exists strictly independently of us; instead, its independence must be reinterpreted. In the realism-anti-realism debate in metaphysics and ontology, such variants of realism have been tested in recent years by (cf. esp. Haslanger 2012; Koch 2006b). In a radical reading as a methodological naturalist, Sellars can be seen as a defender of the prototype of such a realism in ontology and philosophy of science, as I will show. Koch's reflections will also be thematised at that point.

The core of Chap. 6, however, is my further development of Quentin Meillassoux's concept of correlationism itself. I argue that it can be more precisely defined than Meillassoux succeeds in doing. I call this further developed form *systematic correlationism*. With this more precise concept, the next step is to develop what I consider to be a convincing argument as to why the analysed options of correlationist realism and correlationist anti-realism fail in the metalocal debate in philosophy of science. Such positions argue with correlationist arguments and premises and lead either to epistemological scientific anti-realism (ESA) or to ontological scientific anti-realism (OSA). Alternatively, they can reinterpret the ontological realist partial thesis of independence for the domain of the natural external world and assume an asymmetrical co-dependence of the natural external world and subjects in order to make a concession to the anti-realist counterarguments. Both tactics represent a form of correlationism, which is why they concern what Meillassoux has called the *problem of ancestrality*. All of the above positions of ESA and OSA, as well as the realisms without subject-independence, cannot be comprehensively reconstructed within the scope of this chapter. In order to defend correlationism against the frequently voiced strawman objection and to draw lessons from the failure of systematic correlationism for the new scientific realism (NSR), it suffices that I demonstrate systematic correlationism in four exemplary positions: in

[1] Otávio Bueno's anti-realist answer to OSR, structural empiricism, is briefly addressed in the concluding Chap. 8, but does not tap into the same trap as the other options.

[2] The natural external world is often simply called nature or world. In this book, however, I retain the term "nature" as my designation for the alternative of new scientific realism to the concepts of the single real *and* the natural external world and avoid the imprecise concept of world in this context altogether.

Stanford's instrumentalist OSA, van Fraassen's and Wray's constructive-empiricist ESA, Koch's hermeneutic realism, and Sellars, in the radical interpretation as a methodological-naturalist realist hinted at in Chap. 4.

The reconstruction and differentiation of the problem of ancestrality becomes the central task in Chap. 7. Both concepts, correlationism and ancestrality, have been subject to fundamental criticism since their introduction by Meillassoux. I will defend them against several attempts to refute them in Chaps. 6 and 7 and I will adapt them in the process. I argue this defence succeeds, although some points of criticism are justified. My task will then be to free the respective concepts from their theoretical baggage and thus turn them into useful tools for philosophical analysis.

What is the problem of ancestrality at the core of Chap. 7? Basically, the ancestral challenge for systematic correlationism arises from the fact that certain scientific and natural-world entities are simultaneously assumed to be causally or naturally necessary for the existence of subjects, but conversely are themselves logically, constitutively or ontologically dependent on subjects, according to the correlationist. This asymmetrical co-dependency relationship, which is at the heart of correlationism, leads to philosophical positions in the interpretation of the natural sciences that are difficult or impossible to maintain. On the one hand, this emerges from fundamental considerations; on the other hand, all four examples of correlationism can be used to demonstrate why they are unable to resolve the ancestral challenge fully. As a consequence, I can show why systematic correlationism is not a convincing way out of OSR. The first main goal of the second part of this book is therefore to prove the following: *Correlationism is just as incapable as objective realism of providing a convincing position in the realism-anti-realism debate in philosophy of science.*

The argumentative path to this negative thesis—along with the results from Part I—enables a positive turn and thus leads beyond the critique of systematic correlationism. The steps of the argument allow me to see what shape a new scientific realism beyond OSR on the one hand and correlationism on the other must take. *The basic features of any new scientific realism (NSR) can be determined by accepting the ancestrality of nature, its unfixed structuredness, the further development of the criterion of subject-independence and the abandonment of the idea of a single real or natural external world.* This is the second main goal of this second part.

To this end, I will show how nature can be viewed in NSR so that it is neither elevated to a single real external world (as in OSR) nor trapped in a correlation with the sciences and us subjects (as in correlationism). This leads to a change in the image of philosophy of science in NSR: among other things, it requires a *new philosophy of nature* that is in close contact with both the ontology of NR and current results, and with questions of interpretation of the various specialised natural sciences. In this stance of viewing nature, not only is anti-reductionism inherent, but nature will also turn out to be radically plural. In addition, the *natural-scientific interrelation* that exists between nature, its plurality of natural contexts, our sciences with their theories and domain ontologies and our historically evolved scientific practices will take centre stage. The questions of objectivity, critique, plurality of methods and the fundamental disunity of the sciences can be answered in a more appropriate way in NSR in view of the changes in the image of nature, the restructuring of the metalocal debate and the reassessment of the realist criterion of independence. It will also emerge that there is more than one way of articulating such a pluralism in a new philosophy of science. I will argue that the metalocal debate in philosophy of science is reorganised by any conceivable NSR, since temporal and modal realism, among others, always play a role. The task of my last Chap. 8 is not to present a fully formed variant of such a NSR, but to make clear why the basic form to be developed of any NSR should include these philosophical aspects and how this point of departure is made possible by overcoming both OSR and correlationism.

Chapter 6
Correlationism

This chapter introduces Meillassoux's concept of *correlationism*. He uses it to try to identify his philosophical opponents. Not only is Meillassoux's use of the term reconstructed and criticised in what follows, but the concept of correlationism is also decisively developed further. To this end, I will make a distinction between an *epochal concept* of correlationism, a *reasoning scheme* and a *systematic position*. While the epochal concept is controversial in the debate and will not be examined in depth in the present work, the systematic position can be further crystallised, especially with regard to the previously obtained criticism of OSR and against the background of my map of the debate. I argue here in favour of separating correlationism from Meillassoux's historical theses. The reasoning scheme will play a central role in my development of the systematic concept.

My project in this chapter is to recast *systematic correlationism* as a *tool of analysis*. By the latter, I mean a systematic differentiation that is useful for analysing existing debates and sorting positions without reducing them to their status as (non)correlationist. In particular, this analytical tool is useful for better understanding antinaturalist and antireductionist alternatives to anti-realism *and* realism in philosophy of science. This already articulates my thesis that the identification of correlationism and anti-realism often assumed in the research literature on Meillassoux is wrong: In my reading, correlationism is instead *anti-reductionism and anti-dogmatism*, which can be articulated in both realist and anti-realist stances.

In order to create the context for my recasting of the concept and to distinguish between the epochal concept, reasoning scheme and systematic position, I will first discuss Meillassoux's use of the term and its main problems (6.1). By discussing more nuanced criticisms, I argue in the next step that correlationism, with its central reasoning scheme, always opposes the strict subject-independence criterion and argues in favour of either specific epistemological or ontological propositions (6.2). Through my further interpretation and adaptation of the concept of correlationism, the points of criticism and ambiguities of Meillassoux's original proposal can be

J. Voosholz, *Towards a New Scientific Realism*, Synthese Library 518,
https://doi.org/10.1007/978-3-031-98886-8_6

removed. To show that it is not a straw man concept of Meillassoux's opposition—this is the most serious objection from the discussion—I offer some case examples of correlationism (6.3): Stanford's instrumentalist OSA, van Fraassen's and Wray's constructive empiricist ESA, Koch's hermeneutic realism, and the radical interpretation of Sellars as a methodological naturalist realist. With the sharpened *systematic correlationism* that is developed in this way, core views of scientific anti-realisms and correlationist realism are revealed and hidden dynamics of the realism-anti-realism debates become apparent, which can be subjected to the critique of ancestrality in the following Chap. 7.

6.1 What Is Correlationism?

In what follows, I will briefly reconstruct Meillassoux's concept of correlationism, and clarify his arguments and conception with recourse to the central counterargument, the straw man objection, in order to redefine correlationism in the next subchapter on the basis of this analysis. Meillassoux characterises as correlationisms those forms of contemporary philosophy which, in his view, follow in the tradition of Kant in denying the independent nature, "le *Grand Dehors*" (Meillassoux 2006, p. 21, original emphasis), i.e. the independence of the natural external world from our subjectivity, and instead postulate a form of co-dependence. According to Meillassoux, Kant teaches us that any theory about primary qualities, i.e. properties that are attributed to an entity independently of other entities and in particular independently of our reference to or preception of them, is hopeless as a theoretical option (cf. Meillassoux 2008, pp. 1–2), since "it is resolutely *pre-critical*" (Meillassoux 2008, p. 3, original emphasis). In Meillassoux's view, Kant's philosophical turn therefore refutes dogmatic metaphysics, the philosophy that asserts necessary and strictly subject-independent entities. Meillassoux—at least in *After Finitude*—sees current philosophy as being committed to correlationism. This means that it must renounce the radical subject-independence and necessity of the absolute.[1] In his speculative realism, however, Meillassoux aims to re-establish contact with nature—the aforementioned *Grand Dehors*—via primary, subject-independent qualities.[2] He

[1] I will deal with these philosophical vocabularies used specifically by Meillassoux—the absolute, dogmatic metaphysics and similar expressions—and will clarify them each in turn in the course of this subchapter. At this point, they can be read as they are in the broader debate.

[2] The use of the terms *speculative realism, speculative materialism* and *speculative philosophy* varies in different phases of Meillassoux's work. Meillassoux (2006) uses realism and materialism seemingly interchangeably and commits himself to speculative philosophy. In Meillassoux et al. (2007), speculative realism is introduced as a term for the common intersection of the four authors and is subsequently also frequently used by Meillassoux, whereas Meillassoux (2016), p. 133, introduces the following definition, which has since been maintained by Meillassoux: "I call "realism" any position that claims to accede to an absolute reality—thus any speculative position in the wider sense. In this sense, every true materialism has to stand the charge of realism; because it is indeed a realism. But inversely, one can see that not every realism is necessarily a materialism:

strives to prove that our thinking is in contact with a subject-independent nature. Considering this as a philosophical task or challenge seems surprising in light of the analysis of OSR in the first part of this book. Meillassoux does not completely overlook naturalism and similar forms of philosophy, but considers these positions to be vulnerable to correlationist arguments and thus inferior in principle. To better understand correlationism, it is worth looking at how Meillassoux arrives at this concept.

6.1.1 Correlate, Correlation, Correlationism and the Epochal Concept in Meillassoux

The first step towards an improved understanding is to define correlationism more effectively as a family of philosophical positions. The central concept for understanding Meillassoux here is that of *correlate* and *correlation*. He uses CORRELATE to denote the dyad, triad, etc. of two, three or more relata that are only knowable or can only exist if the entire relation and the relata are present together. The fact that both the possibility of being known and the possibility of existing are mentioned here is not considered further by Meillassoux. It should therefore be noted whether a particular correlation involves an ontic or epistemic correlate or even whether both levels are linked. Meillassoux usually uses the ontic phrase of the correlate in his formulations.

Meillassoux often uses the word "CORRELATION" seemingly synonymously with "correlate", but the former always denotes the core of a correlationist position. A correlation can combine several correlates, as can be seen from the examples chosen for analysis later on. An important example for Meillassoux is the correlation that he imputes to Kant: We never have access to the dyad of subject and thing-in-itself individually, but only in their indissoluble relation (cf. also for to the following Meillassoux 2008, pp. 3–7. In addition, if—controversially— Kant can be interpreted in the sense of Meillassoux's correlationism, he could be assumed to have further correlates such as *concept and intuition*). In addition, there is the connection between intuition and thing-in-itself as well as that of the object of intuition and subjects. Together, these relations form jointly with the two co-relata, subject and thing-in-itself, the one central correlation in Meillassoux's reading of Kant. Various pairs, triplets etc. can be considered as relata of such indissoluble relationships, as possible CO-RELATA. Examples besides subject and object are mind and world, referent and language, noesis and noema, or being, *Dasein* and

for anticorrelationist realism can be materialist as well as subjectalist." For the meaning of the other terms, see the further explanations in this subchapter and the commentary in Longo (2014a). In order to emphasise that Meillassoux's position is not only a realism and but also a new realism in my sense, and that it is in his own emphasis primarily a speculative philosophy, I use the term *speculative realism* for his philosophy in what follows.

world. For Meillassoux, these connections are different correlates. The question of how exactly correlates or correlations should be understood in Meillassoux's writings has not been fully answered by what has been said so far (I will come back to this later), but it is possible to begin to redefine the idea of correlationism.

The concept of *correlationism* is derived from correlation. For Meillassoux, CORRELATIONISM refers to the philosophical paradigm that a correlation is philosophically primary. This means that no entity can escape that correlation—everything is always in an ontological or epistemological dependence on this correlation. Everything in our contexts of life and thought is integrated into the respective preferred correlation or requires it as a prerequisite. The philosophies that fall under the concept of correlationism can be heterogeneous, as Meillassoux's main examples from transcendental and analytical philosophy, phenomenology and the majority of post-structuralism and postmodernism show. Meillassoux maintains that current philosophy owes this paradigm to Kant (cf. Meillassoux 2008, p. 3 and pp. 10–11; for Meillassoux, Kant's idea of epistemology as First Philosophy is of particular import, in that Meillassoux claims that Kant's position could not be circumvented by any philosopher writing after the pulication of: Kant 1998a,b).

The definition I have reconstructed thus far differs from Meillassoux's, although his definitional attempts also differ from one another (cf. esp. Meillassoux 2008, 2011b, 2016). Several aspects come together in the definition just outlined, which are justified by the following, more precise analysis to come. For all the differences in Meillassoux's descriptions, three uses of "correlationism" can be distinguished, which constantly play a role: First, Meillassoux uses the term to refer to a philosophical-historical epoch that is ushered in, above all, by Kant's transcendental philosophical turn, but to some extent was already present in Berkeley's so-called master argument, an epoch which continues to this day in Meillassoux's opinion. Here, correlationism functions for theoretical philosophy, read with Meillassoux, just as a paradigm in Kuhn's sense functions for a science.[3] I consider this epochal thesis, in the globality in which Meillassoux sometimes advocates it, to be both false and dispensable. As to the falsity of this thesis, it suffices to point to the fact that reductionist naturalism was advocated in philosophy of science in the 20th century, as I have shown in my Chap. 4. If such positions, which often attempt to reduce any

[3] Cf. esp. Meillassoux 2016, p. 121: "What we might call "The Era of the Correlate, or of Correlation", opened up by Berkeley—and which, in my opinion, continues to dominate Continental philosophy today—can thus be described as consisting in *two opposite movements* in regard to the absolute: on the one hand, correlationism, which, in the form of scepticism (especially), transcendentalism, phenomenology or postmodernism has denied thought all access to the absolute. And on the other hand, a new metaphysical tendency which, given its importance, I shall from now on designate with a neologism that aims to specify it in terms of what it is at its deepest level: *subjectalism*." and. p. 122: "I call the "Era of Correlation" the antimaterialist, post-Berkeleyan era that has shut us up inside correlation, *either through an antispeculative gesture*—which *alone* I call correlationism—*or through a speculative gesture*—which I call subjectalism—that absolutises the correlation of thought and the world through the choice of various traits, all of which are present in human subjectivity." (Emphasis as in the original.)

form of subjectivity, enjoy unbroken popularity to this day, the global epochal thesis is wrong: contemporary philosophy exists that is not correlationist in any sense.

If Meillassoux is interpreted fairly in philosophy, i.e. *charitably*, the epochal thesis could be weakened and correlationism could be re-evaluated as a concept of epoch. Much has already been argued about this (cf. e.g. Davies 2018). I do not think that the epochal concept of correlationism is appropriate. This conviction, as well as the opposing one of Meillassoux's supporters, are mere philosophical-historical opinions. As far as I know, there is no detailed and meaningful historical investigation to substantiate or refute this broad epochal thesis, at least not by Meillassoux. This is less problematic than many believe, as the epochal concept of correlationism will not prove to be essential for the other two dimensions of meaning that I will elaborate. The epochal thesis concerning the era of correlation is not dealt with further in this book and the debate is not taken up here in its philosophical-historical aspects. I leave this to others working on the modern history of western philosophy.

In addition to the epochal concept, Meillassoux defends correlationism as a package of epistemological and ontological theses on the meaning of correlations, which result in three principal forms here one could also speak of philosophical families: weak correlationism, strong correlationism and subjectalism. This is Meillassoux's systematic concept of correlationism, which is more differentiated than many commentators concede (cf. Golumbia 2016). In what follows, I will discuss the advantages and disadvantages of Meillassoux's systematic concept of correlationism.

First, however, I will consider Meillassoux's third way of using the term. Linked to the systematic concept, according to which a philosophical position is to be classified as (a form of) correlationism if it advocates a certain combination of epistemological and ontological theses about a correlation, Meillassoux also uses the term correlationism for the architecture of a pattern of argumentation or rather philosophical reasoning. I will call it a REASONING SCHEME. This does not refer to a predicate-logical rule of inference, but Meillassoux gives several classifications of a philosophy as correlationism when it advocates a variant of two argumentative moves, which he calls the "correlationist circle" and the "correlationist two-step". Both are not precisely defined arguments, but ideal-typical forms of argumentation that can be used to motivate the systematic theses. The existence of these reasoning schemes leads to the systematic theses of weak correlationism. At no point does Meillassoux make it clear whether both, a correlationist reasoning scheme *and* at least one correlationist thesis, must always be present in order for a given philosophical position to qualify as correlationism.

In the context of the following development of a constructive reading of the term in Meillassoux, I will show that some aspects of his argumentation are rightly subject to criticism. Reflecting on this criticism will lead me to reconceptualise correlationism in a novel way as an analytical tool in my next Sect. 6.2. If the conclusion that this new defensible form of systematic correlationism is justified, then Meillassoux's project can be understood as an attempt to use this tool to criticise the whole of so-called Western philosophy from Kant to prominent postmodern, post-analytical, post-structuralist and phenomenological currents of contemporary philosophy. This would make the systematic dimension (correlationism as a combination of theses and as a reasoning scheme) a prerequisite for the historical dimension (the epochal concept), but not vice versa. For Meillassoux's project to succeed, the systematic dimension would have to be combined with a subdivision of current trends in contemporary philosophy into subjectalism, with weak and strong correlationism based on it. In this way, a weakened version of the epochal thesis (á la "all convincing philosophy in the tradition after Kant is correlationism") might be justified. My work does not pursue this goal, as I consider it to be misguided. Instead, I will analyse and criticise correlationism in Meillassoux first by classifying the function of correlationism within Meillassoux's speculative project, then by illuminating the correlationist reasoning scheme, and finally by presenting the combination of theses that Meillassoux associates with correlationism in its three principal forms (weak, strong, subjectalism).

6.1.2 Correlationism in Meillassoux's Theoretical Project and the Straw Man Objection

In *After finitude*, Meillassoux lists a number of intentions that he pursues with the project of speculative realism. The first central intention, which can gain sympathy from the perspective of the present work, is to reorganise the relationship between the natural sciences and philosophy. In contrast to the reason presented in the context of this book up to this point, Meillassoux does not argue from a motivation within philosophy of science. According to him, a different form of philosophy is necessary than has been practised since Kant, i.e. correlationism, in order to grant scientific discourse its full validity. In his view, the difficult relationship between philosophy and the natural sciences stems from the fact that the Copernican revolution of philosophy—for him, this consists in the empirical proof of the mathematisability of nature provided by physics, which is thus separated from the subjective human being—was answered with a "Ptolemaic counter-revolution" (Meillassoux 2008, p. 118). The latter not only criticised the "spontaneously realist" (Meillassoux 2008, p. 114) discourse, but negated the ability of the natural sciences "to accede to knowledge of a world indifferent to [our] thought's relation to the world" (Meillassoux 2008, p. 118). At the same time, philosophy demands of the natural scientist that she alone "provides us with theoretical knowledge of nature" and superficially recognises "science's promotion over philosophy as guarantor of knowledge" (Meillassoux 2008, p. 120). Subcutaneously, however, correlationism assumes that natural scientists do not understand the deeper meaning of their own discourse. In order to address his challenge to such philosophy fruitfully, Meillassoux casts it in the form of the *problem of ancestrality*.

In the, according to Meillassoux, predominant currents of current philosophy—analytic philosophy and phenomenology—he perceives some similarities in the construction of the arguments against the absolute of any dogmatic metaphysics (cf. also for the following Meillassoux 2008, pp. 29–36). Current correlationism always argues in one form or another with the primacy of a human relation to the world.[4] Meillassoux links this to his first concern by assuming the common ontological-epistemological basic patterns to be the starting point for the gap between the natural sciences and current philosophy.

As a solution, Meillassoux proposes to overcome these basic patterns of correlationism so that his ancestral problem becomes solvable (cf. also for the following Meillassoux 2008, pp. 39–53). He therefore begins to develop his own ontological ideas, which are intended to lead to a new *speculative realism* (Meillassoux et al. 2007). His claim in *After Finitude*, however, is not to "tackle this resolution as such" (Meillassoux 2008, p. 128). In subsequent publications, too, he elaborates on several aspects of a possible speculative realism, but does not claim to have yet presented a conclusion to his project.

Meillassoux aims to open up the entirety of so-called Western philosophy since Kant with the research concept of correlationism in order to prove that it includes certain theses and reasoning schemes that are inherently problematic. His own philosophical approach aims to resolve these difficulties by going beyond the paradigmatic position. Correlationism, therefore, is a concept that has much to contribute within his project. As Meillassoux is aware of the magnitude of his aspiration, a philosophy only needs to have a few basic characteristics in order to count as correlationism. Consequently, there are several variants of correlationism. Meillassoux attempts to categorise them by distinguishing between three basic forms: weak and strong correlationism and subjectalism.

This is where the first objection to any form of a systematic concept of correlationism arises, which is the most discussed issue in the debate surrounding Meillassoux's philosophy, but also in the surrounding speculative realism more generally: Is correlationism as an opponent not in fact a straw man, a position that is not held at all? It is related to the fact that Meillassoux presents highly simplified interpretations of Kant, Ludwig Wittgenstein and Martin Heidegger.[5] Critics attack Meillassoux for reducing various positions beyond recognition. But even if one rejects the epochal concept, the question arises as to whether this systematic family of positions fairly classifies existing philosophies. Christian Thorne argues that in order to make the historical thesis seem probable, Meillassoux reads all examples of correlationism as a form of post-structuralism (Thorne 2012). This misreading subsequently makes it impossible for Meillassoux to situate *himself* correctly in

[4] For an argument that concedes Maurice Merleau-Ponty's phenomenology to be correlationist but tries to argue for its justification cf. Booth 2018.

[5] Against the tenability of Meillassoux's interpretation of Kant cf. Thorne 2012, pp. 280–281; Robin 2021; Schäferling 2022, on Hegel, Wittgenstein and Heidegger cf. e.g. Thorne 2012, pp. 286–288; Gabriel 2022; John and Weidman 2023; Leidlmair 2024; Longo 2020; van den Hoonaard 2023. Meillassoux's main points of reference of the classical authors are Hegel 1979, 1989; Heidegger 1997, 2006b; Kant 1998a,b; Wittgenstein 2006c but one should also consult Heidegger 1976a,b, 2006a,b; Kant 2000, 2009; Wittgenstein 1989, 2006a,b.

this very same tradition. The straw man objection is in the background here, even if Thorne does not deny that various correlationisms can be found in poststructuralist positions.

Another variant of the straw man critique is expressed by Paul Ennis. However, it goes in a completely different direction to Thorne's (cf. also for the following Ennis 2011): One objection that Ennis repeatedly raises is that correlationism as a concept is too broad. He proposes—almost in a reversal of Thorne's argument—to reduce correlationism to its core, which is still demonstrable in large parts of current philosophies: the transcendentalism of Kant and Edmund Husserl.[6] Ennis' move results in a version of correlationism as transcendentalism, which could then be successfully criticised and possibly overcome by Meillassoux.

I would like to emphasise two aspects: First, Ennis thinks that the "*correlationist nexus*" (Ennis 2011, p. 38; emphasis as in original; for the conclusion, cf. p. 46) introduces the transcendentalist distinction between an empirical level, on which we as an intersubjective community can explore nature, and a transcendental level, on which we can recognise the conditions of possibility of the empirical level through philosophical reflection. Secondly, he therefore considers correlationism to be synonymous with epistemological anti-realism in so-called Continental philosophy (cf. Ennis 2011, p. 47, whereby the answer to the question of what Continental philosophy is supposed to be is left to the reader). In contrast to Thorne, for Ennis Meillassoux's critique and classification of Kant, Husserl and (implicitly) German Idealism as correlationism is fully correct.[7] However, the criticism of so-called Continental contemporary philosophy is only justified insofar as it represents epistemological-anti-realistic transcendentalism in the sense of Ennis' level distinction.

Both Ennis and Thorne do not separate Meillassoux's historical and interpretative theses from the systematic nature of correlationism. This is not an unfair treatment of Meillassoux, as he makes just as few distinctions in many places himself. In what follows, this will become clear in the analysis of Meillassoux's variants of correlationism. One task that emerges from this for me is to show how a sober, precise and systematic concept of correlationism avoids this error, which gives rise to such different assessments as those of Ennis and Thorne. The point here is not to show that some authors who are generally categorised as post-structuralists or postmodernists, such as Jacques Derrida, do (or do not) advocate correlationism (cf. on such a discussion Hägglund 2011). Instead, I will use my examples in the subsequent subchapters to show that correlationist ESA, OSA and realist correlationism are advocated, and why they are not suitable as an alternative to OSR in philosophy of science due to the ancestral argument against correlationism.

A form of justified criticism of Meillassoux can be distilled from the aforementioned and other straw man accusations: The concept of correlationism is too imprecise. This is underpinned by Gabriel's objection that Meillassoux leaves core

[6] Cf. for an assessment of Husserl as a correlationist Bös 2024; Romano 2023.

[7] For research on Fichte and Meillassoux, cf. Longo 2015a, pp. 151–154.

questions about his position, such as "what exactly is correlationism meant to be?" (Gabriel 2023, p. 74) unanswered. In addition to the confusion of historical, interpretative and systematic considerations and the ambiguous formulations mentioned above, the deeper ambiguity of Meillassoux's concept is due to the "ism" problem already briefly alluded to in the introduction to my second chapter: his use of philosophical *labels* is too abstract. The "empirist", the "subjective idealist", the "speculative philosopher" (Meillassoux 2008, p. 25, 55 and 56) and others form the circle of participants in Meillassoux's dialogical argumentations. However, precise definitions of these figures are lacking, as is a genuine link back to the analysed positions that goes beyond the citation of a work or a name. The level of abstraction in *After finitude* is so high that the straw man objection arises. Meillassoux sometimes recognises this problem himself:

> By showing how it is possible to counter this argumentation, we shall outiine the contours of a contemporary speculation, which would consider that any problematisation of this model would give us reason to hope for the consecutive probiematisation of any philosophy whatsoever that would reject the possibility of thinking what there is when there is no thought. [...] But the counter-proof of this postulate could not consist in anything other than a specific confrontation with the main representatives of these various currents, although this cannot be undertaken here. Meillassoux 2011b, pp. 13–14.

This excerpt from Meillassoux's 2011 essay *Metaphysics, Speculation, Correlation* also applies to *After Finitude* and other publications. Even if my aim is different from Meillassoux's, a consideration of the systematic position of correlationism in his work and its subsequent development can be understood as an impetus for defending systematic correlationism as an analytical tool against the straw man objection. Since it becomes clear that the concept I have further developed is appropriate for my examples, mainly drawn from philosophy of science, I can reject the straw man objection at least metalocally, as I will show in Sect. 6.3.

6.1.3 Correlationist Circle and Two-Step

Before the concept of correlationism can be developed further, it must be reconstructed with regard to its systematic definition in Meillassoux and criticised with literature. For this task, it is useful to shed light on the reasoning schemes that, according to Meillassoux, lead to the systematic position of correlationism in its three principal forms and which thus underlie it, as will be shown in the following sections. The term *reasoning scheme* does not refer to a specific logical pattern of argument, but to the similar architectures of certain argumentations that make use of the same epistemic circumstances of the human being in order to attack the ontological criterion of subject independence. In *After finitude*, Meillassoux calls them the correlationist circle and the correlationist two-step.

I begin with "the correlationist circle" (also for the following Meillassoux 2008, p. 35). Meillassoux sometimes gives it the form of a critical question: "For by what miraculous operation is [...] thought able to get out of itself in order to verify that what is unthinkable for us is impossible in itself?" According to his position, the correlationist has in mind that every argumentation in favour of an absolute being only makes statements about this being from our perspective. It is therefore only knowable as an "absolute" for us. However, if it is bound to us and our perspective, it is no longer an absolute in the sense of dogmatic metaphysics, which would have to exist subject-independently. In the vocabulary I use here, Meillassoux's *absolute* can be described as a strictly subject-independent entity or as a strictly subject-independent principle or law that exists ontologically necessarily. He introduces the two reasoning schemes in *After finitude* as follows:

> If one calls "the correlationist circle" the argument according to which one cannot think the in-itself without entering into a vicious circle, thereby immediately contradicting oneself, one could call "the correlationist two-step" this other type of reasoning to which philosophers have become so well accustomed – the kind of reasoning which one encounters so frequently in contemporary works and which insists that "it would be naïve to think of the subject and the object as two separately subsisting entities whose relation is only subsequently added to them. On the contrary, the relation is in some sense primary: the world is only world insofar as it appears to me as world, and the self is only self insofar as it is face to face with the world, that for whom the world discloses itself [...]". Meillassoux 2008, p. 5; the quotation cited by Meillassoux is from: Huneman and Kulich 1997, p. 22.

This consideration is the outline of a train of thought that can be pursued in different ways to support different theses in different debates. The core of the circle is the assumption of a self-contradiction of strict subject independence: "If I say or simply think (formulate to myself) that something exists absolutely, that is to say independently of me [...], then this is a thought content linked to the speech and/or thought I have just produced." (Meillassoux 2016, p. 118.) The two-step consists in following on from the first argumentative step, according to which there is a two-way dependency between the subjective side of access and the side of the dependent entities, however exactly this is realised. At the point outlined above, Meillassoux cites a phenomenological consideration by Philippe Huneman and Estelle Kulich, which, however, cannot be applied to individual cases.[8] He formulates this more generally elsewhere:

> Correlationism rests on an argument as simple as it is powerful, and which can be formulated in the following way: No X without givenness of X, and no theory about X without a positing of X. If you speak about something, you speak about something that is given to you, and posited by you. Consequently, the sentence: "X is", means: "X is the correlate of thinking" in a Cartesian sense. That is: X is the correlate of an affection, or a perception, or a conception, or of any subjective act. To be is to be a correlate, a term of a correlation [...] That is why it is impossible to conceive an absolute X, i.e., an X which would be essentially separate from a subject. We can't know what the reality of the object in itself is because we can't distinguish between properties which are supposed to belong to the object and properties belonging to the subjective access to the object. Meillassoux et al. 2007, p. 409.

[8] Cf. for a reading of correlationism as central to phenomenology Alloa 2023.

Only the correlationist circle is depicted here, not the two-step. At other points, such as in the excerpt from *After finitude* quoted above, Meillassoux's description of the circle appears as though it were a purely epistemological thesis or a thesis of conceivability is being substantiated by it: We must limit our knowledge claims to what is graspable by our perspective and thinkable by human limitations of cognition. However, this is not the ultimate goal of the correlationist-circular reasoning scheme, because Meillassoux does not oppose only this thesis. Rather, every ontological assertibility and conceivability of a strictly subject-independent entity is denied with reference to an epistemic, linguistic and/or mental, in any case subjective, limitation. This is the human finitude that Meillassoux wishes to counter. Levi Bryant's description in his lexical presentation of correlationism is therefore also correct: "Correlationism is not merely the thesis that we must relate to something in order to know it. Obviously we had to discover dinosaur fossils to know anything of the past existence of dinosaurs. For Meillassoux, it is not relation per se that makes a position correlationist, but rather the assertion of a very specific, unsurpassable relation." (Bryant 2015, p. 47.) What is therefore decisive lies in the fact that, according to correlationism, the epistemic restriction, the subject-dependence of the known, must be inscribed in every ontological statement about a given entity. This is why correlationism contradicts theories that operate with primary qualities; therefore, according to it, the absoluteness of an entity is not something that we could know.

Meillassoux's correlationist circle finds an equivalent, for example, in Demmerling's characterisation of phenomenalist and constitutional anti-realism: "The relevant fundamental thesis is that reality is only accessible via the conscious states of the subjects of knowledge (or in Berkeley's words: *esse est percipi*), and that the subjects of knowledge can never step out of their (conscious) experiences in order to encounter reality directly." (Demmerling 2004, p. 32, my translation.) This is the fundamental form of epistemological anti-realism in the variant that attests a correlation (of the subject of knowledge and the object of knowledge). In philosophy of science, reference can be made here to forms of verificationism, e.g. in Carl Gustav Hempel, but also to instrumentalist proposals such as Duhem's (cf. Duhem 1914, p. 185; Hempel 1950). The basic idea is that all these positions use the correlation circle as a critical, epistemological reasoning scheme in order to invoke a correlation of access between the subject of knowledge and the object of knowledge in science.

> Phenomenalist anti-realists assume that propositions or beliefs that refer to the world can only be justified internalistically, but not externalistically. In generalised form, the idea underlying this position can be outlined as follows: Whatever is actually known of the world is known solely on the basis of one's own sensations, perceptions and experiences. A reality beyond these experiences is unattainable or unknowable for us and in this sense it is our own experiences or justification practices that are constitutive for what is called "reality".
> Demmerling 2004, p. 33, my translation.

Applied to philosophy of science, this could mean that the natural external world as the goal of our theorising practices is inextricably intertwined with them. The natural external world can be understood as a heuristic fiction or as an underdetermined hypothesis per se, which plays an important role in our theorising practices as an instrument for dealing with phenomena that are only ever given individually. Conversely, a phenomenon that we locate in the natural external world in this way is constitutive for us as biological beings. This is only one exemplary application in philosophy of science; others are easily imaginable.

Penelope Maddy introduces the "comparison problem" in philosophy of language with a question similar to Meillassoux's correlationist circle: "how can we compare language with raw reality?" (Maddy 2007, p. 101; as the creator of the term Maddy credits Marino 2006) or also "how can we compare our terms with raw, unconceptualised reality?" (Maddy 2007, p. 176.). For Maddy, the comparison problem, based on Mark Wilson's argumentation (cf. Wilson 2006), is the starting point for arguing in favour of a moderate and historically informed deflationism in the theory of truth and meaning and against a correspondence theory. The correlationist circle in the guise of the comparison problem is thereby turned against a linguistic-philosophical form of representationalism in which the natural external world is strictly subject-independent and must be correctly or less correctly represented by our language. According to Maddy, this strict subject-independence makes it impossible to reach the natural external world entities linguistically. Maddy utilises the comparison problem by demonstrating, using the causal theory of reference, that here the verification of successful reference is only possible if we could step out of our theories about the world in order to test the relationship between (scientific) language and extralinguistic reality by direct comparison (cf. Maddy 2007, p. 126).

Maddy defends her moderate, historical deflationism in *Second Philosophy*, a work the aim of which is a new justification of naturalism and a realist position in logics, philosophy of science and philosophy of mathematics (cf. Maddy 2007, esp. pp. 303–391). It demonstrates two things: The observation takes the comparison problem as a stepping stone can be taken as a first indication that the correlationist circle, when used in a metalocal debate—as here, in philosophy of language—need not entail a correlationist position in all metalocal debates. This is true on the assumption that Maddy's position in logics, philosophy of science and philosophy of mathematics is not a correlationist one. The fact that the correlationist circle occurs in this debate and with this author can be interpreted as a refutation of the objection that correlationist lines of thought are unknown in so-called analytic philosophy and—more importantly—that the correlationist circle cannot be demonstrated outside of Meillassoux's (questionable) interpretations of Kant, Wittgenstein or Heidegger (cf. Blok 2017).

Further uses of the correlationist circle or two-step could be cited and Meillassoux's own example of the event in Heidegger could be discussed. However, neither of these would significantly advance Meillassoux's analysis of correlationism as a systematic concept. To summarise, the correlationist circle has been reconstructed as the reasoning scheme that derives from the problem of comparison, according to which by asserting a strict subject-independence of a given entity, the asserting philosopher runs into a self-contradiction or similarly serious tenability problems with his own theory. The correlationist two-step presupposes the circle and continues it by stating that there is a reciprocal dependency relationship between the given entity and the knowing subject, e.g. a certain entity depends logically, epistemically or ontologically on the subject, but the subject depends causally on the entity.

6.1.4 Weak Correlationism: The First Decision

According to Meillassoux, correlationism uses these correlationist reasoning schemes of circle and two-step to motivate its first guise, which he calls *weak correlationism*. According to his position, the dialectic that then unfolds is due to the weaknesses of this first form, which make it seem possible to recover the ontological absoluteness lost from the point of view of dogmatic metaphysics.[9] However, this absoluteness is not the natural external world, but the correlation itself. Meillassoux calls such positions *subjectalism*. Now, according to Meillassoux, a more radical variant of the correlationist reasoning scheme was devised in the further course of dialectics in order to say goodbye to the subjectalist absolute. Meillassoux calls any philosophy that radicalises itself in this way *strong correlationism*. As with weak correlationism, the question here is which systematic theses characterise such positions.

Meillassoux calls correlationism any form of philosophy in which it is claimed that we only ever have access to the *correlation* of subject and object, language and referent, mind and world, consciousness and being, or a similarly fundamental dyad, triad, etc.: "We will henceforth call *correlationism* any current of thought which maintains the unsurpassable character of the correlation so defined." (Meillassoux 2008, p. 5.) Meillassoux means this philosophical unsurpassable or inviolability when he writes that the chosen correlation is *primary*. Any form of pre-critical metaphysics that is dogmatic in the sense that it presupposes the existence and knowability of strictly subject-independent entities is to be rejected from the perspective of correlationism. Meillassoux calls the existence and knowability of a strictly subject-independent entity an *absolute*. In his usage, however, an *absolute being* denotes a strictly subject-independent, concrete entity that is independent of "the "naïve" stance of dogmatic metaphysics" (Meillassoux 2008, p. 3)—thus naïve realism—in the sense of an *unjustified and accepted* dogma.

He establishes the difference between weak and strong correlationism according to how relentless its "most radical refutation of any attempt to think an absolute" (Meillassoux 2008, p. 30) is in the end. For him, this means nothing other than: How consistently is the primacy of the correlate defended? According to Meillassoux, weak correlationism argues against dogmatic metaphysics with the primacy of the correlate, but the correlationist arguments are not fully developed. Kant serves as an example of weak correlationism in *After Finitude*, as Kant's position forbids any

[9] Cf. for a lexical overview of weak and strong correlationism Austin 2015, however, the interpretation of strong correlationism presented differs significantly from my reading.

positive knowledge of the thing-in-itself, "but maintains the thinkability of the in-itself." (Meillassoux 2008, p. 35.) Strong correlationism criticises this inconsistency in the weak model and thus develops it further.

Meillassoux attempts to demonstrate this in order to make clear the difference between weak and strong correlationism. To this end, he explains how Kant—as a weak correlationist—argues against René Descartes' ontological proof of God (cf. Meillassoux 2008, pp. 29–30). Descartes attempts to prove the existence of an absolute, namely God, with the ontological proof of the existence of God.[10] He tries to prove that a non-existent God is a contradictory concept. By definition, God is absolutely perfect. The predicate of existence, i.e. the property of existing, is something that counts as perfection. If an object A has all the properties S_1 to S_n, but lacks the property S_x to exist, it is less perfect than an existing object B with the properties S_1 to S_n and S_x. Since God is completely perfect, he must exist.

The weak correlationist Kant, in Meillassoux's reading, argues against Descartes' ontological proof of the existence of God by accusing him of a "sophistical" (Meillassoux 2008, p. 30) move. Descartes uses an illicit argumentative manoeuvre to prove God: The property of existence is not a predicate at all.[11] Existence is attributed to an object independently of its other determinations. God's existence is presupposed within the proof, so the argument is therefore viciously circular. A contradiction is only ever possible between a presupposed object and one of its predicates. The existence or non-existence of God cannot lead to a contradiction with the perfection of God, as existence and non-existence cannot lead to a contradiction with any predicate whatsoever.

Why does Kant invalidate Descartes' argument in this way? Meillassoux diagnoses it as follows:

> The nub of the Cartesian argument lies in the idea that the notion of a non-existent God is inherently *contradictory*. [...] But in order to disqualify this argument, it is imperative for Kant to demonstrate that there is in fact no contradiction involved in maintaining that God does not exist. For if there was, it would be necessary to concede that Descartes has effectively attained an absolute. Why? Because although the author of the *Critique of Pure Reason* maintains that the thing-in-itself is unknowable, he also maintains that it is *thinkable*. For Kant effectively allows us the possibility of knowing a priori that logical contradiction is *absolutely* impossible. [...] Consequently, for Kant, the following two propositions have an absolute ontological scope:
>
> 1. The thing-in-itself is non-contradictory.
> 2. The thing-in-itself exists, otherwise there would be appearances without anything that appears, which for Kant is contradictory. Meillassoux 2008, p. 31; on the conceivability of the thing-in-itself he refers to CPR/KrV, B XXVII.[12]

[10] Meillassoux takes as a basis here: Descartes 1996a, 6th Meditation, pp. 71–90; and Descartes 1996b, 2th Part, 1st Article, pp. 40–41.

[11] Cf. Kant 1998a,b, A 594/B 622.

[12] For a more detailed analysis of this particular point in Kant, which is quite similar to Meillassoux's, cf. Kreis 2023.

Meillassoux claims that Kant emphasises the "sophistical character" (Meillassoux 2008, p. 31) of the ontological proof so that he can refute the dogmatic absolute but continue to grant absolute scope to the forms of logic in order to determine the thing-in-itself and prove its existence. According to Meillassoux, Kant's refutation goes "well beyond its specific critique of the Cartesian proof" (also for the following Meillassoux 2008, p. 32). Rather, it is the prototype of the argumentation of weak correlationism in "refuting every proof that would presume to demonstrate the absolute necessity of a determinate entity." Thus, weak correlationism could rebut any form of dogmatic metaphysics—the rationalism of a Descartes' or ancient atomism, Platonic idealism, philosophy of substance in the sense of Baruch de Spinoza or John Locke's empiricism.[13]

According to Meillassoux, weak correlationism can argue in the way described because of a common flaw in these different philosophies: All these philosophies claimed that a being, "that this or that – i.e. some determinate entity – must absolutely be, and be the way it is" (Meillassoux 2008, p. 32). The natural external world, for example, is necessarily subject-independent: "Idea, pure Act, atom, indivisible soul, harmonious world, perfect God, infinite substance [...] viz., in the claim that this or that entity must absolutely be *because* it is the way it is." (Meillassoux 2008, p. 32–33.) For this reason, it is possible for weak correlationism to repeatedly use the reasoning scheme of the correlationist circle, reconstructed by Meillassoux as Kant's counterargument, as *the scheme* for a rebuttal of such positions.

With the reconstruction of the refutation of the Cartesian ontological proof by Kant, Meillassoux makes clear, regardless of whether his interpretative and historical interpretations are to be agreed with, which part of an ontology still eludes the primacy of the correlate in weak correlationism: at the least, the laws of logic. It is also conceivable that laws of nature and mathematical truths would be declared to be exceptions. According to Meillassoux, this is a possible point of criticism of weak correlationism.

Meillassoux's systematic considerations, which were reconstructed using his example, show his view that weak correlationism—and the dialectical development of correlationism as a whole—always requires both the correlationist circle as a refutation of strict subject-independence and, later, the two-step as an argument for reciprocal conditionality or dependence. This answers the question raised in the previous section concerning whether correlationism could already exist from a systematic point of view if only the correlationist circle is advocated or whether further theses or arguments are necessary. The second option is correct, since circle and two-step together justify at least a weak correlationism—it needs the correlationist circle, the two-step and the primacy of correlation. This means that even if completely different correlations can be imagined, there is always a common systematic core in weak correlationism, the primacy of correlation.

[13] Cf. Meillassoux 2008, p. 32–33; for these philosophies cited by him, cf. especially de Spinoza 2002a,b, 2018; Locke 1975; Plato 1977a,c,d; Taylor 1999.

The distinction between weak and strong correlationists is already implied in this demarcation. Meillassoux's analysis of both correlationist models reveals certain main theses. He calls these central propositions the "two decisions" (Meillassoux 2008, p. 36) of correlationism. Weak correlationism only takes the first decision, whereas strong correlationism is characterised by a positive attitude towards both decisions. With the term *decisions*, Meillassoux emphasises that these two theses are embedded in different concepts (consciousness, language, mind, discourse, etc.) in the various correlationist philosophies and are therefore heterogeneous. However, they can be determined functionally: The first decision has the function of forming the conclusion of the correlationist circular reasoning scheme against the assertion of the necessity of an absolute and a radically subject-independent entity. This is the starting point for the different variants of the correlationist two-step that are connected in order to secure the first decision. Through the culmination of both reasoning schemes, the first decision gives the most complete basic definition "of all correlationism" (Meillassoux 2008, p. 36) in Meillassoux. Every philosophy that belongs to correlationism in Meillassoux's sense takes the following 1st *decision*:

> [T]he correlationist perspective insists that it is impossible to abstract from the real the fact that it is always-already *given* to an entity, and that since it is not possible to think anything that would not always-already be given-to, we cannot think a world without an entity capable of receiving this givenness; that is to say, without an entity capable of "thinking" this world in the most general sense, i.e. intuiting it and discoursing about it. We will refer to this first decision [...] as [...] "the primacy of the correlate". Meillassoux 2008, p. 37.

The primacy of the correlate is the quintessence of any weak correlationism in Meillassoux. The thesis of primacy states that nothing is conceivable that is not always already part of the correlation—whereby a wide and diverse range of options are available here for such a correlation. Hereby a thesis of global or metalocal ontological or epistemological subject-dependence of all corresponding entities is advocated. Exactly what the argument for this looks like depends on the correlation for which it is supposed to provide a justification.

Meillassoux does not elaborate on the consequences of this 1st decision. Instead of providing an interpretation, I will reconstruct Meillassoux's dialectic based on this version of weak correlationism, which makes it clear that Meillassoux sees this model only as a step towards a stronger one. Together with the reflection on the points of criticism of Meillassoux's approach, my analysis will offer the possibility of further developing the concept of correlationism. To this end, I will later return to the interpretation of the 1st decision.

6.1.5 Subjectalism: The Return of the Absolute

The second decision only becomes intelligible if we follow the dialectic path of argument laid out by Meillassoux. According to him, the defenders of absoluteness and of the necessity of the subject-independent attack weak correlationism. Meillassoux describes attempts to think the absolute with a correlation in view as *subjectivist metaphysics* (cf. Meillassoux 2008, p. 38) and in later writings as *subjectalism* (Meillassoux 2016).[14] The basic endeavour was on the one hand to take weak correlationism and the argument of the correlationist circle seriously, and on the other hand to try to arrive at an absolute.

The argumentative tactic used for this purpose "consists in *absolutizing the correlation itself*." also for the following Meillassoux (2008), p. 37; emphasis in original. The subjectalist, according to Meillassoux, now argues that a certain correlation *is* absolute. The subjectalists differ from Meillassoux's dogmatic metaphysics in that they do not assert an absolute being, but an absolute correlation. Unlike in OSR, it is not the natural or the only real external world that should be regarded as subject-independent in the first instance. Instead, in subjectalism the correlation itself is necessary and subject-independent. According to Meillassoux, the argument always runs something like this: The in-itself, the subject-independent entities of weak correlationism, is "not only unknowable, but also unthinkable." It is therefore abandoned, but its place is filled anew "so that only the relation between subject and object remains, or some other correlation deemed to be more fundamental." This results in the inception of a new kind of correlationist metaphysics, in which a necessary, absolute correlation takes centre stage:

> A metaphysics of this type may select from among various forms of subjectivity, but it is invariably characterized by the fact that it hypostatizes some mental, sentient, or vital term: representation in the Leibnizian monad; Schelling's Nature, or the objective subject-object; Hegelian Mind; Schopenhauer's Will; the Will (or *Wills*) to Power in Nietzsche; perception loaded with memory in Bergson; Deleuze's Life, etc. Meillassoux 2008, p. 37.

The question of whether these interpretations would hold up if Meillassoux had realised them is not part of my investigation, for the reasons already explained.[15] Here, what is relevant is how Meillassoux describes subjectalism. Every subjectalist argues for the following theses, according to Meillassoux:

"1. Nothing can be unless it is some form of relation-to-the-world [...].

2. The previous proposition [1.] must be understood in an absolute sense, rather than as merely relative to our knowledge." (Meillassoux 2008, p. 37).

The strategy of justifying such an ontological thesis need not always be the same. Proposition 1. emphasises a correlation. In subjectalism, as in weak correlationism, it is to be asserted that every kind of relation of a human to the world is grounded in this correlation. It forms the core of every access to reality, which takes precedence over ontological assertions about reality. However, this proposition goes beyond the 1st decision in that it not only assumes an epistemological condition for the thinkability of an entity, but also imposes an ontological restriction. It applies to

[14] Cf. Meillassoux 2016, p. 121–122: "Subjectalism always consists, beyond its many variants, no less numerous than those of correlationism, in absolutising subjectivity or certain of its traits, so as to maintain post-Berkeleyan antimaterialism without renouncing the speculative status of philosophy."

[15] In order to substantiate or refute Meillassoux's interpretative assertion, more detailed analyses of key works with regard to the possibility of a subjectalist reading would be necessary: Bergson 1991, 2013; Deleuze 1994; Deleuze and Guattari 1987; Hegel 2010; Leibniz 1965; Nietzsche 1966, 1978, 1998, 2001; Schelling 2000, 2004, 2012; Schopenhauer 2010, 2018.

all entities that they correspond to a certain type of relationship to the world, i.e. that they must be part of a correlation in order to exist. Meillassoux gives alleged examples of such propositions 1 in the quotation on the various philosophies from the 19th and 20th century.

In Meillassoux's view, proposition 2 is the unique feature of all subjectalisms. It states, firstly, that the preferred correlation *must* necessarily be the foundation of any possible access to reality and not just that it factually *is* ours; and, secondly, that the ontological restriction of all entities to the correlation is necessary and subject-independent. This means that all subjectalism goes beyond weak correlationism. A subject-independent absoluteness of correlation is postulated and justified. These justifications differ from each other just as much as the preferred correlations. Meillassoux, therefore, does not reproduce an ideal argumentation or even another reasoning scheme, but restricts himself to the reproduction and explanation of the thesis of subjectalism in propositions 1 and 2.

6.1.6 Strong Correlationism: The Second Decision

In the next step of his dialectical argument, it becomes clear that for Meillassoux correlationism does not consist in anti-realism, as Ennis claims, but is always argued from an anti-absolutist motivation, for subjectalism cannot remain unchallenged in Meillassoux's narrative, since subjectalism justifies an absolute, something necessarily independent of the subject. Meillassoux assigns to strong correlationism the task of refuting subjectalism by radicalising the primacy of the correlate and removing the boundaries of the correlationist circle and two-step. Against the background of the critique of Descartes's ontological proof of God, Meillassoux once again explains the extent to which the strong model takes this further step.

According to Meillassoux, the strong correlationist's objection to Descartes can be summarised as follows: "Or to put it more straightforwardly: because absolute necessity is always absolute necessity for us, necessity is never absolute, but only ever for us." (Meillassoux 2008, p. 31.) This is what strong correlationism argues against the ontological proof of God, since Descartes is concerned with establishing the necessity of an absolute, but the attempt is doomed to failure.[16] What distinguishes this argument of the strong correlationist from weak correlationism's attempt at the proof of a performative self-contradiction or sophism is the fact that it utilises the *form* of the Cartesian statement. According to Meillassoux, this happens when one succeeds "by unmasking the vicious circularity inherent in every attempt at absolutization, without regard for the substance of the arguments put forward." (Meillassoux 2008, p. 31.)

[16] I thank Jens Rometsch for pointing out that the historical Descartes did not believe that the existence of God depends on his proof, but instead only wanted to show that atheists are irrational.

Strong correlationism does not stop at the formerly *absolute*, i.e. subject-independent and necessary, forms of logic: They too cannot escape correlation. This circumstance limits their validity. The laws of logic, causality, and even the laws of nature are only laws *for us*, in the eyes of Meillassoux's strong correlationist.

The argument for this is simple: No one can step outside of oneself (one's own thinking, own language and culture, the prevailing discourses, one's own consciousness) in order to compare a world without a relationship to us with a world with a relationship to us (cf. Meillassoux 2008, p. 35). It is impossible to stand beside oneself and examine the world for real contradictions, for example. That there are no real contradictions is thus a truth *for us*—a limited truth (cf. for further critical thoughts on this example e.g. Moss 2024; Priest 2005). Here, again, the comparison problem and the correlation circle are present in an intensified form.

According to Meillassoux's strong correlationist, when we speak of ontological, logical, natural necessities in general, we never mean an absolute necessity of the world as a subject-independent reality or natural external world, but only an intersubjective necessity. The correlation completely encloses us and the knowable, intelligible world. Meillassoux uses various metaphors for this. A strong example, with more than a nod to Wittgenstein, is when Meillassoux's strong correlationism "confines itself to thinking the limits of thought, these functioning for language like a frontier only one side of which can be grasped." (Meillassoux 2008, p. 41.) In this picture, the correlationist is always on the inner boundaries of her own access to the world. For her, nothing can be said about what lies outside of the boundaries—not even that the unthinkable cannot be there. Meillassoux's talk of the *Grand Dehors* also refers to this. The strong correlationist recommends quietism towards a subject-independent reality and agnosticism towards fundamental laws of nature or logic that apply independently of our practices.

From Meillassoux's point of view, it is clear that the argument of strong correlationism against the ontological proof of the existence of God is the prime example of a radical refutation of any form of the absolute. Once again, we are dealing with a variant of the correlationist circle: The problem of comparison forms the starting point of the argument. However, this reasoning scheme is not only applied to an entity of the natural external world or the natural external world as a whole, but, with strong correlationism, to *all* domains, including logical laws. This dissolution of the correlationist circle shatters more than just dogmatic metaphysics:

> According to Kant, we know a priori that the thing-in-itself is non-contradictory and that it actually exists. By way of contrast, the strong model of correlationism maintains not only that it is illegitimate to claim that we can *know* the in-itself, but *also* that it is illegitimate to claim that we can at least *think* it. Meillassoux 2008, p. 35.

Thus, the correlationist argument itself turns against weak correlationism. As a consequence, the in-itself of weak correlationism has become untenable. For it is contradictory to claim that we can think the in-itself, even if we cannot know it:

The in-itself, a strictly subject-independent entity, cannot be thought according to Meillassoux's strong correlationist, since thinking always already stands in relation to it. Strong correlationism no longer allows us to presuppose the law of non-contradiction in order to conclude the absolutely necessary existence and unknowability of the thing-in-itself (or of another "in-itself"). Meillassoux summarises the argument as follows: "for there is no way for thought to reject the possibility that what is meaningless for us might be veridical in itself." (Meillassoux 2008, p. 36). For even the law of non-contradiction as a law of thought or a law of reality cannot claim subject-independent validity for the strong correlationist. It applies to us as a law of logic, but cannot be justified as such from the outside, because we cannot step outside the laws of logic and our perspective. Therefore, it becomes impossible to *infer* an in-itself a priori. Only a subject-dependent in-itself would be justifiable, but this is neither in line with weak correlationism—as there appears to be a contradiction here—nor is it necessary from the perspective of strong correlationism.

Meillassoux's differentiation of correlationism into a weak form, to which he assignes Kant, and a strong form, which he currently considers predominant in analytic philosophy, post-structuralism and phenomenology, draws a line between two levels of radicality in contemporary philosophy. In the weak model, the correlationist circle and two-step is applied to the natural external world and the humans in it, but the possibility of an in-itself of the correlation, i.e. subject-independent and necessary entities such as absolute laws or principles (be they logical, ontological, or natural) is admitted. The strong model is given in every philosophy that subjects *everything* to the primacy of the respective correlation, including all laws and principles. Strong correlationism "prohibits most decisively the possibility of thinking what there is when there is no thought." (Meillassoux 2008, p. 36). Meillassoux uses Martin Heidegger and Ludwig Wittgenstein as examples for strong correlationist positions. Since I am not attempting here to examine which philosophical positions fall under Meillassoux's concept of correlationism in a critical reading, I will refrain from such discussions, despite the straw man objection.[17] Meillassoux attempts to describe a common structure of these diverse

[17] Many proponents of (post-)structuralism and so-called postmodernism are explicitly or implicitly assigned to strong correlationism by Meillassoux: Jean-François Lyotard takes up ideas from the later Wittgenstein in his reflections on language games, but also draws on phenomenological considerations and can therefore be considered to belong to the strong model, cf. Lyotard 1984, esp. pp. 6–22. Jacques Derrida's considerations on the indissoluble correlation of signifier/signified, which are related to each other by *différance* and in which he sees the concepts of sense, thing, *noëma*, reality, trace and presence suspended; in my opinion, Meillassoux should assign this to strong correlationism, cf. Derrida (1982, 2016). Meillassoux clearly characterises the thinking of Emmanuel Levinas when he writes that current philosophy would not "dismiss[...] Levinas' thought of the "wholly Other" as absurd on the grounds that it is refractory to logic" (also for the following Meillassoux 2008, pp. 43–44) only because Levinas' discourse could break with these forms of logic as a strong correlationism, cf. on this Levinas (1987, 1991). Michel Foucault's concept of discourse, that of subject constitution through subjugation and his correlation of power and knowledge make him a strong correlationist in Meillassoux's analysis, as would no doubt Judith Butler, who, following Foucault, advocates strong correlationist positions in queer theory. On Foucault's concept of discourse (cf. e.g. Foucault 1972, pp. 21–87), on his theory of the constitution of the subject as a "correlative history of the modern soul and of a new power to judge" Foucault 1977, p. 23; on the power/knowledge correlation cf. e.g. ibid., pp. 27–31 and Foucault

philosophies when he ascribes to them the shared core idea of strong correlationism: "*it is unthinkable that the unthinkable be impossible.*" (Meillassoux 2008, p. 41) This corresponds to the second core thesis, which I reconstruct as follows:

2nd Decision
The laws of logic, the ontological principles, causality and the fundamental laws of nature do not have a subject-independent and necessary validity, i.e. they are not absolute. They fall within the correlation. There is therefore only a facticity of the (onto-)logical principles, of causality, of the laws of nature and of the correlation.

The second decision concerns the "facticity of these forms" (also for the following Meillassoux 2008, p. 38) and "no longer pertains to the correlation, but rather to the facticity of the correlation." This is my specification of the already mentioned thesis of strong correlationism according to Meillassoux: "*it is unthinkable that the unthinkable be impossible.*" (Meillassoux 2008, p. 41.) What does this mean?

Facticity initially refers only to the fact that the laws of logic, causality, laws of nature and correlation have the forms that they do for us. However, according to the reasoning scheme of strong correlationism, the boundless correlationist circle, these are no longer absolute facts and truths. They are objectively valid because we can determine their validity intersubjectively. It is even inconceivable to us that our reality is different. However, Meillassoux's strong correlationist would not be able to conclude from this fact that something wholly different is not a real possibility. Quietism, scepticism or agnosticism in relation to questions about what is or could be are, in any radically subject-independent scenario, reactions of strong correlationism to the imperative to remain within the boundaries of our thought and our world, for even these scenarios are developed by us, and just because something is unthinkable, this does not mean anything for the actually unthinkable radically subject-independent content of any such scenario. Conversely, this means that the logical forms, causality, laws of nature and correlation factually *exist*, but do not *exist necessarily*. In this respect, according to Meillassoux, the unthinkable is possible within strong correlationism.

6.1.6.1 Outlook on Meillassoux's Further Argumentation

I conclude my reconstruction of Meillassoux's concept of correlationism with a brief overview of how a positive philosophy follows from this designation of philosophy, which he tries to overcome. The following summary can only be a sketch and I will not use the results of his further argument in my discussion. Nor will I defend it against other assessments (cf. e.g. Benéitez 2022; Habdankaitė 2021). However, the

(1978). On Butler's concept of body and discourse cf. Butler (1990, pp. 79–106), on her theory of the body and subjectification cf. Butler (2011, pp. xii–xx). On Meillassoux's position with regards to Foucault and Deleuze cf. Badiou (2008, p. vii); on the question of Deleuze is a correlationist Novak (2020); Pachilla (2022).

background is helpful for the development of a new scientific realism in philosophy of science. Starting with the dialectic described above, Meillassoux develops some arguments to refute correlationism, especially its first decision. Meillassoux aims to achieve this with a variant of the self-application of the correlationist circle (cf. esp. Meillassoux 2008, pp. 50–63). In the constellation of the dialectical tripartite division into weak, subjectalist and strong correlationism, the strong model can only emerge successfully if a "radicalization of the correlation" (Meillassoux 2008, p. 44) is assumed. Consequently, the objection to subjectalism is that the necessity of this or that correlation cannot be proven, since the means for this—laws of logic, causal relations etc.—in turn presuppose the correlation. This means that the mere facticity of the correlation and the boundlessness of the correlationist circle, which we can ascertain from our perspective, cannot be circumvented.

In order to arrive at an absolute, even if not an absolute being—that is, an absolute entity—, Meillassoux asks the strong correlationist whether her statement about the facticity of correlation is necessarily and subject-independently valid, i.e. has absolute status, or is only subject-dependent. "Accordingly, we must try to understand why it is *not the correlation but the facticity of the correlation that constitutes the absolute*." (Meillassoux 2008, p. 52.) Meillassoux thinks this leads to a dilemma for the strong correlationist. If the proposition about the facticity of the correlation is only true subject-dependently, this would be a relapse into subjectalism, for in that case, the strong correlationist would have to admit, contrary to her rebuttal of the subjectalist, that the correlation might be absolute after all; we just do not know this. Meillassoux's analysis of this dialectic is intended to show that strong correlationism should therefore insist on the absolute validity of the factuality of correlation. But what follows if this horn of the dilemma is pursued?

According to Meillassoux, we thus arrive at speculative realism, i.e. his variant of a new realism in ontology.[18] If the thesis of facticity of the correlation is assumed to be absolute, then, Meillassoux contests, the so-called *principle of unreason* (Meillassoux 2008, p. 60) is derivable: "We must convert facticity into the real property whereby everything and every world *is* without reason, and is thereby *capable of actually becoming otherwise without reason*." (Meillassoux 2008, p. 53). This idea of turning the facticity of the correlation into an absolute, i.e. a necessary and subject-independent principle, forms the foundation of his entire positive proposal in ontology:[19]

> The absolute is the absolute impossibility of a necessary being. We are no longer upholding a variant of the principle of sufficient reason, according to which there is a necessary reason why everything is the way it is rather than otherwise, but rather the absolute truth of a *principle of unreason [principe d'irraison]*. There is no reason for anything to be or to remain the way it is; everything must, without reason, be able not to be and/or be able to be other than it is. Meillassoux 2008, p. 60.

[18] Cf. for a book-length defense of a consequent correlationism Bitbol 2019, and: Thomas-Fogiel 2022, for a subjectalist option.

[19] This is obviously a contested move, and even sympathetic readers such as Khamis 2023, do not often agree that is has the intended consequence.

This seems to leave no stone unturned ontologically. Yet according to Meillassoux, a form of hyper-chaos follows from his principle of unreason, which generates some rules of self-preservation, such as the law of non-contradiction (cf. Meillassoux 2008, p. 62).[20] The reason for this is not readily apparent without considerably more reconstructive work, but it is borrowed from the idea that a contradictory entity would imply its own necessity, which would violate the principle of unreason. Meillassoux's self-diagnosed problem is that the few laws of hyper-chaos cannot easily explain why our reality, especially nature, appears to be stable. Meillassoux therefore poses the crucial question of why nature produces seemingly causal relationships through the local, contingent and limited laws of nature, so that our scientific knowledge is possible through observation and mathematical calculation. This needs to be explained, as these local laws of nature themselves exist contingently and could change in principle, as the principle of unreason leads to the insight that, necessarily, every existing entity is contigent.

On the one hand, Meillassoux thus poses David Hume's problem of induction in a different form (cf. also for the following Meillassoux 2007). On the other hand, this stance allows Meillassoux (in his own opinion at least) to prove the absolute reality of the mathematical properties of natural entities, albeit at the price of giving up any ontological guarantee of stability in both directions of the absolute timeline. These mathematical properties are strictly independent of us, but they are also knowable: They should be viewed as primary qualities.[21] Everything that does not result directly from them and the rules of hyper-chaos falls into the domain of correlation. In other words, it enters the world with us humans and thus remains constantly dependent on us, at least in a weak sense of dependency. According to Meillassoux, this ontological minimalism enables speculative realism to respond to the problem of ancestrality, which I will discuss in my next chapter (for a speculative proposal of his metaphilosophy cf. Orensanz 2020). Before I turn to this, it is time to criticise and refine the concept of correlationism as a analytical tool.

6.2 Systematic Correlationism

Now that the concept of correlationism in Meillassoux's philosophy has been reconstructed, it needs to be advanced and evolved. To this end, I will first discuss some points of criticism aimed at Meillassoux and some attempts at interpretations in the following two Sects. 6.2.1 and 6.2.2. These points lead me in the direction of a refinement towards a *systematic correlationism*. I will outline this concept, which is tailored to the context of this book, in the last Sect. 6.2.3 of this chapter.

[20] Cf. for an alternative stance of proceeding from hyper-chaos Meyerson (2023). For an insightful elaboration of hyper-chaos as Meillassoux's ontological meta-time and thought as an emergent outcome, cf. Longo 2014b.

[21] Cf. de Freitas 2021.

The most widely discussed point of criticism of Meillassoux's correlationism has already been mentioned in the previous subchapter: the straw man objection. The entire preceding reconstruction of Meillassoux was aimed at showing that the distinction between weak and strong correlationism remains comprehensible if it is presented separately from the historical-interpretative theses, the epochal concept, that Meillassoux conflates with correlationism. My previous discussion of Meillassoux's use of the term can therefore already be understood as a clarification and limitation of correlationism. In order finally to dispel the straw man objection and the accusation that the level of abstraction is too high, I will give examples of correlationist realism and correlationist ESA and OSA in the next Sect. 6.3. This can be understood as a practical test of the recast systematic correlationism, showing that it is indeed championed by some.

6.2.1 Critique of Finitude: Is Correlationism an Ontological or Empistemological Thesis?

My refined concept of systematic correlationism should also reflect the pressing criticisms and readings of the debate surrounding Meillasoux's speculative realism. A first, important distinction is as follows. In *Meillassoux's Dilemma*, Joshua Heller and Jon Cogburn mainly criticise his philosophical proposal of speculative realism. The mode of their critique is a dilemma argument, in which they claim: "Meillassoux's own speculative metaphysics is inherently unstable, reeling between a fully correlationist finitsm, on the one hand, and an anti-correlationist infinitude on the other." (Heller and Cogburn 2017, p. 164.) I will not consider their actual argument here, which transfers a consideration from Priest's *Beyond the Limits of Thought* to Meillassoux's work. What is interesting is their interpretation of the two-part definition of correlationism: "(1) Verificationism, the proposition that since we cannot think coherently of reality as being unthought, truth must in principle be knowable; and (2) Finitude, the idea that we cannot think coherently of self-subsistent totalities or absolutes, such as reality as a whole." (Heller and Cogburn 2017, p. 163.) My systematic interpretation has produced a very different result, and I will briefly show why I do not consider this reading to be tenable.

First of all, it should be noted that a verificationist theory of truth and meaning and verificationism in philosophy of science, as proposed in early logical positivism, for example, usually become the basis of an ESA:[22] The truth of a proposition consists in the scientific method of its verification. We can only establish the truth of a scientific hypothesis or theory insofar and to the extent that it can be verified by correct methods. This commonly used definition is not what Heller and Cogburn call verificationism. Instead, they call it the thesis that it is impossible to think reality as unthought, and the further thesis that truth or truths are in principle knowable.

[22] Cf. e.g. Voosholz 2017.

Both of these tenets do not follow straightforwardly from standard definitions of verificationism. However, I evaluate the thesis independently of its (admittedly misleading) name, as it can give an indication of why so many commentators equate correlationism with a form of anti-realism.

The first of the two theses of (1), that it is impossible to think reality as unthought, follows from a correlationism with the correlate of reality and thought. Here, therefore, reference is made to an *example* of a correlationist argumentation. This interpretative step by the authors does not initially seem to be problematic and is also shared by other commentators, such as Peter Gratton: "correlationism is a straightforward view: there is no thinking of reality without *thinking* of it." (Gratton 2014, p. 6; emphasis in original.)

The second thesis of (1), that truth or truths are in principle knowable, seems to me to be a misinterpretation: In Meillassoux's words, strong correlationism entails the thesis that it is unthinkable that the unthinkable is impossible. This implies that there are truths or facts that are not only beyond our current knowledge, but that the possibility must be conceded that an unthinkable—so also unintelligible entities—can exist that we can never know in principle. Even if Heller and Cogburn leave underdetermined what exactly they mean by "truth" and the phrase "truth must in principle be knowable" at this point, this premise does not seem to be a fair reconstruction of Meillassoux's correlationism.

Thesis (2), on the other hand, according to which self-contained and absolute totalities are unthinkable for us, could possibly result logically from certain correlationisms. If such totalities would require that they are absolute in Meillassoux's sense of being necessary and subject-independent, the thesis of their unthinkability (or rather non-existence) would indeed follow from correlationism. Heller and Cogburn, however, use a different sense of absolute: they are concerned with totalities that are all-encompassing. Caution is required here. Although there may be correlationisms that hold antitotality arguments,[23] it does not follow from the more cautious interpretation I have given of weak and strong correlationism, as well as of subjectalism that the rejection of totalities is necessary for them or always follows. Therefore, the thesis that every correlationism contains (2) is not tenable without explicit arguments. However, these are missing from their article.

Thus, Heller and Cogburn largely misunderstand correlationism. What they reconstruct correctly is the example of the correlation of reality and thought, which suggests a form of subject-dependence of everything real. This point needs to be explored further. I mentioned above that Heller and Cogburn roughly follow Gratton's interpretation. Therefore, I do not want to ignore his sympathetic reconstruction:

[23] Some of the antitotality arguments I cite in Sect. 5.3, p. 141–158 may originate from correlationists, but this would need to be attested by corresponding readings of e.g. Nelson Goodman or others.

> While "weak correlationism" argues that we can know only what is phenomenal or in language, while the "in-itself" is thinkable, [...] [s]trong correlationism doesn't just deny that we know some "in-itself" independent of us, but that, simply, there is no "in-itself" independent of us. [...] The correlationists think [...] that there is some X out there, but it's always "for us" [...]. As Meillassoux describes it, they [the strong correlationists; J. V.] argue that we are so trapped on this side of the correlation that [we; sic! J. V.] cannot even think the *noumenal*. All we ever have is the pure givenness of the phenomenal, or, for supposed social constructivists, linguistic or cultural structures. Gratton 2014, p. 26

Gratton's rather detailed discussion of examples, which follows on from the cited passage, serves as a defence of Meillassoux's concept of correlationism. Gratton discusses Kant and Dummett as examples of weak correlationism, and Husserl, Maurice Merleau-Ponty and Heidegger as examples of strong correlationism. For Gratton, weak correlationism is the epistemological thesis that we never have access to a radically subject-independent sphere. This makes it clear that Gratton's weak correlationism largely corresponds to epistemological anti-realism with regard to subject-independent nature and the natural external world, which is often paired with ontological realism, as in his reading of Kant. In his discussion of Dummett's linguistic-philosophical anti-realism, Gratton (cf. Gratton 2014, pp. 22–26) shows that, on the one hand, he considers epistemological anti-realism to be necessary as a basis for Dummett's theory of truth, and, on the other hand, that he thinks weak correlationism actually wants to end or undermine the discussion about ontological realism, since it is pointless to want to say anything substantial about the subject-independent parts of reality if they are unknowable.

Although this reading of weak correlationism has many parallels with mine, it differs in important places: It is true that weak correlationism is directed against far-reaching knowledge claims about the sphere of the subject-independent, and often tries to end or undermine the discussion about ontological realism. Subject-independence as a criterion for the natural external world is attacked with the correlationist circle and two-step, thus rejecting any realism that claims to think of the natural external world as strictly subject-independent. According to Meillassoux, weak correlationism is thus always directed against OR, and thus OSR. It also rejects the assertion of the necessity of the existence of certain entities, in particular abstract and concrete objects. However, this does not mean that every form of realism must be rejected and that an epistemological anti-realism must necessarily be advocated by weak correlationists. The thesis of the primacy of correlation, the rejection of necessary existence and subject-independence as an ontological criterion, does not necessarily imply the rejection of the epistmological thesis (RE) of any realism, that the existing entities can be known by us directly or indirectly in some of their properties. The possibility of a correlationist realism is not excluded in Meillasoux's conception, nor the possibility of a correlationist quietism with regard to the question of realism. I therefore disagree with Gratton's assertion that correlationism is *equivalent* to epistemological anti-realism.

Gratton goes on to impute a similar anti-realist attitude to Husserl, Merleau-Ponty and Heidegger. In his reading, their strong correlationisms are characterised by the fact that they moved beyond the debate about ontological realism—in an even more radical form than Kant or Dummett. This means that in Gratton, as in my analysis, the basic thesis and construction of weak correlationism is preserved in strong correlationism. Gratton's idea of a correlationist epistemological anti-realism and the rejection of the ontological realism question is also a core feature of his reconstruction of the strong model. In Gratton's case, these two aspects fulfil the function assigned to the first decision and the application of the correlationist circle in my reconstruction, i.e. they are transferred from weak to strong correlationism and radicalised there. For Gratton, strong correlationism also involves an ontological anti-realism with regard to nature or the natural external world as a radically subject-independent sphere ("there is no "in-itself" independent of us" Gratton 2014, p. 26). However, this is not an ontological anti-realism in the sense of a stance that proves that nature cannot possibly exist as a radically subject-independent sphere, but it consists in the assessment that it is impossible to think of such a nature or natural external world, that its conception cannot be produced without contradiction. Gratton's strong correlationism consists centrally in the prohibition of ontological speculation about the subject-independence of a nature or the natural external world as such.

On the one hand, this reading matches my reconstruction of strong correlationism, since Gratton emphasises the limitations and finitude of human thought and human knowledge of reality, which according to Meillassoux characterises strong correlationism. On the other hand, even in this case I do not think that this thesis of the prohibition of speculation about the radically subject-independent domain necessarily presupposes an epistemological anti-realism. I even maintain that the categorisation as ontological anti-realism is wrong. The ontological prohibition of speculation also refers to the idea of claiming that we can know that the (RO)-thesis of every realism—certain entities have an independent existence and reality—is *true or false*. Strong correlationism in Meillassoux in my reconstruction contradicts the logic of this debate.

Gratton's account also leaves largely untouched Meillassoux's important point, that in strong correlationism the correlationist circle is boundless and applied to everything in order to justify the second decision. This implies that the laws of logic, ontological principles, causality and fundamental laws of nature cannot have a subject-independent and necessary validity and are therefore not absolute in Meillassoux's sense. It is only from their facticity that the ontological prohibition of speculation about how reality is independent of us arises. Gratton leaves aside the deeper philosophical determination in order to concentrate on the anti-speculative result in his analysis of the authors he picks as his examples. I consider this argumentative background to be an important aspect, for it escapes him that the strong correlationism for the ontological prohibition of speculation does not have to assume epistemological anti-realism, even if it does so in his readings of his five selected authors. Instead, correlationist realism is also conceivable here. This explains why correlationist quietism is an obvious option with regard to the question of ontological (anti-)realism (since there is an easily conceivable correlationist argument that we can neither confirm nor rebut RO). In any case, strong correlationism works against the logic of the dichotomous debate.

In my description of systematic correlationism in Sect. 6.2.3, it is important not to stipulate implicitly or explicitly epistemological anti-realism as a criterion, as Gratton, Heller and Cogburn attempt to do. The mistake here is not primarily that Meillassoux cannot be interpreted in this way (his descriptions of correlationism fluctuate, so I do not doubt that alternative readings to mine may open up, and there is textual evidence that Meillassoux himself equates correlationism with anti-realism). The reason why I want to keep these stances seperated is that otherwise one negates the *specific advantage* of correlationism of being a concept transverse to the dual categorisation of realism and anti-realism. By this I mean that these categorisations are independent of each other. In accordance with philosophical honesty, my interpretation concedes the option that correlationism could in principle undermine the debate, which Gratton emphasises several times as a philosophical secondary goal in his readings of Dummett and Heidegger. The finitude of us humans in correlationism is always also our finitude to produce true and reliable statements about the ontological constitution of nature or even reality.[24]

6.2.2 *The Role of Circle, Two-Step and the Two Decisions*

Toby Lovat's description of correlationism comes closer to my reading of correlationism than Heller's, Cogburn's and Gratton's:

> Correlationism, therefore, has two central theses. The first is that our knowledge (or "grasp") of the world is always relative to us. The consequence of this is that we have no means of determining the extent to which our knowledge and experience of the world are generated by, or correspond to the way things are in themselves. The second thesis is that, despite this limitation, the nature of the thinking subject cannot be understood independently of the world of objects and its encounters as an object among other objects. Lovat 2017, p. 15

According to Lovat, this description applies to all correlationism. In particular, however, it characterises Meillassoux's weak correlationism and his reading of Kant. This is also correct from my interpretation insofar as Lovat emphasises several aspects of the reasoning schemes correlationist circle and two-step as a condition of the first decision: The comparison problem is invoked to argue that our knowledge of an entity domain, specifically the natural external world, is always relative to our access, which makes knowledge of strictly subject-independent things a contradictory idea.

In the last sentence of the quotation, however, Lovat points out the peculiarity of correlationism that the correlationist two-step is followed by another argumentative manoeuvre. This shows why we and our conditions of access cannot be understood

[24] For further considerations on this, especially on issues arising from the correlationist Kant and Wittgenstein, cf. Pier (2023) as an excellent collection, especially the papers by Benoist (2023); Kreis (2023); Priest (2023).

as independent of the natural external world either—it shows our role as objects among other objects. Lovat's definition demonstrates how closely the correlationist circle and two-step are linked to the idea of correlationism. He formulates as theses what I have tried to reconstruct as reasoning schemes in Meillassoux. Lovat ignores the two decisions of correlationism—which is not problematic in the context of his argumentation, a defense of Kantianism. For my development of systematic correlationism in the next step, this means that I must not neglect the special role of correlationist circle and two-step in all considerations about the positive and negative core theses. In contrast to Lovat (2017, 2018), I will try to include them as reasoning schemes in the description and not to reduce them to theses or conclusions.

G. Anthony Bruno focuses on the primacy of correlation, i.e. the first decision, in his defence of correlationism: "It is Meillassoux who coins "correlationism" to name the view that we can only access the mutual dependence of thought and being and that, consequently, neither is explicable in isolation, even by the natural sciences." (Bruno 2018, p. 1.) Dan Zahavi chooses a very similar definition (cf. Zahavi 2016, p. 293). Niki Young's investigation of correlationism is also based on the central point of the first decision, which he attempts to break down into three related theses:

> The first claim [...] indicates that correlationism refers to any philosophical position which (implicitly or explicitly) asserts that it would be impossible to attain knowledge of what being might be like independently of thought or vice-versa. [...] Meillassoux further asserts that, at its most basic level, all correlationism rests on an argument he calls the "correlationist circle", which may in turn be formulated as follows: against the realist's allegation that one can make positive claims about the nature of being as it exists independently of thought, the correlationist would claim that the realist is essentially guilty of circular reasoning, since there can be "no X without givenness of X, and no theory of X without a positing of X", such that X "cannot then be separated from this special act of positing, of conception". The correlationist thus insists that one can only have access to the (cor)relation between thought and being, rather than being in itself. Young 2020, pp. 43–44; the quotation is: Meillassoux et al. 2007, p. 409.
>
> Following from the first claim, Meillassoux's second claim [...] emphasises that, for correlationists, the relation between "thinking and being" is "unsurpassable", such that it would be impossible to move beyond the strict boundary limits of the correlation in order to speak of things as they are independently of the way they are given. Young 2020, p. 44.
>
> Finally, the third claim [...] implies that correlationism represents the implicitly dominant anti-realist dogma of Kant and his successors. Correlationism, as understood by SR, represents a prevalent, implicit, yet specific form of (transcendental rather than subjective) idealism which disavows realism by primarily negating its mind-independence component. To this, one may possibly object that many thinkers are not flat out anti-realists since they do not explicitly deny the existence of whatever may exist independently of thought. Young 2020, p. 45.

There are two similarities and one important difference between my reconstruction and Young's account. Like Bruno, Young starts with the first decision, which is always supported by a variant of the correlationist circle. This agrees with my reconstruction of weak correlationism, according to which the first decision is still present and radicalised in the strong model. Young goes beyond Bruno when he emphasises as a second thesis that the first thesis should be understood as primary. However, the above-quoted realisation of this idea can be interpreted in two ways:

Young either means that the first decision is not only meant to assert the thesis of the respective correlation, but also to defend its philosophical primacy. Young would then claim that the respective correlation would not only be defended, but would also become a fundamental epistemological-ontological principle of a certain philosophy before more specific epistemological or ontological theses define it further. This reading would not contradict my reconstruction of Meillassoux. However, it seems questionable why Young separates the two theses—the thesis of correlation and the thesis of primacy. Since Meillassoux does not differentiate here, and Young does not assert a clear advantage of this differentiation, it does not seem obvious.

Alternatively, it can be meant—in the second reading of Young's second thesis—that correlationism always forbids speaking of entities outside the correlation. In my reading, however, this only applies to strong correlationism; after all, according to Meillassoux, Kant speaks of things-in-themselves and the primacy of correlation only curtails our knowledge of subject-independent things. However, restricting the conceivability or expressability of the subject-independent is specific to Meillassoux's strong model. Since Young's theses are supposed to apply to all forms of correlationism according to Meillassoux, in my view the first reading is recommended.

Finally, Young's third thesis states that correlationism is always an idealistic anti-realism. Young does not define what exactly he means by this, but it seems to mean that an idealist anti-realism in my vocabulary is an epistemological or ontological anti-realism that opposes the independence criterion in the form of subject-independence.[25] The fact that correlationism is directed against subject-independence as a criterion of the real is a conclusion that I was also able to distil from my reconstruction of Meillassoux's use of the term. Therefore, this circumstance will form a core of my redefinition of systematic correlationism.

The objection to correlationism raised in the last sentence of Young's quotation can be resolved if it is conceded that anti-realisms do not necessarily follow from the rejection of subject-independence, since realisms do not have to understand the independence criterion as strict subject-independence. Young's definition of realism, however, implies that this is the case, as in OSR or OR. He thus overlooks the option of correlationist realism. In this way, like Ennis and many others, he not only glosses over the subtleties of Meillassoux's presentation of his concept, but also fails to do justice to the role that correlationists often play in realism-antirealism debates, as explained above in the context of the speculation prohibition. Graham Harman summarises this role as follows: "Whenever disputes arise in philosophy concerning realism and idealism, we immediately note the appearence of

[25] Cf. Padui 2011, who undertakes a defence of correlationism, also interpreting it as epistemological anti-realism, and tries to show that this position does not exclude materialism, which in his view, however, Meillassoux does not recognise. Since Padui's concern is not central to the subject of my present inquiry and he is quite unclear in his definition of materialism, I will not pursue his remarks further here.

a third personage who dismisses both of these alternatives as solutions to a pseudo-problem. This third figure is the correlationist." (Harman 2011, p. 7.) It is true that my examples in the next Sect. 6.3 will show that correlationists can speak in favour of both realism and anti-realism. Harman's observation is correct, however, that there is nothing to be said against using correlationism to try to undermine the debate about realism and anti-realism. Whether this succeeds in individual cases or can succeed in general are different questions. Young's interpretation overlooks the fact that the decision in favour of correlationism is not made on the same level as the decision in favour of realism or anti-realism.

This aspect should be adequately reflected in my new recast of the term, even if I agree with Young that subject-independence must play a role. However, it is necessary to specify this role—as well as the whole definitional practice around the concept of correlationism. This also emerges both in Emanuel Rutten's recent embrace of a strong correlationist stance (in an interpretation close to Harman's cf. Rutten 2024, pp. 4–7) and from David Golumbia's sharp criticism of the concept of correlationism:

> What jars the philosophically informed reader is that the issues thereby raised turn out to be central ones in disciplinary philosophical discourse, which Meillassoux is understood to have eliminated the need to address. What Meillassoux's work appears to do is therefore to license a kind of philosophy ex nihilo and without connection to existing practice, and what it thus engenders is a philosophical discourse that repeatedly fails to respect most of the methods of that practice: to state clearly its contentions, to define its terms, to distinguish between philosophical issues (particularly epistemology and metaphysics), or to demonstrate textually its historical-philosophical assessments. As such, it creates a spectral philosophical orthodoxy that is largely incoherent and that few if any practicing philosophers actually endorse; for not only is analytic philosophy not correlationist insofar as we can make sense of what Meillassoux says about it, but the issues Meillassoux claims have been overlooked in post-Kantian philosophy are instead often the main topics of philosophical discourse. Golumbia 2016, p. 4.

On the surface, this can be understood as a justifiable criticism of Meillassoux's sometimes loose use of words, his strong focus on self-introduced opponents, his unclear formulations or his methodological errors, such as the absence of precise textual work when it comes to evidence of current or historical positions. But this goes further than that. Behind the scenes, Golumbia is suggestively concerned with denying that anyone since Kant has actually adopted a correlationist position. However, this now familiar straw man objection obscures a new point: Given the breadth of Meillassoux's formulations and the multitude of levels of description, it remains unclear what the core definition of correlationism *is*. Meillassoux must allow himself to be asked this question in my reconstruction. The answer should be that in every correlationism the reasoning schemes of correlationist circle and two-step are used to justify the primacy of correlation and to deny subject independence as an ontological criterion.

As much as Golumbia's technical criticism of individual formulations and his complaint about a lack of engagement with current and historical literature is true (cf. Golumbia 2016, pp. 3–6), the renunciation of the *principle of charity* as a hermeneutical tool in the evaluation of a philosophical text is just as unhelpful. This applies all the more to terms such as correlationism, which resonate with a part of the debate (cf. e.g. Backman 2014). Golumbia's radically critical approach seems inappropriate to a genuine philosophical debate.

This does not prevent me from emphasising valid points of criticism of correlationism from Golumbia's text. For example, he states: "Meillassoux paints correlationism as incompatible with realism." (Golumbia 2016, p. 6). This hypothesis can be supported by some passages in Meillassoux: "The virtue of transcendentalism does not lie in rendering realism illusory, but in rendering it astonishing, i.e. apparently unthinkable, yet true, and hence eminently problematic." (Meillassoux 2008, p. 27). But here an interpretative caution is required that Golumbia, Ennis and others do not muster. Correlationism turns against the position that Meillassoux calls dogmatic metaphysics or naïve realism: any *realism of the absolute* in his sense. Correlationist circle and two-step question subject-independence as a criterion, so correlationism must—in my terminology—turn against OR and OSR. Again, this does not mean that every realistic position is ruled out from a correlationist point of view. For Meillassoux, the crucial point is that correlationism opposes any position that defends primary qualities, that is, subject-independent properties of natural entities. Golumbia makes the same mistake here that he diagnoses in Meillassoux: If realism is fixed on subject-independence as a criterion, the judgement that correlationism would be incompatible with realism is correct. If, on the other hand, realism is understood as a more complex family of ontological-epistemological positions, as I propose in my map of debates, correlationism need not be hostile to *all* realisms.

Most of Golumbia's other, possibly valid, criticisms concern either Meillassoux's interpretations of Kant and Wittgenstein, or the validity of Meillassoux's assessment of the Western philosophical tradition since Kant. To a lesser extent, it also concerns objections to Meillassoux's own position, which, according to Golumbia, is too naïve compared to the extensive criticism of metaphysical realism. According to him, Meillassoux would defend the latter because he does not separate metaphysical and epistemological questions and arguments clearly enough. Because both levels of consideration are irrelevant to my systematic reflections on correlationism, Golumbia's further arguments against Meillassoux are not central to my investigation.

6.2.3 Systematic Tenets and Reasoning Schemes: Redefining Correlationism as an Analytical Tool

6.2.3.1 Lessons from the Research Literature: Interpreting Meillassoux

The lessons to be defined for the recast *systematic correlationism* can now be drawn from my discussion of the criticism of and interpretations of Meillassoux's systematic considerations. My critique of the interpretations of Heller/Cogburn, Gratton and Young shows that the correlationist rejection of strict subject-independence as an ontological criterion (in the strong model) or as the domain of possible knowledge (in the weak model) of really existing entities results from epistemological considerations of the boundlessness and finitude of human beings (a similar, albeit undeveloped version of correlationism, understood as this first, negative thesis, can be found in Giraud and Millière 2017, p. 110). Thus correlationism always argues against OR and OSR. The more radical a correlationism is, the more clearly an ontological prohibition of speculation is derived from our epistemological finitude. In correlationism, it can be argued in particular that the truth value of the (RO)-thesis of ontological realism with regard to certain entities cannot be established in principle, and that quietism, scepticism or agnosticism can be derived from it (for one possible link between scepticism and correlationism cf. Longo (2018), esp. pp. 48–52). This is underpinned by my consideration of Harman's (and Rutten's) reading. Correlationists are thus classifiable on a *spectrum of radicality* according to how comprehensively they attempt to undermine the ontological question in a given realism-anti-realism debate.

These negative theses of correlationism are always achieved by combining the reasoning schemes of the correlationist circle and two-step. Here, my analysis of Lovat's interpretation has shown that there is a second *spectrum of strength* of correlationism: While every correlationism uses the correlationist circle and two-step, weaker versions use them as restricted to a local entity domain, while stronger ones extend them to metalocal domains, such as to all concrete entities or to some fundamental principles of the natural external world of OSR, such as causality, fundamental laws of nature and/or logical laws, ontological principles, etc. Meillassoux's strong model represents the reasoning schemes in maximum strength and extends them to everything—that amounts to a positioning in the global debate. Meillassoux treats positions on this spectrum of strength as if they neatly divide into two separate species (weak and strong). However, given the gradual differences in the strength of the reasoning schemes, in my view this is a problematic idealisation.

Two further problems with Meillassoux's version of the systematics of correlationism are inherent in the dialectical tripartite division into weak and strong correlationism and subjectalism. First, Meillassoux claims implictly that the spectrum of radicality and the spectrum of strength always have parallel values for a given philosophical position. He believes he sees a weak and less radical correlationist in Kant, and two maximally strong and maximally radical correlationists in Wittgenstein and Heidegger. Even if these interpretations were consistent, there is no argument why an expansion of the entity domain or an increase in the number of entity domains (spectrum of strength) should necessarily be accompanied by a more vehement rejection of the ontological realism-anti-realism question and by a more comprehensive ontological prohibition of speculation (spectrum of radicality). In my further development of the concept, I will therefore not follow Meillassoux in this assumption of coupling strength and radicality.

On this basis, Meillassoux continues to claim that the fundamental variants of correlationist philosophy fall into either his strong, his weak or his subjectalist model. This assumption is also abandoned in what follows, as it only results from the dialectic of those positions assumed by Meillassoux, which, however, might only be covered by Meillassoux's epochal concept. The systematic considerations of my interpretation of Meillassoux and the criticism of this dialectic from the research debate have shown that the complexity of the philosophical positions that fall under correlationism is not sufficiently taken into account in this way. Since there are gradual differences on both spectrums, it is not appropriate to assume natural types of correlationism.

My analysis of the readings of Heller/Cogburn, Gratton and Young has also shown that the classification of a philosophical position as correlationism does not imply a commitment to the categories of realism or anti-realism: although correlationism rejects OR and OSR, realisms with other independence criteria are still conceivable in correlationism on the one hand, and on the other epistemological or ontological anti-realism does not necessarily presuppose correlationism. In principle, correlationist realism, anti-realism and quietism are all possible.

In reflecting on the interpretations of Bruno and Young, I have uncovered that the first decision on the primacy of correlation must be present in every correlationism and is derived from the correlationist circle and two-step reasoning schemes. As with the rejection of subject-independence as the criterion of realism, it has emerged from my analysis of Golumbia's critique that this must be done more precisely than in previous versions of correlationism. The chances of the success of systematic correlationism as an analytical tool are measured by this precision.

Now that the combination of systematic theses and reasoning schemes has become clear, the next step is to adapt correlationism as an analytical tool for use in the realism-anti-realism debate, especially in philosophy of science. To this end, the concept must be reshaped by working out its relationship to the concepts of epistemic and ontological anti-realism as well as to OR. With this step, I finally go beyond Meillassoux's use of the term and modify it. The sole purpose of this is to make correlationism as useful as possible as an analytic-reconstructive tool. I do not claim that systematic correlationism in my sense is the only meaningful interpretation of correlationism.

6.2.3.2 New Version of Systematic Correlationism

SYSTEMATIC CORRELATIONISM in my sense denotes, *firstly*, any philosophy that derives from the comparison problem the reasoning schemes of correlationist circle and two-step in order to argue against all strong forms of subject-independence of certain entities, and in particular against strict subject-independence of the natural external world. The problem, which I call the comparison problem, says that no one can compare reality from our perspective with a subject-independent reality postulated in ontology. Systematic correlationism derives from this negative conclusion and the two correlationist reasoning schemes, *secondly*, either the positive ontological correlationist tenet (OCT) or the positive epistemological correlationist tenet (ECT):

OCT The natural external world and/or certain entities $E_1, \ldots, E_n$(of a local, metalocal or global entity domain to be specified) are always ontologically dependent on their relation to subjects (in a sense to be specified). Conversely, there is also a form of dependence of subjects on the natural external world and/or the entities $E_1, \ldots, E_n$ (often assumed as a causal, natural law-like or logical dependence).

ECT A specific form of knowledge (in the extreme case: any possible knowledge) about the natural external world and/or certain entities $E_1, \ldots, E_n$ (of a local, metalocal or global entity domain to be specified) consists only in a form of intersubjective consensus and/or is dependent on our interests and practices. Quietism, scepticism or agnosticism is therefore recommended as a philosophical attitude towards the natural external world and/or the entities $E_1, \ldots, E_n$.

For the philosophical evaluation of nature and the natural sciences, this means that CORRELATIONIST PHILOSOPHY OF SCIENCE (i) claims that it follows from the comparison problem that all strong forms of subject-independence of the natural external world (and thus also of OSR) are false. Correlationist philosophy of science therefore argues (ii) either in favour of the positive ontological correlationist tenet about the natural external world (OCTNEW) or the positive epistemological correlationist tenet about the natural sciences (ECTNS):

OCTNEW The natural external world and/or some or all theoretical entities[26] are always ontologically dependent on their relation to the research community and/or our practices of theorising. Conversely, there is also a form of dependency of the research community and/or our practices of theorising on the natural external world and/or the theoretical entities (often expressed as causal, natural-historical or natural law-like).

ECTNS Scientific knowledge about the natural external world, laws of nature, natural entities and/or some or all theoretical entities is only a form of intersubjective consensus and/or is dependent on our interests and practices. The philosophical stance advised in this domain is therefore quietism, scepticism or agnosticism.

In the case of philosophy of science, there are different correlations that can be used to introduce the comparison problem as grounds for the correlationist circle and two-step arguments against subject-independence. In the following Sect. 6.3, I present some instructive examples from the wide range of ontological, epistemological and linguistic correlations. They illustrate some of the diversity of what is correlated with each other. On the one hand, the natural external world, laws of nature, different theoretical and natural entities; on the other, the research community, researching subjects and the scientific discourse with its practical components.

---

[26] Often these are to be restricted to unobservable theoretical entities, sometimes laws of nature are included, etc.

According to my definition, systematic correlationism always charactarises an antipode to philosophies of OR or OSR, since subject-independence is rejected as a criterion of independence for the natural external world or for certain entities. The suspicion that correlationism could be equated with anti-realism is averted by showing that realism can be combined with correlationism if the independence criterion from the (RO)-thesis of realism (*certain entities have, or an entity domain has, independent existence and reality*) is understood differently and at the same time a correlation in the sense of OCT or OCTNEW is defined, which describes the co-dependence of the natural external world and subjects in the metalocal domain of science.

Before I pursue the examples, the concept of systematic correlationism described here should be distinguished from Meillassoux's concept: My definition results in a four-part division of correlationism into correlationist realism, correlationist quietism, correlationist epistemological anti-realism and correlationist ontological anti-realism. The quitist alternative—for the reasons already laid out in my Chaps. 2 and 4—is not apt for the metalocal debate, and therefore not thematised in the following section. My four-part devision is different from Meillassoux's dialectical thesis on weak and strong correlationism and subjectalism. My categorisation results from purely systematic differences of philosophical stances regarding the combination of the correlationist thesis with realism, epistemological and ontological anti-realism and their application in philosophy of science, as subdivided by my definition and debate map in Chap. 2. Since the categorisations of realism, epistemological and ontological anti-realism on the one hand, and correlationism on the other run transversely to each other, the possibility of a different three-part order of correlationism arises for systematic reasons. Each correlationist in philosophy of science can be questioned as to whether they accept or reject the realism theses.

With my proposal, I remain true to Meillassoux's basic intention of defining systematic correlationism as a position that arises from the reasoning schemes and the resulting positive and negative tenets. However, I reject Meillassoux's dialectic of weak and strong as inappropriate, at least to the current debate in philosophy of science. Instead, it can usefully be asked of any systematic correlationism in my reading how radical and strong it is, and where it locates itself on these two spectra. Moreover, my systematic correlationism allows for a clear classification of whether an ontological or epistemological tenet is being advocated. Depending on whether OCT, ECT or both positive tenets are defended, this question can be answered differently. In what follows, the example analysed will focus on this second aspect, because the arguments for the tenets, together with the particular appropriation of the reasoning schemes and the rejection of the SI-criterion, turn a philosophy of science into a systematic correlationism. The evaluation of the respective radicality and strength is a deeper aspect that is not decisive in the context of my argumentation. It will be important in other investigations of correlationist philosophy when it comes to evaluating the stringency of any particular correlationism.

Correlationism in my systematic version and the realism-anti-realism distinction are incongruent categorisations of philosophical positions. But there is an important connection to the realism question through correlationism's treatment of the independence criterion and the status of the natural external world and other entity domeins. Alternatives to OSR can be developed in both dimensions. When considering the debate of speculative and new realism in metaphysics and ontology, the question of how a given position relates to correlationism can usefully be combined with the question of whether it is a realism, an ontological or an epistemological anti-realism. As I will show below, the philosophical question of whether the alternatives to OSR are other realisms or anti-realisms can be linked to the question of whether the respective proposal is correlationist or not. The positions for which the second question can be answered in the affirmative run into the problem of ancestrality. Thus these attempts to develop an alternative to OSR in philosophy of science fail. This is where I see the main benefit of this categorisation as an analytical tool in philosophy of science. However, in order to dispel the straw man objection to my version of systematic correlationism, I will first outline the aforementioned examples of correlationist epistemological and ontological anti-realism as well as correlationist realism.

6.3 Correlationist Philosophy of Science

The consideration of some correlationist *realisms*, correlationist epistemological scientific anti-realism (ESA) and correlationist ontological scientific anti-realism (OSA) fulfils two functions in the context of my argument. First, it shows that the straw man objection cannot be levelled against my developed concept of systematic correlationism: such philosophies are out there. To this end, I show how the selected philosophers use the circle-two-step reasoning schemes, based on the comparison problem, to reject strict subject-independence (and sometimes the necessity) of certain entities and to defend a variant of the theses OCT, ECT, OCTNEW or ECTNS.

Second, the chosen examples in the next chapter will show that correlationisms are confronted with the problem of ancestrality and that this is sufficient to cast doubt on them. At least in the metalocal debate in philosophy of science, the correlationisms under consideration do not manage to convince against OSR alternatives, as I will argue. This subchapter introduces the positions that will serve as my main examples.

I begin by discussing correlationist anti-realisms. Since they deny subject-independence, it stands to reason that many correlationisms propose an ontological or epistemological anti-realism. OSA can be motivated by correlationism: Instrumentalism[27] as it is often attested to Duhem and defended in a contemporary variant by Stanford, pursues a correlationist strategy that can lead to an OSA. The comparative problem here serves to justify the fact that although we gain useful knowledge in the sciences about the manipulability of our environment, the entities

[27] For a working definition, see my Sect. 5.1.1, p. 110.

of the theories are only fictional, heuristic tools—instruments that cannot be said to be subject-independent.

But the correlationist strategy can also be used to justify ESA: Wray and van Fraassen, in their variants of constructive empiricism, argue for the correlationist and epistemological anti-realist idea that all knowledge of the fundamental laws of nature and the supposedly fixed structure of the natural external world depends on our theoretical practices and phenomenon-guided interests. They conclude that we should take an agnostic view of the ontological status of postulated theoretical entities and fundamental laws of nature. This is not to deny that unobservable entities or fundamental laws of nature exist or could exist, but contests that the sciences provide us with real insight and knowledge about them. According to van Fraassen and Wray, we cannot possibly know whether a theoretical entity corresponds to an unobservable, natural entity.

I then turn to correlationist realisms to support my argument that correlationism need not lead to anti-realism. I begin with a reconsideration of Sellars' philosophy of science, some readings of which I have already outlined in Sect. 4.2.1. McDowell's interpretation of the radical Sellars that I reconstructed there becomes my subject again here: If Sellars is read only as a methodological, and not as an ontological realist naturalist, he should be understood as a correlationist. I will briefly discuss why correlationist realism is not a common position in philosophy of science. This chapter concludes by examining a contribution to the realism-anti-realism debate in metaphysics and ontology. Despite its focus on the global debate, this position is interesting for the purpose of my analysis because it is a particularly clear case of correlationism, in which explicit statements about the subject-dependence of the natural external world are made from a realist perspective. I am referring to Anton Friedrich Koch's hermeneutic realism, which—to simplify—argues that the existence and reality of the natural external world precedes the existence of subjects chronologically in alignment with fundamental laws of nature and is thus necessary for their genesis in natural history. At the same time, however, the emergence of subjects is understood as a retrojective, logically and ontologically necessary condition for the existence and reality of the natural external world itself. The consideration of this theoretical option serves me as a springboard to the problem of ancestrality, which is my topic in the next chapter.

6.3.1 Correlationist Ontological Anti-Realism in Philosophy of Science

The main task of this section is to show how instrumentalist philosophy of science that advocates an OSA corresponds to a form of correlationism. In particular, I will discuss Stanford's instrumentalist position, which follows from his unconceived-alternatives argument, already reconstructed in my Sect. 5.2.3. Stanford's position also draws on considerations by the operationalist and physicist P. W. Bridgman, logical positivists such as the early Carnap and authors read by Stanford as instrumentalists, such as Ernst Mach, Duhem and Poincaré (cf. on the influence of the first two Stanford 2006, p. 192; on the latter pp. 189–190). Other instrumentalists could also have been discussed here, but Stanford's position is easier to understand, since I have already reconstructed his main argument.

A second dominant form of OSA, besides instrumentalism, can be found in constructivist and relativist philosophy of science, also sometimes described as post-structuralist or postmodernist. All these collective categories are vague "isms" that would require comprehensive clarification if they were to be used here as examples of correlationism. This is particularly true because positions on science that are often orientated towards Kuhn, Feyerabend, Foucault, Rorty and/or Derrida are not only firmly anchored in the corresponding subdiscipline of philosophy, but also in other *science studies*, such as the sociology of science, and require appropriate consideration.[28] However, constructivist-relativist examples can also be found in the narrower debate of philosophy of science. These include Martin Kusch's communal relativism, Lorraine Daston's historical constructivism and Hans-Jörg Rheinberger's deconstructivist philosophy of science. All three seem to defend an ontologically turned co-constitution—and thus ontological correlation—of the intersubjective research community with natural entities.[29] Kusch argues for a limitation of scientific knowledge and its ontological implications to the respective scientific cultures and thus defends an epistemological-ontological anti-realism (cf. esp. Kusch 2002, 2017, 2018, 2020). Daston attempts to conceptualise scientific ontologies that are constantly dependent on intersubjective observations and, to this end, draws on the concept of a historically changeable objectivity developed by her and Peter Galison, which always undergoes changes in tandem with a new form of subjectivity (cf. esp. Daston 2008; Daston and Galison 2007). Rheinberger, on the other hand, argues that the *epistemic things* that appear in our theories and experimental systems are ontologically co-dependent with the socially created, albeit real, laboratory environment (cf. esp. Rheinberger 1991, 1997, 2011, 2015). The thesis of the co-dependence of the intersubjective research community on natural or scientific entities is defended and interpreted differently by each of these three authors. However, one could argue that it always presents itself as a correlationist OSA. For the reasons mentioned and due to limited space, I will focus on instrumentalism in Stanford's variant and provide a brief demonstration of its correlationism.

Stanford's unconceived-alternatives argument essentially consists in bringing together the contrastive UTE and a historical premise: Stanford highlights the possibility that in the past theories were not proposed and therefore not accepted, even though they could have been supported by data and evidence. Thus, our

[28] For examples of the connection between the fields of philosophy, history and sociology of science with a Foucauldian character, cf. Barnes et al. 1996; Kusch 1991; Martín Alcoff 2005.

[29] This seems to be the kind of case that Ennis and other commentators have in mind when they postulate an identity of correlationism and anti-realism.

epistemic situation in the empirical sciences is always precarious. We are not in a position to weigh all relevant, empirically different theories against each other because we never generate all of them due to our epistemic limitations. As should be clear at this point in my investigation, this is a variant of the comparison problem: we can never compare all relevant theories with each other and with the natural external world, and so, due to our epistemological finitude, we are doomed to select only the best historically proposed theory. Due to our experience with rejected theories, according to Stanford—and this is where the further argument for his instrumentalism begins—we should see our theoretical entities, which together form the natural external world of the sciences, as useful instruments for manipulating our environment, but not as elements of a subject-independent reality.

Stanford distinguishes between two instrumentalist approaches: Mach's reductive approach on the one hand, and Duhem's and Poincaré's sceptical-syntactic approach on the other. According to the first (in Stanford's interpretation), scientific discourse and theoretical entities do not signify anything ontologically, but are merely tools for predicting sensory experiences; according to Duhem and Poincaré, by utilising theoretical entities scientific discourse attempts to grasp a subject-independent natural external world, but this fails in principle, so that theories are neither true nor false, but are only useful syntactic tools for drawing conclusions from observations for future observations and possible manipulations of the environment (cf. Stanford 2006, pp. 191–192). Instrumentalism in general can therefore correspond to an ontological, epistemological or epistemological-ontological anti-realism in my vocabulary. Stanford's own proposal differs from the two historical approaches discussed, but can nevertheless be categorised:

> On this alternative, we might allow that our scientific theories really do make just the descriptive claims about inaccessible reaches of nature that they seem to make, and even that the fruitful investigation of nature by human beings typically *requires* that we propose and explore claims of this character, but nonetheless insist that we can make perfectly good practical use of the claims of such theories without believing what they say about the natural world. This approach might be called "epistemic" [...] because it restricts the set of beliefs to which we regard ourselves as entitled by the dramatic empirical successes of our best scientific theories, rather than restricting or reinterpreting the very claims that those theories are understood to be making in the first place. Stanford 2006, p. 194.

> Identifying the various ways in which a theory's concrete consequences can be grasped independently of the theory itself is important only when we seek to specify the beliefs about the world to which we remain committed once we recognize that we are not epistemically entitled to belief in a particular theory even though we know of no better way to pragmatically engage a given aspect or domain of nature than by means of it. [...] Of course, if a theory makes claims about entities or events to which we have no access independently of the theories that describe them or claims for which there is no sense to be made in a way that is independent of those descriptions, the instrumentalist will not believe them at all, but will simply make use of them in the process of generating further predictions and recipes for intervention whose content does not depend exhaustively on the theory's own descriptive apparatus in this way. Stanford 2006, p. 198.

In the ensuing paragraphs, Stanford refers to all scientific entities, not only to unobservable entities or laws of nature. Moreover, he does not only mean theoretical entities, but also observational entities (cf. Stanford 2006, p. 201). Stanford speaks here of nature as something real and does not deny that the sciences, especially in the form of their theories, attempt to make statements about nature. It therefore initially seems questionable whether my characterisation of Stanford's epistemic instrumentalism as an OSA can be correct. But a step-by-step analysis resolves this question: It becomes obvious here that the unconceived-alternatives argument is an argument of the correlationist circle-two-step scheme based on the comparison problem highlighted above. The natural external world—Stanford's *nature*—, however it is constituted independently of us, may, according to Stanford, be a causal precondition of our scientific community and thus of our theories. As the first quotation above makes clear, the sciences attempt to describe the natural external world. But they fail—scientific theories never describe the natural external world correctly with certainty, as is clear from the unconceived-alternatives argument. I. e. with regard to the natural external world, ontological realism is combined here with epistemological anti-realism. With regard to scientific theories and their entities as well as scientific ontologies, on the other hand, Stanford argues an ontological anti-realism:

This is true even though Stanford sensibly chooses the label of epistemic instrumentalism for his position, since scientific theories, their domain ontologies and scientific entities are epistemic instruments for observational predictions, measurements and manipulations within the natural external world in his conception. They give us knowledge about future observations or manipulative possibilities, which in turn are subject-dependent. But epistemic instrumentalism is an ontological scientific anti-realism in the metalocal debate, since theories, domain ontologies and scientific entities have no ontological weight. Stanford's epistemic instrumentalist does not believe in theories, does not consider their scientific ontologies to be domain ontologies of the natural external world, and denies that scientific entities exist in a subject-independent (or even concrete) way. Stanford takes the position on scientific theories in general that he assumes today's scientific realist has for Newton's mechanics: a heuristically decent tool the entities of which (forces, absolute masses, absolute space and absolute time, etc.) do not exist in this way in the natural external world (cf. Stanford 2006, pp. 204–205). Stanford does not avoid the question of what then delineates the boundary of the scientific domain in which we are supposed to be epistemic instrumentalists. After all, our everyday understanding of lifeworld entities, can also be depicted as a form of theory that is also subject to historical change. Stanford answers as follows:

> If we allow that our grasp of the world is, in this sense, "theories all the way down," it might well be impossible to coherently adopt instrumentalism as an epistemic attitude toward any and all theories whatsoever. That is, if we grant that all of our efforts to understand the world around us and our place in it are fundamentally theoretical in the broadest sense of the term, then it may well be that there is no coherent way to adopt an epistemic instrumentalist attitude toward all of our theories simultaneously. It is fortunate indeed, then, that our epistemic position in no way demands this. Stanford 2006, p. 202.

This means that he cannot defend epistemic instrumentalism in the global debate. In fact, he argues that it is not even necessary to champion it in the whole metalocal debate of philosophy of science. On the contrary, what is special about

Stanford's correlationist ontological anti-realism is that it applies only indirectly to the metalocal debate in philosophy of science. He argues for local scepticism with regard to all scientific theories and scientific ontologies that generate such complex models that we cannot be sure that we have formulated all meaningful alternative theories (cf. also for the following Stanford 2006, pp. 35–37 and pp. 202–205). However, this applies *de facto* to all current theories and domain ontologies of the natural sciences. Stanford aims to insist that realist objections to his epistemic instrumentalism now require a local case-by-case review, not a metalocal argument. If we were able to rule out unconceived alternatives for a given local theory, we could assume local realism, epistemologically and ontologically.

The correlation Stanford advocates in the metalocal debate is therefore a variant of OCT about science and ECT about the natural external world: According to him, scientific knowledge about the natural external world is only a form of historically unstable, intersubjective consensus, since theories about it can always be regarded as untrue according to the unconceived-alternatives argument, which is why scepticism is recommended with regard to them. However, our practices of theorising are causally conditioned by the natural external world. Scientific entities and the entire scientific ontologies *de facto* of all scientific theories are, according to OCT, ontologically constantly dependent on us producing and reproducing them as tools. They would disappear as soon as they were no longer useful to us collectively. Therefore, they could not be considered ontologically robust and subject-independent, as wrongly assumed by OSR. In this case, Stanford's OCT about science leads to the stance of OSA.

6.3.2 Correlationist Epistemological Anti-Realism in Philosophy of Science

Now that I have shown how correlationist OSA is advocated by Stanford, it is time to show correlationisms compatibility with ESA using an example. For this, I will refer to constructive empiricism in the variants of Wray and van Fraassen. The reasons for this choice are similar to Stanford's: On the one hand, I can easily introduce the position through my previous reconstructions of central arguments, in this case those of the bad-lot and model argument.[30] On the other hand, constructive empiricism is widely discussed. Many contemporary empiricists within the debate use similar considerations to support their version of an ESA.

Accordingly, many of the other possible examples that could be analysed on the suspicion of being a correlationist ESA often have a reference back to van Fraassen's constructive empiricism. Many also go back to Laudan's historical anti-realism or an interpretation of Duhem's, as a purely epistemological instrumentalist. These include Helen Longino's contextual empiricism. Her position builds on constructive

[30] Cf. Sect. 5.1.4, pp. 115–118.

empiricism and feminist philosophy and argues that science is a type of social knowledge that is both logically and cognitively conditioned by pragmatically anchored values of theory production (cf. also for the following Longino 1990, esp. p. 5, Longino 2002, 2009). Longino does not deny the natural external world, but the results of our natural sciences are logically and cognitively dependent on pragmatic decisions and social configurations. This means that their statements do not provide any reliable ontological information about the natural external world that is independent of us. In contrast to Stanford, for example, she does not deny scientific entities any ontological weight, which is why, in my opinion, it is at least not obviously an OSA, but an ESA.

Superficially, Longino's argument and constructive empiricism are similar in the variants of van Fraassen and Wray. They seem to argue similarly to Stanford: Our theoretical knowledge of the natural external world does not seem certain enough to serve as an ontologically binding description. On closer inspection, however, despite the attempts at appropriation by ontological-antirealist instrumentalists such as Stanford (cf. 2006, p. 194), some deep disparities emerge:

> After deciding that the language of science must be literally understood, we can still say that there is no need to believe good theories to be true, nor to believe *ipso facto* that the entities they postulate are real.
>
> *Science aims to give us theories which are empirically adequate; and acceptance of a theory involves as belief only that it is empirically adequate.* [...] [A] theory is empirically adequate exactly if what it says about the observable things and events in this world, is true—exactly if it "saves the phenomena". A little more precisely: such a theory has at least one model that all the actual phenomena fit inside. van Fraassen 1980, p. 12; emphasis as in original.

The meaning of model here is identical to van Frasssen's use of the term in the model argument: A van Fraassen-model is an abstract structure allowed by a given theory, which can be embedded as such in that theory and in which all relevant parameters of the theory have specific values. Thus, if a van Fraassen-model of a theory is compatible with all the actual phenomena, then it is an empirically adequate theory. With van Fraassen, we should accept it as such in every respect, without being entitled to draw conclusions that this theory is true in a realistic sense and without deriving ontological consequences for unobservable domains of the natural external world. For van Fraassen, and subsequently also for Wray, the aim of the sciences is thus different from what OSR takes it to be; it is not to provide a correct, ontologically valid description of the natural external world (van Fraassen 1980, pp. 8–9). This highlights another fundamental difference between constructive empiricism and Stanford's instrumentalism: From Stanford's point of view, realists are right that the sciences *try* to provide the correct, ontologically binding description of the natural external world. However, this fails for reasons of principle, which is why his instrumentalist variant of OSA is appropriate. In constructive empiricism, however, the error already lies in this basic assumption: Only according to the assumption of the realist of such an ontological intention does an unattainable goal set for the natural sciences result. Wray and van Fraassen suggest that scientists are primarily concerned with developing theories that are compatible with the data, observations and evidence and have the strongest possible empirical implications, i.e. with explaining the phenomena. From the perspective of constructive empiricism, both the realists and Stanford fail to recognise this.

I have already examined two of the main arguments that van Fraassen introduces in favour of constructive empiricism in Sect. 5.1.4. In the bad-lot argument, van Fraassen uses a form of the comparison problem (cf. van Fraassen 1989, pp. 142–143): We cannot compare our theories, either in the sciences themselves or in philosophy of science, with those theories that have not been actually proposed contingently. So our choice for the best theory in terms of realism may be the best of a bad set. If we believed this to be the literally true description of the natural external world, this presupposes that the literally true description of the natural external world is among the alternative theories we are considering. According to van Fraassen, however, the realist always fails to provide us with an argument as to why this should be the case. But assuming the quality criterion of constructive empiricism—empirical adequacy (and strength) of a theory—it is easy to accept that the sciences are successful because our theories are demonstrably empirically adequate, as van Fraassen and Wray argue (cf. also Wray 2018, pp. 49–50).

The bad-lot argument demonstrates only one way of how the comparison problem leads into a kind of limited correlationist circle with respect to the theoretical knowledge of the natural sciences. In arguing for his variant of constructive empiricism, Wray offers separate analyses for holistic and experimental UTE on the one hand and contrastive UTE on the other, which use different variants of the same argument: "When scientists conduct tests to determine which of the competing hypotheses is superior, they are not in a position to infer that the hypothesis that survives such a test is likely true or approximately true." (Wray 2018, p. 41.) Wray does not argue that all UTE arguments use this line of reasoning. But he does want to argue that those UTE arguments that strengthen the case for constructive empiricism proceed from this line of thought.

Put simply, this means that, according to Wray, a problem of comparison is created here for the scientist or the realist: Reality cannot be directly compared with its description by the best current theory, which is why its truth cannot be conclusively shown in an experimental context and why it cannot be ruled out contrastively that our best current theory is false. Here it becomes clear that these UTE arguments in the sense of the correlationist circle strictly separate the scientific ontologies in the unobservable domain from the true ontology of the accessible natural external world, since our scientific field of knowledge is bound to our perspective and is therefore finite. From this van Fraassen concludes that we should take an agnostic view of the ontological status of the postulated theoretical entities. This applies all the more clearly to fundamental laws of nature, which are easier to explain as artefacts of our scientific interpretations than as elements of the natural external world (cf. esp. van Fraassen 1989, pp. 183–214). In contrast, we should adopt a realistic attitude towards phenomena, and this includes the observable theoretical entities in van Fraassen's work. Here—at least if constructive empiricism is not interpreted as epistemic instrumentalism according to the *principle of charity* (as does Stanford 2006, p. 194)—it is therefore not denied that unobserved entities or fundamental laws of nature could exist. Instead, the view that the sciences provide us with real insight and stable knowledge about them and that we can know whether a theoretical entity corresponds to an unobservable natural entity, is questioned. According to van Fraassen and Wray, this is the core of the comparison problem in the empirical sciences, which becomes the basis of the correlationist circle. Conversely, both also advocate a form of correlationist two-step, since the realistically interpreted, observable natural entities are in part causally necessary for our emergence as a natural biological species, even if we are not informed about the actual unobservable processes of this development by our theories, since the agnostic stance also applies to the latter.

According to constructive empiricism, unobservable theoretical entities are dependent on the theories in which they occur and are therefore dependent on subjects. Whether they correspond to unobservable natural entities is in principle hidden from us, which is why an agnostic stance is appropriate. For observable natural entities that also occur in theories, on the other hand, we could adopt a realist stance. Constructive empiricism thus represents the ECTNS thesis: scientific knowledge about fundamental laws of nature and unobservable theoretical entities is dependent on our phenomenon-guided interests and practices of theorising. Agnosticism is therefore recommended as a philosophical stance in constructive empiricism. Thus, I have shown that correlationist ESA is also advocated.

6.3.3 Correlationist Realism

In correlationist realism, SI is abandoned in favour of various correlations between the subject or the sciences and the natural external world. Instead, the independence criterion is articulated differently or mitigated in another fashion. In addition, the comparison problem is used to argue in favour of this conclusion by means of the correlationist circle and two-step reasoning schemes. Furthermore, realistic variants of the theses OCT, ECT, OCTNEW or ECTNS are advocated as a consequence. In what follows, I will show that these aspects, and thus a correlationist realism, are present in McDowell's reading of Sellar's philosophy of science and in Anton Friedrich Koch's hermeneutic realism. I could also cite a number of other examples here to argue in favour of a correlationist-realist reading:

Worrall's EpiSR, which I presented in Sect. 5.4.1, might be an appropriate position. After all, Worrall opts for the subject-dependence of scientific ontologies and argues that the inner nature of concrete entities of the natural external world (for him, this means the nature of concrete objects) is never knowable to us. Instead, the images we form of them, our theoretical entities, are dependent on the scientific approach we choose. However, two things speak against an interpretation of Worrall as a correlationist: First, he does not use the comparison problem in a classical variant, but relies primarily on the incommensurability argument and the PMI. However, a subject-independent realisation of structural properties is possible in Worrall's case, for which a form of realism is to be defended. This is not to say that a correlationist reading of Worrall's EpiSR is impossible, but it seems to me that this case would prove to be too controversial to pursue further.

Another example, this time from the ontological debate, which I have already introduced and for which a correlationist reading may be appropriate, can be found in the multimundialist metaphysical relativism that Rovane discusses. Here, however, it is questionable whether Rovane really wants to defend multimundialism in the metalocal debate of philosophy of science, as was evident from my sketch of her position in Sect. 5.3.3. It also remains unclear whether this position is ultimately realist, which is why I do not pursue this option further.

In the area of philosophy of science, Joseph Rouse's multilevel project seems to articulate a correlationist realism. Rouse's aim is to develop a new naturalism that leaves behind classical forms of strict SI and incorporates an adequate, power-critical and scientifically proven understanding of the human being into the focus of philosophy of science (cf. Rouse 1987, 2002, 2015, 2016, 2018). In order to achieve this goal, Rouse addresses, among other things, the fact that humans are a natural species that is constantly in a process of biological niche construction, as well as the fact that our actions and insights are always guided by interests and have a political dimension. Scientific knowledge and scientific practices are therefore always also instruments of power, with which Rouse aims to introduce a Foucauldian intention into the field of general philosophy of science, without giving up at least methodological naturalism.

Further forms of pragmatist naturalism that attempt to attack the strict SI of theoretical entities in OSR could be added here. Huw Price's pragmatist naturalism "without Mirrors" (cf. Price 2011) is a prominent attempt. In *Quining Naturalism*, Price explicitly argues against classical criteria of SI as the basis of an ontological characterisation of reality through different (OSR-)variants of naturalisms according to Quine, Smart, Armstrong and Field (cf. Price 2007). However, Price's considerations extend beyond the metalocal context of philosophy of science and are conceived as a critique of the global metaphysical debate and as a productive proposal for a solution to the metalocal debate in philosophy of language, which is why an analysis as a philosophy of science would be time-consuming.

This is also the case for Hasok Chang's proposal of a pragmatist variant of scientific realism. However, Chang argues against classical variants of scientific realism in claiming that the ideal of knowledge it presents is neither redeemable in practice, nor helpful as an ideal to strive for because it cannot be made operational (cf. Chang 2022, pp. 1–3). In the course of his argument Chang uses the comparison problem and case studies from different scientific practices to argue for a subtle correlationist solution. What is not so clear is the question of whether he is arguing any longer for a scientific realism in my sense, because of his pragmatist and operationalist leanings. For him, a natural entity "is real to the extent that there are operationally coherent activities that can be performed by relying significantly on its existence and its properties." (Chang 2022, p. 121). This suggests that the reality of natural entities, what Chang calls *real-ness* (cf. Chang 2022, pp. 121–130), is domain-specific and dependent on those researching activities. For me, this makes it questionable if Chang is best read as an OSA, correlationist scientific realist or even as a new scientific realist.

For now, the reasons why I will go into the positions of Sellars and Koch in more detail below, and not those of other authors such as Rouse, Price or Chang, are manifold and mainly of a practical nature. Interpreting Sellars as a correlationist realist in McDowell's reading is not relevant only because Sellars was one of the most influential naturalists of the last century. Interpreting him in terms of correlationism is also obvious at this point of my argumentation, as evidence of his realism and a more detailed reflection on his position has already been provided in Sect. 4.2.1. Sellars is also a central figure to whom many contemporary philosophers of science refer, which explicitly includes Rouse and implicitly Price and Chang.

A possible reason against integrating Koch in this chapter is that he brings his hermeneutic realism to the global ontological debate, not the metalocal debate in philosophy of science. Also, some of his contributions have not yet been translated into English. However, this aspect is mitigated by the fact that Koch's ontology has direct correlationist consequences for the natural external world and for space-time. Koch also aims explictly to advocate a realism for physical entities. The main reasons why I have decided in favour of Koch's position are twofold: In terms of content, his recent philosophy is one of the most explicit and clearest examples of a position in which correlationism is advocated in tandem with realism. Moreover, his ontological proposal does not need to be reconstructed in its entirety by me in order to provide this evidence. The argumentation of his so-called subjectivity thesis is sufficient. A second reason for choosing hermeneutic realism arises from practical considerations: in the next chapter, in the course of reconstructing Meillassoux's argument for ancestrality, I will deal at a central point with one of the possible answers of correlationism, the so-called retrojection strategy. There, Koch will again serve me as a well-argued and explicit example of this argumentation strategy.

6.3.3.1 Sellar's Two Images as Correlationist Realism

The first example is that of a philosopher of whose stance I have already given an interpretation as OSR, more precisely as non-reductive naturalism (in Sect. 4.2.1): Wilfrid Sellars. There I proposed two readings of Sellars' two images, the manifest and the scientific: The first reading, according to deVries and Rosenberg saw a future fusion of both images as the task of current philosophy and Sellars as a representative of a non-reductive naturalism—and thus of OSR. According to this reading, there is a natural external world that is strictly independent of the subject, and we describe it mostly successfully, albeit always only in part, in the scientific image of reality. The manifest image, which includes subject-dependent entities and is the historical precondition of the scientific image, must in this reading be fused by philosophy with the scientific image. The fused image of reality then in turn contains OSR.

McDowell's reading has been distinguished from this. Here, Sellar's *scientia mensura* dictum is less emphasised than other aspects that lead to a different interpretation: Knowledge acquisition, *Erkenntnis*, whether in the sciences or in everyday life, is always practical and processual. This means that ontological postulates of scientific theories are always based on practices, which in turn can only be given meaning in the manifest image and can only be learnt there. Therefore, the dependence of the scientific image on the manifest image is also interpreted differently by McDowell, namely *not only* historically, but also practically. In McDowell's interpretation, this results in a *constant co-dependence* and indissoluble entanglement: The scientific image describes reality partially objectively, as it is independent of us humans, and this partly successfully. This shows that natural-historical processes have led to the emergence of humans. There is a natural necessity here. Conversely, the manifest image remains epistemologically and practically necessary for scientific practices and knowledge. According to this reading, the subject-independent scientific ontologies and their entities seem to depend epistemologically on our scientific practices. This is what I called the radical Sellars reading.

My criticism of McDowell's radical Sellars was that he does not succeed in escaping from the dialectic of realism and anti-realism and can not successfully undermine the debate. I pointed out that the ontological and epistemological status of a natural entity (e.g. the cosmic background radiation) in Sellars' radical view depends on an almost arbitrary choice of the respective image: In the scientific one it exists independently of the subject and is knowable, in the manifest one it is not. However, our choice cannot really be arbitrary; after all, according to Sellars, the sciences make real *discoveries* about the existence of the cosmic background radiation. Therefore, I concluded, the radical Sellars must assume a higher level of ontological anti-realism: The images are contradictory, indispensable, and therefore both not literally true. Since McDowell likewise does not wish to assign Sellars this position of higher-order ontological anti-realism, but a realism, according to my argument in Sect. 4.2.1, McDowell should agree with the interpretation of Sellars' position as non-reductive naturalism.

However, following the account of systematic correlationism, another option seems more appropriate to interpreting McDowell's radical Sellars. I propose that McDowell's Sellars is better understood as a correlationist *and* realist. The manifest image and the scientific image of reality would then describe two indissolubly intertwined parts of reality. SI would be appropriate as an ontological critirion for the natural external world as the totality of objects within the scientific image, but not for the human world of mind, language and culture. Both images of reality would be mutually dependent—natural dependence in one direction, epistemological and practical in the other. This is the correlationist circle-two-step reasoning schemes in action, based on the epistemologically couched comparison problem. Epistemologically, a correlate of the two images would result without reality becoming inaccessible. The independence criterion would therefore be divided and would no longer be strict: SI would exist for the natural external world in the scientific image, but the scientific image itself would be dependent on intersubjective (scientific) practices. Both images would exist independently of each individual subject, although not independently of subjectivity. For a comprehensive interpretation of the two images in this correlationist reading, I would need to go into more depth, but the following quotation, which in *EPM* directly follows Sellars' *scientia mensura* dictum, can establish some credibility for the interpretation of the radical Sellars as a correlationist realist that I have just outlined:

> An integral part of this conviction is [...] the *positivistic conception of science*, the idea that the framework of theoretical objects (molecules, electro-magnetic fields, etc.) and their relationships is, so to speak, an *auxiliary framework.* In its most explicit form, it is the idea that theoretical objects and propositions concerning them are "calculational devices", the value and status of which consist in their systematizing and heuristic role with respect to confirmable generalizations formulated in the framework of terms which enjoy a direct ostensive link with the world. One is tempted to put this by saying that according to these philosophers, the objects of ostensively linked discourse behave as *if* and *only as if* they were bound up with or consisted of scientific entities.
>
> But, of course, these philosophers would hasten to point out (and rightly so) that
>
> X behaves as if it consisted of Y's
>
> makes sense only by contrast with
>
> X behaves as it does because it does consist of Y's
>
> whereas their contention is exactly that where the Y's are *scientific* objects, no such contrast makes sense.
>
> The point I am making is that as long as one thinks that there is a framework, whether of physical objects or of sense contents, the absolute authenticity of which is guaranteed by the fact that the learning of this framework involves an "ostensive step", so long one will be tempted to think of the authority of theoretical discourse as entirely derivative, that of a calculational auxiliary, an effective heuristic device. It is one of my prime purposes [...] to convince the reader that this interpretation of the status of the scientific picture of the world rests on two mistakes: (1) a misunderstanding [...] of the ostensive element in the learning and use of a language—the Myth of the Given; (2) a reification of the *methodological* distinction between theoretical and non-theoretical discourse into a *substantive* distinction between theoretical and non-theoretical existence. Sellars 1956, §43, pp. 303–304; emphasis as in original.

Here Sellars argues against ESA, in particular against logical positivism and related positions such as instrumentalism. According to Sellars, the two errors that lead to the failure of this image of science and scientific discourse are that the stance of logical positivism assumes the certainty and epistemic independence of direct observations through its verificationism (error 1, what he calls "Myth of the Given") and that there the language-analytic, methodological separation into seperated scientific languages, which are different from everyday language, leads to the ontological error of separating real existence (as it can be determined in direct observations) from theoretical existence (of scientific entities) (error 2, ontological separation of the scientific discourses). Sellars contrasts this with the idea of the two images and his scientific realism. According to Sellars, the Myth of the Given is false, the manifest image must first be learnt and our practices, descriptions and observations of reality are permanently and inescapably linked to it. Nothing is simply elementarily given. In the scientific image, a methodological distinction can be made between domains of discourse, but it is nonsensical not to attribute existence and reality to scientific entities and ontologies.

With these arguments, Sellars advocates variants of the positive correlationist theses ECT and ECTNS: The natural external world of the scientific image, as part of reality, is, like the theoretical entities, indissolubly intertwined with the epistemological-practical access that only arises through the manifest image. Within the scientific image, natural entities and the natural external world are subject-independent, but the entire image is inextricably correlated with our manifest scientific practices and particular interests. Independently of this epistemic complex, we can therefore make no statements about the natural external world and the status of our theoretical entities, which is why McDowell's Sellars recommends quietism here. Within our images, however, we should be scientific and everyday realists.

I am not claiming that this interpretation of the radical Sellars is superior to the naturalist interpretation of deVries and Rosenberg or others. Nor do I claim to have presented a deep or even exhaustive reading of Sellars as a correlationist. I merely believe that I have demonstrated the correlationist circle-two-step reasoning schemes, the rejection of the *strict* SI of both images and a form of the ECTNS tenet. It therefore seems possible to me to read Sellars as a correlationist. The McDowell interpretation, which I criticised on the meta-level in Sect. 4.2.1 as ontologically anti-realist, can be better defended and understood as correlationist scientific realism, oposing ESA, OSA and OSR.

The radical Sellars as philosophy of science is thus best read as correlationist realism. It may be objected that this does not support my conclusion that correlationist realism is *really* advocated, since Sellars' position can also be interpreted as OSR and non-reductionist naturalism. I counter this objection with a fully explicit correlationist realism that stems from the global debate of ontology in the next section.

6.3.3.2 Koch's Hermeneutic Realism as a Correlationist Realism

Anton Friedrich Koch's hermeneutic realism is a complex ontological position that I cannot present in its entirety, much like Sellars.[31] I will focus on demonstrating that it is a systematic correlationism in my sense by reconstructing Koch's subjectivity thesis and his argument in its favour. I will then be able to show that this corresponds to systematic correlationism. At the end of the section, I will show that Koch's ontology is not only a realism in its own terms, but also in my nomenclature (cf. for an overview of neighbouring variants of hermeneutic realism in the field of philosophy of science Ginev 2017).

Possibly the core, but at least the most provocative part of Koch's ontology is the so-called "*subjectivity thesis*": "A material spatio-temporal system is possible only if at some time and at some place in it at least one spatiotemporal subject exists ("an I") that relates to objects in thought and perception." (Koch 2006a, p. 8.)

Koch bases the subjectivity thesis on the following premises:

[31] For his departure from the positions of Heidegger and Theodor W. Adorno cf. Koch 2020.

(1) Individuals, meaning mesoscopic concrete things of our life-world, must be individuated: If something is first an individuated thing, then no longer an individuated thing, it does not transform into a general thing, but ceases to be: "The individuality of individuals must therefore be guaranteed with metaphysical necessity [...]" (Koch 2016, p. 24; my translation); Koch excludes contingency as the ground of individuation.
(2) In the case of descriptions using purely general terms, it is not clear and can therefore never be ruled out whether they also apply to something unknown (cf. Koch 2016, pp. 24–25).

    As an intermediate conclusion, it follows from (1) and (2) that purely general descriptions do not individuate the individuals. For if it can never be ruled out that they also apply to something unknown, it would be coincidence if they did not, which was ruled out by the first premise. This makes the assumption of the following premise necessary:
(3) Life-world individuals are ontically individuated by their position in space and time (cf. Koch 2016, p. 25).
(4) The principle of *identitas indiscernibilium*, the identity of indiscernibles (PII), should apply logically (cf. also for the following Koch 2016, p. 26). This states according to Koch: $\forall F(Fx \Leftrightarrow Fy) \Rightarrow x = y$. This holds in every second-order predicate logic with identity and exists in every second-order predicate logic without identity "as a possibility to introduce the identity predicate by definition, provided that one accepts the unproblematic indiscernibility of identicals—"$x = y \Rightarrow \forall F(Fx \Leftrightarrow Fy)$": $x = y \Leftrightarrow \forall F(Fy \Leftrightarrow Fx)$." (Koch 2016, p. 26; my translation.)
(5) Possible worlds that do not violate the PII from the inside, but only from the outside, should be allowed.

Koch constructs his argument from these premises, which leads him in a first conclusion to the necessity of indexical properties of individuals in order to prove his subjectivity thesis in the final conclusion.

> Indexical properties are extrinsic properties, on closer inspection relations of individuals to subjectivity as such. This also gives subjectivity ontological rank. This is what Rosenberg has in mind when he describes the consideration of indexical properties as the ontologisation of subjectivity. What is meant is not the reification of the subjective into an immaterial substance, but on the contrary the thesis that there could be no physical things if they did not also include "things" that are cases of subjectivity, i.e. persons. Koch 2006b, p. 330; my translation.

From this it becomes clear why, in Koch's opinion, the subjectivity thesis can be accepted as soon as it is admitted that the individuation of individuals necessarily presupposes indexicality, and is itself also necessary according to premise (1). Both together would guarantee the subjectivity thesis, which itself also contains an assertion of necessity if it is transformed: In order for our, and every spacetime system to exist, it is necessary that at least one corporeal subject exists in it somewhere and at some time, which refers to individuals in a thinking and perceiving way.

The decisive factor is how the line of argument leads to the preliminary conclusion, according to which indexical properties are necessary for the individuation of individuals. This can be reconstructed as follows:

First, Koch states with regard to (3) that this is a provisional solution for the individuation of life-world individuals, since it only postpones the problem. Now, the question remains what individuates the spacetime positions. If an attempt were made to individuate them by determining at which points in time a certain point in space has which property (empty or occupied by a part of a certain individual), this would be viciously circular reasoning, according to Koch.[32] The question therefore arises as to how else the spacetime positions can be individuated.

Koch justifies (4), the validity of the PII, via the absurdity of assuming its invalidity: If it were not valid, the groundless difference of indiscernible individuals in the life-world would have to be assumed (cf. for the following Koch 2006b, pp. 315–316). Instead of one chair, there would be a collection of 17 identical but different chairs. They would all have the same weight, which is why it would be uncertain whether the chair I weigh and the chair I see are identical. Referring to the location would not help, because the 17 chairs would have 17 indistinguishable locations, as these would only be individuated depending on time and their occupancy by individuals at certain points in time. According to Koch, if PII is dropped, this absurd multiplication of individuals in the life-world cannot be prevented.[33]

Koch justifies premise (5) by comparing two possible-world scenarios to problems of PII. First, he considers Max Black's thought experiment of the twin sphere world in *Versuch über Wahrheit und Zeit*. This is a world in which only two identical spheres of solid iron measuring one mile in radius exist.[34] In Black (1952), a denier and a proponent of PII argue concerning whether the twin sphere world is compatible or incompatible with PII. The details of the dialectic between the denier and the proponent can be reduced to the following argumentative core: The only

[32] Here, a possible objection is that Koch seems to disregard the debate about the ontological status of spacetime and the individuation of spacetime points. Cf. e.g. Brading and Castellani 2002; Calosi and Duerr 2021; Dasgupta 2015; Dieks 2000, 2014; Lehmkuhl 2018; Pooley 2013; Teitel 2022, for the sake of the argument, I will assume that he can answer this query.

[33] The question could be raised here as to why 17 chairs, places etc. should be assumed just because it is not ruled out that two or more non-identical indiscernible objects exist. First, for any conceivable explanation, 16 items seem superfluous and can therefore be eliminated according to Occam's razor. Second, the chairs have no influence on each other and are causally isolated in this example—why? Furthermore, each chair has its own relation to other individual things, e.g. to my body. These relations are always the same and change in parallel, because otherwise the chairs would be discernible. The scenario is less absurd than trivial and leaves important questions open. However, discussing such an objection is pointless and leads away from the topic of whether correlationism exists. Moreover, Koch has potentially further arguments in favour of PII, e.g. its derivation as a rule of second-order predicate logic from other logical rules and innocent premises.

[34] Cf. In Koch's nomenclature of so-called *problem worlds* Koch 2006a, p. 9, devided into those that 1. in time or 2. in space exhibit either a. two-sided infinite repetitions or b. universal symmetries, Black's twin sphere world is a 2b.

option or distinguishing the two spheres is to give them names such as Castor and Pollux. However, since a certain perspective on the twin sphere world is necessary for the baptism, which is not available in the absence of an internal perspective within the twin sphere world and the uselessness of our external perspective on the twin sphere world, the individuation of the spheres fails, which together with (1) allows only one conclusion for Koch: "So we may conclude that the twin sphere world is indeed incompatible with *identitas indiscernibilium*, but precisely because of this, contrary to first appearances, it is logically impossible for some as yet unknown reason." (Koch 2016, p. 30, my translation).

Koch presents a further consideration in the form of a possible world in which the PII is apparently violated. It consists in the world with two-way eternal recurrence introduced by Lewis (cf. Lewis 1986c, p. 63).[35] In this scenario, everything is as in our world, with the added condition that the entire natural history of the universe repeats itself eternally, both in the past and in the future. The universe ends and then begins with a new big bang that is exactly the same as the previous one. Everything repeats itself and runs the same course until a repeated end. Moreover, there is no starting point, as this cycle is eternal. Consequently, there are an infinite number of duplicates of you and me in the past and future.

In the scenario of the world with two-way eternal recurrence, each of my actions has an infinite number of duplicates, as do you, I and the solar system. I and my duplicates are the same in all properties; we are only numerically different in ontological time—there is a Jan Voosholz preceding me and a Jan Voosholz following me. Koch accuses Lewis and other theorists of considering "spacetime points to be magically individuated" (Koch 2016, p. 25; my translation). For Koch, this is not only unsatisfactory, but also fails if the logical-metaphysical principle of *identitas indiscernibilium* is to be upheld: The Lewis example gives rise to the problem that, assuming PII, I am simultaneously identical and not identical with all my duplicates. On the one hand, all essential properties would be the same, therefore I would be identical with all my duplicates. On the other hand, I have different spacetime points (points in different spacetimes, to be precise) than my duplicates and am therefore different from them. According to the magical theory of individuation, given spacetime points are not essential properties of individuals. According to Koch, this is where their mistake lies: As soon as I could indexically mark my spacetime point with *here* and *now*, the apparent contradiction would be resolved.

The case of the two-way eternal recurrence therefore differs radically from the twin sphere world: In the first instance there are subjects with perspectives, they can individuate individuals in their spacetime and in other spacetimes before and after them. "Only ontic individuation is initially problematic and impossible when

[35] In Koch's nomenclature, Lewis' two-way eternal recurrence falls into the 1a category. Examples for 1b, such as symmetric time expansion and contraction, as well as for 2a and others for 2b, such as Strawson's chess-board universe, cf. Strawson 1959, pp. 122–134, are also considered by Koch, but carry no weight in his argument.

viewed from the outside. For the external view [...] the imaginary world violates the *identitas indiscerniblium* and would therefore be impossible for a reason that we still have to find. But in this case no independent reason will be found." (Koch 2016, p. 32; my translation). Koch instead concludes from this that the hypothetical external view is impossible, but not the imaginary world with two-way eternal recurrence. In this respect, everything is in order with it, there is the individuation of the mesoscopic individuals of the life-world desired by Koch: "My desk has the indexical property of now standing here in my study, the desks of my doppelgangers do not, they not only do not stand in my study, but do not yet exist at all or no longer exist." (Koch 2016, pp. 32–33; my translation).

This in turn means that, according to Koch, his *problem worlds* are possible iff there is indexicality and perspectives. This of course means they are impossible if they lack indexicality and perspectives. He thus concludes that there must be indexical properties, and necessarily so:

> Among their relational characteristics, individuals themselves must possess such characteristics through which they are related to thinking and speaking beings in such a way that they can be epistemically individuated by them. Such relational characteristics are indexical properties. But these only exist in worlds in which at some point in time and somewhere (not necessarily always or everywhere) a priori self-localisation occurs, which enables the use of indicators, i.e. only in worlds in which at some point in time and somewhere at least one corporeal, a priori self-localising subject occurs. However, since the individuality of individuals is a categorical, essential property that cannot depend on coincidences, there must be corporeal subjects in all possible worlds sooner or later, here or there. Worlds with symmetries and repetitions are possible because space and time are principles of separation for the concepts of qualitative and numerical identity; worlds without subjects, however, are impossible. The difficulties that space and time bring with them for *identitas indiscernibilium* must also be healed by space and time. Therefore, space and time are impossible without subjects embodied in them, i.e. spatiotemporal subjects. Space and time on the one hand and bodily, corporeal subjectivity on the other refer essentially to each other; they stand in a relation of reciprocal, interrelated, essential dependence, in what I will briefly call a interrelationship (*Wechselverhältnis*). Koch 2016, p. 33; my translation.

I read this as Koch's explication of a correlation. Spacetime and bodily subjectivity form Koch's core correlate, which he demarks with the subjectivity thesis (this goes back to Koch 1990, §3). Koch's entire correlation is far more complex: for one thing, the subjectivity thesis implies four corollaries for him. One is the so-called *epistemic personhood thesis* (*epistemische Personalitätsthese*), which states that subjectivity knows itself a priori as a physical and temporal being, i.e. as a person (cf. Koch 2006b, pp. 331–333).[36] On the other hand, Koch adds further ontological-metaphysical main theses to the subjectivity thesis. A consideration of

---

[36] The other three corollaries are the ontological person thesis (subjectivity is necessarily embodied as a person among potentially many persons), the epistemic perspectivity thesis (the complete description of the real is impossible, since individuals are ultimately only knowable perspectivally and cannot be transformed into a neutral totality or into one another) and the ontological perspectivity thesis (individuals are *appearances* insofar as they are essentially related to spatiotemporally embodied subjectivity without being reducible to subjective states, which is why their subject-reference belongs to their objective being); cf. Koch (2006a).

the overall correlation of his hermeneutic realism is neither possible nor expedient here. The ontological core is described with the subjectivity thesis, as all other elements are embedded in the correlation based on this correlate of spacetime and bodily subjectivity. Koch's description of the correlation as a "relation of reciprocal, interrelated, essential dependence" clearly presents the ontological-natural-historical correlation between space and time and the bodily subject as such. Although the cited formulation speaks of *at least one* spatiotemporal subject or bodily subjectivity as an abstract term, there must inevitably be several in the case of human subjects, since Koch adds a further correlate of individual (perceptual) consciousness and discourse or discourse community.[37] This is necessary for the constitution of spatiotemporal, i.e. corporeal, and truth-knowing human subjects.

As mentioned, this idea is concretised in the two successive corollaries of Koch's personhood theses: "Subjectivity knows itself a priori as a corporeal and temporal being, i.e. as a person" (also for the following Koch 2006b, pp. 331–332; my translation) (epistemic) and "subjectivity is necessarily embodied as a person among potentially many persons" (ontological). When these two theses are combined with the subjectivity thesis, Koch takes a position that will appear surprising against the background of the *retrojection strategy* of realist correlationism reconstructed in the next chapter (Sect. 7.3.3). Meillassoux uses this term to describe any attempt by correlationism to affirm the chronological and natural-historical antecedence of the natural external world, while at the same time preserving the primacy of correlation (or, in my proposal, one of the tenets of systematic correlationism) by defending a logical or ontological necessity of correlation, or a subjective correlate for the natural external world or the passage of time. Koch must now assert the following on the basis of his personhood and subjectivity theses: *In order for a concrete spacetime system to be possible at all, there is the ontological necessity that a discourse community of spatiotemporally embodied subjects arises within the spacetime system at some point in time and somewhere and that discourse community refers, in thought, perception and speech, to individuals (i.e. other spatiotemporal elements of the system – meaning concrete natural entities of the natural external world).* This thesis is also a special variant of the retrojection strategy, as it seeks to be able to uphold scientific realism. Although Koch gives his project the name of hermeneutic realism, he does not completely abandon idealism, but explicitly distances himself from it. His position

> is not a *metaphysical realism*, according to which the real is independent not only of each of our opinions, but also of the fact that there are opinions at all; nor is it a reductive *scientistic realism*, according to which natural science and ultimately physics alone set the standard for what is the case and exists. It is also not a *neutral realism* that answers the question of which categories of objects exist in a highly inclusive way, as Markus Gabriel advocates; rather, it takes a courageous stand in favour of life-world individuals and persons and, on the other hand, also allows physics to come into its own. Koch 2016, p. 8; my translation.

[37] For a more detailed account of what follows in Koch's view, cf. Koch 2023.

In this sense, he describes himself as a hermeneutic realist. Since physics comes into its own and describes the preconditions of the discourse community that are necessary under laws of nature, Koch's position implies a form of mild scientific realism. However, the subjectivity thesis contradicts the strict SI of the natural external world, as, according to it, the latter is ontologically dependent on subjects and the discourse community. Koch thus contradicts OR and OSR. According to his own statement, he attempts to be a realist in relation to life-world objects without becoming a naturalist (cf. Koch 2016, p. 8). In the case of hermeneutic realism, the problem of comparison takes the form of the argument about PII: Black's twin sphere world scenario is ruled out because, according to him, there is no internal perspective in this world. This is a variation of the usual comparison problem, as it is not related to our natural external world, but to imaginable external worlds in general.

It is therefore crucial that Koch argues for a clear and explicit form of the OCT thesis with the correlate between subject/discourse community and natural external world/spacetime: In hermeneutic realism it is claimed that the discourse community is necessarily dependent on the natural external world in terms of natural history and laws of nature, that subjects are causally dependent on the discourse community and that the natural external world, understood as spacetime, is logically dependent on individual subjects. Koch attempts to combine this with a realism about scientific and life-world entities that are epistemologically and causally independent of the subject. Since, according to his argument, the ontological dependence of subjects also applies to them indirectly (via individuation of spacetime points), I have shown that a correlationist realism is advocated in the case of Koch.

6.3.4 Correlationisms, Anti-Realisms and Objective Realisms

Koch and McDowell's radical Sellars advocate a correlationist realism, Stanford a correlationist OSA and van Fraassen and Wray a correlationist ESA. I hinted at other likely candidates; it can be assumed that there are other such realisms, even if the tandems of correlationism and either ESA or OSA are the more obvious combinations. In my view, this is mainly due to the fact that SI and the thesis of the single real external world can easily be linked argumentatively, as I have argued in Sect. 3.2.[38] However, the insistence of many OSR proponents on the SI criterion as a hallmark of a robust realism provides another strong reason why a departure from the objective realist position is often understood as an anti-realist opposition, even though it may only be a correlationist one. I have argued here against the claim that correlationism and realism cannot be combined in two respects: Systematically, the impossibility cannot be justified, since the theses of realism and the tenets and reasoning schemes of correlationism can be combined without

[38] Cf. Sect. 3.2, p. 52.

contradiction. Moreover, as the examples show, it is appropriate to understand Koch and McDowell's radical Sellars as realists and correlationists who present different, softened forms of the independence criterion.

If there is sympathy for ESA or OSA in today's debate, I believe it is for the following reasons: As I have argued throughout the first part of this book, the three challenges of UTE, historical and anti-totality arguments cannot be met by OSR. Now, this chapter shows that the comparison problem and the correlationist circle and two-step can also get some OSR variants into trouble. The reasoning schemes serves as the ground for a variety of arguments that shake SI as an ontological criterion. I believe that these two impulses motivate many contemporary anti-realisms. However, I am not advocating correlationist anti-realisms here, since I think the position is ultimately mistaken, which is the topic of my next Chap. 7. The conjecture concerning this mixed motivation is nonetheless instructive in providing an impetus for some of the louder voices (at least in philosophy of science) of realism: The consideration and evaluation of the philosophical difficulty behind anti-realisms deserves to be taken seriously in its complexity and strength. Another valuable result of the preceding sketch of correlationist positions is that, with systematic correlationism, we have gained a concept that sheds new light on already established analyses of anti-realism, such as Demmerling's distinction between hermeneutic forms of anti-realism (to which he counts Heidegger, Gadamer, Derrida and Rorty) and constitutional forms:

> What is experienced as reality or identified as a thing within it is the result of individual efforts of understanding and interpretation. Unlike the phenomenalist vote in favour of anti-realism, the hermeneutic variant also problematises the notion of something given (be it ideas or sensory data). Rather, the supposedly given is considered to be the product of a historical development, in the course of which different forms of thought and language develop, which only provide the framework (changeable at different times and places) within which reference can be made to reality at all. In this way, hermeneutic anti-realism cuts the solipsistic roots that belong to the traditional versions of phenomenalism [...]. The individual consciousness is replaced by a community of speakers and interpreters. In the terminology of the newer philosophy of mind: Compared to the phenomenalist variant, hermeneutic anti-realism is externalist, at least in a social sense. Demmerling 2004, p. 34; my translation.
>
> The objections that are raised against realism in this context [on the part of constitutional anti-realism; J. V.] essentially refer to the fact—and here there is a certain proximity to the position that I described above as "phenomenalist"—that we cannot conceptualise the world without our own acts of understanding. As a result, we cannot stand beside ourselves in order to judge "from the outside" whether and to what extent the connections we have created between the elements of the manifold correspond to reality. For us, there is no perspective that would allow us to take God's point of view and look at the world from it. The structure of things-in-themselves, as Kant would have put it, remains unknowable to us. Because it is unknowable to us, it is—I would now conclude—also irrelevant to us. The nature of the world that is possibly independent of us, if there is such a thing, can also be removed here—as already in phenomenalism—from our mental housekeeping without loss, since nothing changes as a result. A constitutional view of things shares with the position of hermeneutic anti-realism the idea that it is we who have a particular form of structuring influence on what we can experience. However, it is not so much about the idea of a (historically or culturally predetermined) conceptual scheme with the help of which we structure the raw content of experience. Rather, the catch of the constitutional alternative to

> the hermeneutic approach lies precisely in the fact that, according to it, object and concept are produced simultaneously and are firmly interwoven. Demmerling 2004, pp. 39–40; my translation.

Demmerling's description of the fundamental reasoning scheme for hermeneutic anti-realism and for constitutional anti-realism are two differently radical forms of the correlationist circle, based on the comparison problem. In the latter case, this amounts not only to a refutation of OR, but also to a correlation of object and concept, which are necessarily related to one another. In Demmerling's sketch, forms of agnosticism and quietism towards a subject-independent natural external world are inscribed in the constitutional anti-realisms, and not just scepticism, as in phenomenalist variants. These attitudes originate from the same source as the positive, anti-realist forms of the OCT, ECT, OCTNEW and ECTNS tenets in constitutional anti-realism: A higher degree of radicality on the first correlationist spectrum (compared to hermeneutic realism) and a minimum degree of strength (at the level that such anti-realisms relate at least to a metalocal, if not global, domain) on the second spectrum result in a qualitative difference in the forms of possible anti-realisms. Only from this starting position is it possible for these forms to defend the correlation of concept and object.

At this point in the consideration of correlationism, I could try to explain why only a realistic correlationism has a chance of success in philosophy of science and how it deals with the three challenges of the first part. But every correlationism faces Meillassoux's problem of ancestrality and I will show in the next chapter that none of the proposed correlationist solution strategies is convincing. In doing so, my further development of the problem of ancestrality will counter the widespread objections to Meillassoux's formulation, and will show that the further development provides indeed a fundamental objection to correlationism.

But first, the preliminary definition of new realism from the previous part of this book can be specified: NEW REALISM is any philosophical stance in a metalocal or global realism-antirealism debate that advocates realism (i.e. the RORE theses) but rejects OR *and* correlationism. In doing so, any new realism takes the latter two options seriously by considering the comparison problem and correlationist reasoning schemes without completely rejecting subject-independence. Before discussing in my final Chap. 8 how a new scientific realism understood in this way might be introduced into the relevant debate, in Chap. 7 I will reconstruct and develop further the central argument that Meillassoux introduces against correlationism: the problem of ancesterality.

Chapter 7
The Problem of Ancestrality

The central task of this chapter is to show why systematic correlationism—whether as correlationist realism, ESA and OSA—runs into unavoidable problems in philosophy of science. To this end, I will reconstruct and modify Meillassoux's arguments, which he introduces in the context of the *problem of ancestrality*.[1] To this end, I will integrate into my analysis the extensive criticism that has been levelled at the ancestral question in the course of the debate. In addition, my consideration of ancestrality can help to defuse the sometimes scathing criticism of Meillassoux's concept in essential respects by modifying and clarifying it at vital points. A key to the successful reformulation of the problem of ancestrality lies in the differentiation of two levels of the problem that I will discuss in what follows: It arises in the correlationist treatment of scientific discourse on the first, *scientific-discourse level* and in the question of the genesis and enabling conditions of each correlation on the second, *ontological level*. In the course of my discussion of both, I will again take up the example of correlationism from the previous chapter and show why the inherent problems of these positions become apparent when they are confronted with the ancestral question.

I begin my investigation with Meillassoux's introduction of the problem (7.1) by outlining the basic intention and explaining my reinterpretation on the two levels in more detail (7.1.1). I then discuss Meillassoux's definitions of ancestrality and arche-fossils (7.1.2). In the second subchapter, looking at the first level, I highlight the more superficial way of formulating the problem of ancestrality, namely as the question of how statements about ancestral entities such as the Big Bang can be understood in scientific discourse within the framework of systematic correlationism (7.2). The argument here is reconstructed as a dilemma at the scientific-discourse level, including some points of criticism (7.2.1), in order to then deal with the, in my opinion, central objection to it. This consists in

[1] The interpretation and presentation of Meillassoux's philosophy presented in this chapter builds on considerations from previous works; Voosholz (2014, 2019).

J. Voosholz, *Towards a New Scientific Realism*, Synthese Library 518,
https://doi.org/10.1007/978-3-031-98886-8_7

the accusation of a pseudo-problem (7.2.2), which is discussed by Meillassoux from a Kantian perspective, but which I will also present from the perspective of constructive empiricism as a stronger and clearer example of a correlationist ESA. My conclusion will state that the problem of ancestrality cannot be refuted in this way, but that certain forms of correlationism only really run into problems on another level (7.2.3).

This second, ontological level of the problem is dealt with in the concluding Sect. 7.3. The second core of the ancestral challenge is to explain the genesis of the respective correlation and to allow the ontological enabling conditions of the correlates and co-relata within one's own correlationist position. In a first step, I explain why Meillassoux argues that this is only to be achieved in correlationism by means of a *creatio ex nihilo*, including further points of criticism, for which I briefly consider this *terminus technicus* (7.3.1). I then discuss the core strategy of systematic correlationism in order to solve this second level of the problem of ancestrality with ontological arguments: the retrojection strategy (7.3.2), which can also be demonstrated by the examples of the radical Sellars, Koch's hermeneutic realism and Stanford's instrumentalist OSA. Finally, I explain why a synthesis of the two levels always succeeds in putting systematic correlationism into explanatory difficulties (7.3.3). From this synthesis, I conclude that correlationist alternatives to OSA are not fully viable in my view, but that lessons can be learnt from correlationism and the problem of ancestrality in order to develop a new scientific realism.

7.1 The Problem of Ancestrality in Meillassoux

7.1.1 The Ancestral Question in Meillassoux and My Reinterpretation

In *After finitude*, Meillassoux claims that correlationism runs into insoluble ontological and philosophical difficulties because every absolute is considered unthinkable and unattainable (cf. e.g. Meillassoux 2008, pp. 28–29). According to him, it is true that correlationists are right in assuming that absolute beings—necessary and subject-independent entities—and with it dogmatic metaphysics had finally proved untenable. The error of correlationism, however, consists in considering the absolute to be generally unthinkable and unattainable on this basis.

Meillassoux challenges correlationism by formulating the *question of ancestrality*. What Meillassoux understands precisely by this is the topic of the next section. From his considerations and their criticism, an argument can be reconstructed that can be applied to any correlationism. To this end, a distinction is made between two argumentative levels on which the ancestral problem can be posed. Through this reinterpretation and extension of the challenge, I will show that much of the criticism levelled at the problem of ancestrality arises from a one-sided and therefore truncated reading, which Meillassoux's treatment of ancestrality, however, invites. In contrast to the previous chapter on correlationism, I begin my presentation with the distinction between the two levels that I have developed and successively introduce Meillassoux's considerations and its criticism.

I would like to call the first level the *scientific-discourse* and the second the *ontological* level of the problem. Jointly they encompass what I see as all the essential points of Meillassoux's attempt to attack productively the many different positions of correlationism. My proposal to distinguish two levels of the ancestral problem will show that the *systematic correlationisms* analysed by way of the examples in the previous chapter can either be attacked on one of the two levels, or encounter explanatory difficulties by synthesising both levels. In the latter case, I make clear how the levels enrich and support each other argumentatively.

I cannot and will not address the question of whether and to what extent the problem of ancestrality can fulfil the task that Meillassoux assigns to it in his theoretical project, as this lies outside the scope of my research question here.[2] Nevertheless, it is important to point out that Meillassoux does not stop at the concept of the ancestral, but introduces an extended form of ancestrality in the course of his train of thought, which he calls *dia-chronicity*. This concept, and the considerations that give rise to this extension, can best be understood if they are discussed against the background of alternative philosophies of physical time. However, in the context of my research question, this consideration cannot even be made cursorily, because the focus here is not on Meillassoux's conception of physical and ontological time or his ontological proposal, but on the implications of his negative insights into correlationism and ancestrality for the correlationist philosophy of science. In the final chapter of this thesis, the topic of the philosophy of time and Meillassoux's ideas for an ontological proposal are taken up again, albeit not exhaustively. My task in the following is a reinterpretation of the question of ancestrality, a reading that looks beyond Meillassoux's context and a systematic refinement of the ancestrality argument.

7.1.2 Definitions: Ancestrality, the Problem of Ancestrality and Arche-Fossils

The rhetorical gesture with which the ancestral critique of correlationism is presented in *After finitude* makes it clear: Meillassoux introduces his problem—*the question of ancestrality*, which later becomes the question of dia-chronicity—as a provocation. With the correlationist circle, Meillassoux claims to have invalidated

[2] Some interpreters and critics even argue that it is not even intended to refute correlationism (and in no way does); cf. Rutten 2024, p. 8: "So ancestrality seems not really an *argument* against correlationism. It doesn't refute it. At most it results in a situation of *parity* between correlationism and realism, thus arriving at an *aporia*."

any dogmatic metaphysics. In the context of this book, dogmatic metaphysics can be read as including OR in ontology and OSR in philosophy of science. With ancestrality, he now means to present correlationism—on his model, especially the strong variant—with a challenge that it cannot overcome.

To construct his argument, Meillassoux introduces his concept of *ancestrality* (cf. also for the following Meillassoux 2008, esp. pp. 9–27). Ancestrality refers to the period *before* the first subject(s) and thus before the emergence of the phenomena of consciousness, mind, cognition, conceptual perception, knowledge and thinking. If it is assumed that humans are the only subjects in the universe, then the formation of our solar system, for example, is an *ancestral event*. For Meillassoux, the time before each subject as conceived by the current natural sciences—and generally by any cosmology that does not assume a spiritual-subjective origin of the cosmos—is a time before the existence of the decisive correlations:

- I will call "ancestral" any reality anterior to the emergence of the human species – or even anterior to every recognized form of life on earth.
- I will call "arche-fossil" or "fossil-matter" not just materials indicating the traces of past life, [...] but materials indicating the existence of an ancestral reality or event; one that is anterior to terrestrial life. Meillassoux 2008, p. 10.

In order to criticise and overcome correlationism, Meillassoux introduces the *problem of ancestrality*. He argues, in a nutshell, that correlationism on the one hand declares ancestrality to be impossible, but on the other hand ascribes objective or intersubjective validity to scientific statements about it. In this way, correlationism creates a contradiction: true statements with impossible referents. Correlationism thus negates the question of whether it is possible for the time of correlation to lie in the absolute, scientific time of the ancestral. According to Meillassoux, this firstly forecloses the possibility of comprehending the discourse of the natural sciences, as it necessarily contains such ancestral statements. Secondly, the *origin* of the respective correlation, its coming into existence, is only explainable by a *retrojection of the past from the present*, which is not only insufficient, but also leads to self-contradictions.

Arche-fossils are, as can be seen from the quotation above, objects, processes or events that exist in the present and which require us to assume that nature is older than any concrete subject, and even than subjectivity as such. Meillassoux uses our solar system as an example of an arche-fossil. Let us assume that the biological beings on our planet are the only entities in nature as a whole that possess subjectivity. According to everything we know about astrophysics, geology, chemistry and evolutionary biology, this planet must already have undergone certain stages of formation, i.e. certain natural processes, in order to come into existence. Arche-fossils then point to the necessity that the genesis of the solar system precedes subjectivity in time, i.e. that this process had properties that were not bestowed on it by our intervention, this process had *subject-independent* properties.

Thus one can state: Ancestrality is the time before any concrete subject and the emergence of subjectivity in nature. Arche-fossils are empirical evidence for and indications of ancestrality. Meillassoux's problem with ancestrality lies in the fact that correlationism recognises the ancestral explanations of the current natural sciences for arche-fossils, but does not grant subject-independence to ancestral entities, and therefore runs into contradictions or has to interpret the scientific discourse in a revisionist way.

7.2 Surface: The Scientific-Discourse Level

7.2.1 Correlationist Dilemma: Self-Contradictions or Revisionist Reinterpretation of the Scientific Discourse

This sketch of the problem already shows the direction that Meillassoux is taking. One way of approaching the first level, which I call the *scientific-discourse* level of the ancestral problem, is to follow the most common way of describing the argument. Meillassoux introduces his problem of the ancestral as an objection that could be raised from the position of a scientist with a good knowledge of current philosophy as well as by a philosopher of science with naturalistic inclinations. The problem arises when a correlationist attempts to determine the meaningfulness of a scientific statement about an event before the emergence of life (or the modern human or the mind) without being able to acknowledge its strictly subject-independent status.

I have shown in the previous chapter of this book that even in all the correlationisms analysed for the use of the correlationist two-step, a law-of-nature-like, causal and/or natural-historical dependence of the subjects or subjectivity on a temporally preceding natural external world is assumed, whereby at the same time strict SI of that world is called into doubt in each case. At this level, the problem of ancestrality, together with the scientific discourse on arche-fossils and ancestral time, indicates that no realistic, literal interpretation of the language of the discourse seems possible without running into contradictions. As Meillassoux himself recognises, this problem arises not only from ancestral statements about the pre-human past, but also from statements about a future without subjects or spacetime domains that are otherwise separated from present observers for principal reasons. But it will become clear from the context of the deeper, ontological level why Meillassoux argues primarily with the distant past.

But back to the problem of the impossibility of a language-philosophical realist, literal interpretation of the scientific discourse without contradictions: In the realisms analysed here, scientific statements, including those on the Big Bang and the accretion of our solar system, are not to be reinterpreted. According to them, these sentences and descriptions have no hidden meaning, but are to be understood literally. In this context, Meillassoux poses the question of how it is possible that the ontological or logical conditions for the existence of ancestral entities or the ancestral natural external world were not given, yet these form the law-of-nature-like, causal or natural-historical precondition for subjects. Meillassoux suspects a dilemma: either the correlationists have to reinterpret the discourse in a revisionist way after all, or they end up in a self-contradiction.

7.2.1.1 Counterargument I: Alien Subjects

Following this way of formulating the problem, I will refute two obvious and similar counterarguments: The first is an objection that argues against ancestrality by introducing potential extraterrestrial subjects that could have observed the emergence of life on our planet (even the accretion of the solar system or of our galaxy). This is formulated, for example, by Harald Wiltsche, who notes that only temporal distance and spatial proximity *together* would result in an ancestral entity (cf. Wiltsche 2017). Thus, the accretion of the solar system is an unobservable event for us; however, due to the time required for the propagation of light, this does not apply to a potential observer that is 4 568,2 million light years away, as the light from the formation of the protosun and solar-planetary accretion disc has just reached her (cf. on the age of the solar system, etc. Bouvier and Wadhwa 2010). Wiltsche aims to use this as a defence of phenomenology (as a form of correlationism):

> On Meillassoux's view, ancestral things confront phenomenology with a dilemma because of the impossibility of them becoming present in conscious experience. But if the argument in this section is correct, then Meillassoux's objection is simply not sound. Given the appropriate spatial location, we can and in fact do "observe" ancestral things. For instance, many of the objects and events we "see" through the Hubble Space Telescope ceased to exist or occurred prior to the existence of life and consciousness. And yet it is possible for these objects to become present in conscious experience. Wiltsche 2017, p. 817.

Wiltsche's empirical-physical argument against real ancestrality posits that due to the connection between spacetime, the relativity of time and the speed of light or information, which Meillassoux ignores, there are no real ancestral entities. However, this argument only works insofar as it is legitimate to introduce a counterfactual conditional by postulating fictitious subjects in order, for example, to make the accretion of our solar system a non-ancestral event. Here, Wiltsche benefits from the fact that Meillassoux allows counterfactual conditionals as a means to understand actual, in principle observable entities that are factually unobserved, such as the tree in the forest or my desk when I briefly go to the kitchen, as part of the correlation.

However, even if the postulation of fictitious subjects were legitimate, Wiltsche's argument against ancestrality would be misguided: After all, alien subjects also once began to exist. Their existence must similarly be preceded by natural processes, since as observers of our genesis they are necessarily part of our natural history. Like us, the alien subjects require a body for the observation of our solar system, which must have been created beforehand through the interaction of objects in natural processes. Some natural processes, then, preceded the existence of the alien subjects and are therefore ancestral.

Moreover, in the currently prevailing theories of cosmology about the origin of the universe—that is, those that assume a Big Bang as an ancestral primordial event—it seems evident that at the beginning of the universe no organisms existed that could be considered as carriers of subjectivity. This is due to the empirical fact that no complex particle compounds existed for a considerable time after the Big Bang, and nor even did atoms for a few hundred thousand years.[3] Cosmic microwave background radiation can be cited as an arche-fossil for these ancestral events in this context.[4] This argument has not yet been put forward by Meillassoux, but in view of the definitions it seems obvious to me.[5]

7.2.1.2 Counterargument II: The Ancestral Witness

A second objection, which at first glance is quite close to the first, is explicitly dealt with and refuted by Meillassoux: If the correlationist concedes that there was no biological being as a subject at ancestral times, there could nevertheless have been an "ancestral witness" (also for the following Meillassoux 2008, p. 11) present, a (fictitious) subject without corporeality. This quasi-divine entity would then have the function to "turn[.] every event into a phenomenon" for herself, which would lead to a justification of our natural history as a phenomenal history of the ancestral witness.[6] But how can this metaphysical scenario be reconciled with correlationism? Meillassoux argues that the scenario of the ancestral witness cannot be introduced with consistent correlationism. The ancestral witness would be an absolute and necessary being, which is ruled out by strong correlationism and also by my systematic correlationism. Instead, Meillassoux's subjectalism must be adopted in a special form in order to create the conditions for this scenario. This can be imagined as follows: The ancestral witness is the subjective side of the chronologically and ontologically first correlation, since it stands in a phenomenal, indissoluble relationship of entanglement with the entirety of natural history at all times. An external world event may also be phenomenally given to me, but it is in any case given to the ancestral witness and can thus always be defined by its dependence on at least one subjective instance. To this end, it must be assumed that the correlate of natural history/ancestral witness is eternal and exists by itself, independently of the existence of human subjects. This presupposes that this correlate is absolute—necessary and subject-independent—, which according to Meillassoux is the hallmark of subjectalism: a correlation is hypostatised to

[3] Cf. e.g. on the proof of the ion HeH^+ as the first molecule Güsten et al. 2019; on the possible starting point of chemical-biological evolution cf. Kauffman 2020.

[4] For a general presentation, cf. e.g. Scott 2006; On the probative value of cosmic microwave background radiation against the background of primordial nucleosynthesis, cf. e.g. Mathews et al. 2018.

[5] On why I believe that Meillassoux's strategy of isolating his own position from the current hypotheses and theories of the natural sciences in order not to bind himself to their paradigms is fundamentally misguided in NSR, cf. my Chap. 8, pp. 315–337.

[6] Cf. e.g. for such a speculation about a *god of the philosophers* Tetens 2015, 2017a.

something absolute; in particular a vital, subjective, mental and/or conscious aspect of the correlation is treated as *absolutum*. Here Meillassoux objects:

> But correlationism is not a metaphysics: it does not hypostatize the correlation; rather, it invokes the correlation to curb every hypostatization, every substantialization of an object of knowledge which would turn the latter into a being existing in and of itself. To say that we cannot extricate ourselves from the horizon of correlation is not to say that the correlation could exist by itself, independently of its incarnation in individuals. We do not know of any correlation that would be given elsewhere than in human beings, and we cannot get out of our own skins to discover whether it might be possible for such a disincarnation of the correlation to be true. Meillassoux 2008, p. 11.

With strong correlationism in all its varieties, Meillassoux has, in his opinion, a means at hand against the subjectalist hypostasis, which introduces the correlation between natural history and ancestral witness as an absolute correlation. Central to my further argument is that systematic correlationism can in no way apply this metaphysical trick in the domain of philosophy of science without explaining why such knowledge of the ancestral witness is possible. But nobody in philosophy of science seriously attempts such proofs of a god-like being as a means of an argument from the metaphysics of nature. Against this background, the scenario of the ancestral witness cannot possibly be used by a consistent correlationist against the scientific-discourse version of the ancestral problem.

7.2.1.3 The Correlationist Dilemma Realised: Revision of the Discourse or Self-Contradiction

The ancestrality problem cannot be solved in either of these ways—especially since neither of the available strategies of the correlationist objection is strong, as I will show. But in order to present the problem introduced above in all clarity, it is first worth asking how correlationism can be argued against from the standpoint of current science:

As already explained, ancestral events become inferable through arche-fossil materials. Arche-fossiles can be analysed using scientific methods—such as thermoluminescence or measuring the absolute age of a rock according to the decay of certain isotopes (cf. Geyh 2005; Stokes 1999). The results show that it makes sense to assume a period of natural processes before humans. With this statement, which is taken for granted in geology and geophysics, Meillassoux strives to expose a contradiction in correlationism: Only world *and* human, referent *and* language, object *and* subject etc. allow for a correlation. The ancestral statement, however, speaks of a *subject-independent* world that precedes it, a world without humans, referents without language and objects without subjects. This strategy seems dubious at first. Meillassoux asks more precisely what status this statement can have for any correlationist:

On the one hand, the ancestral statement is *true* for the correlationist, or at least in principle *truth-apt*: It is a hypothesis that has been tested with scientific methodology through intersubjective research practice and that has been put forward in scientific discourse. Since it does not depend on the exact content but only on the general form of the hypothesis, *this* statement of Meillassoux is permissible and reasonable. It is irrelevant, for example, that the object *X* is *n* years older than the emergence of life/modern humans, it is only important that many concrete objects, processes and events can be dated to a time before the emergence of life/humans, i.e. when they needed to be subject-independent. Meillassoux writes that for the correlationist, the "ancestral statement is true insofar as it is founded upon an experiment that is in the present—carried out upon a given fossil-material—and also *universalizable* (and hence by right verifiable by anyone)." (Meillassoux 2008, p. 16). Here one can object that present experience, even in correlationism, is only supposed to provide evidence for the ancestral statement, while the truth-aptness (and possibly the truth) of the statement is independent of it.[7] But it is possible to propose a philosophically honest reinterpretation of Meillassoux, since he only has to claim that the scientific discourse contains factual claims about ancestral entities and that the correlationist ascribes truth-aptness to this discourse. This would then mean that the correlationist recognises at this level that events occurred before the emergence of consciousness, and that processes took place before the existence of the mind, life, and so on. This, in turn, would require the prior, subject-independent existence of entities from a temporal perspective, as any direct access to past, ancestral entities is blocked to us in time.

On the other hand, according to Meillassoux, the ancestral statement is, strictly speaking, correlationally impossible and thus not permissible in the eyes of the correlationist: Here, the actual existence of entities would not only be assumed *de dicto*, but *de re* outside of the correlation, which is also claimed to be indirectly knowable.[8] However, due to the correlationist circle, this is impossible or (depending on the variant) at least nonsensical, unthinkable or unknowable.

Even if Meillassoux's concept is replaced by the more specified notion of systematic correlationism, there is still a problem here. Since systematic correlationism cannot admit that ancestral statements in scientific discourse are to be interpreted literally, are true or truth-apt and imply the subject-independence of natural, ancestral entities, this part of scientific discourse appears at first glance to be almost impossible to interpret without contradiction. I will only consider whether the apparent contradiction can be resolved from the perspective of the four examples of correlationisms once it is clear how this scientific-discourse level of the problem can be further refined and reinterpreted in the light of the criticism it has attracted.

Meillassoux uses this consideration of the contradiction between the attribution of existence and the denial of existence for a critique of certain current philosophies of time: Even time itself, which is supposed to form the antecedent, is contra-

[7] I have Jens Rometsch to thank for this criticism of Meillassoux.

[8] In this context, by *de dicto* I mean a purely linguistic attribution of a modal property such as actuality, by *de re* a resilient truth assertion of this modal property with regard to the entity. I do not try to make a similar use of these terms as Rutten 2024, which I take to be an unfavourable reconstruction couched in the problematic terms of "world-for-us" and "world-in-itself".

dictory in strong correlationism (cf. Meillassoux 2008, p. 17).[9] According to all correlationists, the concept of time does not exist independently of the one who thinks it.[10] Even if we refrain from assuming this form of theory of the subjective intrinsic proper time for all correlationisms, a contradiction seems to emerge in correlationism if it is *not* to result in ESA or OSA with regard to the scientific discourse with its explanations, models and theories. Correlationist realism seems to run into difficulties here, and Meillassoux asks how this conflict can be resolved:

> *[T]he ancestral statement is a true statement*, in that it is objective, but *one whose referent cannot possibly have actually existed in the way this truth describes it.* It is a true statement, but what it describes as real is an impossible event; it is an "objective" statement, but it has no conceivable object. Or to put it more simply: *it is a non-sense.* Meillassoux 2008, pp. 16–17; emphasis as in original.

The point criticised here of declaring a statement to be true, partially true or approximately true, and at the same time claiming that the referent of the statement is ontologically impossible or epistemologically unknowable, describes a self-contradiction of correlationism, if Meillassoux's analysis is correct. However, it is reasonable to assume that hardly any correlationist would claim that objective statements without possible referents should be allowed and truth-apt. Since ancestral statements refer directly to concrete entities such as the solar system, it cannot be claimed, by analogy with an argument in favour of the truth-aptness of the sentence "*All orcs in Tolkien's works have legs.*", that ancestral statements are truthful fictional statements about fictional contexts, as in fictionalist instrumentalism. Meillassoux's considerations can be translated into the following dilemma, taking into account the criticism already mentioned:

P1 Within scientific discourse, statements are asserted as knowledge the referents of which are ancestral entities.
P2 Some of these statements posit that some ancestral entities are necessary natural conditions for the existence of subjects and thus for the emergence of subjectivity in general.
P3 Scientific discourse should be interpreted literally to attribute it full validity.
IC1 The existence of some ancestral entities is a necessary, natural, temporally prior condition for the existence of subjects and subjectivity in general and we have scientific knowledge about this connection.

[9] Cf. Benoist 2023, for a critical assessment of Meillassoux's chances of succeeding in overcoming such relativities himself.

[10] Meillassoux's historical example here is again Kant, although his interpretation is not convincing at this point. Cf. e.g. CPR, A 34–36/B 51–53. Kant seemingly binds time to humans here. This has not gone unquestioned by thinkers who follow Kant positively. Sellars, for example, proposes a distinction between *time* and *time∗* in order to distinguish between space and time as forms of the intuition (without ∗) and the forms of the world that structure sheer receptivity (with ∗); cf. Sellars 1968, p. 4. This indicates that Sellars does not aim to represent a consistent correlationism, but rather understands time in the manifest image possibly like a Kantian, but in the scientific image he does not regard time∗ as a form of the intuition.

P4 Assumption of systematic correlationism: Strictly subject-independent existence of scientific and natural entities is impossible and/or knowledge about such strictly subject-independent existence is impossible.

P5 Ancestral entities do not have a strictly subject-independent existence (a) or we have no knowledge of the strictly subject-independent existence of ancestral entities (b).

P6 If the existence of A is a necessary and temporally prior condition of B, then A must be strictly independent of B.

IC2 In the case of (a), the assumption of systematic correlationism leads to a contradiction between IC1 and P6.

IC3 In the case of (b), the assumption of systematic correlationism leads to a contradiction between IC1 and P6.

C Systematic correlationism becomes self-contradictory if it attributes full validity to scientific discourse by interpreting it literally.

Meillassoux takes the truth of P1 and P2 for granted, and the scientific sources already cited in this chapter also speak in favour of these premises about scientific discourse. That P5 is a permissible application of P4 to ancestral entities also seems unproblematic; P4 is what correlationists wish to defend. They can therefore deal with the argument in three ways: Either P3 must be reinterpreted or denied in such a way that IC1 does not follow. This is the option of revisionism with regard to the scientific discourse. I will address whether this option can be defended later when discussing Stanford's instrumentalism. Meillassoux thinks this is impossible and would be undesirable. The second option is to deny P6. This amounts to the *retrojection strategy* already mentioned in the last chapter. I will discuss this in my Sect. 7.3 on the ontological level of the ancestral problem, but here it is relevant that, according to Meillassoux, the strategy leads to a self-contradiction. Finally, and this is the third way of defence, a correlationist might admit that her position leads to contradictions in her interpretation of the scientific discourse. This would involve learning lessons from correlationism, but giving up P4 and developing a form of NSR that is able to grant scientific discourse its full validity even in the domain of the ancestral. This is one of the fundamental considerations in the final Chap. 8 of this book.

7.2.2 *The Ancestral as a Pseudo-Problem: The Transcendental Counterarguments*

First, a further objection can be formulated against the validity of the scientific-discursive variant of the ancestral argument. This essentially states that although premise P6 is valid, in its application to ancestral entities the mistake is made of formulating a natural necessity in the antecedent (if ancestral entities are a natural necessity for the existence of subjects, ...), but in the consequent strict independence of ancestral entities from subjects is required and not only natural or causal independence. This counterargument accuses Meillassoux of introducing a pseudo-problem for correlationism with ancestrality by ignoring the distinction between a purely ontological level (on which ancestral entities depend on subjects) and a purely empirical level (on which subjects depend on ancestral entities), which resolves the paradox of the ancestral.

Meillassoux deals with this objection explicitly, albeit less rigidly because of his dialogical method, which I reconstruct in a first step. Since his rebuttal again uses Kant as an example and, as before, I will not attempt to justify this interpretation, I will draw on two different, current interpretations of Kant. In doing so, I show that Meillassoux's counterargument retains its effectiveness despite these two readings, and that the burden of argumentation lies on the side of the correlationist who wants to separate a purely empirical from a purely ontological level. In a second step, I formulate and invalidate a similar counterargument that can be articulated from the perspective of constructive empiricism. It attempts to separate a pure level of natural-scientific discourse from a pure meta-level of philosophical reflection on the sciences.

Meillassoux considers the *transcendental counterargument to ancestrality*: The correlation is *ontologically primary*, but it does not follow, and especially not necessarily, that at the level of natural causality the occurrence of the correlation coincides *chronologically* with the occurrence of the natural external world (cf. Meillassoux 2008, pp. 22–26). In Meillassoux's treatment of the transcendental counterargument, it is argued from a Kantian perspective that the transcendental subject as part of the Kantian correlation is a time-independent, universal condition of the possibility of subject-consciousness: "But a condition for objective cognition cannot be treated as an object" (Meillassoux 2008, p. 23). This statement from the Kantian prohibits the conflation of transcendental and empirical-scientific levels of observation. Meillassoux's Kantian opponent expresses this as follows:[11]

> As for the transcendental conditions of cognition, they cannot be said to arise or to disappear – not because they are eternal but because they are "outside time" and "outside space"– they remain out of reach of the scientific discourse about objects because they provide the forms for this discourse. Meillassoux 2008, p. 23.

In this reading, the ancestral reveals itself—claims the Kantian—as a pseudo-problem. From her perspective, there is no such problem if a distinction is made between scientific empirical inquiry and transcendental philosophy: "on the side of the object, bodies are born and die; while on the other side, conditions provide the norms for knowledge of the object." (Meillassoux 2008, p. 23). Although there is a logical connection that ontology or epistemology is the foundation of natural science, there is no reverse connection. Here, Meillassoux attempts to begin refuting this conviction of the Kantian opponent—as representative of every correlationist who attempts the "immunizing" (Meillassoux 2008, p. 24) of ontology or epistemology against the empirical in this way. A Carnapian suspicion seems to

[11] Cf. for a critique of ancestrality expressed in the debate, which essentially corresponds to such an argument Saldanha 2009, esp. pp. 317–320.

impose itself on Meillassoux at this point:[12] The binary separation between genesis and validity, which is in the background of this consideration, cannot be sustained.

Before this idea, which Meillassoux develops into an argument, is presented in the next section, the question arises as to whether it is legitimate to postulate a Kantian who argues in this way or whether she is a straw man—especially since it was indicated in the last chapter that Meillassoux's historical readings of Kant, Wittgenstein, Heidegger and other authors are subject to persistent criticism in the research literature. I consider it legitimate to reconstruct Meillassoux's transcendental counterargument and to regard his answer as meaningful, since this hypothetical objection to the problem of ancestrality from a Kantian perspective also arises from other readings of Kant, as I will briefly sketch.

7.2.2.1 The Argument of the Kantian

The technical weaknesses of Meillassoux's reading of Kant become irrelevant if the transcendental counterargument also arises from sound contemporary interpretations of Kant. To this end, it must be shown that current Kant-interpretation (or better, a family of Kant-interpretations) advocates three points: (1) the transcendental level of consideration of reality and the empirical level of consideration of reality are categorically distinct; (2) on the transcendental level, relationships of dependence exist between principles of knowledge in an unchanging, extra-temporal context that are necessary for subjectivity and knowledge of objects; (3) temporality, time and events of the natural external world are to be located exclusively on the empirical level of consideration. Together, these three points are sufficient to put forward the counterargument against the ancestrality problem reconstructed above.

All three tenets are explicated in clearly heterogenous, contemporary interpretations of Kant. As examples, I will cite the readings of Lucy Allais and Kiyoshi Chiba. In her book *Manifest Reality*, Allais establishes an interpretation of Kant according to which: "Kant's transcendental idealism includes idealism about spatio-temporal objects, i.e. that there is a substantial sense in which such objects are mind-dependent." (Allais 2015, p. 101.) Following prior work by Rea Langton (cf. esp. Langton 1998), she argues further: "Kant also thinks that spatio-temporal reality is incomplete in a way which could make sense only if it were not metaphysically fundamental" (Allais 2015, p. 207 and cf. also the arguments in Part II, pp. 176–204). According to Allais' Kant, the conditions for the existence of spatio-temporal reality thus lie on a transcendental level, which itself cannot be spatio-temporally constituted. In Allais' Kant's view, spatio-temporal reality is not metaphysically fundamental, but time and space are forms of the intuition—namely intuitions a priori—, so that the existence of spatio-temporal properties

[12] Cf. for a critique of any philosophy that immunises itself against the processes on the side of the empirical Carnap's arguments for the impossibility of an intersubjective language without reference to the empirical and physical Carnap 1931, esp. pp. 447–450.

is bound up with the existence of subjects that have intuitions.[13] Thus both points 1. and 2. are fulfilled: 2. is fulfilled, since according to the interpretation the conditions of subjective experiences themselves cannot be spatio-temporal. For this, the transcendental level on which these connections and dependencies are elaborated must be categorically distinct from the empirical level, since the latter always describes a state or a change of a spatio-temporal object or the experience of a (temporal) phenomenon in an intuition; that is, 1.

Point 3. can also be demonstrated in Allais' reading: It follows from the manner of the aforementioned interpretation, but becomes particularly illuminated when it is emphasised how Allais relates Kant's transcendental idealism to empirical realism. According to Allais, Kant is neither an epistemological anti-realist regarding the transcendence of experience (cf. Allais 2015, pp. 209–217), nor an anti-realist in philosophy of science regarding unobservable objects (cf. Allais 2015, pp. 217–221), but a structural realist regarding empirical reality: "[S]tructural realists [...] hold that science gives us knowledge only of relational features of reality." (Allais 2015, p. 209 and the argument on pp. 221–230.) Following the distinction made by Worrall and Ladyman in Sect. 5.4, Allais distinguishes between epistemic and ontic structural realism. Roughly speaking, the former maintains that scientific knowledge is only possible from structural, relational elements and properties of reality, the latter that reality consists of nothing more than these structural elements and properties. According to Allais, Kant holds both positions. Point 3. can be derived from this and from 1. and 2., according to which temporality and events are located exclusively on the empirical level of observation. If the transcendental and empirical levels are categorically distinct, and only extra-temporal conditions for foundational principles of reasoning and knowledge are described on the former, the structural, relational elements and properties of reality are to be located on the empirical level. One of its essential structural principles is temporality and thus all events of the natural external world are to be located there as phenomena. Allais' Kant thus represents all three points that Meillassoux ascribes to his Kantian opponent.

Chiba proposes a vastly different interpretation, which sees Kant as a decidedly epistemological anti-realist. However, even according to this interpretation, Kant fulfils all three requirements for presenting the transcendental counterargument against ancestrality as a pseudo-problem in Meillassoux's sense. Chiba argues in favour of understanding Kant's philosophy and his coinciding affirmation of transcendental idealism and empirical realism according to the "anti-realistic two-world interpretation" (Chiba 2012, p. 89; my translation), and thus explicitly opposes realist interpretations such as that of Allais:[14]

---

[13] Cf. the interpretation of the conclusion of some arguments from the *Transcendental Aesthetics* (in Kant, CPR, A26/B41) Allais 2015, pp. 186–204, esp. p. 195: "So our representations of the structure of space and time do not present us with a mind-independent feature of reality: they are merely the forms of our intuition and do not represent things as they are in themselves."

[14] Cf. Chiba 2012, pp. 47–56.

> The contrast between the two-worlds and the two-aspects interpretation deals with the question of whether "appearance" and "thing-in-itself" are two aspects of one and the same single thing (which we come to know empirically) or not. The former [...] answers this question negatively and claims that "appearance'" and "thing-in-itself'" are two distinct entities (or at least are not two aspects of one and the same thing). [...]
>
> The two-world interpretation harmonises well with the anti-realist interpretation; accordingly, the "appearance" is understood as something cognition- and knowlegde-dependent [*Erkenntnisabhängiges*] such as imagination, intentional object, logical construct from sense data, etc., whereas the "thing-in-itself" is understood as something existing *independently* of our cognition and knowledge. Chiba 2012, pp. 67–68; my translation.

Accordingly, in Chiba's reading, Kant can be ascribed point 1., according to which the transcendental level of consideration and the empirical level of consideration are categorically distinct. It is clear, given Chiba's anti-realist two-world interpretation of Kant, that a distinction must be made between the transcendental level, on which things exist in themselves but are not knowable, and the empirical level, on which spatio-temporal objects exist that are knowable. According to Chiba's Kant, the connection between things in themselves and spatio-temporal objects is given by a transcendental affection: "Kant's theory of receptivity compels us to regard *every* sensory matter as a product of transcendental affection" (Chiba 2012, p. 349; my translation), where Chiba, with recourse to Jacobi and Vaihinger, reconstructs Kant's argument as follows:

> *Level 1*: For empirical knowledge it is necessary that sensory matter be given to us (and this is indeed given to us).
>
> *Level 2*: Sensory matter is not a product of our spontaneity [implication (I)]. In other words, the process by which it is produced in our mind (namely the affection) is one that is not dependent on our spontaneity.
>
> *Level 3*: From this it follows that this process is also independent of our knowledge acquisition. For since our knowledge acquisition only arises from the interaction of receptivity and spontaneity, something independent of spontaneity must also be independent of knowledge acquisition and cognition as a whole.
>
> *Level 4*: But what is independent of our cognition cannot, according to the anti-realism of the Kantian transcendental idealism, be a spatiotemporal object. Affection must rather be a transcendental process. (And since sensory matter is actually given to us, one can also say that something independent of cognition, namely a thing-in-itself, in the sense of the two-world interpretation, actually exists.) Chiba 2012, pp. 355–356; my translation.

There is thus a conditional relationship between things-in-themselves and subjects with empirical cognition and knowledge acquisition.[15] Thus Meillassoux's second point is also effective in Chiba's Kant: 2. on the transcendental level, relationships of dependence between principles of knowledge exist in an unchangeable extra-temporal context that is necessary for subjectivity and knowledge. This last point also arises from the core of Chiba's reading: 3. temporality, time and events of the

[15] Chiba admits that the possibility exists that things-in-themselves could be identical with subjects of cognition or even be (in my translation) "drastically different from what is normally imagined by the word "thing"; it may, for example, not be a substantial entity, but rather something like a process or structural order, even something completely incomprehensible in itself." Chiba 2012, p. 356, fn. 501.

natural external world are to be located exclusively on the empirical level. Chiba's Kant also supports this, since in this reading objects must be given in order to speak of their spatiotemporal form as events, but they never exist independently of our knowledge acquisition and general cognition, which is Kant's epistemological anti-realism for Chiba.

It has thus been shown that the contemporary interpretations of Kant by Chiba and Allais, as different as they may be, impute to him the same crucial points that Meillassoux assigns to his Kantian opponent, and that these premises are sufficient for the argument. Allais' and Chiba's Kant, despite their differences, must both argue against ancestrality in the manner mentioned. This does not save Meillassoux's reading of Kant. It should be noted, however, that Meillassoux is not arguing against a straw man here, but against interpretations such as those of Allais and Chiba.

7.2.2.2 Meillassoux's Reply to the Transcendental Counterargument of the Kantian

The refutation of the transcendental counterargument to ancestrality begins with the status of the transcendental subject as the sum of conditions of possibilities and the strict distinction of genesis and validity, of transcendental and empirical levels of consideration, that this entails. Using the example of receptivity, Meillassoux demonstrates that the transcendental subject cannot be immunised against objective and concrete discourse without contradiction, as at least Allais' Kant seemingly attempts to do. In order for "sensible receptivity and its spatio-temporal form" (cf. also for the following Meillassoux 2008, p. 24) to exist, each subject must have a very specific "point of view on the world", a single, individual perspective on reality. The entire experimental setup of the Kantian forbids any radical "view from sideways on" (McDowell 1998b, p. 214) as McDowell suggests, or a "view from nowhere" (Nagel 1986, esp. p. 25–27) as Thomas Nagel has called it: According to McDowell, Nagel and also Kant, an external perspective on the ontological situation without subjects and their cognitive and knowledge practices is not to be had. For Kant, the world can therefore only be conceived as the horizon of the subject, as a "regulative ideal" or "regulative principle" (Kant 1998a, A 509/B 537, A 537/B 545, and A 684/B 712), and never as an object that can be fully known. These things—receptivity, fixity in space-time, the horizon of the world and limited knowledge of it—are, according to Meillassoux, necessary parts of what the Kantian calls the transcendental subject; they determine, among other things, its point of view. Thus, "the transcendental subject is posited as a point of view on the world" (Meillassoux 2008, p. 24). Meillassoux uses this fact to substantiate his counter-objection:

> But if the transcendental subject is localized among the finite objects of its world in this way, this means that *it remains indissociable from its incarnation in a body;* in other words, it is indissociable from a determinate object in the world. Granted, the transcendental is the condition for knowledge of bodies, but it is necessary to add that the body is also the condition for the taking place of the transcendental. That the transcendental subject has *this*

> or that body is an empirical matter, but that *it has* a body is a non-empirical condition of its taking place – the body, one could say, is the 'retro-transcendental' condition for the subject of knowledge. Meillassoux 2008, p. 25; emphasis as in the original.

Meillassoux argues that the transcendental subject is inseparable from corporeality. He does not have to claim that precisely these bodies, with only a single evolutionary history, can produce subjectivity. It remains conceivable and possible that higher organisms on our planet[16] are the only living beings with subjectivity—but it is equally possible that there are others in other places that have other evolutionary histories and possibly no biological histories in the narrow sense. But Meillassoux notes that *corporeality in general* exists as a necessary condition for subjectivity. Asserting this for "real" or empirical subjects is still consistent with the Kantian position. It becomes problematic for Meillassoux's fictitious Kantian opponent as well as for Chiba's and Allais' Kant if Meillassoux can extend his point to the transcendental subject; for then the latter loses the status of a bundle of extra-temporal conditions and relationships of dependence—and finds itself fully exposed to empirical, objective discourse.

Meillassoux's argument in favour of the premise that corporeality also belongs to the conditions of possibility of the transcendental subject is that every single empirical subject not only exemplifies the transcendental subject, but instantiates it. *Instantiate* means that a concept is first formed and made possible by a set of entities. For example, the biological species *Oryctolagus cuniculus* is formed by all specimens of European rabbits. If there had never been any European rabbits, no such biological species would exist. *Exemplify* here means that there is a logically independent concept that is merely represented or objectified by certain entities. For example, in some varieties of Platonism in philosophy of mathematics, it can be argued that the abstract geometric figure of the equilateral triangle exists independently of the set of all concrete drawings, representations and natural cases of equilateral triangles. It is only exemplified by these drawings, representations and natural cases. The equilateral triangle is therefore an abstract object that exists outside of space and time.

In Plato, according to Meillassoux, the idea of the human exists independently of the individuals, therefore the human (and the human subject) in Platonism is only exemplified by each one of us (cf. Meillassoux 2008, p. 25). For an empiricist, "the species "man" is instantiated by individual men because this species does not exist apart from the individuals in which it is incarnated" (also for the following Meillassoux 2008, p. 25)—just as with European rabbits. The Kantian is now committed to not allowing the transcendental subject to exist "apart from its bodily individuation". Meillassoux warns that "otherwise, it would be guilty of speculatively hypostatizing it as an ideal and absolute subject." By this he means that such a reifying exaggeration of the subject would lead to an absolute idea of the

[16] Besides humans, here one can mention apes, other monkeys, dolphins, octopuses, crows and a few other species—the whole zoo of suspects.

human (or something similar). The Kantian would change from a transcendental to a speculative, Platonic idealist.

In other words, the Kantian would contradict herself by claiming that the transcendental subject exists *only* as a universal condition of the possibility of actual subjects. If the transcendental subject were a purely abstract object, like an equilateral triangle in Platonism in philosophy of mathematics, then some things would apply to it that Meillassoux's Kantian, but also Chiba's and Allais' Kant must deny. Firstly, it would be a single absolute abstractum; secondly, it would be ideal—and could only be represented by actual subjects in time and space—; thirdly, it could not contain anything that requires a concrete point of view within the horizon of the world, it could not have a perspective. Allais cannot accept these three suggestions because, according to her, Kant begins with the concrete subject and shows that each of us who instantiate the transcendental subject must always have a concrete, empirical point of view, a perspective (albeit always a different one in each case, requiring humility of us all). Allais' realist two-aspect interpretation identifies empirical subjects, which Kant conceives realistically, and the transcendental subject, which Kant conceives idealistically, as two sides of the same entity. Although Chiba interprets this differently, he must also reject these suggestions on his Kant's behalf: The whole idea of transcendental affectation requires every transcendental subject to be equally endowed with spontaneity and receptivity, and thus every transcendental subject possesses a point of view, a perspective, like many other transcendental conditions besides. So, Meillassoux's argument works: For some of these transcendental conditions to hold—those that are tied to point of view like receptivity—corporeality is an unarticulated, assumed precondition. Some corporeality of some potential subject must be given in order for a condition such as receptivity to emerge. This is what Meillassoux means when he writes that the transcendental subject must *take place*.

Receptivity needs an empirical body to take place. Since space and time, according to Meillassoux's reading of Kant,[17] emerge with the mind and therefore with the subject; the Kantian is forced to endeavour a *retrojection of the past from the present* or another ontological *ad hoc* solution in order not to become entangled in contradictions. This boils down to the fact that the ancestral problem catches up with the Kantian, like all correlationists, in the form of its *ontological* expression: *How is the transition of the transcendental subject* "from not-taking-place to taking-place" (Meillassoux 2008, p. 26) *thinkable*? Before this is elaborated in the next Sect. 7.3, it is important to clarify whether Meillassoux's refutation of the Kantian form of the transcendental counterargument can be transferred to the possible counterargument of the constructive empiricist, who also wishes to deny that the ancestral problem exists for her. Finally, I summarise the consequences that arise from the scientific-discourse level of the problem of ancestrality.

[17] Contrary to this interpretation of space and time in the *Critique of Pure Reason*, McDowell, for example, claims that Kant does not advocate this idea, see McDowell 2009c, p. 103. In contrast, the interpretations of Allais and Chiba agree with Meillassoux in this respect.

7.2.2.3 The Agnostic Argument from Constructive Empiricism

As explained above, it is possible to argue a similarly structured counterargument regarding the problem of ancestrality from Wray's or van Fraassen's position of constructive empiricism on the level of scientific discourse, which, in addition to the confusion of levels in premise P6, also offers an adapted interpretation of P3.[18] If the formal reconstruction of the scientific discursive argument from the conclusion of the previous subchapter is taken as a basis, this means that the constructive empiricist's riposte to P3 and P6 should have the effect that no contradiction follows from IC1 and P6 without a revisionist interpretation of the scientific discourse having to be made.

My fictitious constructive empiricist would therefore use two considerations from her correlationist ESA:[19] "The scientific propositions of which Meillassoux speaks are to be understood literally only insofar as observable natural entities are described in them. Ancestral entities such as the accretion of the solar system or the Big Bang are unobservable; they are mere theoretical entities. We should therefore remain agnostic about their existence, even though, with Meillassoux, we interpret the discourse correctly in the sense that it makes statements about these entities, for their ontological postulation is dependent on the best current theories and scientific practices, which can change and are underdetermined by observations. Therefore, the first intermediate conclusion of the scientific-discourse argument from ancestralitya IC1∗ must be reformulated so that the existence of *some* ancestral entities is a necessary, natural-causal precondition for the existence of subjects, and so that we only have provisional scientific theories about their properties and closer determination. Moreover, premise P6 implies a conflation of two levels. It applies at the level of a given theory, but not at the level of reflection in philosophy of science. According to it, "P6∗ If the existence of *A in a given scientific theory* is a natural partial cause of *B*, then *A* must be strictly independent of *B*", but the *real, natural ancestral entities*, about which we have no certain knowledge but remain agnostic, are subject-independent parts of the natural external world, and some of them are probably preconditions for us subjects on earth. This applies at the level of philosophical reflection. We cannot know whether they correspond to the *postulated, merely theoretical ancestral entities*, according to my constructive empiricism. Therefore, no contradiction with assumption (b) of my systematic correlationism follows from IC1∗ and P6∗."

How does this argument and the apparent concession of the constructive empiricist succeed? She admits that the hypotheses are scientifically well-founded, that the existence of a scientific discourse on our planet necessarily requires the emergence of the species *Homo sapiens* and for that some kind of accretion of the solar

[18] This counterargument is mine, but there is a critical debate about Meillassoux's concept of knowledge, which indirectly criticises his understanding of science, cf. e.g. Sacilotto 2013.

[19] The following argument seems to be roughly what Wiltsche 2017, has in mind as his solution of the problem of ancestrality in phenomenology but does not quite give it.

system, the star and the planet is required, among other things. The constructive empiricist trivially concedes that the scientific discourse is empirically conditioned by contingent natural events. However, she believes that hypotheses about the *determined existence* of such-and-such ancestral entities cannot be justified as knowledge about phenomena and that we should therefore adopt an agnostic attitude towards them. The hypotheses would only serve to explain current phenomena, such as arche-fossils. They would only be empirically adequate for our purposes and would not explain how the world came into being independently of the subject. In constructive empiricism, there is an insurmountable difference between the scientific level of observation, e.g. physical cosmology, and the meta-level of philosophy of science. This shows that subject-independent nature is not described by scientific theories, as that is not their aim and purpose.

However, this is where the weakness of this argumentation strategy lies in my view. This form of strategy is not only a theoretical possibility for Wray, van Fraassen and other constructive empiricists. It is basically open to the correlationist E(S)A and is sometimes used directly. For example, Raoni Padui argues that ancestral statements would not be understood literally by most empirical scientists (cf. Padui 2011, pp. 97–98).[20] According to him, this is not a weakness of natural science, because Kant's critical turn proved the limitation of philosophy in arriving at objective certainties. Consequently, the ancestral statements must be restricted. Both Padui and representatives of the agnostic argument from constructive empiricism defend a separation of real or natural ancestral entities, about which nothing can be known, and postulated or theoretical ancestral entities, in relation to which we should be agnostics. The former can be considered on a meta-level, while the latter are negotiated in scientific discourse.

This way of reasoning judges entire sub-disciplines, such as parts of physical cosmology and theoretical biology on the origin of life, as—in the best case—deeply misunderstood by the public and arguably most of its practitioners, or—in the worst case—as essentially meaningless endeavours. This problem seems to go beyond the usual realist objections to constructive empiricism by means of the no-miracle argument, and is also stricter than the remark against empiricism that to be engaged in the practice of science, one cannot be so agnostic about one's subject. This argument pushes further: Parts of the scientific discourse must be assumed to be empty of content or necessarily confused with this counterargument. It amounts to a revision of the discourse of the natural sciences, which should be respected according to the constructive empiricist's own standards. In my view, this hidden revision explains an often vague reservation that realists harbour against constructive empiricism. Its promise to take scientific statements at face value seems

[20] According to Padui, the literal meaning of the ancestral statement (*There were dinosaurs.*) does not have to be endorsed; the correlational statement (*For us: There were dinosaurs.*) is sufficient on the philosophical meta-level. He also accuses Meillassoux of addressing the ancestral problem to the wrong position: "Meillassoux fails to see that the ancestral question itself, as a problem of scientific or epistemic realism, is really a product of the very tradition he is trying to overcome." (Padui 2011, p. 93).

to create contradictions with the distance from theoretical entities in our judgments on the meta-level. This reservation become clear by considering the problem of ancestrality on the level of scientific discourse.[21]

The other part of the argument, the strict distinction of levels into internal scientific discourse and the meta-level of philosophy of science, seems equally untenable: the latter cannot explain the emergence of its own discourse, as it can only take place on the assumption that there are human animals with an evolutionary history and a specific ecosystem on our planet, and constructive empiricism in particular refers to the contingent social and natural preconditions of theories and practices. Their explanatory burden is often passed on to empirical sciences, which makes the distinction an important methodological and thematic one, but not a strict one in the required sense—the levels are not isolated. This part of the counterargument is similar to Meillassoux's retro-transcendental objection to the Kantian: Our concrete practices and theories require human subjects and their genesis requires an explanation, also on any philosophical meta-level, which, according to the rules of constructive empiricism, can only be provided by the discourse of the natural and social sciences.

In my view, both aspects that the fictitious constructive empiricist utilises in her counterargument to the ancestral problem, the hidden revision of the scientific discourse and the strict distinction of levels, cannot be justified. This means that this form of correlationist ESA is also affected by the challenge. I cannot rule out that another argument can be put forward by actual constructive empiricists in order to avoid the ancestral argument, but from now on I assume the burden of proof is on their side.

7.2.3 Conclusion and Outlook: Ancestrality on the Level of Scientific Discourse

To illustrate ancestrality, Meillassoux gives the example of a timeline in a popular science textbook (cf. Meillassoux 2008, p. 9), in which one could read: 13.5 billion years before our era: origin of the universe; 4.45 billion years before our era: origin of the earth; 3.5 billion years before our era: origin of life on earth. With the idea

[21] This hidden way of yielding to the scientific discourse on the ancestral, and explicitly Meillassoux's argument, while still insisting on an epistemological correlation, can be found, for example, in Žižek (2012). For an account of his treatment of ancestrality, cf. Moshe (2013, pp. 6–7). Žižek poits, with Meillassoux, that reality is independent of us (and precedes us), but nevertheless insists that we remain inscribed in the correlation. Confronted with the question of whether dinosaurs existed as science describes them, he tends to answer with "yes": "Before rushing to an answer, we should remember how relative "external reality" is with regard to our point of view, which does not mean that we "created" it, but that out of the infinite complexity of the Real-in-itself a part or slice of reality was selected as correlative to our perceptual apparatus." (Žižek 2012, p. 647).

of ancestrality in mind, it makes sense to indicate possible points in time for the emergence of the correlation (or at least *our* correlation) on this timeline.

The problem of the ancestral at the *scientific-discourse level* is that the natural sciences use ancestral statements that describe the natural preconditions of subjects, but at the same time the correlationist philosophy of science that comprehends this cannot allow the subject-independent assertion of statements about ancestral entities. According to Meillassoux, this is a dilemma, because correlationist philosophy must either interpret the ancestral statements (and thus the scientific discourse) in a revisionist way, or fall into the self-contradiction of characterising the statements as true or truth-apt, and at the same time as ontologically or epistemologically impossible.

The formal presentation of the scientific-discourse argument of ancestrality has shown that two forms of contradiction are possible: The contradiction of the correlationist consists either in recognising that ancestral statements are partially or approximately true, but impossible, or in denying premise P6 of the argument, that if the existence of A is a necessary and temporally prior condition of B, A must be strictly independent of B. Meillassoux calls the second option the *retrojection strategy of correlationism*, since the ancestral entities exist in the past, but are dependent on the correlational present. Meillassoux argues that this always leads to a contradiction; in the next subchapter I will analyse whether this judgement is correct. Since the revision of the scientific discourse is not a real option for philosophy of science, as the consideration of the counterargument of the constructive empiricist has also shown, the other two options deserve a final look.

The various options for arguing against the ancestral problem can roughly be applied to the variants of systematic correlationism discussed in the previous chapter. In the case of correlationist realisms, I consider a form of retrojection strategy to be obvious. That is because it is compatible with the attempt to show that one's own philosophy can evaluate ancestral statements in an ontologically robust sense in a way that is realist in philosophy of science, without becoming entangled in the noose of contradictions laid out by Meillassoux. This is possible by refuting or restricting premise P6, i.e. the retrojection strategy. Such an approach also has the advantage for the correlationist realist that the problem of ancestrality is then only directed against correlationist anti-realisms. In the following subsection concerning the ontological level, I will show that this response strategy underestimates the problem of ancestrality by reducing it to a conflict between correlationism and scientific hypotheses about the past.

The strategy of the ancestral counterargument, and the objection to the narrow reading of premise P3 analysed above, fits the correlationist ESA much more neatly, especially in my example of constructive empiricism, for agnosticism, or the commitment to not knowing about the natural external world or its unobservable parts, is the central feature of these forms of correlationism. The correlationist ESA claims that their preferred position can refute the premise that scientific discourse, when dealing with the ancestral, is fundamentally realist, and therefore no revisionist interpretation is needed. Correlationist ESA must offer an interpretation of this part of discourse, but, as I have shown above, it fails to do so because both the Kantian and the constructive-empiricist argument cannot do without a reinterpretation of the discourse. It could be shown that the deeper reason for this is that the strict distinction between a transcendental and an empirical level (or a philosophical meta-level and a natural scientific object level) is not tenable in principle. Meillassoux assumes that this option is fundamentally hostile to science, but this does not seem justified, either for transcendental philosophy or for constructive empiricism.[22] In synthesising the two levels and reviewing my ontological anti-realist example of correlationism, Stanford's instrumentalism, I will briefly return to Meillassoux's accusation later.

Meillassoux discusses the scientific realist way out of correlationism in relative detail in *After Finitude* and other publications. I will work out an essential argument against the viability of this solution on the ontological level of the ancestral problem, which is the subject of the following subchapter. There I localise the main argument against any ontologically robust, realist correlationism and correlationist OSA.

7.3 Depth: The Ontological Level

Meillassoux's challenge to correlationism can be formulated in a second way, which I will call the problem of ancestrality at the ontological level. This is not a completely separate formulation, but merely emphasises what the conditions for the existence of a given correlation are. This variant of the ancestral argument quickly leads to problems of justification for the various correlationisms in philosophy of science, but also in ontology. At this level, it also becomes clear why the retrojection strategy ultimately fails.

7.3.1 The Genesis of Correlation as creatio ex nihilo

With the corporeality argument against the Kantian, the scientific-discursive way of formulating the ancestral problem gradually transforms into the second way, the *ontological*. The problem of the ancestral can be posed at this level without recourse to a literal meaning of the discourse of natural science. Even realist correlationisms run into difficulties on the ontological level in a way that only emerges through my more differentiated consideration in the present subchapter. In Meillassoux's work, this level remains mostly implicit, even if he makes clear in some places that

[22] For example, Meillassoux writes: "And our correlationist then finds herself dangerously close to contemporary creationists", in: Meillassoux 2008, p. 18. The comparison with the Christian-religious current, which holds the thesis that the universe and the earth are only 6 000 years old and were created by God, despite all the apparent scientific evidence for a greater age, is intended to make such philosophy appear anti-scientific and irrational.

the examples of scientific discourse are not decisive: "Those [facts are of interest; J. V.] which make possible an elucidating mediation—like the problem posed by the science of dating in the first chapter of *After Finitude*. This type of investigation must allow us to throw away the ladder once it has been climbed. [...] [T]his (f)actual complex was useful, but not in principle indispensable to the demonstration [...]." (Meillassoux 2016, p. 149). Meillassoux aims to defend a rational, mathematical-scientific world view and to legitimise the discourse of current and future natural sciences as realist. It is therefore obvious that he does not attempt to explicate a purely ontological formulation. However, uncovering this deeper level of the ancestral problem has two major advantages: Firstly, it allows one to pose the problem of ancestrality without presupposing the theses of, for example, current physical cosmology. Secondly, the philosophical core of the unease can emerge more clearly with the combination of realism and correlationism.

I would therefore like to propose the following formulation of the question of ancestrality on the ontological level: How are the *ontological conditions and genesis* of any conceivable correlation possible without having to assume either (1) a *retrojection of the past from the present*, or (2) the simultaneous *creatio ex nihilo* of the correlation and the natural external world, or (3) the implicit presupposition of subject independence of the natural external world?

Before I address the question of what can be understood by the expressions of a *creatio ex nihilo, the ontological conditions and genesis of every conceivable correlation* and a *retrojection of the past from the present*, the following can already be said about this question, which amounts to another dilemma (or trilemma): It starts at a different point. This critical question is not about the contradiction between an ancestral statement in the form of a scientific hypothesis or theory—*such as: the planet earth emerged 3.5 billion years ago*—on the one hand, and the premise that no subject-independent entities exist outside the correlation, or that we cannot know anything about them on the other. For the question formulated above, no ancestral statement needs to be presupposed in order to create a contradiction with the positive or negative tenets of *systematic correlationism*.

7.3.1.1 What Does "Ontological Condition and Genesis of the Correlation as *creatio ex nihilo*" Mean?

Meillassoux had already mentioned the concept of *creatio ex nihilo* before *After finitude* in *L'Inexistence divine*. Before examining Meillassoux's idiosyncratic use of this formula more closely, it is worth taking a look at the origin of the term. It originates in the Christian theology of creation of the 2nd century and was essentially minted by the apologists Theophilus of Antioch and Tatian, who took up Plato's doctrine of creation (cf. Hubler 1995, pp. 102–126; Schmitz 2010; pp. 61–65; both in recourse to May 1978, 2004). As a thesis, the *creatio ex nihilo* denotes the act of creation of a biblical god who created the matter of the universe out of nothing and did not merely transform what already existed into the matter of the universe. From this usage in Theophilus and Tatian, the theological thesis in

Christian scholasticism of the Middle Ages developed into the dogmatic position of the Catholic doctrine of creation: Thomas Aquinas argues in the *Summa theologiae* that god created the world out of nothing. This argument, which Thomas develops in the *Quaestiones* 45 and 46 of the first part of the *Summa theologiae*, enables a better understanding of why and in what sense Meillassoux takes up the term. Following the creation stories of the Bible and the cosmological considerations of Aristotle, who defends the thesis of the eternal duration of nature (cf. esp. Aristotle 1984, Met. 1069b3–34), Thomas considers the question of the genesis and the ontological conditions of the cosmos:

> Now what proceeds by particular emanation, is not presupposed to that emanation; as when a man is generated, he was not before, but man is made from "not-man," and white from "not-white." Hence if the emanation of the whole universal being from the first principle be considered, it is impossible that any being should be presupposed before this emanation. For nothing is the same as no being. Therefore as the generation of a man is from the "not-being" which is "not-man," so creation, which is the emanation of all being, is from the "not-being" which is "nothing." Aquinas 1920, I q. 45 a. 1 co.

So, when people ask about the cause of the whole, the universe or the cosmos, it is clear from Thomas' proof of god and the revelation of the Bible in the creation stories of the Book of Genesis, that god is the answer. Since he had nothing to fall back on for the creation of all being, the cosmos or the whole must come from nothing. This is the core of the theological-philosophical thesis of *creatio ex nihilo* in relation to the cosmos, defined as the whole of existence, which Thomas develops in the quoted passage. Accordingly, the divine act of creation is not a transformation of a preceding, for example chaotic, being or another world, but goes back solely to the eternal biblical God, who makes something, namely all of creation, out of nothing.[23]

Furthermore, in the quoted passage from the *Summa theologiae*, it should be emphasised that Thomas compares the genesis of all being and thus that of the cosmos with that of an individual human being, i.e.—contemporarily speaking—individual embodied subjectivity, which also emerges from non-human, non-minded and non-subjective being. Meillassoux refers to this dimension: In rejecting any absolute, such as the eternal god of the Bible, the origin of the natural external world as an event becomes a problem for the correlationist, and with it all further events in the natural history of this external world up to the emergence

---

[23] This last point is explicitly defended by Aquinas in the following passage, STh I q. 45 a. 2 co: "I answer that, Not only is it impossible that anything should be created by God, but it is necessary to say that all things were created by God, as appears from what has been said (q. 44, a. 1). For when anyone makes one thing from another, this latter thing from which he makes is presupposed to his action, and is not produced by his action; thus the craftsman works from natural things, as wood or brass, which are caused not by the action of art, but by the action of nature. So also nature itself causes natural things as regards their form, but presupposes matter. If therefore God did only act from something presupposed, it would follow that the thing presupposed would not be caused by Him. Now it has been shown above (q. 44, a. 1,2), that nothing can be, unless it is from God, Who is the universal cause of all being. Hence it is necessary to say that God brings things into being from nothing."

of subjectivity remain dependent on the status of scientific results. This is a more general formulation of the scientific-discourse level of the ancestral problem. Thus, on the second level of the problem, the genesis and the ontological conditions of correlation (and hence also those of the subjectivity contained in the correlation) themselves become an insoluble question, as will be argued below. In this sense, the problem is twofold: How can there be entities before the emergence (giving) of correlation? *And:* How can this emergence of the correlation, its natural genesis and the fulfilment of its ontological conditions be explained without contradiction even within correlationism?[24]

Systematic correlationism may be able to retreat to explaining cosmogenesis through a *creatio ex nihilo*. But if the natural-historical genesis of the correlation itself is also to be explained by a *creatio ex nihilo*, correlationism violates the principle of sufficient reason: Everything that exists has a reason why it exists. If the genesis of the correlation is *not* granted the status of a *subject-independent* event and is understood as *creatio ex nihilo*, as the causeless emergence of the correlation in the natural external world, it becomes inexplicable in correlationism. The *creatio ex nihilo* of an entity *S* removes it from any rational discourse when thinking about logical or ontological enabling conditions of *S*—we can ask neither for reasons nor causes because there are not supposed to be any. If the correlation appears spontaneously and without reason out of nowhere, then nothing can be said about its necessary, let alone sufficient conditions. Agnosticism or quietism may be

[24] The term *creatio ex nihilo* is already introduced by Meillassoux in *L'Inexistence divine*, his PhD thesis. As the multi-volume version of this work has not yet been published, we can only refer here to the excerpts from Harman's *Quentin Meillassoux. Philosophy in the Making*. There, the first excerpt (*Advent* ex nihilo *is a Rational Concept*) deals directly with the *creatio ex nihilo*: "This indicates in the most striking fashion that if we think advent in its truth, it is an advent *ex nihilo* and thus *without any reason at all*, and *for that very reason* it is without limit. In revealing the contingency of laws, reason itself teaches that becoming is ultimately without reason. [...] By advent [in the original: *surgissement*; J. V.] *ex nihilo* we do not mean that being arose entirely from original nothingness. What we mean, stated in classical terms, is that there is more in the effect than in the cause: that this "more" therefore has no reason at all for its advent, and hence nothing (no law) can limit it." (Meillassoux 2011a, pp. 176–177.) Two aspects of this passage are noteworthy: The fact that Meillassoux seems to be willing to make use of explanations based on an *ex nihilo* in his philosophy and the fact that Meillassoux does not seem to rely on the dominant usage as explanation or creation *from nothing* following Thomas Aquinas, but on the ancient usage before and among the apologists, who in turn rely on Plato's treatment in the *Phaedo* and the *Politeia* (cf. esp. Plato (1977b), Phaedo 96A–97B, 100E–101E, Republic 508E–509B, 517C): A given effect cannot be reduced to the particular and constituent causes alone, but can only be explained by genuine emergent transcendence of causes, the emergence of a novelty that Meillassoux calls *surgissement ex nihilo*. Meillassoux's own speculative use of *surgissement ex nihilo* is a positive one. From speculative realism, he argues that *surgissement ex nihilo*, genuine novelties and events without some form of *bottom-up* causality, are possible and a rationally justifiable principle. This can be expressed in the vocabulary of the debate on strong emergence in such a way that, according to Meillassoux, there are processes in nature that produce strongly emergent properties in which the complexity of a modular system exceeds the combined and added complexity of the modules. This is to be distinguished from a *creatio ex nihilo*, a creation out of nothing, which, according to Meillassoux, is not open to correlationism as a theoretical option.

acceptable in the case of cosmogenesis, but they are not acceptable in the case of the genesis and the ontological conditions of the correlation.

7.3.1.2 Correlationist Trilemma: Self-contradiction, Quietism or Retrojection of the Ancestral

For systematic correlationisms, the question of cosmogenesis is generally not a genuinely philosophical question, but is referred to the empirical sciences (or left to religious belief, mythical narrative, etc.). However, the decisive question is: *How is it possible that the respective central* correlate *is ontologically formed in the preferred* correlation*?* How does it arise, and what necessary and jointly sufficient conditions must be fulfilled to allow its existence? Here it becomes clear that two aspects must be answered either separately or together: the genesis of the respective correlate as a process of coming into existence or emergence and the necessary ontological conditions that allowed this to happen and allow the correlation to exist. The example of Kant's correlationism in the interpretations of Chiba and Allais shows why these two aspects cannot be coherently answered separately for many correlates: genesis and validity, empirical and transcendental levels cannot be neatly and clearly separated, as Meillassoux argues, and there is no principled distinction in many cases. The analysis of the constructive-empirical ESA has also shown that the strict distinction of the scientific and meta-levels cannot be maintained in the way that would be necessary to answer the aspects mentioned fully separately from one another. It is not simply possible to assign the question of genesis to the domain of the empirical—and thus indirectly to the natural sciences—and to address the question of ontological conditions fully independent in the domain of the transcendental or philosophical meta-levels. The ontological conditions include the actual circumstances of the initial co-relata and the correlate.

Every correlationism must be able to answer the question of genesis and ontological conditions for each correlate it uses, or at least show that an answer is possible. The explanation of the ontological validity and natural-historical genesis of a correlate through a *creatio ex nihilo* leads to the following problem: If, for example, the correlate of subjectivity and objects of the natural external world were to arise through a *creatio ex nihilo*, either the existence of this correlation would be subsequent in time, but ontologically necessary for the genesis and existence of the natural external world, or the emergence of the correlation would *coincide* with the emergence of both co-relata and thus the (objects of the) natural external world. The first assumption is the subjectivity thesis of Koch's hermeneutic realism and an instance of the retrojection strategy, which I will examine in more detail in the next Sect. 7.3.2.

The second assumption of simultaneity or joint emergence, on the other hand, either leads to contradictions within correlationist ontology or is an empty postulate. This is because dealing with a primordial event that produces the correlation and its co-relata, especially the natural external world, equally out of nothing is either not an explanation at all (if the *ex nihilo* is taken seriously) or the primordial event must be understood as an absolute event that must have taken place necessarily and subject-independently. The latter option of an absolute origin is ruled out by systematic correlationism, as otherwise a contradiction with the rejection of SI would arise. In the simultaneity assumption, the contingent primordial event, in which the co-relata and the primary correlate arise spontaneously out of nothing, must be dogmatically posited. In order to be conceivable as an event, it must itself already be part of the correlation. Apart from its status *ex nihilo*, however, nothing can be said about it. Therefore, such an explanation is empty. The primary correlation, and thus reality, has suddenly and contingently begun to exist. There are therefore no genesis or ontological *preconditions* for us to explain or know, the correlation is inexplicable, and the gesture towards this primary ur-event is thus without meaning. In addition, a contradiction arises between the strong thesis of the primordial event and the scientific discourse. According to the latter, subjectivity and the natural external world cannot be traced back to a simultaneous primordial event, which in turn leads to a revision of the discourse, i.e. the problem returns to the first level.

The question, especially for correlationist realisms and correlationist OSA, is therefore: How to explain that the ontological conditions are met and the genesis of an ontologically primary correlate and its co-relata without resorting to a true *creatio ex nihilo*? If the correlate is ontologically primary, it must be given for natural history to be possible. In a correlationist realism or a correlationist OSA, the ways to answer are limited, depending on how the natural external world is understood and which philosophy of time can be appealed to.[25] However, irrespective of this, the aforementioned critical ontological question gives rise to a number of problems for any correlationism:

First, it can be stated that every correlate and every co-relatum—and thus every correlation—itself has ontological conditions. In the Kantian correlation mentioned above, for example, Meillassoux draws attention to the ontological condition of the corporeality of every possible subject as the basis of Kantian receptivity. In the correlate of scientific practices and postulated theoretical ancestral entities of the constructive empiricist, I have argued in favour of the hidden precondition of real, natural ancestral entities, which, however, lie beyond the reach of philosophical and scientific discourse. Stanford, on the other hand, assumes a correlation from an unknowable nature that necessarily conditions the emergence of subjects, which in turn necessarily precedes the eventually resulting correlate of scientific entities and scientific community in time. An ontological condition for Stanford's correlation, therefore, unlike in constructive empiricism, is that an unknowable nature must exist that is not identical with the natural external world (as the sum of scientific entities, our objects of knowledge). In Koch, on the other hand, I have already dealt with the ontological condition of the natural external world, which develops in roughly the same way as physics describes it, that subjects must exist in it over the course of time that relate to individuals in the natural external world by thinking and acting indexically. This can be repeated for each correlate, since, as was shown in the

[25] Cf. my Sect. 2.2.1, pp. 19–21.

previous chapter on correlationism, correlates are inconceivable without co-relata and each co-relatum remains in need of explanation even within a correlation. To be sure, in itself, the ontological conditionality of correlates and co-relata is not a problem.

On the basis of the aforementioned argument, in order for the correlation to be possible, it is necessary to explain the genesis of these co-relata and the primary correlate as well as the ontological conditions as an inseparable whole. The causes or the natural history of all the correlates, co-relata and the ontological conditions involved must be analysed. It may be that these are empirically inaccessible and philosophically unobtainable—here we may refer to the emergence of the first consciousness in animals or the emergence of the first fully linguistic community. But even in such cases, at least hypotheses or narratives must be conceivable in order to explain this genesis: For the emergence of human self-consciousness, for example, a group of animals forming consciousness via processes of evolution could be conceivable,[26] or a world in which gods create us as self-conscious beings. Such descriptions, whether in a particular worldview or independent of worldviews, operate with emergence as an "Event *Y*" (Meillassoux 2008, p. 13), for which a genesis must at least be possible, i.e. must also be ontologically possible. The problem of correlationist OSA and correlationist realism is that the correlation is also ontologically primary for an event *Y*, and the latter cannot therefore be said to exist subject-independently. The two aspects of ontological conditions and genesis in their combination are a problem for every correlationist, since the disclosure of ontological conditions makes any even hypothetical genetic event impossible. Confronted with the question of the origin, ontological conditions and genesis of the correlation, the only option besides arguing for its strict subject-independence and giving up its own position, it seems, is to resort to a *creatio ex nihilo*, which is not convincing. This can be formulated formally as an argument that can be adapted to any correlation as follows:

- P1 Each correlation contains at least one primary correlate with at least two co-relata.
- P2 Each correlate and each co-relatum has ontological conditions that must be fulfilled for its existence.
- P3 The fact that the primary correlate exists with its co-relata requires an explanation.
- IC1 At least one hypothetical event *x* must be possible without contradiction, which would explain the genesis of the primary correlate with its co-relata and the fulfilment of its ontological conditions.
- P4 Assumption of systematic correlationism: Strictly subject-independent existence outside of the correlation is impossible and/or knowledge about the possibility or impossibility of entities outside the correlation is impossible.

[26] Cf. for such philosophical speculation based on scientific results Tomasello 2014.

- P5 Event x has no strictly subject-independent existence outside the correlation (a) or we have no knowledge about the possibility or impossibility of event x (b).
- P6 If the hypothetical possibility of the existence of A is a necessary and temporally prior condition of B, then the hypothetical A must be possible and strictly independent of B.
- P7 If event x simultaneously generates the primary correlate with its co-relata and all their ontological conditions, it is a *creatio ex nihilo*.
- P8 Either event x simultaneously produces the primary correlate with its co-relata and all their ontological conditions, or event x is a necessary and temporally prior precondition of the primary correlate and the co-relata as well as all their ontological conditions.
- P9 The assumption of x as a *creatio ex nihilo* is an empty explanation or contradictory under the presuppositions of systematic correlationism.
- IC2 It follows from P6, P7, P8 and P9 that the hypothetical event x is possible and must be strictly independent of the primary correlate and its co-relata.
- IC3 The assumption of systematic correlationism leads with (a) to a contradiction with IC2.
- IC4 The assumption of systematic correlationism leads with (b) to a contradiction with IC2.
- C Systematic correlationism leads to self-contradictions.

This argument can be used and adapted as a reasoning scheme to be applied to any correlation, in the way that ontological conditions for various correlations have already been superficially listed in the previous chapter. This is the ancestral argument on the ontological level. In contrast to the scientific-discourse form, it is not necessary to invoke the scientific discourse with its ancestral statements in order to create a problem that leads systematic correlationism into self-contradictions. Instead, the hypothetical ancestral event x is sufficient, which only has to be possible. However, as the intermediate conclusions IC3 and IC4 show, in systematic correlationism it is either impossible that such an ancestral event x exists or we cannot know or say anything about even the possibility of such an ancestral event x, because even the assertion of sheer possibility would be too much to ask of the correlationist.

Of course, the conclusion only follows if all premises can be defended. I have presented a dilemma argument for the validity of P9 in this subchapter: Either the desired explanation is defunct because of the claim of a *creatio ex nihilo*, or event x becomes a necessary and subject-independent primal event, an absolute event in Meillassoux's sense, and thus a contradiction within correlationism. This secures P9 and the *creatio ex nihilo* of the correlation with its ontological conditions is not an option for systematic correlationists.

Hence, the argument stands and falls with P3. Only if the correlationist admits that the existence of the primary correlate with its co-relata is in need of explanation does the contradiction follow. If P3 is not recognised by a systematic correlationism, I call this a QUIETISM REGARDING THE CORRELATION. Quietism here refers to the philosophical attitude that a given domain of philosophical discourse is unnecessary and consequently that contributions to it are superfluous, misleading or even harmful and that it is better to refrain from philosophical discussion in this domain altogether. The given domain can be a single philosophical problem (e.g. that of ancestrality or the genesis and ontological conditions of the correlation) or a particular debate (e.g. the realism-anti-realism debate in philosophy of science or ontology). Quietists do not try to solve the problem under discussion, but to dissolve it; not to answer the question, but to make it disappear; not to win the debate, but to undermine it.[27]

Quietism with regard to the correlation is essentially based on rejecting the question "*How is the correlation possible?*", delegitimising it and showing it to be nonsensical. I could begin an elaborate argument here as to why I consider this to be philosophically dishonest in this particular case of correlationist scientific realism and anti-realism. However, in order not to prolong my discussion unnecessarily, I would like to point out that this strategy has little chance of success in philosophy of science, since no positive claims could then be made about the enabling conditions of scientific ontology and theoretical entities, the status of scientific theories and the relationship between scientific statements about reality and the natural external world as part of reality. This follows because the argument for the futility of the critical question of how a certain correlate is possible also applies to these elements of philosophy of science that are co-relata in the correlation(s). The quietist, with respect to the correlation, retreats to the position that the ancestral question is misleading and does not require any philosophical answer. In particular, any attempt to give a philosophical theory of the genesis and ontological enabling conditions of the correlation would be pointless; this must be the conclusion of her argument. What seems to follow would be a philosophy of science that would essentially consist in the recommendation that the discussions within philosophy of science are largely meaningless. This is the basic problem of any *negative local quietism*. According to Spiegel (2021, pp. 109–170), three forms of quietism can be distinguished: In *general quietism*, all philosophical activity is quietistically rejected. All philosophical problems, questions and debates are therefore not solvable, but only resolvable. According to this attitude, the only conclusion that can be drawn from philosophising is to step back from the activity of philosophising. The second and third varieties both belong to local quietism, i.e. the philosophical attitude that the questions and problems of an unnecessary or meaningless domain of some

[27] Here, there is no discussion about the answer, rather, the ontological question is treated therapeutically. McDowell can be quoted as representative of this approach when he writes about the late Wittgenstein: "Wittgenstein's quietism is not a refusal to engage in substantive philosophy in the face of what everyone has to accept as genuine problems. It is an activity of diagnosing, so as to explain away, some appearances that we are confronted with genuine problems. The supposed problems disappear, leaving no need for theory construction to make things less mysterious. [...] Quietism does indeed urge us not to engage in certain supposed tasks, but precisely because it requires us to work at showing that they are not necessary. And it is indeed work. Therapeutic philosophy is designed to spare us the travails of positive philosophy, but it has its own difficulties" (McDowell 2009i, pp. 317–318).

discourse is delimitable and not comprehensive. They should therefore be rejected and be shown to be in such-and-such a way malformed or misleading, without dismissing all other philosophical discourse. Either local quietism is positive, which is the second type according to Spiegel. Positive local quietism demonstrates the unnecessary nature of a particular problem or debate with philosophical arguments by building on premises that it positively defends (and thus engages in at least some degree of "theory production", if you will, elsewhere in the philosophical landscape). The third variety is a negative local quietism, which does not attempt to produce any commitments, positive argument or philosophical theory itself and does not defend any premises of its own, but only points discursively to the unnecessary nature of a problem or a debate without proving it. Here it becomes clear why there is no fourth variety according to this nomenclature: General quietism must be negative, otherwise it would commit a self-contradiction.

For the correlationist quetist, this means that for every assertion such as: "*Nothing can be said philosophically about the status of scientific theories or the existence of theoretical entities*", the correlationist in turn invites the question of how this assertion is justified. If the answer refers to the natural sciences as the only authority in this area, the problem of ancestrality again arises on the scientific-discourse level and the quietist would fall in a regress. If no reference is made to the natural sciences, it would have to be shown in a positive quietism that all philosophical questions associated with the ontological conditions and the genesis of the respective correlation are based on a general error, which in turn requires positive philosophy—which would be a general surprise to me, as nobody, to my knowledge, has tried this argument. So more likely, the correlationist quietist would need negative local quietism, but that position, if combined with correlationism, contaminates all things in her correlation, meaning she should not say anything at all about, for instance, the instrumental value of our fictitious theoretical entities, and so forth. If the question about the genesis and ontological conditions of the correlation is meaningless or malformed, and we should therefore adopt negative local quietism, it also seems to be the case for the historical and practical consequences of the correlation and the ontological (and epistemological) determining, i.e. all the things for which the correlationist in philosophy of science introduced a correlation in the first place.

For the examples of systematic correlationism discussed here, quietism with regard to correlation and the simple rejection of P3 is not an option. The only sensible way to escape the argument on the ontological level, therefore, is to refute premise P6: *If the hypothetical possibility of the existence of A is a necessary and temporally prior condition of B, then the hypothetical A must be possible and strictly independent of B*. It is similar to premise P6 of the ancestral argument on the scientific-discourse level. This rejection of P6 leads to the retrojection strategy of systematic correlationism. It is the task of the next section to explain this strategy in more detail, and to demonstrate in my examples that it is explicitly or implicitly advocated and to argue for its failure.

7.3.2 The Retrojection Strategy of Systematic Correlationism

The RETROJECTION STRATEGY consists in the argumentative attempt of systematic correlationism to affirm the chronological, causal or natural-historical antecedence of the natural external world, but at the same time to defend one of the tenets of systematic correlationism—OCT or ECT—by arguing for a logical or ontological necessity of the correlation or a subjective co-relatum for the existence of the natural external world or ancestral entities. On the scientific-discourse level, this means that premise P6, *If the existence of A is a necessary and temporally prior condition of B, then A must be strictly independent of B.*, can be denied. If the systematic correlationist succeeds in defending the possibility that *A* is a naturally necessary condition of *B*, while *B* can be a logical or ontological condition of *A* (or a part of *A*), the premise would be refuted. This also applies to the similar premise P6 on the ontological level of the problem, which states *If the hypothetical possibility of the existence of A is a necessary and temporally prior condition of B, then the hypothetical A must be possible and strictly independent of B.*

Meillassoux begins his formulation of the retrojection strategy as follows: Apparently, the ancestral problem can be solved by "something like a simple codicil, always the same one, which he will discretely append to the end of the phrase:" (also for the following Meillassoux 2008, p. 13.) The event *x*, for example, took place so and so many years, ages etc. before the "emergence of humans—*for humans* (or even, *for the human scientist*" or *in our myth*). Without directly intervening in the statement about ancestral entities or the genesis of the correlation, the integrity of the positive and negative correlationist tenet is seemingly preserved in the eyes of the systematic correlationist with this qualification of the statement: After all, the ancestral statement or thesis for the genesis of the correlation seems to be restricted to the subject-dependent and correlational context. Meillassoux describes this attitude as the foundation of his formulation in the voice of the correlationist of the retrojection strategy:

> [B]eing gives itself (occurrence 1) as anterior to givenness [= genesis of the correlation; J. V.] (occurrence 2). At the first, superficial level, I forget the originary nature of givenness, losing myself in the object and naturalizing givenness by turning it into a property of the physical world, one that is liable to appear and disappear in the same way as a thing (being gives itself *as anterior to givenness*). But at the deeper level (being *gives itself* as anterior to givenness), I grasp that the correlation between thought and being enjoys logical priority over every empirical statement about the world and intra-worldly entities. Thus I have no difficulty reconciling the thesis of the *chronological* anteriority of what is over what appears – this being the level of meaning that is superficial, realist, derivative – with the thesis of the *logical* priority which givenness enjoys *vis-á-vis* what is given in the realm of givenness (to which the aforementioned chronological anteriority belongs) – the latter thesis corresponding to the deeper, more originary level, which alone is truly correct.
> Meillassoux 2008, p. 15.

As with the transcendental counterargument, according to Meillassoux the correlationist strategy consists in introducing a level distinction: The primary logical or ontological level is demarcated from the secondary empirical and chronological level in such a way that the hypotheses or narratives of the second are subordinate to the philosophical insights of the first, which means that they need to be distinct. A factual emergence of life, minds, subjects, thought, etc. (whatever the subjective co-relatum is or what the subjective co-relata are) as part of the primary correlate of one's own preferred correlation would thus be something that only shows itself from a *given present perspective*, i.e. only subject-dependent. The present perspective would only be possible within the existing correlation. Meillassoux calls this a retrojection strategy, since, according to him, the present point of view introduced by consequence cannot be transcended, so that the ancestral entities only appear to us as such: According to the retrojection strategy, in order to comprehend any ancestral statement about the cause of arche-fossils in scientific discourse or about the genesis of the primary correlate, its co-relata and its ontological conditions in a deeper sense, the systematic correlationist "should not proceed from the ancestral past, but from the correlational present. This means that we have to carry out *a retrojection of the past on the basis of the present.*" (Meillassoux 2008, p. 16) Peter Hallward advocates a similarly formulated retrojection strategy in response to the ancestrality problem:

> Almost no-one actually balks at ancestral statements because correlationism as Meillassoux defines it is in reality an epistemological theory, one that is perfectly compatible with the insights of Darwin, Marx or Einstein. There's nothing to prevent a correlationist from thinking ancestral objects or worlds that are older than the thought that thinks them, or indeed older than thought itself. [... All] that the correlationist demands is an acknowledgement that when you think of an ancestral event, or any event, you are indeed thinking of it. I can think of this lump of ancient rock as ancient if and only if science currently provides me with reliable means of thinking it so. [...] What's less obvious is how we might think such a [ancestral; J. V.] world without thinking it, or how we might arrive at scientific knowledge of such pre-given objects if nothing is given of them. Hallward 2011, p. 137.

Hallward insists that the natural external world is older than the first consciousness or the first living creatures, even for the most hardened systematic correlationist. To read this treatment of the problem according to the *principle of charity*—and not to imply that Hallward is guilty of an inappropriately shallow interpretation of the problem—it must be assumed that Hallward is here defending a form of the retrojection strategy. According to this, the statements and assumptions about the ancestral past necessary to solve the problem would be unproblematic as long as the epistemological restriction is asserted that these statements and assumptions depend on contingent epistemic circumstances as well as intersubjective interests and practices (thus assuming a form of ECT as compatible with ancestrality). Nathan Brown argues against Hallward's solution:

> When the correlationist thinks the ancestral object *qua* correlate of thought, she effects a temporal retrojection of the past from the present, such that "it is necessary to proceed from the present to the past, following a logical order, rather than from the past to the present, following a chronological order". For the correlationist [...] "the deeper sense of ancestrality resides in the logical retrojection imposed upon its superficially chronological sense". Thus, stricto sensu, the correlationist cannot think ancestral objects as prior to the thought that thinks them. Meillassoux's argument is simply that if we accept the priority of logical over

> chronological succession [...] we will be unable to assess scientific statements regarding ancestral phenomena without destroying the veritable meaning of those statements, which concern the chronological priority of that which came before thought, regardless of any temporal retrojection performed by thinking. What is at stake here apropos of "thinking and being" is a disagreement regarding the priority of the logical correlation between thinking and being over the chronological disjunction of thinking and being. Meillassoux's point is that the correlationist's insistence upon the priority of the former eviscerates the proper import of the latter. Brown 2011, pp. 142–143; emphasis as in the original; quotations from Meillassoux 2008, p. 16.

Brown argues that Meillassoux's core argument against Hallward's retrojection strategy consists in rejecting the *priority* of logical or ontological dependence of the ancestral on the correlation, which is placed above the subordinate chronological (or natural-causal or natural-historical) dependence of the correlation on the ancestral. Hallward believes that he can assume such a dependence without running into problems. This could already be refuted on the ontological level with the ancestral argument presented.

In what follows, I will argue in favour of reconstructing the retrojection strategy in a more general way and therefore also of presenting the counterarguments more generally. This subsection is devoted to describing and rebutting the retrojection strategy of systematic correlationism. Hallward responds explicitly to Meillassoux, but I will show that, at least in my examples of Stanford's correlationist OSA, Koch's hermeneutic realism, and the radical Sellars, a form of retrojection strategy is implicit.

By analysing Meillassoux's rejoinder, I aim to reveal how the anti-realist variant of the retrojection strategy can be argued against. However, the realist-correlationist solution following Koch's argument for the necessity of an intersubjective knowledge community in every possible spacetime, and a fictitious, similarly situated argument of the radical Sellars, will require a further argument from me, since Meillassoux does not deal with this option. I will argue against any ontological and logical dependence of an entity on other, not yet existing concrete entities by pointing out that such justifications are begging the question.

7.3.2.1 The Anti-Realist Retrojection Strategy by the Example of Stanford's Instrumentalism and Meillassoux's Rejoinder

When Stanford writes that scientific theories make statements about unattainable entities of the real natural external world, including ancestral entities, he aims simultaneously to argue that these statements, through their form, are truth-apt. The following rule therefore applies to his instrumentalist approach to ancestral entities: "[I]f a theory makes claims about entities or events to which we have no access independently of the theories that describe them or claims for which there is no sense to be made in a way that is independent of those descriptions, the instrumentalist will not believe them at all, but will simply make use of them in the process of generating further predictions and recipes for intervention." (Stanford 2006, p. 198.) Following this, regardless of our physical theories, it does not make sense to speak of a Big Bang or the accretion of the solar system as though we knew them.

Regarding the ancestral events described by the natural sciences, Stanford is an ontological anti-realist because he sees them only as useful tools for generating predictions and providing intervention measures. The ancestral entities are part of the correlate of theories or theoretical entities and our scientific practices and interests. This is a secondary correlate attached to a primary correlate: For theories and practices to exist, there must be (human or other corporeal) subjects that stand in a natural-causal dependence to an unknowable, real natural external world within which they exist. Since Stanford here advocates a clear epistemological anti-realism with regard to the real natural external world, we cannot speak of a chronological priority of this real natural external world. We only know of its existence because we as subjects depend on some form of real natural external world in an unrecognisable way, which is why the external world is in a primary correlate with us, to which our scientific theories and knowledge practices form a second correlate; combined, they make up this correlation.

Meillassoux's argument begins from the postulation of a primary epistemological correlate (in our case, the unknowable natural external world and human subjects) with a secondary ontological correlate (the ancestral and other theoretical entities ontologically depend on our interests, practices and contingent theories in the sciences and vice versa). The fact that Meillassoux is not primarily arguing against general ontological anti-realism—as is sometimes attributed to Foucault, Richard Rorty and other so-called postmodernists in social constructivist or relativist readings—in my opinion is due to the fact that Meillassoux considers such positions to be self-contradictory: They would have to assume that everything is produced by discourses or other shared, social practices—be it the stars, the human or the laws of nature and the discourse itself. But if without discourse there would be no spatiotemporal entities, no abstract entities and no subjects, how can there be a discourse? Either it is a *creatio ex nihilo* or a mysterious, unfounded entity, or it is an absolute correlation—which would assign this form of philosophy to subjectalism. Since, without going into more detail here, Meillassoux considers both to be erroneous, in his view such correlationism commits a self-contradiction regardless.

Consequently, Meillassoux's argument is directed against what he considers to be the better variant, of which I cited Stanford's instrumentalism as an example. Meillassoux summarises this strategy as follows:

> This means that we have to carry out *a retrojection of the past on the basis of the present.* What is given to us [according to correlationism; J. V.], in effect, is not something that is anterior to givenness [= genesis of correlation; J. V.], but merely something that is given in the present but gives itself as anterior to givenness. The logical (constitutive, originary) anteriority of givenness over the being of the given therefore enjoins us to subordinate the apparent sense of the ancestral statement to a *more profound counter-sense*, which is alone capable of delivering its meaning: it is not ancestrality which precedes givenness, but that which is given in the present which retrojects a *seemingly* ancestral past. To understand the fossil, it is necessary to proceed from the present to the past, following a logical order, rather than from the past to the present, following a chronological order. Meillassoux 2008, p. 16.

The ancestral entities of the theories adopted in the natural sciences are, according to Stanford, present heuristic tools, parts of the currently produced theory that has a practical, instrumental use and necessity for us. According to Stanford, talk of ancestral entities is to be understood literally, but fails for all complex theories, since we have no knowledge of their truth. Because all current theories of ancestral entities are complex, we have no knowledge of a reality that precedes us and on which we depend; we only have functional theories.

The quoted passage shows that the chronology of events in reality is not doubted by correlationists who argue in this way, but only the primacy of the (onto-)logical over the temporal-causal order. In this case, a primacy of the first epistemological and second ontological correlate over the merely possible, natural-historical dependence of both correlates on the unknowable natural external world is rejected, which Meillassoux criticises:

"Now, why is this interpretation of ancestrality obviously insupportable? Well, to understand why, all we have to do is ask the correlationist the following question: [...] Did the accretion of the earth happen, *yes or no?*" (Meillassoux 2008, p. 32.) In line with his pro-scientific agenda, here Meillassoux uses the example of the formation of the planet Earth, but he could also have chosen a hypothetical ancestral event *x*, provided it is given a meaning. The seemingly simple yes-or-no question boils down to the ontological form of the same self-contradiction that also comes into play on the level of philosophy of science. Stanford must answer: We do not know what really happened. The actual genesis of the correlation and with it the ontological conditions remain obscure to Stanford.

Since the ancestral conditions for the epistemological-ontological correlation of his instrumentalism, about which, according to Stanford, natural science only *seemingly* provides us with information, are rejected, he cannot say anything about the anteriority of subject-independent nature, the ancestral conditions of correlation and any hypothetical event *x*, not even affirm its possibility. The retrojection strategy here conceals the fact that it only postulates that the correlation exists without answering the legitimate question of an explanation for its existence. Stanford adopts from scientific theories the idea that we (and our correlation) originate from a subject-independent nature and real ancestral events and processes, without applying his philosophical premises and conclusions to determine that any epistemic position—including Stanford's own—is better suited than others to know this. Therefore, Stanford's explanation for the correlation is also empty. The ancestral past, which is referred to in truth, is that described by natural sciences. According to Stanford, however, it belongs entirely to the practices and interests of the correlational present. If the question of the genesis and ontological conditions of the correlation is justified, then Stanford's retrojection strategy does not work to escape the ancestral argument.

7.3.2.2 The Retrojection Strategy of Correlationist Realism

The reconstruction in the last chapter has shown that McDowell's radical Sellars argues that the scientific and the manifest image of reality are mutually dependent: On the one hand, the natural history of the natural external world is necessary for the emergence of human subjectivity and for the emergence of the manifest image. Epistemologically and practically, however, the scientific image is constantly dependent on the existence of the manifest image. Confronted with the argument of ancestrality, there are two possibilities for the radical Sellars: Either he rejects the ancestrality question and adopts a quietism regarding correlation. This may be McDowell's own strategy (cf. an interpretation of his quietist understanding of philosophy Demmerling 2014), but it might not be a fair interpretation of even the radical reading of Sellars. The radical Sellars does not dismiss every assertion about the status of scientific theories and ancestral entities as an unnecessary consideration, as a consistent local quietism on this question would require him to. The second possibility of the radical Sellars is a realist rejection of premise P6 and the elaboration of a realist variant of the retrojection strategy.

This step seems obvious: Since ancestral entities appear in the scientific image, scientific realism is appropriate for them, and, within the image, even subject-independence. According to Sellars, there was therefore the Big Bang etc. and our theories are true. This means that the radical Sellars does not run into Stanford's problems, at least immediately. For him, however, the entire scientific image is dependent on our practices and thus on subjectivity, which is why no strictly subject-independent existence can be attributed to ancestral entities outside of the scientific image. In the manifest image, ancestral entities would therefore also be problematic and do not occur. Confronted with the question of the accretion of the earth, he seems to give a clear yes according to the *scientia-mensura* dictum, which would mean a relapse into the OSR and speaks against the *interpretation* of the radical Sellars. But let us assume for the sake of argument that the radical Sellars would hold to the correlationist premise that the question is not clearly decidable, since the images are equal in their validity and the answer must be yes in one and no in the other. In order to formulate a retrojection strategy, the radical Sellars could make use of a thesis by McDowell from the *Unboundedness of the Conceptual* (cf. McDowell 1996, pp. 24–36) and modify it for his philosophy of science. The following argument would be conceivable from the perspective of the radical Sellars, with the addition of McDowell's thesis:

"When we have a veridical experience, we know that the objects of the natural external world are such and such, which we then make the content of a conceptual judgement. The natural external world must itself be conceptually structured in order to exert such an influence on a subject's thinking. In the scientific image, it is true that concrete entities, some of them ancestral, in the natural external world in which they appear make abstract thoughts and mentally constituted, minded subjects possible in the first place. In the manifest image, it is true that the natural external world with its entities must itself be conceptually structured, whereby every conceptual structure must be produced and reproduced by the rational practices of minded subjects capable of judgement. The conceptual natural external world is therefore in constant dependence on the community of minded subjects. A natural external world, which for its part would have no conceptual structure, could not even enter into our rational practice of judgement in the scientific picture. The supposed contradiction disappears if we assume that the community of minded subjects in the present is also capable of conceptually structuring and thus grasping the ancestral natural external world. It could therefore be argued here that it is a necessary condition for ancestrality, the fate of the ancestral natural external world, to be conceptually structured by minded subjects of the present."

This is just one possibility for what an argument from the now completely counterfactual position of a strictly correlationist, radical Sellars might look like. But this form of argument in favour of a realist correlationism finds a representative in Koch with his hermeneutic realism. As shown in Sect. 6.3.3, he adopts the "*subjectivity thesis*": "A material spatio-temporal system is possible only if at some time and at some place in it at least one spatiotemporal subject exists ("an I") that relates to objects in thought and perception." (Koch 2006a, p. 8). Koch argues in favour of this by introducing counterfactual possible worlds in which the principle of *identitas indiscernibilium* is violated from the outside, but which cannot be excluded. The correlate of the individual subject and the natural external world, for Koch a space-time system, is extended in the further course of his argumentation by the discourse community as the enabling condition of the individual subject and the life-world and physical individuals as parts of the natural external world to which subjects refer indexically (by thinking, speaking and perceiving).

The correlation and the corporeal subject as a co-relatum are thus logically and ontologically necessary for Koch so that the natural external world can exist, even if the external world necessarily conditions the individual corporeal subject in terms of natural history. This is therefore a realistic retrojection strategy that corresponds in its essential points to the fictitious correlationist retrojection strategy of the radical Sellars that I outlined above, though my radical Sellars is less demanding. If Koch were therefore to respond directly to the problem of ancestrality, there is a simple answer: *We, corporeal subjets, are a necessary event in time with our partially true knowledge of the ancestral past.* The ancestral past of the natural external world could not have existed if we (or other corporeal subjects) had not been produced at a later time by the natural history of the natural external world. Subjectivity is thus also the fate of the natural external world in Koch's case.

7.3.2.3 The Second Rejoinder: Arbitrary Ontological Predetermination and Vicious Circularity

Meillassoux does not deal with this rather special, realistic retrojection strategy. First of all, this thesis of the logical and ontological enabling of the ancestral past by the present seems implausible: A necessary occurrence in time seems to contradict the cosmic randomness that seems to have conditioned the emergence of human life in our solar system. Moreover, this teleological moment goes against the usual naturalistic (and even "just" scientific) intuitions.[28] However, both are not yet arguments against Koch's and the fictitious Sellars' retrojections.

Ancestral statements of natural science do not seem to cause difficulties for the realist retrojection strategy, which is why Meillassoux's yes-or-no question does not have the desired effect. His reply is directed against the anti-realist variants of retrojection. Nevertheless, for him "[t]here is no possible compromise between the correlation and the arche-fossil: once one has acknowledged one, one has thereby disqualified the other." (Meillassoux 2008, p. 17.) The fact that Koch and other realist correlationists escape the ancestral problem here at first glance is due to Meillassoux's account. Through my reconstructive differentiation of the two levels of ancestrality, a new question can be posed on the ontological level:

Why is a world without subjects ontologically impossible? There is no contradiction in imagining our spacetime system as non-subjective. All we have to do is to assume that life never originated anywhere in the universe, i.e. (counterfactually) not even on our planet. This is not only quite easy to imagine, but also does not violate any logical or ontological principle (such as PII). Why should this possibility not exist? On what level can there be a fate in which lifeless nature necessitates subjects? Here at the latest, the fictitious radical Sellars would need an argument that actively *breaks with* the *scientia-mensura* for good. This means that such a radical interpretation is untenable because such an arbitrary ontological predetermination, a fate, of the natural external world is incompatible with the other commitments that make up Sellars' stance.

The ontological ancestral questions thus posed also concern Koch. For a genetic explanation of how embodied subjectivity and the discourse community have developed in natural history, Koch refers to the empirical sciences. In this way, he avoids the danger of understanding the "emergence" of subjectivity as a *creatio ex nihilo*: A spontaneous self-positioning of the first subject would then be the ontological and causal initial event that retrojectively creates the timeline in which it apparently takes place: a necessary moment as an anthropocentric miracle. Koch's ontological and epistemological realism about the natural external world and time itself prohibits the covert assumption of a *creatio ex nihilo* of correlation.

Koch must ensure that the subjectivity thesis applies to all possible worlds and the real natural external world. In his argument, reconstructed above, he initially only proves the validity of the subjectivity thesis for problem worlds in which the PII is violated from the outside, but not from the inside. More precisely, it seems questionable how Koch can claim that the subjectivity thesis is valid in the actual world, the reality and independence of which Meillassoux is concerned with arguing for, the "Grand Dehors" (Meillassoux 2006, p. 21). The answer is Koch's premise (5): *Possible worlds that do not violate the PII from the inside, but only from the outside, should be allowed.* For from this premise Koch can conclude that the world with the eternal two-way eternal recurrence introduced by Lewis is a possible world,

[28] Cf. a rejection of this teleology by an avowed new realist and thus anti-naturalist Gabriel 2023, pp. 77–78.

and from this in turn that we—assuming modal realism about possible worlds—cannot rule out the possibility that we are in such a world. This suggests that the subjectivity thesis also applies to us and in general.

The idea is as follows: If embodied subjectivity via indexicality ensures the individuation of individuals in some (problem) cases, and we cannot exclude in our reality that we live in a problem world, for example with two-way eternal recurrence or time dilation and contraction, then we necessarily need indexicality and consequently embodied subjects here and now. First, this would mean that the *world of eternal ancestrality* is not possible.[29] This would mean assuming a universe that shares our natural history, but in which no subjectivity arises. This is surprising, because it seems fully possible: the laws of nature, logic and ontological principles allow for a natural history without living beings, and thus embodied subjects. This counterfactual scenario seems no more fantastic than those in which the laws of nature include two-way eternal recurrence or time dilation and contraction. The world of eternal ancestrality is not a problem world in Koch's nomenclature: there are no duplicates of mesoscopic individuals to which Koch grants primacy; PII applies to this context without any issues.

To put it clearly: If premise (5) is taken as a foundation and the rest of Koch's argument is sound in the first place, then premise (5∗) is excluded: *(5∗) All possible worlds that do not violate PII should be admitted.* For with the acceptance of (5), Koch makes the decision to reject the world of eternal ancestrality and to allow the worlds of two-way eternal recurrence and time dilation. This seems at the least arbitrary, if not questionable. In my opinion, Koch would need a strong argument to make this decision convincing. The subjectivity thesis cannot help in the decision between (5) and (5∗), because the decision against the world of eternal ancestrality and in favour of the world of two-way eternal recurrence is necessary in order to be able to argue for the subjectivity thesis at all. In my opinion, the prospects of defending this seem slim.

My fictitious radical Sellars, who in his retrojection strategy uses a variation of McDowell's unboundedness of the conceptual, also relies in his argument on the fact that at some point in the universe a conceptual structuring must take place in order to make it possible. Again, the historical Sellars does not imply this idea. The point rather is to show that this is unavoidable as a form of the realists retrojection strategy. Since the realist theses concerning the chronological order are to be defended, but the negative correlationist tenet of the impossibility of subject dependency must also apply, the only possibility left for the use of the retrojection strategy is to postulate this form of retroactive logical and ontological necessity. However, the realistic retrojection strategy in any case fails.

[29] Gabriel reconstructs this question as the relevant core of the problem of ancestrality: "How, goes the somewhat more modally refined question, can one think of a reality that would also have existed if creatures like us had not existed—a reality, therefore, that cannot have those structures we ascribe to it when we experience it as cognisable *post festum* inscribed within it qua thing-in-itself?" (Gabriel 2023, p. 76).

Here we can correct Meillassoux's sentence, as quoted above: Not between arche-fossils and correlate, but *between ancestrality and the correlation there can be no compromises.* Koch excludes the world of eternal ancestrality in order to be able to admit the other problem worlds under the premises (1)–(4) already considered, and to conclude with (5) what has already been implicitly assumed: The subjectivity thesis is nothing more than the rejection of the possibility of eternal ancestrality. Here, the possibility is dismissed out of hand that it was indeed a real possibility of our natural history not to form embodied subjects; the premises already exclude the existence of the natural external world without our species and all other potential species with subjectivity. Koch's argument and the entire experimental structure of hermeneutic realism are viciously circular; the thesis to be proved is implicitly achieved by the arbitrary exclusion of certain possibilities that directly imply the falsity of the thesis.[30]

It would be ontologically, logically and naturally possible that there are no subjects in a spacetime system (never and nowhere). Assuming modal and temporal realism, this sentence is true. But Koch and every other representative of the realistic retrojection strategy must reject it. However, since the realist retrojection strategy serves to defend one's own correlation against the ancestral problem on an ontological level, they all argue in a vicious circle.[31] The correlationist realist must postulate a retroactive necessity, a fate of the natural external world, according to which the ancestral entities must become part of a future correlation. However, this is arbitrary, since every justification already implicitly presupposes this premise. No realistic correlationism escapes the contradiction of the ancestral argument on the ontological level. Thus the realist retrojection strategy fails.

This section has shown what the deeper, ontological level of the ancestral problem consists of. It represents the critical demand on every correlationism to leave a way open for a reflection on and statements about the ontological conditions and the genesis of the respective correlation. This variant of the question

[30] This is an implicit begging the question, not the circular horn of Agrippa's trilemma of the impossibility of final justification (cf. the classic passages in Sextus Empiricus (1994). The presentation in (Williams 2001, esp. pp. 60–64) is informative, as are those as *Münchhausen trilemma* in (Albert 1985, esp. pp. 16–21) and (Krockert 2016, esp. pp. 1–12) on the justification trilemma between infinite regress, circular justifications and dogma).

[31] I will not explain in detail why such a form of circular argumentation actually undermines the conclusion. Here the burden of argumentation lies on the side of the realist correlationist. Some attempts at argumentation that could be interpreted in this way even speak of an "*virtuous circle*" (Dickie 2015, chap. 3). But it seems unconvincing that such a retrojection strategy can be defended in such a manner. There are attempts in so-called dialetheism and noneism to deal with the circularity of one's own philosophical justification or contradictions in one's own ontology (cf. e.g. Beall 2004, 2009, 2014; Beall and Restall 2006; Priest 1998, 2000a,b, 2014; Routley 1979; Routley (Sylvan) 1980, which build on da Costa (1974), da Costa and Alves (1977) advancements in paraconsistent logics, among others). However, it is both unclear whether correlationism and dialetheism can be meaningfully brought together and questionable whether a dialetheistic solution to the problem of ancestrality is possible, either by "defusing" the circularity of a realistic retrojection strategy or by "containing" a contradiction at the end of the argument from ancestrality.

of the ancestral can be posed with reference to a hypothetical, possible event *x* without presupposing the statements of the scientific discourse. In this respect, the ontological level is argumentatively stronger and philosophically deeper than a reduction of the problem to the level of philosophy of science.

7.3.3 The Synthesis of the Two Levels and Conclusion

At the conclusion of the section on the scientific-discourse level and at the beginning of the section on the ontological level, it was already demonstrated that the two types of formulations can be converted into one another; Meillassoux does not differentiate conceptually between different forms of his problem. However, the separation of the issues allows the challenge to be presented more clearly: It is thus *on the one hand* a *critical question* addressed to systematic correlationism concerning how those ancestral theoretical entities that are assumed in scientific discourse as preconditions of our physical and biological emergence are to be understood. *On the other hand*, it can be understood as the *specific question* about the genesis and the ontological conditions of the respective correlation.

Some major criticisms of ancestrality, such as those already outlined by Hallward, Wiltsche and Padui, stem from the weakness and ambiguity of Meillassoux's formulation of the problem. Often only the surface of the problem is considered, which is why hasty rejections such as Hallward's might seem appropriate. Meillassoux's staging as an advocate of supposedly threatened scientificity against a fideism obscures the profound philosophical relevance of his argument. The conclusion that systematic correlationism runs into insoluble difficulties in view of dinosaurs and star formation seems correspondingly exaggerated. Through my differentiation on the two levels, I have not only tried to modify and clarify the problem, but also to argue in favour of a more modest conclusion, but one that is valid.

My considerations in the preceding subchapters and sections were designed, on the one hand, to present the ancestral problem in sufficiently general terms on the ontological and scientific-discourse levels to demonstrate its challenge to systematic correlationism and to defend it against counterarguments. On the other hand, I have posed some variants of the ancestral question to my exemplary positions of systematic correlationism in order to show that their answers are not convincing.

The conclusion of this chapter cannot therefore be that correlationist realism, correlationist ESA and OSA are untenable per se. In my view, the evaluation of philosophical positions must always proceed in detail. But what my discussion of the correlationist dilemma, the genesis of correlation as *creatio ex nihilo*, the insufficiency of quietism in philosophy of science and the failure of the retrojection strategies shows is the ability of ancestral questions to put on the spot a wide variety of correlationisms. By analysing the four strong philosophical options chosen, I have shown that it is worthwhile to engage with correlationism in philosophy of science. As an analytical tool, ancestrality points to a structural problem of all positions that advocate systematic correlationism. This is not to say that no such position can solve the problem posed in some way. From my perspective, I would assess the chances of correlationist ESA as much better than correlationist OSA, let alone realism, due to the fact the former is not yet convincing with regard to the scientific discourse, but that the latter runs into deep contradictions. However, the suspicion nevertheless arises that an alternative to correlationist philosophies of science offers greater prospects for a successful solution. Therefore, after this discussion of systematic correlationism and ancestrality, it is advisable to look for a new scientific realism that avoids not only the errors and difficulties of OSR, laid out in part I of this book, but also the pitfalls of correlationism.

In order to draw lessons from the failure of systematic correlationism in the face of ancestrality for this new scientific realism (NSR), it is worth considering both levels of the problem, first separately and then in combination. On the level of scientific discourse, the agnostic counterargument from constructive empiricism fails as much as does the transcendental counterargument. The strict distinction of an empirical and transcendental or inner-scientific object-level and philosophical meta-level required for these arguments cannot be convincingly justified in my view. This leaves the correlationist ESA, which is particularly vulnerable at this level, with no choice but to interpret the discourse of the natural sciences in a revisionist way or to admit a contradiction. This poses a real dilemma, because at least Wray and van Fraassen, as constructive empiricists, refuse to undertake such a reinterpretation. Together with the improved no-miracle argument,[32] this dilemma on the scientific-discourse level of the problem of ancestrality results in a large argumentative basis against the correlationist ESA in philosophy of science. In this light, NSR must present a sound realism with regards to the ancestral statements of the natural sciences and respect the realist impetus of the discourse.

On the ontological level of the problem, it became clear that within correlationist realisms and OSA it is difficult to think of the ontological conditions and the genesis of correlation without contradictions. One trap is to assume a *creatio ex nihilo* of correlation together with its ontological conditions. This would result in a conflict with scientific, ancestral statements—and thus back to the level of scientific discourse. But it also would lead to the embarrassment of having to resort to a dogmatic, metaphysical manoeuvre that cannot be further justified. Therefore, none of the variants of correlationism discussed take this path. For them, the correlationist trilemma arises on the ontological level: either the respective position runs into self-contradictions, or it has to advocate a negative quietism regarding the correlation that is untenable in philosophy of science, or has to apply the retrojection strategy. Stanford's retrojection strategy in defence of his correlationist instrumentalism, which implies an OSA with regard to ancestral entities, fails in the face of Meillassoux's reply: The strategy cannot be upheld, since nothing more could therefore be said about ancestral nature, which Stanford, however, must assume to be conditional for subjects. The retrojection strategy of the correlationist realisms of Koch and the fictitious radical Sellars also leads to an unsolvable

[32] Cf. the present work, p. 131.

problem: They attempt to prove that our natural external world necessarily had to produce subjects, but this fails; it is begging the question.

If both levels of the problem are considered in their synthesis, further lessons can be drawn for the NSR: The failure of realist retrojection and the inadequacy of the revision of scientific discourse show that the ontological conditions must hold subject-independently—or more precisely: be maximally modally robust, i.e. whether we know them or not. It further shows that the genesis of the correlation—or the interrelation of reality and science that takes its place in NSR—results from the interaction of ancestral entities that exist maximally modally robust and, in that sense, independently. The complete rejection of the subject-independence of (ancestral) scientific entities or knowledge about them, as found in correlationist OSA, is therefore false. The combination of realism and correlationism in the field of philosophy of science is not appropriate, as the examples examined, when confronted with the problem of ancestrality, led to self-contradictory positions.

Including the considerations on the ontological level, it turns out that the realistic interpretation of the scientific discourse must make it possible to draw on the theories of physical cosmology concerning time and spacetime, the accretion of the solar system, the origin of life on earth and the basal social systems in the animal kingdom in order to be able to answer the questions about the genesis and, in part, the ontological conditions of the interrelation of nature, human subjects and the sciences. It would be a mistake, especially committed by reductive-naturalist OSR, to deny the existence and philosophical relevance of such a complex interrelation. What one can take onboard from correlationism, therefore, is the need to discuss this interrelation in NSR, which will be one of the topics discussed in my next chapter. Interpreting the relevant theories and scientific ontologies realistically also means that temporal realism, which was briefly considered in my map of debates, belongs to the metalocal realm of NSR and must be closely linked to the results of the natural sciences, especially those of physics. Otherwise, the genesis of the intellectual, cultural and historical forms of scientific knowledge acquisition is inexplicable. In this respect, the most radical position of NSR would be to set aside theory production in philosophy of time entirely in favour of new realist interpretations of physical spacetime theories (but of course NSR is not monolithic, and one could also argue for combining classical debates in philosophy of time with interpretations of current spacetime physics and cosmology). Proponents of NSR should also take a realistic view of the robust possibilities of a nature without (human) subjects, especially if these possibilities concern not only a distant ancestral past, but also the possibly not-so-distant future of nature due to climate change, species extinction and many other volatile conditions of our planet. Not only because of this need to consider such robust possibilities of a reality without us, modal realism, even if not regarding possible worlds, is an indispensable component of NSR.

The synthesis of both levels also shows that philosophy of science, as soon as it answers fundamental ontological or epistemological questions, can never be immunised against the discourse of the natural sciences and the insights of the empirical sciences in general, which includes not only the natural sciences but also the social, historical and cultural sciences. In NSR there can therefore be no strict distinction of an empirical-scientific and a philosophical meta-level. This does not mean that NSR must follow the lead of Quine and other naturalists in order to postulate or strive for a reduction, integration or equation of empirical natural, social and cultural sciences with philosophy. It should simply be clear that in ontological or epistemological considerations in philosophy of science (and, I would argue, in philosophy in general), the premises and conclusions of other sciences must be taken into account in order to preserve their insights and avoid inappropriate revisionist interpretations.

With NSR based on the results of my analysis of the double stalemate of OSR and the ancestral challenge to systematic correlationism, I can meet a final critique of the problem of ancestrality: *It might be the case that ancestrality is indeed a problem for systematic correlationism, which is in fact advocated, but which nonetheless remains ineffective.*[33] This would be the suggestion, akin to sceptical challenges in epistemology, that systematic correlationism could simply continue without really addressing ancestrality. Here, I would hold, the ancestral problem in the synthesis of its two levels enables one to provide strong arguments against current positions of systematic correlationism. However, one could fear a similar fate in the debate as that encountered by the anti-realisms: to remain ineffective despite having convincing *counter*arguments, since no convincing positive proposal can be presented. I concede that Meillassoux's speculative-realist counter-proposal is not entirely convincing. All the more reason why I see the need to develop an NSR that represents a genuine alternative to the OSR and correlationisms in their current state in order to advance the debate. My final chapter outlines the framework within which such positions can be developed, draws lessons from the critical examination of OSR and correlationism and identifies some of the tasks that any NSR must solve.

[33] I owe this point of criticism to Philipp Wüschner.

Chapter 8
Towards the New Scientific Realism

In what follows, I deploy the results of my enquiry to achieve the second goal of the present book: Laying the groundwork for a new scientific realism that points to a way out of the double stalemate of OSR without falling into the pitfalls of systematic correlationism. This enables the development of innovative realist solutions in the metalocal realism-anti-realism debate in philosophy of science, thus answering the question of this study. In contrast to Chap. 2, which consists in remapping a familiar territory—the realism-anti-realism debates of the last decades—, my following remarks can instead be viewed more as a nautical chart drawn by observations made from a distance. Nevertheless, I do not want to imply by this that the following outlook is uncertain or random. Rather, the results of the small-step analyses of the arguments, theses and philosophical proposals can be used to take a final step towards the goal of the present study, to provide answers to the following questions: If everything relevant from the preceding study is taken into account, what variants of a new scientific realism (NSR) are then possible? Within what framework must any conceivable NSR orient itself? What are the key lessons from the failure of OSR and systematic correlationism, and how might different approaches lead to different types of NSR?

Section 8.1 addresses the question of the general framework for any NSR. To this end, the first section draws lessons from the double stalemate of OSR and the refutation of systematic correlationism (8.1.1), then presents ontological (8.1.2) and epistemological (8.1.3) grounding assumptions as well as challenges that every NSR will necessarily share. Section 8.2 addresses one of two absolutely essential task of NSR: the development of a new philosophy of nature. This step is necessary to replace the untenable conceptions of a single real external world and a systematically correlated natural external world in the context of philosophy of science. I give an overview of the aspects that such a new philosophy of nature can take into account and I draw on (8.2.1), before discussing two key issues: the inherent plurality of the ontological domain of nature (8.2.2) and the possibilities in the theoretical landscape for local laws of nature and other core concerns (8.2.3).

J. Voosholz, *Towards a New Scientific Realism*, Synthese Library 518,
https://doi.org/10.1007/978-3-031-98886-8_8

The final Sect. 8.3 discusses the ways in which we, our scientific practices, and the plural domains of nature are multiply connected. It considers what an epistemology of NSR might look like (8.3.1), how central the issues of scientific objectivity and critique of the natural sciences are anchored in NSR (8.3.2), and why a conception of scientific transdisciplinarity is recommended for any NSR (8.3.3).

It should be clear from what has been said so far that I cannot offer a comprehensive philosophy of science here, that gives elaborate answers to all these issues. Nor will I describe any individual position. When I write of *the* NSR, then, as with *the* OSR or *the* systematic correlationism, I am referring to a family of positions that share key theses, features and arguments. In this respect, the following is not merely a sketch, it is the sketch of a branched and not easily surveyable stretch of shoreline, to continue the nautical metaphor. That also means that at this point I will be unable to review current versions of scientific realism that might be considered or could be (re-)interpreted as new scientific realism as developed here. An example that comes to mind is Michela Massimi's perspectival realism (cf. Massimi 2012, 2018a,b, 2019; Massimi and Mccoy 2019; and the seminal Massimi 2022).[1] I will briefly comment on the connections of some of her ideas in my section on the natural-scientific interrelation, as well as some of the criticism that is levelled against her proposal. However, in-depth analysis of perspectival realism as a potential new scientific realism is not my concern in this book. The same goes for other proposals that could be considered as such (cf. Chang 2001, 2012, 2017, 2018, 2022; Dalton 2022; Lyons 2025; Mizrahi 2020; Rock 2021).

It is also not the goal of the present chapter to defend the thesis that my drafts for a NSR delineate the outlines of any possible philosophy of science worth defending, not even as concerns the kind reader who has followed all of my arguments so far. I claim to have shown in a fair argumentative debate that OSR and systematic correlationism in philosophy of science should be exceeded by a kind of NSR. However, I will concede that in my account the possibility remains that there is a form of non-correlationist ESA or OSA that is similarly motivated to NSR and can be effectively argued for. This possibility is not even underdetermined in the current theoretical landscape of philosophy of science: Otávio Bueno's structural empiricism, like the NSR, is not yet a fully elaborated theoretical option, but it cannot simply be classified as systematic correlationism either (cf. Bueno 1999a,b, 2000a,b, 2008a,b, 2011, 2015, 2016). But I will not deal with this or other non-correlationist—if you like *new scientific*—anti-realisms in the context of the present work (such as modal empiricism in Quentin Ruyant 2021). Using the analytical tool of systematic correlationism, I have merely shown that the anti-realisms in philosophy of science that utilise and defend systematic correlationism, including those of Wray, van Fraassen and Stanford, do not represent a convincing alternative. To reiterate, I have *not* shown that there could be no anti-realism that provides a

[1] Of course, there are other perspectivalisms out there, owing to Giere (2006) and on some historical accounts go back to Ernst Mach (cf. Gori 2023). For a slightly different approach cf. Gensollen and Jiménez-Rolland (2021).

convincing philosophy of science in the face of the problems of OSR and systematic correlationism. What I defend as a conclusion in the context of the present book is that the first stalemate between OSR and anti-realist counterarguments *does not provide a convincing motivation* to become an anti-realist in philosophy of science.[2]

8.1 A Starting Point for New Scientific Realism

First, I will summarise the core insights that follow from the double stalemate of OSR and the ancestral criticism of systematic correlationism for NSR. Then I will discuss the ontological and epistemological thesis in their new scientific realist form. In doing so, I will build a bridge between two local realism-anti-realism debates that I briefly considered in my map in Chap. 2, and I will argue in favour of modal as well as temporal realism as integral parts of the position of any NSR. Finally, I argue that neither the problem of ancestrality nor the problem of subject-independence should be overestimated in NSR.

The new scientific realism is committed to the two theses of realism in the metalocal debate. I will present modified theses, the components of which will be considered one by one and in context in the course of this chapter. The advantage of cutting the position down to theses is that a framework is set for any possible NSR, which can be clearly delineated in contrast to the theses of objective scientific realisms, EOSR and OOSR,[3] and the tenets of ESA, OSA as well as correlationist scientific realisms. Each NSR argues for the EPISTEMOLOGICAL (ENSR) and the ONTOLOGICAL THESIS (ONSR) OF NEW SCIENTIFIC REALISM:

ENSR We can know some entities of nature in some of their properties and their local, structural dynamics by means of scientific methodology and theorising. Our best theories about nature are partially true and should be understood literally.

ONSR Our best natural scientific theories provide plural, correct domain ontologies of nature with their true parts, the entities of which are maximally modally robust and exist independently of their scientific-theoretical description.

---

[2] This also applies to positions not examined, such as the already mentioned pragmatism in philosophy of science, which can be interpreted as OSA or ESA, such as Healey (2021) and Price (2011, 2013, 2019).

[3] These were: *EOSR: We can know the subject-independent entities that together constitute the single real external world, at least in some of their properties and some of their connections to natural laws, by means of scientific methodology and theorising. Our best theories about the external world are therefore approximately or partially true and should be understood literally. OOSR: Our best theories provide in these parts or in approximations a correct ontology of the single real external world, as the latter is subject-independent.* Cf. the present work, p. 103.

The nature referred to in both theses corresponds neither to the natural external world of systematic correlationism nor to the single real external world of OR. How this term is to be understood, and that philosophical reflection on nature forms an essential part of NSR, will be discussed in Sect. 8.2. It is no coincidence that the formulation of the partially correct theories and domain ontologies of nature resembles selective scepticism along the lines of semirealism with a nod to structural realism. I will discuss this aspect, as well as related epistemological questions concerning methodology, theorisation and critique of the sciences, in Sect. 8.3. In the following sections, I will first emphasise why and how each NSR distinguishes itself from OSR and systematic correlationism with these two theses and which other differences should be taken into account. I highlight the ontological and epistemological perspectives this opens up in order to solve the problems of the previous realist options.

8.1.1 Beyond Objective Scientific Realism and Systematic Correlationism in Philosophy of Science

In the metalocal realism-anti-realism debate of philosophy of science, NSR presents an alternative starting position to OSR and systematic correlationism. NSR distinguishes itself from both by neither accepting the subject-independence criterion in its general form nor—and this point is more decisive—embracing the thesis of the single real external world. The latter is rejected as false. In contrast to systematic correlationism, NSR does not assume a correlation in which the co-relata are in a state of necessary reciprocal dependence. Instead, NSR assumes nature and natural entities that exist—roughly speaking here—independently of their interrelation with us as subjects and our scientific practices.

But even in this delimitation of NSR it becomes clear that insights from the two families of philosophy under scrutiny are also integrated; for example, the interrelation of nature (and natural entities) with subjects and scientific practices. How nature relates to the rest of reality, and what place within the plural and complex nature falls to us as natural entities, deserves philosophical attention. However, some form of independence of nature from our scientific modes of knowing and acting is required. A realist view of our best theories as partially true is also necessary. All of these are pages taken out of the books of either OSR or systematic correlationism.

NSR can derive some core insights and open questions from the double stalemate of OSR. To this end, it is worth briefly revisiting some of the conclusions I drew in my Sect. 5.5: A positive lesson was taken from French's OSR that philosophy of time or spacetime and philosophy of mathematics have an important role to play in the metalocal debate in philosophy of science. The question of how spacetime relates to nature in NSR is virulent and the relationship between the natural sciences and mathematics needs to be explained soundly. In this context, each NSR must provide insight on how its own position relates to (mathematical) abstracta, as these fulfil one (or more) specific and central function in most natural sciences and this is necessary for the interpretation of their respective languages and theories. However, the structuralist approaches to solving interpretation problems in philosophy of quantum physics can also be pursued further in NSR in case they appear promising. (Which I take them to be, inspite of some commentators' concerns that OntSR not delivering fully what it had promised; Berghofer (2018), Chen (2024), Ladyman and Lorenzetti (2023), Lehmkuhl (2011). In this light, it might make sense to look at not only tweaking a little the way scientific theories are regarded, as proposed by Benitez (2023), but at giving up on OSR in one's structuralism and taking the structuralist position to NSR.)

A negative aspect that stood out when considering the various examples of OSR at the end of the first part of this book was that the antitotality arguments prohibit a fixed structure of nature in the form of fundamental laws of nature as well as reductionism to a *single real* external world or a fundamental level of nature. The debate about the status of laws of nature is thus a necessary part of the metalocal debate of NSR as well. We cannot have NSR without an innovative position regarding laws of nature. Importantly, any form of reductionism of the ontology of nature to a single scientific ontology and to fundamental laws and symmetries was also ruled out as an option for NSR. It is thus clear that nature is in itself plural and nonuniform in a sense that needs to be defined more precisely. In the same instance, nature can never be equated with reality, but can only be understood as one of its ontological domains, the internal structure of which becomes the object of research in NSR in close cooperation with the natural sciences. What these insights mean for the concept of nature in NSR will be a key question that any such philosophy will have to ask itself. The reorganisation of the debates on realism mentioned above should also not be underestimated. In particular, it follows from these conclusions that NSR cannot be a metaphysics of science in the sense of French and Chakravartty, since no philosophical theory of reality as a whole is offered and the separation between global ontological and metalocal debate stands firm. The world or whole is not to be confused with the topic of nature.

Moreover, my consideration of OSR has shown that no anti-realist option has been offered. None of the counterarguments to OSR sufficiently motivates anti-realisms (I will come to the alternative motivation of correlationist anti-realisms in a moment). In the interpretation of scientific theories, it had been shown that the most favourable, realist method is selective scepticism, according to which our best scientific theories are partially true. This method of interpretation can already be read from my definitions in theses ENSR and ONSR. The analysis of Chakravartty's semirealism and French's variation of similar arguments revealed that this entails the task for the respective selective realist sceptic of clearly naming which aspects of a theory, description, hypothesis or model are selected for the realist or sceptical stance. With this strategy, however, it is fundamentally possible to respond to the UTE and historical arguments, as has also become clear, if one does not fall into the trap of relying on the fixed structure of the single real external world for justifying one's selective scepticism.

The analysis of the problem of ancestrality and systematic correlationism also gives rise to questions that should be addressed in NSR and to some insights for its conception. The comparison problem and the correlationist reasoning schemes of circle and two-step present NSR with the task of clearly naming how the NATURAL-SCIENTIFIC INTERRELATION of nature, natural entities, subjects with more or less contingent scientific practices in their communities and the scientific theories, hypotheses, models, languages and explanations is to be understood in a realist manner, without downplaying its complexity. How is it possible for successful referential relations to emerge between parts of the theoretical results of our disparate scientific activities with natural entities? If one accepts nature to be plurally structured internally, this needs to be explained. The question of the correct understanding of the sketched scientific interrelation also brings other open issues into focus, e.g. the connection between our epistemological interests, the sciences as a historically evolved social system, the ethical, political, economic and generally practical consequences of our scientific results, technologies and products, individually and collectively, with nature on the one hand and our theories and beliefs on the other. How can the methods and practices of the sciences be meaningfully and fairly criticised from a subject-specific and philosophical perspective, without losing a realist grip on nature?

Any NSR must also be able to reflect on the ancestral preconditions, the genesis and the ontological conditions of the natural-scientific interrelation.[4] This means that it must be open to the insights of history and sociology of science, evolutionary biology, cognitive science and psychology regarding this interrelation, for example, without adopting any naturalistic or correlationist presuppositions. This consideration must also be incorporated into the reflection on the transdisciplinarity of philosophy of science itself. How the sciences are to be categorised and characterised in comparison to other forms of human knowledge is an open question, not only for NSR but with a distinct perspective derived from the negative lessons of systematic correlationism, and the necessity of criticism and critique of the sciences.

Since the realist theses ENSR and ONSR directly assert the independence and unconditionality of ancestral natural entities, the problem of ancestrality is of little relevance to these new positions themselves. However, even in NSR it follows from this consideration that a temporal realism is necessary, in which time or space-time must be ontologically and logically independent of perspectives within it. Whether a genuinely philosophical position on spacetime can limit itself to presenting a new realist interpretation of physical spacetime theories is an open question that I will only briefly address later.

Another central result in contrast to systematic correlationism is the insight gained from my analysis that NSR cannot be immunised against the results of the empirical sciences, be they natural, social or cultural sciences (under which

[4] This conception is not to be confused with the *intrarelation* as an explanation of phenomena in agential realism, cf. Barad 2007; Murris and Bozalek 2023.

I subsume the rigorous work of the humanities). The focus on the trans- and interdisciplinary context, and an interpretation of the disunity of the sciences as appropriate to the plurality of nature, and the other complex domains of reality besides nature are therefore consequences of this core insight. This not only means that NSR cannot operate without transdisciplinary work and interdisciplinary procedures, but that some of its tasks can arise directly from the current results of specialised sciences. Therefore, the by-now classic idea of general philosophy of science as a purely meta-level discipline is called into question.

8.1.2 NSR's Ontological Moves: Nature Instead of the External World, Modal Robustness Instead of Subject Independence

A first striking difference between the realist formulas of OSR and NSR is the replacement of the formulation of the *subject-independent entities that together constitute the single real external world* with *natural entities that exist maximally modally robust and independently of their scientific-theoretical description*. This means that two aspects are approached differently on the ontological level by NSR: The criterion of SI is not invoked (in parallel to systematic correlationism); moreover, there is no talk of a single real external world, but instead of nature.

I would like to address the latter aspect first. The rejection of the single real external world is motivated by the antitotality arguments. Equating the target domain of the natural sciences with reality, and reality with the single real external world fails because the problem of pseudo entities (PPE), or the *placement problem* as Price (2004, p. 7) calls it, immediately arises when the existence of subject-dependent entities (such as subjects) is recognised, but these cannot be part of the single real external world *per definitionem*. The other world- and totality-problems and set paradoxes mentioned in Sect. 5.3.1 also speak against such an absolute conception of reality as the single real external world. Gabriel's main ontological argument and Rovane's argument against unimundialism have shown that even more restricted variants of the single real external world, such as the totality of objective facts or the target of objective scientific consideration, run into internal contradictions. Finally, I had reconstructed some of Cartwright's arguments against fundamental laws of nature as a fixed structure of the single real external world. Together with my consideration of the failure of French's OSR, this result means that NSR can no longer assume that there is a unified overall structure of all concrete or spatiotemporal entities, especially not one that derives overall from fundamental symmetries of physics or from fundamental laws of nature that would apply to everything and always. Thus, NSR has no choice but to reject the single real external world with its fixed structure, and with it ontological naturalism. But the strategy cannot be to regain something like the single real external world, namely the natural external world of systematic correlationism, by assuming a fundamental epistemological or ontological nexus of this natural external world with other co-relata.

Instead, both mistakes can be avoided by conceding that there is no such thing as a unified domain constituted by fundamental laws of nature. Letting go of this unified picture of nature, the naturalist world-view, is no easy feat yet one can be consoled by the fact that nature remains wide open to us, that laws of nature still apply, and that our theories are still as valid. What one has to see is that nature is essentially ontologically plural, i.e. at least, (1) not everything in nature is related to everything else; (2) we must not assume fundamental laws of nature that necessarily fix all the different domains of nature; and (3) causal links and natural relationships between different natural entities are not organised in a unique, uniform or fixed way. Points 1. and 2. have been sufficiently clarified by me, point 3 follows, but requires explanation: A classical picture of nature in reductive naturalism is that causal relations and quantum probabilities decisively structure the single real external world, in that the entities of one causally closed and fundamental level (such as the particles of the Standard Model) causally necessitate and sufficiently condition the entities of the next higher level, these levels are in a hierarchical order (cf. for additional arguments against this level-picture Voosholz 2021b), and science consists primarily in revealing these mechanisms and offering reductive explanations of phenomena and entities at as fundamental a level as possible. If such assumptions of order and the presupposition of fundamental laws of nature, with which everything can be explained relatively simply, are dropped in NSR, this picture is ruled out. As the scientific results of our increasingly disparate natural sciences show, nature is far more complex, plural and disorganised than such philosophical models admit.

This is not to say that nature eludes our thought or complex structures of order, nor that there are no successful reductive explanations of entities, especially phenomena, at a basal or less complex level. However, the presuppositions of a totalised order of nature, a complete or necessary knowability of all natural entities and a global applicability of reductive explanations to all areas of nature must be opposed in any NSR. I will consider which understanding of such a plural nature without fundamental laws is appropriate to being a central question of all forms of NSR.[5] To present a philosophy of nature is thus the *first task* for every NSR.

The second aspect that is understood differently on the ontological level in NSR compared to OSR is the realistic-ontological criterion of independence. SI is not championed here; instead, my definitional thesis speaks of natural entities *that are independent of their scientific-theoretical description and maximally modally robust*. It could be argued here that this criterion belongs to the family of SI criteria, and that some of the representatives of OSR that I have cited give similar criteria. Perhaps surprisingly for the reader, I agree with both observations. The

[5] This deviates clearly from the idea of a renaissance of natural philosophy that is widespread in parts of speculative realism; cf. Gironi (2017). For a criticism of the political dimension of speculative realist philosophy of nature cf. Malabou (2020) and a defense cf. Harman (2021).

main problem that arose with the SI criterion *in the context of philosophy of science* through my counterarguments in the first part and through the correlationist circle reasoning scheme in the second part of this work was that it led to insoluble difficulties with the assumption of the single real external world. Moreover, as I emphasised when introducing the criterion, SI is too vague a criterion for independence, which is nonetheless widely used in OR and OSR. Outside the context of philosophy of science, SI quickly leads to difficulties, especially in the global debate. This is because, as Price correctly points out (cf. Price 2013, pp. 6–8), the PPE comes back to haunt the scientific realist who embraces SI: If only what exists independently of subjects really exists, subjects *trivialiter* do not really exist, which leads to a spiral of problematic auxiliary assumptions to support some *ad hoc* solution.

The way out that Gabriel emphasises helps: "a fact is called *natural* if it is maximally modally robust, i.e., if it would also have obtained had nobody noticed it, and it belongs to the universe." (Gabriel 2023, p. 80). I will not import the notion of a natural fact here, rather the notion of maximal modal robustness as that which would also have obtained if other conditions would have been different, especially regardless of its epistemic situation. Applied to the metalocal debate in philosophy of science and for NSR, this means that natural entities are maximally modally robust in my sense. Therefore, the definition states that they are independent of how we describe them scientifically-theoretically; the first part about their maximal modal robustness implies the independence of descriptions. This latter part of the ONSR thesis can be formulated more specifically: Natural entities are independent of whether we describe them, whether we describe them one way or another, and whether our description is completely correct. In particular, this means that it cannot be ruled out that there are several correct descriptions of them and that they would also have existed if they had never been described. In this respect, natural entities are maximally modally robust with respect to us, our scientific practices and theories. With the restriction that they belong to the universe—which designates the epistemic field of the natural sciences (cf. Voosholz 2024c)—Gabriel implies that there are also maximally modally robust facts that, in his understanding, are not natural. Mathematical facts, for example, can also be shown to be maximally modally robust, at least in certain Platonist philosophies of mathematics, but are not necessarily thought of as parts of nature. Every NSR has two options here in my estimate: Either the new scientific realist claims that nature includes all maximally modally robust entities, which would also interpret all maximally modally robust mathematical, logical and other entities as parts of nature, or she restricts nature to maximally modally robust entities with a certain characteristic, such as the ability to enter into causal interactions. In the remainder of the chapter, I do not presuppose either decision, but I discuss possible differences in other issues arising from this CHOICE OF THE SCOPE OF NATURE. However, it is already clear here that NSR must either present an argument that nature is not a totality—in the case of unrestricted nature—in order to avoid the conclusions of its the antitotality arguments or NSR mus motivate its own restriction criterion, which with the maximal modal robustness becomes a second mark of the natural.

NSR will inevitably face a counterargument due to the definition of maximal modal robustness: Does this mean that a laboratory-bred guinea pig would have existed even if we were not describing it in scientific-theoretical terms? It clearly seems to depend on our scientific practices. Or does this consideration mean that the guinea pig in NSR is no longer a natural entity? Here, NSR gives an immediate answer: Our laboratory and general scientific practices are themselves the natural behaviour of animals, because they are part of our behaviour and we are animals. No cultural divide from nature should be allowed such that human behaviour and human subjects are considered unnatural or non-natural. The assertion that humans are natural entities in no way implies that once humans have been sufficiently researched and understood in the various fields of nature (e.g. as objects of spacetime physics, human biology or neuroscience), everything about humans that can be sensibly and scientifically said has been said. This is clearly wrong-headed, as people also act and live in domains of reality that are not, or at least not only, domains of nature, such as in literature, art or politics. A poem, a symphony or a revolution may have mereological parts that are natural entities (paper, soundwaves, fires), but the former can never be reduced to the latter. This is also true of scientific theories and practices—our mental and social capacities are necessary for them to obtain. From this perspective, it becomes clear that the guinea pig or the generated boson in the particle accelerator are natural entities (in contrast to the theories describing them) and accordingly maximally modally robust. If we set them free, both behave independently of us, as the respective local domain of nature allows. The generated boson and the bred guinea pig exist regardless of whether we observe, describe or ignore them.[6]

In addition, this also reveals that our various scientific practices can not only become the subject of social, historical, cultural and philosophical research themselves—which is obvious anyway—, but that as the natural behaviour of animals it can be a target system for the natural sciences themselves. This does not mean that social, historical, cultural and philosophical research is dispensable for achieving a comprehensive understanding of our scientific endeavours. On the contrary, as will be explained in Sect. 8.3.3, trans- and interdisciplinary research involving all these perspectives is necessary for such an understanding from a new scientific realist perspective. Just because humans have access to other ontological domains (for instance via political, artistic or mathematical practice) does not mean that we are also always part of nature and thus maximally modally robust. However, we are *also* a natural phenomenon that can be investigated scientifically. In

[6] A lot has to happen in the particle accelerator for the generated boson to appear, which in fact is only the case when humans operate the particle accelerator. Theories are also needed to build particle accelerators. However, these points are irrelevant to the question under consideration. This paragraph also says nothing about the measurement problem of quantum physics (cf. Harrison 2022). It only claims that a generated boson can exist without us having or giving a description of it or documenting its existence in any way. My understanding of new realism thus turns away from Maurizio Ferrari's idea of documentality, which I consider to be a form of ontological anti-realism, especially in the context of philosophy of science, cf. Ferraris (2012).

environmental science, physical and ecological methods can be used to investigate, for example, how early experiments with radioactivity produced certain isotopes in their environment that are detectable in the atmosphere and soil today and have certain ecological and physiological consequences (cf. i.e. Hamilton and Robison 2004). This also emphasises that we and our actions, and their consequences for our global ecosystem, are maximally modally robust: Once we have radioactively contaminated an environment, it remains so regardless of whether and how we describe it.

There is a crucial addition here to be made in my view: Natural entities are *not only* maximally modally robust with regard to their scientific-theoretical description. They are also robust with respect to most other natural entities. This follows from the first insight into the plurality of nature described above, namely that not all natural entities are related to all other natural entities: The planet Jupiter will continue to orbit the sun even if, in Douglas Adams' science fiction scenario, a highly technologised alien species blows up the planet earth because it got in the way of an interstellar construction project. The orbits of the planets in our solar system are also maximally modally robust in the counterfactual scenario that we are dealing with a cosmos of eternal ancestrality and that life will never develop anywhere. In this sense, natural entities and many of their properties are modally robust, even if certain modalities change in an overlapping natural domain. The distant stars do not influence our creation right now, being and passing away—and as a rule a star still exists even if its neighbouring star would never have formed. Certain stars, such as our sun, may be part of our natural history, but not all. Our sun will not be part of the natural history of aliens in the Andromeda Galaxy. Natural entities are generally also characterised by maximum modal robustness in this sense.

More decisive for philosophy of science than this general modal robustness, however, is the independence of natural entities from our scientific-theoretical descriptions, our measurements and our epistemic access to them. The ontological realist criterion of maximum modal robustness with respect to our scientific-theoretical description might be *intended* by some OSR proponents, but it is not precisely formulated and its philosophical consequence is rarely named: Such a conception of the independence criterion requires modal realism, which, moreover, should not appear as a realism about (external or possible) worlds, so as not to evoke the aforementioned problems of OSR with totalities.[7] This means that, as a prerequisite of any NSR, a realism about possibilities must also be advocated in order to fulfil the ontological independence criterion in philosophy of science without binding it to possible worlds, as is attempted, for example, by Gabriel, Boghossian and Meillassoux.[8]

[7] Additionally, modal realism in NSR is required because of the importance that is attributed to the reasoning with models and the fact that sometimes counterfactuals and counterpossibles play a crucial role in scientific models; cf. Frigg 2023; Sterpetti 2024.

[8] Cf. my presentation of modal realism in the present work, pp. 22–27, and the reasons given in Benovsky (2021).

I have already mentioned which arguments pose a difficulty for such modal realism and which paths could lead out of them. Gabriel and Meillassoux develop their modal realisms in the course of their ontologies. It is clear from the projects of both authors that they would represent a form of NSR, though not necessarily in the specific sense that I give to this family of positions in the context of this book. In this respect, it would be immediately possible for both authors, who so far have been quoted most extensively as representatives of new realism with their arguments against O(S)R or (systematic) correlationism, to implement this demand for NSR in my sense. But any NSR, if it is not based on an ontological position, can alternatively develop and defend a modal realism within the framework of philosophy of science. This offers the possibility of developing NSR that is not necessarily based on Gabriel's, Meillassoux's or other preliminary work from the global debate in ontology and metaphysics. In both cases, the necessary condition would be met if modal realism without possible worlds is defended. Presenting or connecting such an argumentation is thus a *second task* of any future NSR.

But this is not to say that discussion of this form of robust modal independence of nature from scientific-theoretical descriptions is a *central* element of NSR, on the contrary: It is merely a way of countering correlationist or anti-realist criticism of SI. Once this form of independence is satisfactorily articulated, it does not seem to justify a particularly interesting stance or thesis within NSR. This suggests that the question of the independence of scientifically analysed entities is less relevant than other aspects. With recourse to my Chap. 3, it can thus be stated: The axis of existence is more central in NSR than the axis of independence.

Moreover, the maximally modally robust independence of natural entities is not yet a sufficient characterisation of nature, even if it is true that natural entities must be maximally modally robust. A narrower scope of nature means that there can be other maximally modal entities in an ontology of NR (and hence NSR). The criterion of maximal modal robustness allows entities that are investigated by—leaving the question of mathematics aside here—the social sciences, humanities, cultural sciences, and philosophy also to be attested relative modal robustness, so that this form of independence is related to the later consideration of objectivity in NSR: States, historical events, mental illnesses, struggles for monetary resources, poems or chauvinistic beliefs are also relatively modally robust as social, cultural, historical and/or intellectual-mental entities with respect to their scientific-theoretical description, even if not maximally robust, like natural entities. It is a spectrum with a limit, not a dichotomy.

8.1.3 NSR's Epistemological Moves: Partially True Theories and Temporal Realism about Ancestrality

Speaking of the epistemology of science in NSR in the singular can lead to a false conclusion: That of assuming that at the level of the general philosophy of science with regard to all sciences—or at least all empirical sciences or sciences of nature—it could be fully explained how we arrive at knowledge about the respective part of reality under investigation, which and in what way methods are used, hypotheses, models and theories are generated, deployed and tested, and how we then arrive at explanations, descriptions, predictions, discoveries, applications, technologies and other goals of the sciences, to which new research then follows in an ascending circular motion, a spiral of knowledge about reality. As tempting as such an image is, it is an inadmissible simplification. On the one hand, at the level of the general philosophy of science, where the metalocal realism-anti-realism debate takes place and where the present work is situated, some basic observations can be made about how we evaluate and criticise methods and practices, how hypotheses, models and theories in the natural sciences relate fundamentally to their target systems in nature, and what goals might be appropriate for the sciences in the exploration of different domains of a plurally understood nature. On the other hand, it is not possible to say anything about specific methods and practices of certain current or historical research communities—this falls into the areas of various specialised philosophies of science (of astrophysics, quantum chromodynamics, ecology, systems biology, etc.) and the respective scientific methodological critique—; nor can certain current or historical theories and their models, hypotheses, descriptions, explanations or internal and external areas of application be addressed. In addition, NSR does not allow for a critique of science that has *the* natural sciences per se as a target. If the plurality of nature and the disunity of the natural sciences are taken seriously, such generalisations are out of the question. In addition, I can only name the tasks that are open in NSR at the level of general philosophy of science with an eye to the epistemology of science within the framework of this book and point out possible answers, not present a comprehensive proposal.

In the conclusion of the first part, one theoretical option of OSR in epistemology of science proved to be superior to the others: selective scepticism. This is defended in different variants by Chakravartty and French. I consider this element of an epistemology of science to be similarly useful and important in NSR, although a separate way of denoting theories, hypotheses and descriptions as *partially true* must be found. The selective-sceptic attitude has many advantages: Natural entities can be direct referents of theoretical expressions. For specific theories, hypotheses and descriptions, it is possible to indicate clearly which of their parts are to be understood in a realist or in a sceptical manner. This also makes it possible to appreciate the instrumental use of outdated theories—such as Newtonian mechanics—in current research practice. In a given theory, a model always represents a possible state of a certain target system. Firstly, however, it usually contains a reduction in complexity compared to the target system(s), i.e., it disregards properties or even whole subsystems; secondly, a model always stands for the possibility of how a natural system could be, and in this sense is always merely permitted and not required by a theory. UTE ensures that several alternative models can always be derived from a theory.

In addition, selective scepticism makes it possible to respond appropriately to the UTE and historical arguments. However, Chakravartty's variant is based on a subdivision of detection and auxiliary properties that can be criticised, whereas French's variant is based on a global structuralist thesis. Therefore, it is a task of any future NSR to develop a selective scepticism based on the ONSR and ENSR theses, the debate about modelling practices and—last but not least—the insights of current sciences. I think it likely that, in contrast to the OSR variants, no singular subdivision criterion can be given for which parts of theories require which attitudes. This is because, at least at first glance, it is not clear why the same epistemological rules should apply in different domains of nature. Why, for example, should it not be possible to pursue a structuralist thesis of local symmetries locally in the field of quantum chromodynamics (i.e. to be a sceptic about objects in theories of quantum reality or take a thin concept of objects, but to be a realist about the structural properties of those systems), yet in the field of systems biology to assume causal detection and auxiliary properties close to the definition of Chakravartty's semirealism? This is only a hypothetical idea, but if the ontologies of nature are not unified, why would a unified criterion approach to selective realism and scepticism be the default idea? This third task (if one counts those two that concern the ontological move) can be tackled in different ways. One possibility is to start with the general philosophy of science and deduce from the scientific interrelation which plural domains of nature are accessible to us and why these require different theoretical approaches. Selection criteria for the respective ontology of science could then be formulated for each of these theoretical approaches.[9] Another possibility opens up if the process is carried out bottom up: Here, we would have to start with individual historical and current theories about a local domain of nature and find one or more selection criteria for these theories, which could then in turn be collected in the respective specialised philosophies of science for a particular natural science and fed, in a third step, into the general debate of philosophy of science. I consider the latter approach to be superior, as with the former I see the danger of imposing a selection criterion on certain theories without first considering the individual theory. However, this suspicion must first be examined as part of this *third task*.

As I have already argued in the first section of this subchapter, the definitional theses ONSR and ENSR essentially solve the problem of ancestrality: There once were ancestral entities independent of us, some of which were preconditions for us and the genesis of, among other things, our scientific interrelation within nature. They are maximally modally robust and independent of the scientific-theoretical description we give of them. In this way, NSR is able to avoid the epistemological problems that systematic correlationism encounters on the scientific-discourse level of the problem of ancestrality. However, this has a direct consequence: Temporal realism is necessary for this solution. Like modal realism, temporal realism is a prerequisite for a successful NSR, so that this local debate at least overlaps with NSR (if it has already been solved, e.g. in the context of the global ontological or

[9] Also, in this way problems with empirical confirmation such as that raised in Allzén 2021 can be avoided.

epistemological debate, as in the case of Meillassoux), but otherwise becomes a proper part of the metalocal discussion in philosophy of science.

From my presentation of this debate, it emerges that there are three realistic options that are advocated by the majority:[10] Eternalism, which attributes the order of past, present and future to the subjective perspective and treats concrete entities independently of their temporality; gradualism, which adheres to the objective non-existence of the future and assumes in one way or other a growing block of space-time; and realist presentism, which defends an ontological primacy of the present. There are also attempts to develop alternative philosophies of time, such as realist theories of proper time. Realist presentism appears untenable in view of the ontological level of the ancestrality problem, as it is committed to a realist retrojection strategy as part of philosophy of science. However, this fails. It seems to follow from this that NSR must advocate eternalism, gradualism or a correspondingly sensitive realist alternative theory in order not to fall prey to the problem of ancestrality. However, through the interpretation of the physics of spacetime, especially gravitational, relativistic spacetime, but also information and quantum gravitational theories, there is the further possibility of judging the realism debate of philosophy of time as too insensitive to current physics and—instead of extracting arguments from (supposed) physical results, so as to argue in favour of eternalism or gradualism—not to argue a separate, metaphysical-ontological theory of time, but to present a NSR-interpretation of current spacetime theory that identifies ancestral entities as unproblematic for one's own position.[11] No matter which of these theory options is pursued, the incorporation of a suitable temporal realism into NSR is always a *fourth task* for it.

When arguing from the position of non-correlationist realism, ancestrality is not a problem, but is part of our knowledge about the past of certain systems, be it the universe, our galaxy, the solar system, biological evolution on our planet or the emergence of complex social behaviour in certain animal populations. The positive significance of ancestrality should therefore, as Gabriel (cf. Gabriel 2023) rightly points out, not be exaggerated. In particular, I think it is unjustified to accuse the proponents of systematic correlationism in general or Kant in particular of a "*Ptolemaic counter-revolution in philosophy*" (Meillassoux 2008, p. 118; original emphasis). Meillassoux is not the first to bring such an idea into play; Bertrand Russell had already written: "Kant spoke of himself as having effected a "Copernican revolution", but he would have been more accurate if he had spoken of a "Ptolemaic counter-revolution", since he put Man back at the centre from which Copernicus had dethroned him." (Cf. Russell 1976, p. 9). The scientific and philosophical-historical questions about the correct evaluation of Copernicus' heliocentrism, about how diverse the philosophical reactions to it were and how Kant's metaphor about the "first thoughts of Copernicus" can be better interpreted

---

[10] Cf. the present book, pp. 19–21.

[11] One possibility here would be to present a realist interpretation of radical super-substantialism, such as exploratively in Lehmkuhl 2012, 2018.

do not lead me any further here.[12] However, these questions have nothing to do with the criticism of systematic correlationism I have reconstructed through the two levels of ancestrality and therefore cannot be answered by them.

* * *

Four tasks arise for any NSR as a consequence of the ENSR and ONSR theses: (1) propose a new philosophy of nature; (2) advocate modal realism without possible worlds; (3) offer an interpretation of the scientific interrelation and introduce a suitable variant of selective scepticism (or another epistemology of the natural sciences) that goes with it; (4) introduce a temporal realism into the metalocal debate. In the remainder of this chapter, I will apply the results of my previous work to the 1st and 3rd tasks in order to develop a perspective on what a future NSR might look like. I will not discuss the 2nd and 4th tasks further here. There are several reasons for this, which I have already touched on. The new philosophy of nature, its plurality and the absence of a fixed structure are genuine tasks of NSR. In addition, as has already been seen, there are other aspects to this question, which I will explore briefly in the following Sect. 8.2. Asking which conception of the scientific interrelation would be appropriate and how selective scepticism, voluntarism or other epistemologies of science should be included in NSR is also specific to NSR. This includes reflections on the significance of NSR for trans- and interdisciplinarity, objectivity and critique of science, the understanding of theories, models, explanations and methods, and the question of how the relationship between mathematics and the natural sciences is organised (this also answers the last concern in an analysis of the ontological turn of speculative realism in Pucciarelli 2022, p. 112). These fields—new philosophy of nature and scientific interrelation—are too broad to be fully surveyed in the context of the present book, let alone to be

[12] Kant, CPR, B XVI. Meillassoux's thesis is discussed positively in Gironi 2011, esp. pp. 34–37. However, even the question of whether there was anything revolutionary about Copernicus' theory is controversial: Kuhn doubts this with regard to the hypothesis of heliocentricity in *De Revolutionibus Orbium Coelestium* (cf. Kuhn 2003, pp. 134–137). Meillassoux imagines a straightforward path from Copernicus via Galileo Galilei, Tycho Brahe and Johannes Kepler to Isaac Newton, whereby he fails to recognise the intricacies of this history: Beginning with the unauthorised preface to *De Revolutionibus* penned by Andreas Osiander, which "rejects any scientific claim to truth in the sense of the ancient definition" (Blumenberg 1965, p. 41; my translation), to the inconsistencies of the heliocentric theories before and including Kepler, to the various religious, mathematical, political and philosophical levels on which the "Copernican" turn was prepared and the spread of the new natural philosophical model took place (cf. e.g. Kuhn 2003, esp. pp. 141–265 and Blumenberg 1965, esp. pp. 37–47 and 94–121). Russell's and Meillassoux's understanding of Kant's philosophical innovation as a Ptolemaic counter-revolution is *firstly* uninformed to the extent that both imply that Kant took "his own Copernican revolution" (Meillassoux 2008, p. 119). Hans Blumenberg points out that the Copernican metaphor is used in the two passages of the second preface to the Critique of Pure Reason (B XVI, B XVII and B XXII, fn. 1) in meanings that have nothing to do with a radical change of models or theories, let alone the dethronement and banishment of the human to the periphery (cf. Blumenberg 2000, pp. 595–615). I would venture that Kant implied that his CPR should be read as thinking through a hypothesis without a fundamental theory behind it, as Copernicus did, but this is beside the point here.

dealt with comprehensively. Nevertheless, it is worth giving an outlook on them in order to incorporate the results of my previous work into these considerations of any future NSR.

I will not deal further with the questions of modal and temporal realism as local debates that must be included in the metalocal debate in NSR, as there are many ways of meeting these demands. Gabriel and Meillassoux, for example, already argue modal and temporal realisms in the global discussion as new realists, so that all preconditions for a respective NSR would be fulfilled. The realist interpretation of spacetime theories in philosophy of physics also seems to me to be an obvious way of meeting this requirement. The decisive factor in my decision not to deal with these questions any further is therefore that the options for realising these aspects are numerous and do not appear to be genuinely problematic to me, even if they remain important and large jobs for the new scientific realist. I also find the independence criterion of NSR as maximum modal robustness of natural entities straightforward, especially in the sense of independence from our scientific-theoretical descriptions. The ancestral problem can thus be solved immediately despite the assumption of a scientific interrelation.

However, one important result can be noted here for any NSR—regardless of how it is carried out individually. With the fact that

1. modal realism must either be integrated into NSR or already be part of a global ontological NR as a prerequisite of NSR,
2. and that temporal realism—whatever strategy is pursued for it—must always be a local debate linked to NSR,
3. and as shown above, the local debate about mathematical entities must be resolved within NSR,
4. and these necessary partial relations or necessary preconditions do not all exist for all the debate positions of either OSR or systematic correlationism

C. the NSR is always accompanied by a restructuring and expansion of the metalocal debate in philosophy of science.

I have thus demonstrated the thesis that any NSR is necessarily accompanied by a restructuring (and expansion) of the debate in philosophy of science, given my map of the discourses thus far. This is a feature that unites all forms of NSR. It is not the only feature, but I consider it relevant because it gives an indication of the innovative power of NSR. Another key innovation that is a central feature of every NSR is the claim to a new philosophy of nature.

8.2 New Philosophy of Nature

In the first section of this subchapter, I address what I consider to be the central aspects that any new philosophy of nature must address. The reasoning from the previous subsection, why a NSR without a philosophy of nature is an empty position, is taken up again here in the first section, leading to the discussion that this also implies a non-trivial positioning towards abstract-mathematical and historical-social entities. I further discuss two aspects, as they are particularly difficult and urgent, in more depth: The second section addresses the idea of the plurality of nature in its various dimensions. In the third section, I discuss why issues of laws of nature, causality and complexity theory need to be reconsidered in order to do justice to nature from the perspective of any NSR.

Before I embark on these tasks, a few notes on the choice of words. I call this subchapter *New Philosophy of Nature*, not *New Natural Philosophy*. This is because the term *natural philosophy* (quite closely linked to *Naturphilosophie* in German) would possibly lead some readers to believe that the mode of philosophising about nature outlined here is linked to romanticism, German Idealism or early modern natural philosophy in Newton's sense[13] or even be considered a term as used by historians of science to imply a broadly philosophical study of natural entities and "nature before it could easily be identified with what we call 'science' today. This is done to avoid the modern and potentially anachronistic connotations of the term 'science.' But 'natural philosophy' (and its equivalents in different languages) was also an actor's category, a term commonly used throughout the early modern period and typically defined quite broadly as the study of natural bodies." (Blair 2006, p. 363). I do not want to suggest any of this. The reasons for the urgency of a new philosophy of nature are independent of whether there is a philosophical sympathy with Schlegel, Schelling or Newton. Moreover, the framework within which such a philosophy of nature can operate cannot simply be found in one of the positions usually labelled natural philosophy. This leads to another reason why I avoid this designation, as it suggests to those familiar with the debates on new and speculative realism that Iain Hamilton Grant's *Philosophies of Nature after Schelling* (cf. Grant 2006) is being alluded to here. This is not the case. This philosophical proposal is not central to the perspective of my present work, and nor do I think that its interpretation and the further development of Schellingian philosophy of nature is a solution to the philosophy of nature proposed here. My concern is for the deep systematic issues at stake in the contemporary debate in philosophy of science.

One could also ask why I use the old and loaded concept of nature and not a more neutral word with less of a philosophical-historical bias (cf. for a discussion about dropping the concept of nature e.g. Briadotti and Dolphijn 2017). My answer here is very simple: I do not want to leave *nature* to the naturalists. I consider it relevant that the natural sciences attempt to generate actual knowledge about the most diverse domain of nature—and in doing so to understand them, on the one hand, as inexhaustible and, on the other hand, as accessible to our methods and practices of theorising.

[13] I would like to thank Christoph Demmerling for pointing this out to me.

8.2.1 *Aspects of a New Scientific Realist Philosophy of Nature*

If the single real external world with its fixed structure is dismissed by any NSR in the same way as the natural external world as part of a correlation, the question arises as to what such a fragile, incoherent and seemingly alien sphere looks like. However, the idea that the nature of NSR is a dangerous and unstable place connects to tropes of its wildness and savagery, in which it must be subdued in order not to destroy us, rather than considering the consequences of the previous argument. Therefore, I will now concisely assemble the conclusions about nature that have emerged from the arguments so far in order to sketch out the framework within which nature must be thought about within NSR. I organise this on the basis of the tasks for a new philosophy of nature as part of each NSR, to reflect the character of an groundwork-laying chapter.

Before the ontological state of nature can be discussed, I must first do justice to the ENSR thesis already outlined. Nature is knowable in some parts—suggesting that it is not in others. Certain natural entities can be known both in some of their properties and in their local structural dynamics. This is due to the interrelation of nature with subjects and scientific practices. This scientific interrelation will be discussed in more detail elsewhere, but an important conclusion can be drawn from the fact that some parts of nature are knowable and others are not: *Nature neither eludes knowledge nor is nature made to be known.* Nature does not evade our grasp, nor does nature in all its complexity reveal itself to us. These two ideas should be understood carefully because they are formative for the scientific interrelation and new scientific realist understanding of nature. Some aspects of nature are open to scientific methods, theories and practices, others are not. In this sense, nature is inexhaustible from a human perspective. Moreover, we will not come to an end in research in the sense that we understand all domains of nature in every respect. This seems to be a trivial point, but in this context it takes on a further dimension: As human animals, we are designed through evolutionary processes to be able to perceive some of the natural entities of our original biological niche that are relevant to us. In a limited sense, *we* are therefore "made" for knowledge of nature by means of our partial perception of nature, albeit only to a limited extent: The natural sciences produced in our historical processes go far beyond this perceptual niche when they investigate the ocean floor or the interaction of different molecules.[14] The fact that this work is hard and not always crowned with success is not due to any agency of nature that resists our cognitive grasp, but to the fundamental attempt to shift the boundaries of our knowledge into new local domains of nature. If such a shift is successful, we can speak of scientific progress—even if many things in

[14] Jens Rometsch has kindly pointed out to me that evolutionary explanations for perceptual senses may still seem reasonable, but that there is at least one further step between these and our capacity for scientific knowledge, which consists in the need for a capacity for abstract reasoning and reasoning with others that could be understood as decontextualised by environmental conditions in our species development.

the new domain behave differently to what our best current theories suggest. Any form of ontological assertion about the local connections and what I will call natural contexts must always keep this insight in mind.

The first important ontological point about nature, which can be discerned from the ONSR thesis—read carefully—when it only mentions plural, local domain ontologies of nature, concerns its placement in our broader ontology: It has already been made clear that nature does not constitute the whole of reality for any NSR. This resolves the PPE, because it is no longer puzzling how *everything* that exists can occur in nature, although it apparently excludes entities. The problem disappears in NSR, and also in many ontologies of NR; they turn out in some sense to be pseudo problems. Thus Gabriel's loose, metaphorical talk of the "ontological province among others" (Gabriel 2015c, p. 27) with respect to the universe in a way applies here. Abstract entities such as numbers, moral values or mental objects do not (at least apparently) occur in nature, yet they do exist there. Socio-cultural entities such as states, poems and financial crises are also not pure natural entities, even though they—depending on the ontological view—may have natural entities as parts or elements, or are necessarily linked to some natural entities (human animals, for instance). This also makes it clear that knowledge of nature cannot be the only genuine form of knowledge, for otherwise we would have no knowledge of the other ontological domains and their entities—a point that will be important for the scientific interrelation.

NSR must therefore defend a *relative independence* against any reductionism of these other ontological domains to the domains of nature. This cannot be done by assuming that non-natural entities are necessarily fully modally robust in relation to nature. While this may seem obvious for abstract entities such as numbers—at least, assuming a form of abstracta realism or Platonism in philosophy of mathematics—, socio-cultural entities, moral values or mental objects seem to be modally connected to (and even apparently necessarily dependent on) the existence, constitution and activities of us human animals on this planet. Hence the idea of a relative rather than absolute independence of such entities from nature. NSR must therefore clarify why the aforementioned and other non-natural entities can never be traced back to natural entities *alone* and that, as ontological domains, they have their very own dynamics, rules and contingencies, which in turn are not comprised in those of nature or determined by nature alone. This is sufficient as a working definition of RELATIVE INDEPENDENCE, but deserves a more precise theorisation in any NSR. Gabriel's ontology of fields of sense can be read as a proposal that makes this distinction (cf. Gabriel 2024a).

A central consequence of the idea of nature as an ontological province, a set of domains among other domains, is that nature cannot be completely ontologically closed. By ONTOLOGICALLY CLOSED I mean that a domain of reality does not maintain stable relations with other domains of reality. At least one, if not two, stable relations exist between nature and other ontological domains: In any case, nature contains mental, acting, social, culture-creating beings. The factual interrelation between us subjects and nature, which presents itself as a scientific interrelation through the historical genesis of our natural sciences, is a stable, if not ontologically necessary or eternally persistent connection between nature and our socio-cultural and some abstract entities. Through our actions, non-natural entities have causal consequences for and in various domains of nature—here, for example, one could point to the connection between the image of the human as a being designed for the greatest possible satisfaction of needs and security with the global economic system of capitalism and the development of population density and biodiversity in the earth's ecosystems, or to the breeding of the lab guinea pig, or the creation of the boson. In this respect, nature is not only not ontologically closed, it is also not causally closed. It will be an important task of any NSR to understand the consequences that this causal and ontological incompleteness of nature entails and to find arguments that expose the frequent assumption of ontological and causal closure as premature. I make some preliminary considerations in Sect. 8.2.3 of this book.

In Sect. 8.1.2 I have already mentioned another central point as part of the *first task of a new philosophy of nature*: Nature is plural in the sense that there are always only local contexts—such as laws of nature, rules of behaviour, systems, complex structures and dynamics—that cannot be completely unified on a minimum or maximum level or scale. From now on, I will refer to these plural, local domains of nature with their own laws of nature, rules of behaviour, systems, complex structures, dynamics, entities and properties as NATURAL CONTEXTS. In this respect, there is nature, but not *the* nature as a fully comprehensive and unified entity (not even a structure) that exists like other entities. How exactly the complex, partially disjunct and disparate inner structure of the nexus of different contexts of nature is constituted in the ontological domain that I call *nature* must be discovered by the various natural sciences and the new philosophy of nature. For the reasons given above, this endeavour always remains a *work in progress*.

What makes this piece of work difficult is that the complex, partially disjunct structure of nature is dynamic. The order and relations of the natural contexts are not fixed or necessary by any meta-law, but arise contingently as a result of the relationships that arise from the interactions between different natural contexts. They can be discovered and described empirically, but every NSR should raise doubts when scientists or philosophers claim that they can be derived theoretically. Apart from the arguments against the structural thesis—for instance, Cartwright's argument against the existence of fundamental laws of nature—and Meillassoux's considerations, such a dynamic can also be derived from empirical examples if these are considered under the idea of the plurality of nature. One example is the emergence of life in our solar system, which gives rise to new local contexts (e.g. ecosystems) with their very own local laws of nature, systems and natural entities, which then enter into new relationships with each other and with the other natural contexts. For example, the proliferation of photosynthesising single-celled organisms between 2.4 and 2.3 billion years ago drastically increased the oxygen content of the earth's atmosphere, which influenced the geology of the earth (cf. e.g. Lyons et al. 2014). Human activities have also produced and continue to produce new chemical compounds that are distributed around the planet and launched into the solar system as artificial satellites. In all these and many other cases, new interactions and new natural contexts are created.

Neither the thesis of the dynamics of different local areas nor that of the plurality and irreducibility of nature should deny that causality exists and that weakly or strongly emergent structures, systems and phenomena can arise. From the failure of any comprehensive metaphysics of nature as a theory of the whole of nature or its most fundamental basic structure—in OSR as well as in systematic correlationism—it only follows that there is no *superordinate* reason why the contexts of nature are so and not otherwise. In this context, Meillassoux claims that such reasons cannot exist for ontological considerations and that this fact can be proven deductively. This is Meillassoux's "*principle of unreason*" (Meillassoux 2008, p. 60) again. I consider this thesis—without going into the exact argument that leads Meillassoux to it—to be exaggerated in the context of my consideration for two reasons: First, a fixed meta-structure of nature would at least be conceivable in a counterfactual scenario and, in my view, the proof of a contradiction within this idea is still pending. Second, a purely deductive proof, as Meillassoux attempts to present it, is not very credible from my philosophical premises; after all, the existence of fundamental laws of nature, for example, is an empirical question that cannot be answered without empirical research in the natural sciences. I therefore do not follow Meillassoux on this point. In the next Sect. 8.2.2, I offer some subsequent but nevertheless preliminary considerations on the plurality of nature.

In addition to these aspects of the causality, complexity and plurality of nature, there are others that a new philosophy of nature should address:

One of the open tasks is to explain why any reductionism of nature to the whole—the idea that nature is identical with the cosmos or spacetime—or to an allegedly basal level such as that of quanta, or the hypothetical level of strings, *must* fail. This includes an explanation of why no natural context contains and completely determines all other natural contexts so that they could be ontologically reduced to it. An important argument is that there are strongly emergent entities and properties that only occur in a specific natural context (e.g. the global ecosystem of our planet, in galaxies, etc.). These are not reducible to a larger spatiotemporal context (e.g. our spacetime does not possess the corresponding entities and properties, even if it must necessarily exist so that the natural context can develop in it). Here there is a relationship of necessity between two contexts, without them being able to be reduced to each other. The same applies to the reduction to supposedly minimal parts: Matter and energy in the global ecosystem is necessarily dependent on the existence of e.g. molecules. Nevertheless, the structures of the natural context, such as the distribution of matter and energy or properties occurring within it, such as trophic relations, are not reducible to smaller units. The discussion about ontological reductionism within nature is given new impetus and urgency by NSR.

The rejection of reductionism to the whole also raises the question of how the spacetime in which we find ourselves is related to nature. There has always been and still is the idea of equating nature with our spacetime. The idea is that what is called nature is only accessible to us insofar as it occurs in our spacetime, since our senses and our empirical scientific methods are bound to this spacetime. Therefore, philosophical speculation about nature beyond our principled scientific grasp is rejected, often out of a naturalist or empiricist impulse.

The equational strategy has at least two flaws from the point of view of the new philosophy of nature: first, an external one, which consists in the assumption that because we have no method of measuring other spacetimes, our spacetime is the only one. There is a whole group of conceptions of a multiverse—from physical hypotheses to (untestable) conjectures (for an overview cf. Ellis et al. 2004). One of the physically open questions here is whether there are multiple spacetimes (and how we could prove or disprove the hypothesis). We can imagine other spacetimes without contradictions, even if only with some theoretical physical skill, and make statements about their possible properties. It follows that we can neither assume their existence nor rule them out; this is a real, ontological possibility. The second error of the equational strategy is that it does not take into account that our spacetime itself is a natural entity: We have theories about it that explain its interaction with other natural entities, such as black holes or accelerated systems. Gravitational theories, e.g. on the curvatures of spacetime, have been empirically tested in new ways since the detection of gravitational waves (cf. e.g. Abbott et al. 2016; Barish and Weiss 1999). However, it seems problematic that events and objects in nature that lead to local distortions of spacetime thus distort nature. It must not itself be an emergent natural entity. Moreover, spacetime has many other properties that—if what has been established so far about nature is correct—simply do not belong to nature.

A separate question is whether all those natural entities that are not themselves parts of a spacetime occur in a spacetime and in this sense must be spatiotemporal entities. This question is not identical, but related to the question of whether all natural entities are material-energetic, more precisely systems, the subsystems of which are causally dependent on particle structures and interactions of the Standard Model of quantum physics.[15] The last question must be answered in the negative: First of all, it argues against the fact that the empirical test is pending whether e.g. dark energy or dark matter, should they turn out to be natural entities, can be explained by the Standard Model, such as baryonic matter or electromagnetic radiation. Irrespective of these empirical questions, there are irreducible natural entities that are definitely not matter or energy: local laws of nature. Why should these not be granted the status of natural entities? It is not clear if local laws of nature have subsystems. Even if this were the case, it is not clear why they should be causally dependent on the structures of the Standard Model.

Is there an implication here for the question of whether all natural entities that are not parts of a spacetime occur in a spacetime? Possibly, because it seems strange to understand local laws of nature as occurrences (or even events) in

---

[15] Although here I am speaking colloquially of particles, I am not opposing structuralist interpretations of quantum physics, such as Frenchs. The argument does not change if such a stance is assumed, and whether structuralist NSR would be compatible with local theories of emergence, as suggested in Cordovil et al. (2023). Neither does it exclude "common ground" explanations of nonseparability if one assumes an object-oriented interpretation, such as Ismael and Schaffer (2020).

spacetime. But a definitive answer to these questions would require both a better understanding of the relevant physical theories than I can present in this chapter and a deeper consideration of some arguments from the philosophy of gravitational and spacetime physics. The question of whether all natural entities known to us, which are not themselves parts of our spacetime, occur in our spacetime must therefore remain open for now.

Another aspect to be addressed is the relationship between natural entities and mathematical entities. This question is related to, but not identical with, the question of how the current natural sciences relate to mathematics. If it is assumed that mathematical entities are abstract entities (be they abstract structures, objects or properties) that exist with maximal modal robustness in relation to our scientific access to them, the question arises as to whether they could be natural entities. This immediately raises the question of whether they occur in their own natural context or whether other (possibly all) natural contexts also contain abstract mathematical entities, as Meillassoux and (to a lesser extent) Maddy, for example, seem to suggest, albeit without designing a NSR (cf. Maddy 1990; Meillassoux 2008, Maddy has nevertheless since adjusted her position.). However, this does not seem to be the only option open to NSR. Even if it is not disputed that mathematical entities are maximally modally robust abstracta, it can be argued that they exist in a domain of reality other than nature. It then needs to be explained why so many natural contexts seem to be organised mathematically or at least can be so well described theoretically and measured empirically using mathematical methods (which is, of course, a classic question, cf. Wigner 1960). If, on the other hand, a different attitude towards mathematical entities is assumed, such as intuitionism, formalism or fictionalism,[16] the latter question becomes all the more urgent. In any case, it is linked to the question of how current natural science and mathematics are related; after all, it cannot be ignored how fundamentally these disciplines need and influence each other. I will return to this last point in Sect. 8.3.3, in the context of considerations on trans- and interdisciplinarity.

At the beginning of this section, I referred to the scientific interrelation that exists (at least) between nature, us human animals and our scientific practices. There, I only touched on one additional aspect that calls for consideration: the question of the difference between nature and culture or society. In NSR, this can be formulated ontologically as the question of how socio-cultural entities and natural entities are connected. It may be possible to show through a social ontology on which to base an internal differentiation of the former entities that their connections with natural entities are in principle not uniform, but fall into different groups or types. The question of how natural entities (e.g. some crystals and metals) can become parts of socio-cultural entities (e.g. objects of economic value and means of payment) also plays a role here. Several questions are closely linked to this. They include the interplay between the natural sciences and the social, cultural and human sciences, which, however, goes beyond a new philosophy of nature and provokes the question

[16] Cf. my Sect. 2.2, p. 17–19.

of a new philosophical meta-theory of all sciences, which I discuss briefly in my Sect. 8.3.

However, the question of the connection between nature and socio-cultural entities can also be posed outside of such a meta-theory. Gabriel's *Fictions*, for example, contains a social realism that already looks at the relationship between social and natural facts from the perspective of his NR (cf. esp. Gabriel 2024a part III; for the question how this is entangled with general epistemological realism and an account of falsity in various guises, cf. Gabriel 2024c). One task here would be to link the philosophy of nature with the theory of social facts in such a way that sociality and culture can be understood both as processes in nature and as a separate ontological domain with their own entities and rules of behaviour. This seems to make sense not only from the perspective of any NSR in my proposal, but also from Gabriel's own perspective, since his definition of a social fact makes use of natural entities by referring to biological species: "A fact is *social* if its being-so essentially involves the coordinated behaviour of several individuals of a species." (Gabriel 2024a, p. 221).

There are different ways of understanding sociality and culture as components, processes or phenomena within nature, without denying that the domains of the social and cultural (if one wants to separate them) have a *relative independence* from nature. They are thus separate and ontological domains of reality the existence of which is not ontologically subordinate to that of nature.[17] In order to argue that sociality and culture are also natural phenomena without creating a contradiction, it could be argued, for example, that objective scientific knowledge can be gained from natural contexts, which as natural systems are inextricably intertwined with the socially and culturally conditioned activities of human subjects. The vast majority of our planet's ecosystems today cannot be understood without knowledge of processes that arise from human sociality and culture. NSR has the task of determining how the line between nature and the socio-cultural sphere runs in each individual case and whether there is a one-sided encompassing relationship in each case. The idea behind the latter option is that contexts could occur in plural nature that are structurally and causally inextricably intertwined with subjects and their actions and are investigated, for example, by urban ecology or climate physics. These natural sciences investigate natural contexts just as objectively as quantum physics or organic chemistry. How independent or dependent an area on subjects and their socio-cultural activities might be would not be decisive for the question of scientific objectivity, nor for the ontological affiliation of the natural context to nature.

Even if the ontological and philosophy of science aspects were clarified, this would not end the treatment of the questions of difference and the complete interrelation of nature and culture or human society. For human practices and ways of life, especially our interaction with nature in the broadest sense, also emerge here. This raises issues of ethics, political theory, anthropology and social theory,

[17] Cf. the positive ontological proposals in Gabriel 2015a, 2016, 2024a,b,c.

philosophy of culture and theory of action, the investigation of which in NR, alongside questions of NSR or (social) ontology, represents a task in its own right.

This list of aspects that play a role in a new philosophy of nature could be continued and refined. It comprises only some of the important tasks that a new philosophy of nature must solve from my perspective. I will limit myself further in the following, as examining the aspects already mentioned would be enough to fill another long book. I will only deal with the two aspects of causality and complexity theories mentioned above (including the ontological and causal incompleteness of nature) and the plurality of nature. I have chosen these two questions because the others have either already been dealt with more comprehensively in the debates in philosophy of science and the ontologies of NR, or they seem to me to be more straightforward. I will not hide the fact that some of the other tasks and open questions are linked to the points I discuss in the next sections. Since my remarks are merely *preliminary considerations* on any new philosophy of nature, in what follows I consider them in a correspondingly limited sense. Due to the limitations of this work, it is impossible for me to pursue all the connections here or trace all the consequences that arise.

8.2.2 The Plurality of Nature

The previous reflections already imply something about the plurality of nature in every NSR: Nature is ontologically plural, i.e. at least at first: not everything in nature is connected to everything else. In particular, not all natural contexts are directly related to all other natural contexts; moreover, we know of no natural context that encompasses all other natural contexts or is necessarily related to them. It is possible, for example, that in addition to our spacetime there are others in nature that are possibly causally isolated from ours. Second, the ontological plurality of nature means that we cannot assume any fundamental laws of nature that necessarily fix all natural contexts or apply in all natural contexts and situations. Third, the causal connections and natural ties of different natural entities are not ordered in a unique, uniform or fixed way. This means that ontological reduction to a supposedly smallest scale, lowest or most fundamental level is incorrect. Fourth, nature is neither ontologically nor causally closed, but is at least influenced through our actions by abstracta and socio-cultural entities that do not necessarily or even obviously occur as entities in a natural context. The assumption that nature could be otherwise linked to abstract entities arises at least from certain views on mathematical entities.[18]

[18] By this I only wish to point to the surprising mathematical organisation of some natural contexts mentioned above, not to the problematic, reductionist and para-idealist thesis that nature consists exclusively of information, as suggested in Floridi 2008; Tegmark 2007.

Furthermore, nature in NSR should be understood as dynamic, both with regard to the natural entities themselves and with regard to their structures and local laws of nature as well as to the natural contexts. These and their relations change and, as far as we know, are only determined by their local interactions, not by an all-encompassing metastructure. As already mentioned, nature does not elude knowledge, nor is it made to be known. We have good reason to believe that some natural contexts can never be known by us (such as natural entities outside our causal light cones), while others are already well known.

Every new scientific realist has at least two strategies at her disposal to develop an understanding of this plurality: irreducibility and the dynamics of nature. Both strategies are compatible with each other, but can also be used separately. The first is to learn about the inner structure (or disorder) of the ontological province of nature from the global debate on ontology of NR. The second seeks to draw inferences about some of the dynamically linked local contexts of nature from the results of our current natural sciences and the reconstruction of domain ontologies that can be derived from specific, possibly linked theories. For a comprehensive understanding of the plurality of nature, the combination of both strategies seems equally promising and challenging.

If Meillassoux's ontology were taken as a basis, a number of questions would arise: In his philosophy, nature, like everything else, is only made possible by the "hyper-chaos" (Meillassoux 2008, p. 62), the contingent production of reality, which for him follows the principle of unreason. Without being able to go into the exact details of his argument, his conclusion can be summarised as such: Nature was produced by hyper-chaos and is locally stable (cf. for the following Meillassoux 2013b; esp. pp. 150–156; Meillassoux 2008, pp. 82–117, and for a recent system-theoretic reading cf. Zwick 2024). By this Meillassoux means that the locally applicable laws of nature, the existence of matter and energy in our spacetime and many other facts through which natural contexts exist, exist contingently and continue to persist.

He has a form of ontological materialism in his sights, according to which, in the case of our spacetime organised according to matter and energy, spatiotemporal concrete entities with mathematical properties stand in certain natural-law governed contexts. These contexts, like the laws of nature that govern them, are neither totalisable nor obtaining in principle (cf. Meillassoux 2007, pp. 69–76). The mathematical properties of material or material-energetic entities constitute what Meillassoux aims to defend in a comprehensive realism. However, he does not develop a philosophy of science or of mathematics, so it is questionable how this is to be interpreted from the perspective of, for example, the philosophy of physics. According to this view, are mathematical properties, such as relative momentum or locations of point particles, to be understood here as abstract structural relations that determine nature? If so, is Meillassoux thinking of a pluralised and fractured version of French's world structure? Would Meillassoux advocate a selective realism in philosophy of science only for these mathematical properties, and a selective scepticism for all others (because causal properties are too shaky in his philosophy of radical contingency)? Or is my suspicion justified that Meillassoux generally has a sceptical attitude towards our best current scientific theories and accordingly wishes to speak in a realist voice about the mathematically describable properties of directly observable natural objects? These are all questions that a new philosophy of nature would have to address if it were to take Meillassoux's ontology as its starting point.

At first, it seems obvious that Meillassoux would defend a form of ontological monism for nature. His comments on speculative materialism could easily be interpreted as the conviction that nature consists of the material-energetic objects of physics and that all other natural entities are either local and contingent laws of nature or abstract properties of these objects. Complex natural systems, such as the world of the living, could then be understood as a strongly emergent structure, but one that emerges from the contingent dynamics of the simple material-energetic objects of physics, their abstract properties and the local laws of nature. In relation to these systems, the interpretation may be appropriate, but Meillassoux has not yet explicated this (nevertheless, he repeatedly hints at such a reading, cf. Meillassoux 2014, 2016). For Meillassoux, with the principle of unreason, our nature *and* our spacetime are only one of many that are produced and rejected in absolute ontological time without reason. Our nature is therefore stable for the moment, but is not necessary or fixed by anything, hence it can change or cease to exist without reason or cause. In fact, in relatively recent writings, Meillassoux goes so far as to inscribe a radical plurality into nature, despite its ontological uniformity, when he declares his speculative realism to be a clear anti-reductionism:

> There will always be [scientific; J. V.] disciplines in the plural, even if their configuration is modified profoundly from one epoch to another; and this not only, in vague terms, because of the "domains" separable as surveyed lands, all of the same nature, but, we suspect, because there are in fact in our world fissures of beings as such, which resist unification by means of a superior knowledge that would traverse all the others in order to give their ultimate reason. The fissured real is given above all in the irreducibility of the world of contents and qualities of sensible life, to that of the natural sciences, strictly mathematised. To seek to reduce the first to the second at any cost, or to seek to inject the second at any cost into the first, these are two attitudes which in truth are symmetrical and which do not give a chance to the irreducible, to the fissure that runs through the actual (the world that is in fact ours). To affirm that sensible content—whether conceptualised as qualia, or as secondary qualities, or in some other way—are but epiphenomenal lures, and that the real is entirely a matter of the mathematizable basis of physical, chemical or neuronal sciences, is only the inverted version of the thesis according to which there are already infinitesimal contents, sensations, volitions in matter as mathematically rendered by science. These are two inverse reductionisms which tend to have little or no interest in anything that persistently resists their monism. Our materialism is, on the contrary, an antireductionism. Meillassoux 2016, pp. 145–146.[19]

What I have so far designated by the general name *natural context*, without characterising its plurality, diversity or limits, corresponds only in part to Meillas-

[19] At this point in the text, Meillassoux elaborates on his version of a strong emergence theory, which is laid out in the concept of *surgissement ex nihilo*, to which I have already referred in my Sect. 7.3, p. 278.

soux's idea of a *fissured real*. Initially, the domain of the contents and qualities of perceiving life seems to be part of the latter. However, I indicated above that, from the point of view of NSR, depending on how one explicates it, this is a different ontological domain to that of nature, at least insofar as, for example, social entities (if these are to be included at all) have parts that can be natural entities, but do not occur as such in nature. However, Meillassoux also suggests that this is not the only *fissure* and that others criss-cross nature, so that its realisation through a multitude of scientific disciplines is appropriate as an epistemic approach to its ontological plurality. This seems to go in the direction that every NSR should have in mind if my previous characterisation of nature is correct. However, Meillassoux does not specify at any point whether or how this plural form of nature can be described scientifically, nor which *fissures* run through it exactly.

In addition to the question of whether Meillassoux's ontology and the figure of *fissure* are convincing in themselves, which is not answered here, several questions arise for me in relation to nature: Can Meillassoux's absolute time, in which nature arises and passes away, be reconciled with my reflections on spacetime and current physics on the subject? For Meillassoux, nature appears to be in some ontological time, but a spacetime must be understood as a further context of nature, as I showed in the previous section. It is questionable whether and, if so, how this apparent contradiction can be resolved. Must we assume, contrary to realist theories of time and spacetime, that an ontological meta-time must be assumed with Meillassoux? That would make Meillassoux an undesirable starting point for a new philosophy of nature. It is also questionable how anti-reductionism can be reconciled with the claim that the ontologically fundamental entities of nature are material-energetic objects of physics that possess abstract, mathematical properties determined by local and contingent laws of nature. The theory of emergence and complexity that would be necessary for fusion of theories from different disciplines would not only require insights from the general philosophy of science and ontology, but also answers to questions about the emergence of subjectivity in this nature and labour in the particular philosophies of science of physics and biology. Overall, therefore, Meillassoux's speculative materialism, understood as a pluralism about nature, seems to present NSR with a heap of unresolved challenges.

In contrast, Gabriel's ontology of the field of sense does not contain any monistically coloured elements but is thoroughly pluralistic. However, here too there are tasks that any NSR and a new philosophy of nature based on it must solve: Gabriel describes two fields of sense, i.e. ontological domains, which he calls *nature* and *the universe*: "Naturalism identifies the world with the universe, the domain of inquiry of the ensemble of the natural sciences, which one calls 'nature' as well." (Gabriel 2016, p. 279; my translation.) Gabriel argues here and elsewhere with the universe as the field of investigation of the ensemble of the natural sciences against a "micro-foundationalism" (Gabriel 2016, p. 166; my translation) a metaphysical reductionism to supposed elementary particles, a position which I have labelled reductionist physicalism. This contributes to the conclusion that Gabriel's universe is clearly an ontological province, i.e. one field of sense, one domain among many. The universe as the field of investigation of the ensemble of natural sciences therefore cannot be translated into the concept of nature outlined here. For one, it is an epistemic field, thus an epistemologically discerned domain, not an ontological domain in the sense of the nature required here (for a more elaborate interpretation cf. Voosholz 2024c). But, what is more, it is most likely to describe and pick out the areas of nature, some of them natural contexts, that are described partially correctly by our current scientific theories. Gabriel supports this interpretation: "there are non-thoughts, which can be fused epistemically into a field, the universe" (Gabriel 2021, p. 44). The *non-thoughts* mentioned are those entities (objects, in Gabriel's nomenclature) that are not thoughts themselves (cf. Gabriel 2021, pp. 44–45). My arguments in favour of the plurality of nature are transferable to Gabriel's universe: This domain is also radically plural in its internal organisation, so that it is not possible to conceive of it as a single field of sense that then contains neatly natural entities, and only those. Instead, Gabriel's universe forms an extremely heterogeneous domain of several, partly interconnected fields, i.e. ontological domains that also contain theoretical entities of all kinds. Gabriel recognises this, which is why he speaks of the epistemic fusion of this field. Some of these entities are also natural entities. How exactly this relationship is structured, however, is a question of philosophy of science, which Gabriel does not address in his ontological investigation thus far (cf. his response to my argument Gabriel 2024b, pp. 346–348). But he at least hints at the possibility of a new philosophy of nature:

> In this way, field of sense ontology inherits the basic idea of domain ontology. In doing so, it drops the optional assumption of classical domain ontology in favour of ontological realism, that object domains are individuated as universes of discourse delineated by different disciplines or methods. This allows us to assume that nature itself could be organised in a domain-specific way, which leaves open the option of distinguishing between the universe (as the domain of an idealised unified physics) and nature (as the domain of the ensemble of natural sciences). Gabriel 2016, p. 193; my translation.

Gabriel's use of the term *nature* in *Sinn und Existenz* and other texts is often identical with that of *universe*. But this is not always the case. The first option, that nature could be organised in a domain-specific way, not only allows the universe to be restricted to the domain of futuristic physics in the way described—an option that Gabriel does not consistently pursue in his writings—but it does open up the possibility of pluralism in nature. In *How Mind Fits into Nature* a deeper differentiation is made: The term *universe* retains its meaning and is equated with "*anonymous nature*" (Gabriel 2021, p. 41; emphasis as in original), but Gabriel develops the concept of nature further:

> Nature reaches further than the mindless universe. We know it from the inner viewpoint of a minded life-form, the necessary existential conditions of which include natural processes; and we contribute to these processes in a "top-down" manner, through the exercise of our mental capacities. What we think, be it true or false, essentially codetermines how things are with us, and thus with the natural structure that we are. Gabriel 2021, p. 48.
>
> New Realism needs "an explicitly non-scientific philosophy of nature" (Voosholz 2024c), p. 162, one that deals with GEN [= Gabriel's extended nature; J. V.] on an ontological and epistemological level that is only partly amenable to scientific reconstruction. While thus irreducible to scientific knowledge narrowly construed, such a philosophy of nature is by no

> means anti-scientific or revisionary. It can fully embrace scientific knowledge and success, such that we can guarantee that we made it, while constantly insisting on the fact that nature has not fully been unearthed. In spirit, it is thus similar to good old regulative idea accounts of scientific knowledge, according to which it is a justified ambition of the natural sciences to reduce the chaos in GU [= Gabriel's universe; J. V.] (to reduce the heterogeneity of natural phenomena by way of its distinguished modes of explanation) without ever hoping to overcome heterogeneity in a final unified picture of the natural world, maybe even culminating in some fundamental physical theory of everything or other. Gabriel 2024b, p. 348.

Gabriel develops this extended concept of nature in the context of a treatise on mental realism and in his response to my argument, but until now it has remained a sketch. It is therefore hardly surprising that no further statements are made about the structure or properties of extended nature. However, two facts can be deduced from what is quoted: First, the ontological and causal openness of extended nature is given; here by the "top-down" influence of mental capacities on the natural contexts associated with us human animals. Second, these have effects in some natural contexts (like ours) but not in (most) others. It follows that Gabriel's extended nature is complexly structured and has at least two different natural contexts.

The second quotation clearly shows that his extended nature thus encompasses more than the domain of the universe. It is obvious that extended nature can also contain natural contexts that are neither interrelated with us nor enter the purview of the natural sciences. These considerations from Gabriel's NR can therefore be expanded, at least in principle, into a new philosophy of nature as I outline it. Whether that is advisable, given some strong demands on the global debate about ontology and epistemology, is an open question.[20]

I will refrain here from citing further NR ontologies, such as those of Harman, Bryant, Ferraris, Longo or Grant. It should have become clear from the brief considerations of Meillassoux's and Gabriel's proposals that philosophical work must be done in each case in order to gain a new philosophy of nature from the ontology in this way.

The *second strategy* of describing individual natural contexts and their dynamic interactions, building on the results of our best current natural sciences and the special philosophies of science dedicated to them seems to be a useful complement to the first strategy, if not even more central to the issue: Only if these are given adequate attention can the ontological gaps in any new philosophy of nature be filled.

I cannot even begin to take on such a task of synthesising various special philosophies of science and the results of current natural sciences within the framework of this study. Therefore, I am merely addressing the question of what the philosophical method might look like that could enable such a synthesis. It has been shown in this book that the ontology of nature cannot be uncovered according to naturalistic, e.g. Quineian method, in which philosophers of science *simply* read off

[20] For a critical overview, cf. Voosholz 2024b; esp. Azzouni 2024; Benoist 2024; Bueno 2024; Koch 2024; Kolozova 2024; Moss 2024; Stang 2024; Travis 2024; Voosholz 2024a.

the "metaphysical commitments" (Chakravartty 2007, p. xii) from the best scientific theories. The discussion and open question of selective realism and scepticism is closely linked to this problem: The existence of which natural entities do I assert if I consider this or that current theory of gravity to be partially true? The crucial point is that the scientific ontology of a theory (or a whole family of theories) can never be reductive with respect to the ontology of nature, because of its plurality. Even in the ideal case, an ontology of science corresponds at most to one of the local natural contexts, without necessarily indicating the form of its ontological connection to all other natural contexts.

In order to counter the two methodological problems mentioned above, we need to develop our own approach. The aim is to avoid the trap of developing the ontology of nature according to the model of a *prima philosophia* as a reflection of the respective general NR ontology or of assigning this task solely to the empirical natural sciences. Both these options will fail. The scientific ontology of a particular theory must first be developed through fundamental research in the respective discipline and the respective special philosophy of science. Second, two questions must be addressed: How should selective realism and scepticism be applied to it? For which scientific entities do we have good reasons to assume an identity with natural entities? And what clear connections are there between this ontology of science and those of other theories? Philosophical as well as scientific expertise is needed to answer both questions. Only then will an attempt be made to embed the respective natural entities, with their established or possible relations to other natural contexts, in their own ontology of nature, based on the N(S)R's ontology. This can give rise to a network of natural contexts. In it, there may be isolated, disjunctive natural contexts without any connection (or any stable connection) to other contexts; however, it will also be dynamic, because new connections and natural entities begin and others cease to exist due to changing circumstances. Therefore, in contrast to French's or Chakravartty's projects, the aim is no longer to form a unified ontology or even a metaphysics of nature, but rather to show not only the structural bridges and causal networks, but also the fractures, the isolation of some natural contexts and the limited scope of natural laws and symmetries.

This methodological outline is provisional and there are alternatives to it. The discussion about the correct method is an integral part of the new philosophy of nature. Reference can be made to the rich debate on methodology in philosophy of science.[21] The aim and approach, even of alternative approaches, are characterised by the idea that the best current theories and practical considerations in the respective science on a particular natural context or phenomenon dictate which scientific ontology is appropriate. This legacy of naturalism remains, in a modified form: It is not a matter of merely uncovering ontological *commitments* that we have knowingly or unknowingly entered into by holding a theory to be true. In

[21] Cf. for an overview of various methods of special philosophies of science, cf. Part IV in Saatsi 2018b, pp. 277–380; and for the discussion of methods from different perspectives e.g. Hankinson Nelson 2002; Suppes 1993; Wagenknecht et al. 2015.

the philosophical development of a domain ontology for a natural context, the specific ontology of a natural context, including the dynamic links with other natural contexts, is developed in a critical examination and dialogue with the scientists, their practices and current theories. This opens up our own perspectives on certain local and metalocal natural contexts. In contrast to the majority of natural philosophies, the new philosophy of nature does not aim to inform us about what or why something happens in nature. However, it enables us to understand the dynamic, complex and interactive ontology of the plurality of nature in sections, opens up a view of specific local structures of nature, their causal and natural law conditions and gives us an idea of whether and, if so, where connections to other natural contexts exist.

8.2.3 *Laws of Nature, Causality and Complexity: Scales and Natural Contexts*

The question of how natural contexts are connected and delimited from each other can be answered in many different ways in NSR. For example, it is possible to approach this area epistemically. Accordingly, both current empirical and theoretical scientific practices would be analysed via a specific phenomenal area that is understood as a plural natural context. This analysis would have to demonstrate how practitioners of the respective science(s) individuate this natural context and which other contexts play a decisive role in their work. Practical aspects of the scientific interrelations in which a particular natural context finds itself would therefore form the starting point for its conceptualisation.

Another approach is to answer the question posed at the beginning only after using the dual strategy explained in the last section. This would attempt to focus on the ontological interconnectivity and uniqueness of the various natural contexts. The tools of general NR ontology would then be available, as well as the scientific ontologies of specific theories or subdisciplines developed by specialised scientists and philosophers of science. Regardless of whether this or another strategy is chosen to explain how natural contexts are related and distinct from one another, a comprehensive answer can only be achieved through minute work. The two approaches presented presuppose close co-operation between philosophy and the respective disciplines. The question posed at the beginning expresses a legitimate and important concern of NSR, but I will not describe it here in further detail.

This section is concerned with the general treatment of natural contexts themselves, not with the delimitation and linking of specific natural contexts. This endeavour arises from a number of more general questions: How can it be that local laws of nature only apply in a certain natural context? Why do new forms of complexity develop into new natural contexts through processes of causality and emergence? How are probabilistic rules or contingent events embedded in them? All these problems are known from long debates in philosophy of science, the various natural sciences and ontology. In this context, I relate them to plural nature and the natural contexts of NSR. Rich research debates exist on these and other topics. It is therefore necessary to examine whether the established answers are compatible with NSR.

As an example, I consider Nancy Cartwright's philosophy of science, her theory of local laws of nature and her adapted understanding of nature, which she develops in *The Dappled World* from her arguments against fundamental laws of nature.[22] I choose this example because determining the scope of a local law of nature is central to the definition of a natural context. This corresponds to a pluralistic turn of French's OSR, which seeks to defend this proposal globally, monistically and reductively. In contrast, I suggest here how such a theory of local laws of nature allows us to think many plural structures of heterogeneous, yet partly interconnected, partly disjunct natural contexts, rather than the one structure of reality. My aim here is not to integrate Cartwright's theory into a form of NSR or to characterise it as one of its necessary components. I merely aim to show that it can be used for new forms of philosophy of science after a necessary critique (from an NSR perspective).[23]

Cartwright's argument against fundamental laws of nature, which I reconstructed in Sect. 5.3.4, leads her to a discussion of the image of nature. Her reflection runs parallels to my work here: For Cartwright, too, nature is inconsistent and neither reducible to a fundamental level nor ordered by a fixed structure. In *The Dappled World* she says right at the beginning about the laws of nature:

> The laws that describe this world are a patchwork, not a pyramid. They do not take after the simple, elegant and abstract structure of a system of axioms and theorems. Rather they look like – and steadfastly stick to looking like – science as we know it: apportioned into disciplines, apparently arbitrarily grown up; governing different sets of properties at different levels of abstraction; pockets of great precision; large parcels of qualitative maxims

---

[22] Cf. Cartwright (1999), pp. 24–25: "To grant that a law is true—even a law of "basic" physics or a law about the so-called "fundamental particles"—is far from admitting that it is universal—that it holds everywhere and governs in all domains. [...] Return to my rough division of the concrete facts we know into two categories: (1) those that are legitimately regimented into theoretical schemes, these generally, though not always, being about behaviour in highly structured, manufactured environments like a spark chamber; (2) those that are not. There is a tendency to think that all facts must belong to one grand scheme, and moreover that this is a scheme in which the facts in the first category have a special and privileged status. They are exemplary of the way nature is supposed to work. The others must be made to conform to them. This is the kind of fundamentalist doctrine that I think we must resist. [...] Not only do I want to challenge the possibility of downwards reduction but also the possibility of "cross-wise reduction". Do the laws of physics that are true of systems (literally true, we may imagine for the sake of argument) in the highly contrived environments of a laboratory or inside the housing of a modern technological device, do these laws carry across to systems, even systems of very much the same kind, in different and less regulated settings? Can our refugee facts always, with sufficient effort and attention, be remoulded into proper members of the physics community, behaving tidily in accord with the fundamental code? Or must—and should—they be admitted into the body of knowledge on their own merit?"

[23] For a pluralist realist project for certain families in physics, cf. Gabovich and Kuznetsov 2023, albeit that the compatibility with NSR would need to be discussed.

> resisting precise formulation; erratic overlaps; here and there, once in a while, corners that line up, but mostly ragged edges; and always the cover of law just loosely attached to the jumbled world of material things. For all we know, most of what occurs in nature occurs by hap, subject to no law at all. What happens is more like an outcome of negotiation between domains than the logical consequence of a system of order. Cartwright 1999, p. 1.

Cartwright combines her image of nature with her depiction of the disunity of sciences. To a certain extent, this is methodologically sincere: According to NSR, the natural sciences also come to know natural entities, local dynamics, structures and natural laws in certain natural contexts. Cartwright points out that even our current best physical theories should only ever apply to clearly defined target systems, which must also be clearly delimited in practice for empirical research.[24] According to Cartwright, when the natural sciences describe these clearly defined target systems or phenomena in theories, they depict local structures, dynamics and natural laws as scientific entities *and natural entities* through their methods.[25] Cartwright seems to assume that there is an approximate congruence between scientific entities and natural entities, that the domain of a scientific theory corresponds to the natural context it describes. However, from the perspective of selective realism and scepticism in NSR, regardless of its exact formulation,[26] this is too optimistic: Some epistemological reductions of one science or theory to another might succeed before it turns out that both theories or sciences have looked at the same natural context from two different epistemological (e.g. methodological) perspectives. We have reason to believe Cartwright's assumption that there is a close relationship between theories and natural contexts. However, NSR is not entitled to the further conclusion that there is an approximate identity of scientific and natural context ontologies such as Cartwright implies. A central idea is the following:

> The impressive empirical successes of our best physics theories may argue for the truth of these theories but not for their universality. Indeed, the contrary is the case. The very way in which physics is used to generate precise predictions shows what its limits are. The abstract theoretical concepts of high physics describe the world only via the models that interpret these concepts more concretely. So the laws of physics apply only where its models fit, and that, apparently, includes only a very limited range of circumstances. [...] Laws, where they do apply, hold only *ceteris paribus*. By "laws" I mean descriptions of what regularly happens, whether regular associations or singular causings that occur with regularity, where we may, if we wish, allow counterfactual as well as actual regularities or add the proviso that the regularities in question must occur "by necessity". Cartwright 1999, p. 4.

The ideas articulated here initially find their correspondence in plural nature and the scientific interrelation of NSR. However, the broad concept of a law of nature should

---

[24] Cf. Cartwright 1999, p. 2: "Physics in its various branches works in pockets, primarily inside walls: the walls of a laboratory or the casing of a common battery or deep in a large thermos, walls within which the conditions can be arranged *just so*, to fit the well-confirmed and well-established models of the theory, that is, the models that have proven to be dependable and can be relied on to stay that way. "

[25] This is due to Cartwright's idea of entity realism, which I have presented together with its criticism by Chakravartty in Sect. 5.1.5, pp. 119–120.

[26] Cf. my Sect. 8.3.1 for two suggestions of the formulation, pp. 341–342.

be restricted in NSR. The reason for this is that, at least for the above proposal to use local laws of nature as ontological characterisations of natural contexts, it must be assumed that they themselves should be natural entities. Cartwright's definition of a law of nature as a description of a regularity in nature must be read and reconsidered accordingly. If it is the ontologically binding characterisation of a regularity, as suggested in what follows, Cartwright's proposal would be compatible with the idea put forward. A Humean understanding of natural laws, as other passages suggest, would be problematic for the transmission (cf. Bhogal 2020).

At this point, Cartwright introduces a concept that, applied to NSR, can serve to define natural contexts more precisely with natural laws as natural entities: "Metaphysical nomological pluralism is the doctrine that nature is governed in different domains by different systems of laws not necessarily related to each other in any systematic or uniform way; by a patchwork of laws." (Cartwright 1999, p. 31.) Here it is crucial that by deriving a natural context from a single local law of nature, or a group of such local laws, this context is clearly defined both practically and theoretically, but at the same time it is not excluded that it is closely related to other natural contexts.

Stéphanie Ruphy argues against Cartwright's use of the term "domain of a scientific theory." (Ruphy 2016, p. 61.) Ruphy criticises Cartwright's insinuation that there are examples of phenomena that, according to the orthodox interpretation, lie within the scope of a theory and a law of nature, but which nevertheless defy them. One of Cartwright's much-discussed examples comes from Neurath: For a banknote that is blown across the Stephansplatz in Vienna by the wind and for which Newtonian mechanics should actually apply, it is practically impossible to specify where and when it will come to rest (cf. Cartwright 1999, pp. 26–27 with reference to Neurath 1987, p. 13). According to Ruphy, Cartwright concludes from this that nature is *dappled*, i.e. that Newton's laws do not apply in this case. She thus justifies her metaphysical nomological pluralism. Ruphy criticises the fact that the validity of Newton's laws in this natural context depends on which question is asked. If the final distance of the banknote to the ground is to be determined (which is always zero when measured and calculated), the law applies; if we want to know more, it does not apply.

> But what has been shown is that the frontiers of the domain governed by a law are question-dependent. Consequently, the domain of what a physical theory accounts for cannot be construed in the ontological sense of a domain, and "ordered" and "unruly" are predicates that can only apply to the question-dependent sense of domain, not to its ontological sense. In other words, *the motifs of the patchwork vary*, depending on one's cognitive expectations. Ruphy 2016, p. 65.

Ruphy concludes from this an even more cautious, anti-metaphysical stance, in which no natural contexts are determined with certainty by local laws of nature. She assumes a dependence of the laws of nature on epistemic attitudes, which amounts to an ontological anti-realism. Several objections can be raised here: First, it is not said that the banknote comes to rest on the ground at the Stephansplatz at all. Perhaps it sticks to the facade of St. Stephen's Cathedral, perhaps it is blown off the square, perhaps it first comes to rest and is then blown away again. Second, Ruphy's interpretation of this and other thought-experiments cannot change the fact that the behaviour of many natural entities—in the laboratory of grammar school physics or in the movement of the earth around the sun—provide good examples of entities that correspond to Newton's laws. It does not seem to matter which questions we ask about falling stones and feathers in a vacuum tube or planetary motion. Third—this point brings us to the present enquiry—NSR can make itself independent of Cartwright's conception of metaphysical nomological pluralism insofar as there are further reasons in favour of the plurality of nature and the ontological irreducibility of the various contexts of nature to one another that are independent of Cartwright's argument.

However, Ruphy's objection shows that Cartwright's partly epistemological, Humean-sounding conception of laws of nature cannot simply be adopted in NSR. It needs to be thought through—at least if the proposal that local laws of nature essentially (co-)determine the boundaries of natural contexts and their interconnections is to be followed. This is not an assumption that every NSR must necessarily make. However, it does provide an opportunity for differentiating the boundaries and interrelations of the individual natural contexts, the causal, emergent and complex relationships of which can be researched and interpreted philosophically and scientifically.

Let us remember: Local laws of nature in NSR are to be understood as natural entities. Even under this assumption, there are a number of options for a more precise definition—these could include causal rules of a natural context, abstract natural principles, a family of dispositions of certain natural entities or a local concrete structure in French's sense. Cartwright's conception of nomological pluralism is flexible enough to allow for different elaborations of a NSR. Cartwright proposes that local laws of nature divide nature into domains that can be interpreted as contexts of nature in NSR. She follows this with an argument for understanding natural laws as abstract truths the applicability of which to a concrete situation is relative to the existence of a suitable objective, concrete characterisation of the situation. Even if an NSR uses Cartwright's nomological pluralism to define natural contexts, it need not follow her further argument of natural laws as abstractions. Therefore, I will not explain which arguments lead her to this conclusion, how she combines them with entity realism and which objections are raised against them (cf. Cartwright 1999, pp. 47–49). Here it has become clear that the integration of Cartwright's proposal is possible in different ways and answers the question of local laws of nature and the definition of natural contexts. Whether this is, for instance, compatible with the recent work of Michel Ghins is up for debate (cf. Ghins 2024), although I see good reasons for embracing Ghins' take on a realism about local causal powers and the arguments against necessitarianism in NSR and combining it with a reinterpeted conception of nomological pluralism.

Which other possibilities are available for achieving these and related goals of a theory of complexity, causality and emergence in and across different natural contexts is not considered further here. A contrastive evaluation that adapts Cartwright's proposal is accordingly postponed to future research. These steps of analysis would result in the next step that amounts to a positive proposal for a new philosophy of nature and deserves more attention than the sketch I can develop here. But I think I have shown that NSR is suitable for productively taking up the results of different discussions in philosophy of science on these important issues.

Elsewhere, I examine George Ellis' theory of causality, complexity and emergence in the light of a possible NSR (cf. Voosholz 2021b, mainly discussing Ellis 2006, 2008, 2012, 2016; Ellis and Gabriel 2021; Ellis et al. 2012 as well as the additional considerations in Voosholz and Gabriel 2021, esp. Healey 2021, Green and Batterman 2021). If Ellis is read, as I advocate, as a representative of a counterfactual theory of causality, such as James Woodward's asubjective manipulation theory (cf. Woodward 2003, 2007, 2021), some of Woodward's and Ellis's basic insights can be transferred to any NSR which is so inclined. Many considerations on plurality, the existence of strong emergence in new contexts and causal relationships can be harmonised with a NSR. This does not only apply to *bottom-up* approaches from the smallest parts to more complex systems, but also to those that run in all directions. Ellis's idea of a global order of nature into finitely many, discrete levels on which natural entities are organised and which are each investigated by certain disciplines poses a crucial problem. I therefore argue elsewhere in favour of freeing Ellis' transdisciplinary theory of causality, emergence and complexity from this assumption. Adding this to the approach of NSR also has the advantage of dissolving difficulties that arise for scientific realism due to competing theories in physical cosmology (cf. Merritt 2021).

The question about the right theory of causality and laws of nature is of course linked to a topic that goes beyond philosophy of science and is discussed in the debates on mental causation and emergence theories of subjectivity: How are the natural contexts shaped in which subjects occur as natural entities? Epistemically, this is significant in NSR insofar as we only have a very limited space and a very limited period of time in nature, meaning our pocket of spacetime, from which we can access nature and, indiscernibly, many natural contexts. This limitation makes it all the more special that we, as subjects, have gained increasingly greater access to new natural contexts through the historical-social development of scientific interrelation—new ecosystems, more distant stars, smaller units of energy, matter and structures, our inner systems as living beings.

In the new philosophy of nature, the idea plays a role that we have emerged as subjects in natural contexts (often called mesoscopic) with a certain complexity, size and energy. Here, in NSR, we must neither make hasty judgements that all subjects share these natural contexts and the contingently chosen scales appropriate to them, nor that subjectivity must always be based on a comparable scale of complexity, energy and materiality. This does not detract from the embodiment argument of Meillassoux and others that subjects must possess bodies, and that physicality and levels of complexity are relevant to specifically human subjectivity and corporeality. However, we should not fall into an anthropocentric prejudice by insisting that subjects must always be, roughly, like us. Integrating these considerations into a new philosophy of nature, as I have outlined, is a worthwhile task for the future.

The analysis and critique of Cartwright's reflections on local laws of nature have shown that while specific adaptations of this theory are necessary in order to transfer it to a new philosophy of nature, this endeavour can be profitable for answering the questions posed at the beginning of this section. The prospect that this also applies to other theories on crucial topics in the philosophy of nature, such as causality, complexity and emergence, shows that NSR does not lead to a break with previous work in philosophy of science, but instead offers a new and specific perspective on the systematic quality of these questions.

8.3 The Scientific Interrelation

In the following, I outline what NSR means for the epistemic side of philosophy of science of the natural sciences. The consideration of the scientific interrelation, which I introduced in Sect. 8.1, plays a central role here. Accordingly, in the first Sect. 8.3.1, I argue that many core properties of the scientific interrelation can be understood as epistemically appropriate consequences of the new philosophy of nature.

I then turn to issues arising from the results of the previous section: I argue in favour of the objectivity of scientific theories, explanations and results, but at the same time justify the importance of criticising historically evolved scientific practices and some forms of the scientific interrelation. Political and ethical questions pertaining to power, technology and other parts of our knowledge system arise here. I also discuss briefly how objectivity and fallibility are related in the natural sciences.

The last Sect. 8.3.3 is devoted to the question of how the integration of results about nature can succeed in the plural epistemology of the natural sciences on the one hand, and how the study of nature can contribute to a comprehensive study of plural reality on the other. This raises the problem of trans- and interdisciplinarity. Philosophy of science can contribute to integration, and I will argue that this is one of the central tasks to which NSR should dedicate itself.

Extensive research debates exist in relation to all of the topics in this chapter—understanding of theories, discussions of methods, interpretation of models and explanations, objectivity, critique of science, inter- and transdisciplinarity—which I do not cover below. Like the development of a new philosophy of nature, this attempt would be a research project in its own right. However, throughout the following sections I will refer to works from the respective debates whose arguments seem to me to be compatible with the motivation of many options for a NSR. Therefore, NSR's epistemology of the natural sciences cannot reinvent the wheel, but can reorient existing ideas and contributions in a new way and in a new direction. If I succeed in outlining this orientation in what follows, I will have achieved the aim of this subchapter.

8.3.1 Theories, Models and Explanations, a Plurality of Methods and a New Epistemology of the Natural Sciences

In addition to the question of a new philosophy of nature, a central task of any NSR is to gain a deeper understanding of the *scientific interrelation*. I had provisionally defined this as a relationship between plural nature, natural entities in their natural context, subjects, our historically developed scientific practices in our communities and our scientific theories, hypotheses, models, languages and explanations. The scientific interrelation arose from the ENSR thesis: We can know some natural entities in some of their properties and local, structural dynamics—that is, in their natural contexts—by means of scientific methodology and theorising. Our best theories about nature are partially true and should be understood literally. The ENSR thesis not only points to a selective-sceptical understanding of scientific ontologies, but also implies the scientific interrelation: A natural science is in contact with natural entities of a certain natural context, usually on the basis of already better-known natural contexts, by means of hypothesis formation and specific, empirical methods. Through subject-specific and historically evolved practices, theories and models, some of which are partially true, are thus formed in a constant interplay (cf. for one relevant debate Bueno and Krause 2007; Bueno et al. 2012; da Costa and French 2000; Frigg 2023; Knuuttila 2011; Suárez 2008). This means that the underlying scientific ontology for the natural context does not completely correspond to the ontology of the natural context, but does so in a relevant way.

As mentioned in the last subchapter, we use explanations, models and theories to exert our influence in a given natural context and to control interactions of natural entities, among other things, with the aim that they correspond as closely as possible to our models, theories and explanations. This means nothing other than that a genuine, complex and realist interrelation exists that runs in both directions, without limiting the maximal modal robustness and independence of the natural entities.[27] The degree of this tie differs significantly both from natural context to natural context and from scientific practice to scientific practice: Current astrophysics influences its natural context less than organic chemistry; genetics in 1960 had less influence on many of the natural entities it investigated than genetics in 2025. It is a task of historical work and specialised philosophy of science to make statements about the quality of this respective form of scientific interrelation.[28]

In NSR, the scientific interrelation occupies the place that is often occupied in OSR by naturalistic models of the approximately true knowledge of the single real external world and in systematic correlationism by a correlation involving us and the natural external world. Through the new understanding of nature as radically plural,

[27] For an alternative view of representation in the sciences from an NR perspective cf. Kanev 2017.

[28] Cf. for the case of the philosophy of biology e.g. Elliott-Graves 2020; Griesemer 2013; Horan 1988.

irreducible to a single natural context and ontologically and causally incomplete, it quickly becomes apparent that the form of scientific interrelation differs in different natural contexts. The non-uniformity of the natural sciences and the increasing subject specialisation through the development of new natural contexts no longer appears to be in need of explanation or inherently undesirable, but rather as an epistemological consequence of the ontological structure of nature (cf. as a classic idea Dupré 1993; and empirically e.g. Langhe 2010; Leahey and Reikowsky 2008).

This argument applies not only to the plurality of disciplines in the natural sciences, but also to the plurality of methods and the divergent roles that models, and causal and probabilitic explanations play in different disciplines and sub-disciplines. It no longer seems meaningful to speak of *the* scientific method if it does not refer to the rational, discursive and historically sustainable investigation of reality through a division of labour (cf. Stump 1992). These aspects are not sufficient to characterise a practical approach to how the sciences gain knowledge of any domain of reality, e.g. any natural context. When one says that the sciences proceed rationally, in my view this can best be interpreted in the sense of a planned approach; that a specific question, e.g. on a phenomenon, structural context or an entity is assumed, which must lie within the framework of one's own epistemic access. By *discursive and historically sustainable* I mean that the procedure and the results are documented in order to make them accessible to others, intersubjectively verifiable, open to critique and revision and accessible for future generations. The sciences are also based on a division of labour, which means that larger research questions are broken down into smaller sub-questions and can be worked on by different individuals and groups, each of which draws on the results of the others. However, none of these aspects characterises a single method as a universal and sufficient procedure for arriving at facts. This is particularly true between the different natural sciences. Nor is a consensus necessary in order to be able to continue research based on a division of labour (cf. for an interesting argument for an epistemology of the sciences based on dissent Solomon 2001, esp. pp. 65–96).

This also implies that the supposed methodological divide between the humanities, cultural studies and social sciences on the one hand, and the natural sciences on the other, which supposedly has consequences for the objectivity of the results, amounts to nothing more than a metaphor from the perspective of NSR. The suspicion arises that it serves the purpose of gaining distinction and competing for academic funding. This does not mean that there is no possibility of a meaningful distinction between the natural and other sciences based on methods or theories. My characterisation of nature and natural contexts opens more than one way of providing one. But the actual empirical methods and practices of theorising diverge as strongly within the natural sciences: diachronically through the historical development of a subject such as chemistry (cf. Weyer 2018), as well as synchronically between different disciplines and sub-disciplines—e.g. current astrophysics, quantum chromodynamics and solid state physics—which is not surprising given the different contexts of nature under investigation. On this basis, it cannot be claimed that one method connects the natural sciences in such a way that they can be distinguished on that basis alone from all other sciences. This also has the consequence that we cannot expect a great unified science or natural science, neither in the sense of an all-encompassing theory, nor in the sense of a single method for coming to know nature.

Massimi's perspectival realism focuses less on the results of our sciences, such as theories, but more on the process of the scientific investigation, and in that less on different methodologies of special sciences in comparison, and more on the question of how research communities in different fields produce reliable knowledge, such as nuclear physics, environmental climate science and psychology (cf. Massimi 2022, chap. 4). By focussing on the interaction of communities for scientific knowledge production, Massimi can be read as one take on the scientific interrelation in the sense of NSR. She can be read as an anti-realist (for the, in my opinion, both systematic correlationist and anti-realist interpretation of Massimi in combination with hermeneutic realism cf. Khalili 2022) or thoroughly criticised for contaminating scientific realism with perspectivalism (for criticism cf. Panagiotatou and Psillos 2023, for a possible defence cf. French 2024b). Without going into the details required to assess whether the realist reading she offers holds up, I would propose the possibility of reading her as a new scientific realist. What does her shift of attention to the community-driven knowledge production yield? Massimi carefully explains why work in one disciplinary field on a part of a given natural context can lead to new discoveries and new models and theories in other disciplinary fields, undertaken on a different part of the same natural context, as she highlights in the case of the shell model in nuclear physics and its connection to prior work in environmental science on isotopes. Yet Massimi does not solely focus on the perspectives of the disciplinary fields. Scientific practice for her also encompasses "(i) the body of scientific knowledge claims advanced; (ii) the experimental, theoretical, and technological resources available to reliably make those scientific knowledge claims and (iii) second-order (methodological epistemic) principles that can justify the reliability of the scientific knowledge claims so advanced." (Massimi 2022, pp. 5–6.) This can be regarded as an apt description of exactly this slice of the scientific interrelation. I would not argue that Massimi in this and other contributions (Massimi 2018a, 2019, 2022; Massimi and Mccoy 2019) exhausts what can be said about the scientific interrelation with regards to scientific practice and the scientific process in and across different disciplinary communities, but one can clearly regard this as an important contribution to what any NSR needs to capture.[29] Importantly, Massimi in *Perspectival Realism* also attempts to describe how we come to know lawlike dependencies in nature and how this allows us to give accounts of phenomena and natural kinds. This, in turn, provides a stab in the direction of a new pluralist philosophy of nature without fundamental laws—I leave the question open for now whether this account represents all the aspects of a NSR such as I have outlined above or only some as I cannot reconstruct Massimi's

[29] For an exploration to combine Massimi with Jerzy Neyman's methodology cf. Kubiak 2024.

proposal of her "phenomena-first ontology" (Massimi 2022, p. 217) in the necessary detail here.[30]

What I have said so far about Massimi and any NSR's position regarding the scientific interrelation does not result in a neat, concise position in the debate about scientific explanations, models and understanding of theories. I consider NSR to be compatible with a range of proposals (Massimi's included) provided, firstly, that they support the realism of the ENSR and ONSR theses and, secondly, that they allow a form of selective scepticism. Chakravartty's proposal of a division of detection and auxiliary properties in the theories of the natural sciences, which are accorded a decidedly different status within the ontologies of science, is basically compatible with NSR.[31] Chakravartty's characterisation of detection properties as causal properties would make the trans- and interdisciplinary elaboration of a theory of causality, complexity and emergence for different natural contexts mentioned in the previous subchapter all the more urgent. This would entail a focus on relational structural properties, which would make NSR epistemically compatible with structural realist considerations (at least for the field of quantum physics) without falling into OntSR or EpiSR. In particular, the separation of natural and postulated entities emphasised in my interpretation of Chakravartty, namely theoretical entities with and without detection properties, could complement NSR's separation of scientific ontologies and the domain ontologies of natural contexts. Natural entities, which we know about because of their detection properties, and purely theoretical entities, the status of which in relation to the natural context is opaque, both occur in scientific ontologies. The domain ontology of a natural context contains only natural entities, but we usually do not know all of them and their properties. Yet, any NSR gives us good reason to believe that some specific scientific ontologies are similar to the domain ontology of a specific natural context. This sketch shows that a transformation of Chakravartty's variant of selective scepticism into a NSR is possible without a hidden introduction of his brand of OSR. Selective scepticism does not require the assumption of a single real external world, but instead local laws of nature, causality and emergence must be given a place in the new philosophy of nature.

Modifying Chakravartty's theory is of course only one way of arriving at a selective scepticism that is appropriate to the scientific interrelation. In Sect. 8.1.3 I considered the possibility of a bottom-up process that leads from the individual historical or current theories concerning a particular natural context to one or more selection criteria in the respective scientific ontologies. In contrast to Chakravartty's solution of focussing on causal properties with which we can come into direct or indirect contact, this approach would redefine the properties in each scientific ontology. This procedure is more complex and demanding because it requires detailed work in the various specialised philosophies of science. One needs to work

[30] Elsewhere (Voosholz 2024c), I analyse how Gabriel's position on nature and the universe combines with different recent proposals such as Massimi's and Chang (2022).

[31] Cf. Chakravartty (2007, pp. 39–48) and my Sect. 4.2.2, pp. 101–102.

through the different contemporary theories, as one would have to ask individually which properties are directly or indirectly captured by the corresponding methods and how they are linked in models and theories with carrier entities for which the assumption of an identity of scientific and natural entities would be required or likely. The possibility that this also applies to other scientific entities of the given theories is not excluded by such a bottom-up selective scepticism towards them; the interpreter refrains from making a judgement in these cases. By means of this piecemeal process, Chakravarty's criterion of causality can be justified if the same detection properties and natural entities are arrived at in both cases. But it would also be possible that one could arrive at a different kind of selective scepticism and realism, although it would possibly work equally well for NSR.

With regard to scientific explanations and models, I follow Juha Saatsi and Arnon Levy in the conviction that the current discussions and proposed philosophical positions in both topics are not yet complete (cf. for an overview: Levy 2018; Saatsi 2018a). In the opinion of these and other authors, a unified or pluralistic theory in philosophy of science is still missing in both cases. For example, the theory of pluralistic explanations proposed by Christopher Pincock can be taken up in any NSR as well as in any OSR (cf. Pincock 2018). According to this proposal, there are at least three types of scientific explanations: causal (which answer the question *"What are the causes of a phenomenon, process or event?"*), constitutive (which answer the question *"What entities must be present for a phenomenon, event or process to occur?"*) and abstract (which answer the question *"What are the abstract structural presuppositions of the context in which a phenomenon, event, process or other entity occurs?"*). Pincock argues that for natural phenomena all three forms of explanation are appropriate and cannot be absorbed into each other. Any NSR can theoretically undertake this analysis just as well as OSR, and in the case of natural phenomena NSR is even better able to explain what a natural entity or natural context is.

The same applies to the reflections of Frigg, Suárez, Cartwright, Bueno, French, Ladyman, Thomasson and others on scientific models that I already mentioned above. Frigg's considerations with regard to a fictionalism about scientific models (his DEKI framework cf. Frigg 2010a,b, 2023; Frigg and Nguyen 2020, named for "denotation, exemplification, keying up, and imputation" Frigg and Nguyen 2016, p. 226), for instance, combines well with a view NSR in concert with NR in the ontology of fictions. To take a different example from the possible positions, Amie Thomasson argues in favour of an indirect realism towards models by conceiving them as abstract entities created by fictional practices, which are nevertheless capable of incorporating certain aspects of natural entities. They are thus able to give us true explanations, predictions and functional manipulation possibilities for a certain natural context in one respect, if never in all (cf. Thomasson 2020a; for different version of models as abstractions compatible with realism, cf. Bueno and Krause 2007; Bueno et al. 2012; da Costa and French 2000; Psillos 2011). The limited scope and the finite application contexts of many models and the corresponding modelling practices are unproblematic in NSR, in contrast to OSR, as I have already argued in the analysis of Cartwright in the previous subsection.

The question of abstractions, idealisations and the role that models and modelling play in practice in various sub-disciplines, on the other hand, is just as virulent in NSR as it is in OSR. It can be said that the debates about pluralistic explanations and the role of models do not appear in a fundamentally different light as a result of the adoption of NSR. However, this does not apply to every position: In particular, the anti-reductionist and pluralist theses of NSR speak against metaphysically charged conceptions (cf. for such, e.g. Craver 2007).

It can be summarised that any NSR can produce a meaningful epistemology of the natural sciences by analysing the scientific interrelation and contributions to current debates on theories, models, explanations, plurality of methods and scientific practices. At the very least, this endeavour does not appear to be any more difficult than under the assumptions of OSR or systematic correlationism. The options outlined in the debates and their possible links with NSR only represent a small section of the theoretical landscape. I do not want to suggest here that this task is trivial or that no objections can be formulated, for example, against pluralism via scientific explanations or methodological heterogeneity. But it seems to me that the path towards defending NSR in this respect is no rockier than the alternatives.

8.3.2 Objectivity in and Critique of the Natural Sciences

I have already argued that the fact that even if the natural contexts, in which we are embedded, are in some cases determined by subjects and their socio-cultural activities this does not detract from the objectivity of scientific results, theories and disciplines.[32] Following this idea and the absence of a strict methodological boundary between the natural and other sciences, as explained in the previous section, the question arises as to whether NSR wishes to assign a different status to research in the social sciences, humanities and cultural studies. Dilthey's historic demarcation between hermeneutic-interpretive disciplines, ("*Geisteswissenschaften*") which are interested in understanding the contingent individual, and neutral-explanatory disciplines ("*Naturwissenschaften*"), which are orientated towards explaining general laws of nature, is sometimes used. Similar systems distinguish between descriptive and normative forms of science and see objectivity and objective knowledge as the domain of the natural sciences and mathematics.

In what follows in this section, I turn to the questions of division and co-operation between the various sciences. My first concern here is to find out whether, from the point of view of NSR in general, there is anything to be said in favour of characterising the natural sciences as a whole as particularly objective. This will lead me to the conclusion that NSR should include and enable modes of critiques of the sciences.

---

[32] Cf. the present chapter, p. 323.

Due to the plurality of nature, its status as an ontological domain of reality among others, and the heterogeneity of the natural sciences, it does not seem obvious to me that they must possess a fundamentally different and distinguished form of objectivity. A defender of such a thesis of the *excellent objectivity* of scientific research could point to the maximal modal robustness of the natural entities investigated. The argument would then be that the natural sciences *qua* their field of investigation are special and their results are more certain or reliable because natural entities are particularly stable. However, a number of results that have already emerged in the preliminary considerations for a new philosophy of nature speak against this: Not all natural entities are stable; natural laws apply locally, not globally; natural contexts are dynamic; some are even partly dependent on socio-cultural entities; nature is neither causally nor ontologically closed.

The defenders of excellent objectivity may concede that the natural sciences also research humans as natural entities, but differentiate that it considers them in their natural existence, in contrast to their subjective existence, which is characterised by social norms, psychological states and cultural narratives. This is the domain of the other sciences, which are contaminated, as it were, by this subjectivity. The critic of excellent objectivity is certainly right in that the social sciences, for example, investigate phenomena and systems of humans as subjects organised in groups and thus not primarily as an aspect of nature, unlike human medicine, for example. Therefore, an intentional, emotive and mentalistic vocabulary is required in these fields of research. However, empirical methods are also used here to explain spatial and temporal relationships. In addition, observable phenomena within a natural context often provide the starting point for a social scientific research question or a theory of the social science—for example, the regular movement of groups of people in a city system, concrete acts of violence that indicate the outbreak of a social conflict, the disturbed eating behaviour of a group of test subjects, etc. How culture and the social are integrated into nature, which natural contexts are culturally and socially permeated, saturated or even created, and whether the objectivity of the results is given in other ways but to the same extent, can thus be investigated in a new and innovative way in NSR.

With these considerations, I do not wish to suggest that there are no methodological and theoretical differences between different sciences or to reject the usual systematics of the sciences and the academic division of labour. There is no need for this due to NSR. It does not in principle argue against such considerations, as it is clear that much of the success of the sciences is based on them. However, caution is required in claiming a higher degree of objectivity in the natural sciences than in other sciences. If such a conclusion is to be the goal of an NSR—it remains a *possible* position in my view under the aforementioned considerations—it must be argued for differently than via the outlined and known paths. Conversely, in my view, the new philosophy of nature indicates that the burden of proof for the thesis of the excellent objectivity of the natural sciences lies with its proponents. If anything, the argument from stable natural entities suggests a different level of objectivity is possible between sciences, sub-disciplines and theories investigating different natural contexts. Yet proving a principled difference in the general level of objectivity in comparison to other sciences is not an easy conclusion and should not simply be assumed.

As a form of realism, NSR argues in favour of attributing objectivity to the sciences as a whole and in some degree to each and every science. However, unlike OSR, it is committed to linking any objectivity of a statement or scientific theory to the condition of fallibility.[33] In *Limits of Epistemology* Gabriel argues that the difference between truth and holding-true arises for the mere possibility of objective existence and knowledge about it.[34] Scientific knowledge of nature is no exception, especially since it is always guided by interests. Natural sciences investigate phenomena, natural contexts and questions discursively and practically in a social group. The selection of these research entities is neither predetermined by universal curiosity, which is *per definitionem* indiscriminate, nor by the plural structure of nature itself, but is orientated towards social (economic, political, technical, technological) reasons and the historically contingent discourses within a multidisciplinary and interdisciplinary scientific community. Sometimes, it is also due to the random curiosity of one or two scientists, who uncover something that is taken up by the group and then shapes a field, of course, but that taking up happens due to the social reasons and the historical discourse. OSR, especially in the form of naturalism, often leaves this aspect untreated when science is defined as the rational, step-by-step realisation of the single real external world.

This means that NSR enables *specific scientific critique* of the motivations, practices, methods, theories and discourses of clearly defined scientific research projects, programmes or sub-disciplines. In NSR, pluralism prevents critique from turning against the natural sciences as a whole, as it characterises the overgeneralisation of such a gesture—the plurality of nature and the plurality of the social web undermine such gestures. Following NSR, a critique of science of the individual natural sciences, both from within the respective disciplines and from within any particular philosophy of science, must be a vital component of scientific and philosophy of science practice. I have already explained in Sect. 5.2.4 why I consider scientific

---

[33] To be fair, it must be added that this connection is sometimes also seen by proponents of OSR and realist orientated pragmatists. A new contribution to the relationship between fallibilism, relativism and anti-realism is Stump (2021).

[34] Gabriel 2020, cf. esp. pp. 183–184: "If there are to be registerable items, we have to factor in a difference between holding-true and truth, bearing in mind that the possibility of this distinction does not imply that the rules that facilitate our cognitive access to the world (and so our practices) are themselves independent of rules (and thus of our practices). That would be an equally absurd assumption. Our practices are dependent on their respective rules because they constitute them as one set of distinct practices among others." and pp. 225–226: "By pursuing this strategy, Sellars extends the difference between *causes* and *reasons* to a dualism of nature and mind. But he thereby faces the following problem: he has to explain how we can *know* about a nature that presents a purely natural order, a realm of 'is' rather than 'ought', even though our knowledge of this order qua knowledge already brings normativity into play and so presupposes its own fallibility. For, according to Sellars, our knowledge of the existence and operations of the natural order cannot be a priori; it can be acquired and extended only in and through the natural sciences."

critique to be central.[35] That argument concluded that it is necessary to assign historical and social arguments central positions in philosophical work on the natural sciences in order to become aware of errors and unjustified presuppositions through scientific and philosophical analyses of the sciences as a social system, historical factor and worldview, for example. The insight firmly anchored in NSR about the scientific interrelation that the natural sciences, as historically evolved human practices, always also fulfil diverse social, political and cultural functions, opens up the possibility of integrating feminist, decolonial, anti-racist and other ideology-critical forms of scientific critique. Critique is directed neither against the objectivity nor the preciousness of natural scientific knowledge; quite the contrary, it is one enabling condition of both and should be valued also as such.

NSR opens up a new path in the discourse of critique of the natural sciences, as many of these ideas appear to be traditionally linked to anti-realist approaches in philosophy of science (cf. Rowbottom 2019). The reason for this is twofold: The neutrality thesis about natural science, roughly the idea that scientific theories, methods and results are neutral with regard to the social, which in OSR derives from the objectivity—understood as complete subject independence—of the results that describe the subject-independent, single real external world as approximately true, prevents the compatibility of scientific critique in many cases. That is not the case in NSR. But in addition, at least those anti-realisms that build on a correlation (involving the natural sciences) in their reflection appear attractive for a critique of science. The openness to historical and social arguments that easily results from systematic correlationism also helps to increase the attractiveness of developing scientific critique from such perspectives. With its reflection on the scientific interrelation, NSR offers an opportunity to conduct scientific critique within a realist framework without the downside of over-generalising or undermining objectivity.

8.3.3 Integration of Current Research: Interdisciplinarity, Transdisciplinarity and the Exploration of Reality

By asking about the critique of science in NSR, I have touched on a level of interaction between the social sciences, humanities and the natural sciences. I have already criticised the question of the possible demarcation between different areas of the sciences as defined in OSR. These considerations are not the end of the process of reflecting on these relationships and thinking about the connection between the various sciences (including the natural sciences). This includes, in particular, the question of how mathematics and the various formal natural sciences relate to each other. As these questions cannot all be answered fully in this book, they are merely posed here in all clarity, making it obvious that there are various ways of answering them in a given NSR.

[35] Cf. pp. 139–141 of the present work.

From the fact that nature is pluralistic, irreducible and not causally or ontologically closed, it follows that no single science—including physics—is or will ever be able to explain or model the entire ontological domain of nature. I explained in the penultimate section that a plurality of disciplines and methods follows from the ontology of nature as an epistemological flip side. Furthermore, a deeper understanding of the ontological province of nature requires not only the multidisciplinary investigation of different, individual natural contexts, but also transdisciplinary research into the structural dynamics of *the interrelationship* of plural natural contexts. In the context of this deliberation, INTERDISCIPLINARY means drawing on the theories, methods and models of different sciences in order to investigate a specific entity. I use the term TRANSDISCIPLINARY to refer to the scientific practice of drawing on methods, models and background theories from different sciences, as well as knowledge from domains different from scientific knowledge, in order to develop new theories in agreement with them that focus on the dynamic structures of linked (natural) contexts.[36] The practices of transdisciplinary work can lead to the identification of new (natural) contexts and/or the emergence of new sciences. The reason why I have bracketed the word *nature* so far is that there are transdisciplinary approaches not only between the natural sciences, but in principle between all sciences—also between the natural and all the non-natural sciences.

Transdisciplinarity is thus an important mode of scientific research, not only of natural reality. This in no way means that such an interplay between the sciences replaces philosophy. On the contrary, the integration of the results of the most diverse sciences is an important part of a new realist philosophy of science. Furthermore, philosophy, and in particular philosophy of science, can play a useful role as translator and mediator. With the study of metastructures and the sciences themselves, but also with the ability to deal with specialised languages, different ways of understanding and explaining the world and the foundational forms of rational argumentation, such a vision does not seem presumptuous.

This should not be understood as an attempt to present a comprehensive philosophy of philosophy or a definition. Instead, I suggest that the NSR can provide a transdisciplinary, alternative picture of the study of reality that treats the various sciences more democratically. Quine's naturalism aims to subsume philosophy and all other sciences under the banner of a unified science. There are also images of a battle of the disciplines (cf. Brown 2001), of philosophy as the fundament of the sciences in general, as their queen or as an ethically or politically legitimised critic of the entirety of the natural sciences. These cannot be perspectives of NSR if the scientific interrelation is taken seriously as a subject of our philosophical analysis. NSR neither conceals the ability of philosophy to reflect on scientificity and the sciences in a special way, nor does it negate the competitions and power struggles

[36] Apart from this, the term *transdisciplinarity* is often used to refer solely to the introduction of non-academic knowledge into the sciences, as in Koskinen and Mäki (2016). My usage tends to be more confined to e.g. Lynch et al. (2021), Augsburg (2014). That said, questions concerning how technologies interrelate with scientific practice and natural contexts are not easily put aside.

in the social system that the sciences also are. However, I consider the approach of striving for a transdisciplinary exploration of reality in harmony with the other sciences to be a promising attempt to shape scientificity.

Interdisciplinarity and perhaps even transdisciplinary is also required to redefine the studies of technology. Many disciplines, especially the social and historical sciences investigate technology, but take foundational concepts either from philosophy of technology or science and technology studies. In my view, both the disparate field of philosophy of technology, which misses a focal point in at least predominant definitions of its subject and theories that unify it as a philosophical sub-discipline (cf. Franssen 2020; Vallor 2020), and the strong programme of sociology of knowledge and the Latourian framework of actor-network theory widely employed in science and technology studies (cf. Latour 2005; Matta 2014; Sismondo 2004; Virtanen and Silvast 2023; Vora and Atanasoski 2019), with its—at least from the perspective of NSR—unsatisfactory metaphysics of nature in the background, can profit from an infusion of new ideas from philosophy. I suggest that the analysis of the scientific interrelation broadly understood (so not only that of *natural* sciences) can provide just that. Our artefacts, both instruments in the sciences and theory-laden products, cannot be thoroughly studied without explicitly thematising the scientific interrelation. In turn, we need to understand technologies and technology better (in transdiscinplinary research) to account properly for our scientific interrelation in its entirety.

Against the background of such an idea of transdisciplinarity, determining the relationship between the sciences is a worthwhile, albeit difficult, task. In what follows, I will demonstrate this using the relationship between mathematics and physics. I have already emphasised why it is essential for any NSR to determine the relationship between the mathematical-formal natural sciences and mathematics. I choose physics as an example, fully aware that it is not itself unified with the gap between relativity and quantum theories and, to a lesser extent, faces the current crisis about the Hubble tension in physical cosmology and astrophysics that arise from the different measurements of the Hubble constant.[37] This choice is motivated, first, because it is a well-researched subject, and second as it has already been dealt with in the course of the reconstruction of structural realism. In what follows, I briefly argue in favour of the fact that this determination of the relationship can turn out differently in different variants of NSR, leading to decisively different philosophies.

These possibilities for divergences about this subject in NSR have two sources: On the one hand, different answers can be given within the framework of NSR as to what mathematics actually investigates, what the ontological domain of its entities is. In Sect. 8.2.2, I explained that in NSR with Meillassoux's ontology, mathematical properties such as relative momentum or locations of point particles can be understood as abstract structural relations that determine some natural

[37] Cf. Bengochea et al. 2020; Freedman et al. 2025; Hollis 2024; Kollmeier et al. 2014; Planck Collaboration et al. 2018; Riess et al. 2019, 2024; Secrest 2021; Wolf and Duerr 2024.

contexts. If such mathematical properties are attributed to a natural entity in physics, e.g. our solar system, which has a certain location in space-time and a certain momentum relative to other solar systems, etc., then it stands to reason that abstract mathematical structural properties should also be understood as natural entities. From this it could be deduced that although mathematics does not analyse a separate natural context (in any case, a separate argument would be necessary for this conclusion), it does enable research on the abstract structural properties of natural entities. Since these abstract structural properties are in many cases what we try to describe in theories and models, there is a natural connection between mathematical-formal natural sciences and mathematics. If it is true that we often find the same abstract mathematical structural properties in models as between natural entities, it is also clear that mathematics does *not only* investigates abstract properties from nature (cf. Maddy 2007, pp. 305–391), for we can construct models that have no equivalent in a natural context known to us, but also possess these abstract properties. This can help us to develop an understanding of why different axiomatisations of different areas of mathematics are possible, partly with consequences for what is true in them. For it does not seem necessary that we are always dealing with the same abstract properties in plural natural and non-natural domains. If such a form of philosophy of mathematics is developed within the framework of NSR based on the reading of Meillassoux provided here, a close relationship between mathematics and the natural sciences is given and easily explicable. However, the difficult question must be addressed as to how it can be that natural entities have abstract properties in the first place, and what exactly this should mean.

Similar positions follow in NSR even without Meillassoux's position in the background if it is assumed, for example following the early works of Maddy (cf. Maddy 1997), that mathematical abstracta really do appear in nature, and from certain positions of structuralism in philosophy of mathematics. Again, this will of course also depend on whether one wants to adopt a thoroughly structuralist picture of natural contexts in one's NSR that is amenable to an isomorphism explanation via abstract mathematical structures to explain Wigner's question about why many physical contexts seemingly behave according to abstract mathematical rules.

A starkly different picture emerges when an NSR is built on, for example, Gabriel's field of sense ontology and an object-platonist philosophy of mathematics, as defended by Mark Balaguer. Mathematical entities would have to be understood here as abstracta that belong to a different ontological domain than nature. Obviously, there is at least one connection between the two *provinces*: Following object platonism, humans have the possibility of gaining knowledge about the ontological domain of mathematical objects (how exactly this is possible in platonism is a much-discussed problem since Benacerraf (1973); Gödel (1947), cf. e.g. Balaguer (1998b) for a possible answer). This knowledge influences their actions in various natural contexts. It is therefore possible to speak of an indirect causal influence of mathematical abstracta on nature (cf. Bueno 2021). However, mathematics and the natural sciences according to this perspective still investigate two ontologically fully distinct domains of reality. The dependence of the success of physics on mathematical methods and formally expressed theories is thus very much in need of explanation in such a NSR proposal. Several paths seem feasible here: Perhaps there are further connections between the natural contexts studied by physics and the domain of abstracta in general, due to which the mathematical tools of physics are successful and nature behaves mathematically in many contexts. An alternative to this would be to see local laws of nature as abstracta with causal forces, which is why many structural relationships in natural contexts that are governed by local laws of nature can only be grasped by means of mathematical modelling. Further options are conceivable, including a local anti-realism with regard to mathematical abstracta as merely instrumental for physics, for instance. In any form of NSR, it is an unresolved task to explain why mathematics plays such a central and successful role for physics (and other natural sciences). In this respect, a NSR based on Gabriel's and Balaguer's convictions leads to a different philosophy of science of the natural sciences than the approach from Meillassoux's ontology or conceivable others.

The question of the relationship between mathematics and the natural sciences in NSR is open, albeit restricted by its ontological and epistemological framework. Even if the question of this relationship is urgent, so are the determinations of the links between mathematising and non-mathematising sciences, between natural, social, cultural and human sciences, and many other concerns that can be treated differently in each NSR. There is no reason to fear that even if my argument in favour of NSR were to fall on fertile ground, this would reduce the plurality of philosophies of science.

Chapter 9
A Short Conclusion

The previous Chap. 8 has already given an outlook on the possibility, urgency and necessary agenda of a new scientific realism. Here I would like to summarise as concisely as possible the main conclusions of the individual chapters as steps in my argument for why we should strive for a such a new scientific realism.

The aim of this book was to show why the variants of *realism* in philosophy of science that have been and are currently predominantly represented are argumentatively insufficient, why common forms of *anti-realism* are equally unconvincing and how this debate motivates a *new scientific realism* (NSR)—all this in the context of the natural sciences.

In the second chapter, I presented definitions of the central terms, and I developed a sophisticated cartography for the debates about realism and anti-realism. I began by using the term *realism* for those positions that advocate the following two theses: First, the *ontological* thesis (RO) that the entities in question have an independent existence and reality; and second, the *epistemological* thesis (RE) that these entities can in turn be registered directly or indirectly by systems (and that we can in principle know some of them). This resulted in two divergent types of anti-realism: *ontological anti-realism* rejects RO, while *epistemological anti-realism* rejects RE, in part or whole. These definitions were defended against objections by first demonstrating the coherence of the categorisation in a series of debates.

This review of the debates allowed me not only to test the definitions, but also to reconstruct those disputes that had an important role to play in the subsequent chapters: Those concerning mathematical entities, modalities, temporal realism, the global debates of realism, meaning scepticism and idealism in epistemology and ontology, and finally those concerning metaphysical and scientific realism. This analysis led to my *map*: RO and RE are either discussed *locally* for a limited entity domain, all entities are discussed *globally* or several local entity domains are systematically linked *metalocally*. I then identified the realism-anti-realism debate in philosophy of science as metalocal, whereby I discussed the affiliation or exclusion of certain local issues.

J. Voosholz, *Towards a New Scientific Realism*, Synthese Library 518,
https://doi.org/10.1007/978-3-031-98886-8_9

In the third chapter, I addressed the question of whether the definitions I presented in the previous chapter, which have proved to be useful as philosophical cartographical tools in the debates, also capture their philosophical core. I argued in favour of identifying ontology and epistemology with the core questions of existence, independence and knowledge as the defining elements of the disputes about realism and anti-realism. In doing so, I have positioned myself against Dummett's thesis, according to which the theory of truth is the focal point of these disputes. With Brock, Mares and Devitt, I argued that a one-sided focus on the philosophical level of language is untenable. In contrast to those who either do not identify a core in the debates at all, or who further narrow the dispute to only either epistemology or ontology, I have defended a definition that systematically links both levels and places them at the centre of the analysis. In this way, I have striven to refine the notion of realism, hoping to protect it from the accusation of being a platitude, as levelled by Rometsch against Boghossian. In addition, I have further differentiated both the axes of independence and of existence as parts of the ontological thesis.

With the completion of the debate map and in securing the working definitions, objective realism (OR) could be introduced on the basis of three characteristics: First, the thesis of the subject-independence of the entities in question; second, the fact that these entities together form a single real external world; and third, that we possess knowledge of these entities and thus of the single real external world, at least in some of its properties. I have shown what various, often vague formulations of the independence criterion in OR—such as independence from intentional references, our practices and discourses, our mind or mental life, or our intersubjective community— attempt to capture and collectively define the independence of entities from subjects. The single real external world was introduced as a set, sum, system or whole of subject-independent entities. Finally, a minimal epistemological optimism about necessary knowledge is indispensable for OR. In OR, one must argue that we possess knowledge about some properties of the subject-independent entities and about their connection in the single real external world in order to be shielded against sceptical objections. In this way, OR is similar to metaphysical realism, but is less rigid and therefore easier to defend: While metaphysical realism, which I have defined following Putnam, also advocates the external world thesis and usually the subject independence thesis, it additionally asserts that there is a complete and true description of this external world and that we express it in parts in the case of true knowledge claims. This results in a correspondence relationship between the descriptions and the external world in metaphysical realism. I have shown, on the basis of Putnam's and Field's argument, that OR is easier to defend and seemingly more obvious than metaphysical realism.

An important consequence of OR is the structure thesis, according to which the single real external world has a fixed structure. It turns out that this structure is often identified with natural causality, laws of nature, logical laws or comparable fundamental rules of the world. These are themselves regarded as subject-independent and are supposed to apply to everything, always and everywhere in the single real external world in order to guarantee its stability. Like the idea of the external world as a sum, set or whole, the structural thesis was later taken up again and criticised along with various antitotality arguments.

In the fourth chapter, I turned to objective realism as part of the metalocal debate in philosophy of science. To this end, I looked at different varieties of naturalism. I defined it as a realism about the ontology (or ontologies, if a pluralist is at work) of science that are implied by our best or future or ideal scientific theories. In addition, it is characterised by a confidence in scientific methodology, holds the resulting theories to be true or approximately true, and defends the thesis that at the end of research or under ideal conditions all entities of the single real external world will be contained in the scientific ontology or ontologies. Naturalism combines these convictions with a strong theory of truth and often with the idea of assigning the task of a metaphysics of the external world to an appropriately interpreted natural science. Whenever naturalism is understood in this way, the philosophy arguing such a position is also an OR, because it assumes theoretical knowledge of a single real, subject-independent external world. Quine and his self-proclaimed heirs served me as examples of the reductive variety of such a naturalism. The common idea is that of reductive physicalism (RP), i.e. a form of naturalist reductionism according to which everything that really exists is ultimately ontologically reducible to the entities of physics, and epistemologically all real knowledge is ultimately physical knowledge. Since this also applies to all scientific entities, such a philosophy utilises the idea that physics will one day represent the unified science in which all knowledge is integrated and which alone is capable of explaining the single real external world.

In what followed, I pointed to problems concerning Quine's version of RP, namely incompatibilities and internal contradictions with other parts of his philosophy. They explain why many of his followers, who remained faithful to RP, turned away from other aspects of his philosophy. I have presented two problems that concern every RP and even every form of reductive naturalism: The explanatory problem of pseudo entities, which consists in the apparent impossibility of accommodating entities such as beliefs, feelings, values, social institutions, or the human mind in RP or explaining them from the purely physical external world of RP. Neither does an error theory seem possible, nor can eliminativism of all pseudo entities be defended without contradictions. I have also argued against supervenience theories that attempt to explain pseudo entities as secondary properties of physical entities. Thus, this first problem in reductive naturalism seems to be unsolvable. The same applies to the second hurdle: The implicit or explicit presupposition of a unity of science seems inherently problematic, even as a future or ideal unitary science. I have argued with Dupré, Noble, Galison, Ruphy, Demmerling, Fodor and others in favour of a disunity of the sciences, which results, among other things, from the necessary plurality of methods, the impossibility of reducing all *natural kinds* to *physical kinds* and the failure of the generality argument.

I then turned to two versions of non-reductive naturalism. Such positions reject the idea of reductionism, but retain all the basic naturalistic theses. The first example I discussed involved different readings of Sellars' philosophy of science. The *irreducibility of the manifest and the scientific image* in Sellars form a counterweight to his *scientia-mensura* dictum. More important than Sellars' analysis for my ensuing argument was Chakravartty's *semirealism*, which attempts to present a metaphysics of the single real external world based on a cautious and balanced non-reductive naturalism. His concept of *selective scepticism* about scientific theories and scientific ontologies was particularly striking here and proved to be helpful in the subsequent course of my work. With the concept that theories are partially true, Chakravartty finds a way to mediate between pessimism and optimism regarding our knowledge situation in the natural sciences. To this end, he introduces a transdisciplinary method that requires both philosophical and scientific expertise in order to advocate realism regarding some scientific entities and their properties and scepticism concerning others when interpreting scientific theories. In doing so, Chakravartty identifies and solves a problem of the metalocal philosophy of science debate, namely wishing to pass judgement on all theories of the natural sciences and their scientific ontologies at once. Since both Sellars in the readings examined and Chakravartty champion a non-reductive naturalism, they have served me as examples of *objective scientific realism in the philosophy of science* (OSR) in the further course of the book. Despite their achievements, I rejected both positions as untenable in the next step.

In the fifth chapter, the last of the first part, I proved the first main thesis of this book: Two argumentative stalemates exist, one between the best versions of OSR—identified as Chakravartty's semirealism and French's ontic structural realism—and a second between these and the best arguments directed against OSR. This double stalemate explains the apparent paralysis of the current debate. To demonstrate this, I conducted a detailed analysis of the arguments and presented the crucial criticism of OSR in three families:

The first family includes the *arguments from the underdetermination of scientific theories by the evidence*. Most OSR positions cope well with the difficulties of experimental and holistic underdetermination. However, the arguments resulting from these show that individual hypotheses or propositions cannot be tested independently of the rest of the theory via an *experimentum crucis*. Theories require certain scientific ontologies without necessarily entailing ontological obligations, which reductive naturalists often assume. These arguments from holistic and experimental underdetermination can be a reason to reject entity realism. It implies an ontological realism about theoretical entities that is not based on an epistemological realism about theories and is therefore undermined. Chakravartty's semirealism copes better with this form of underdetermination. Additional forms of contrastive underdetermination represented by van Fraassen's bad-lot and model arguments serve to shake up realist arguments and to rebut others, not to refute OSR. I have shown that they always impose a defensive manoeuvre on OSR—be it the unfortunate combination of holism and naturalism that leads to self-contradictions as in Quine, selective scepticism, or the argument of the *easy way out*, i.e. the insistence on scientific progress to get around the problems of current theories posed by underdetermination.

The *historical arguments* of the second family are either directed against the no-miracle argument or attempt to refute OSR directly. My reconstruction of Kuhn's incommensurability argument has shown that it attempts to refute epistemological realism about theories. I found that this argument alone is not strong enough to succeed against more sophisticated variants of OSR. I then considered the pessimistic meta-induction (PMI) and showed that it is understood to be a conglomerate of related arguments, only some of which are true inductive inferences, while others are deductive arguments with empirical premises. While I argued that the genuine inductions fail to achieve their argumentative goals, I finally determined that the variants of Wray, Psillos and Lyons successfully refute common variants of the no-miracle argument. They show that approximate truth is not the only reason for the empirical success of theories and that even in successful theories not all theoretical terms have to refer. This means that our best current scientific ontologies cannot be completely identical to the domain ontologies of the respective natural contexts and that we should expect further theory change. This is one of the important reasons why I have recommended a form of selective scepticism for NSR. Using the PMI, I reformulated the no-miracle argument so that it is not shaken by the former, but at the same time suggests only moderate realism. Later, however, I added further reasons for realism in philosophy of science in the form of the ancestrality argument and considerations from scientific practice, which provide motivation independently of the anti-realist countararguments. Finally, I dealt with Stanford's unconceived-alternatives argument, which combines empirical-historical premises about the natural sciences with contrastive underdetermination. This undermines the plausibility of the epistemological thesis of OSR, but Chakravartty successfully shows that a sound selective scepticism is able to escape the troublesome consequence of the argument, since any solution to the valid PMI variants is also a solution to Stanford's argument.

A common consequence of the valid historical arguments is that they get in the way of the *easy way out* of the underdetermination arguments: Pointing to future science does not help, since there is nothing to suggest that anything will fundamentally shift the course of the historical process of theory change. Furthermore, these arguments support Chakravartty's moderate approach, especially his selective scepticism. Finally, I have addressed why a complete rejection of historical arguments would be a mistake on the part of NSR: Without them, no sophisticated critique of science could be conducted, that is, an analysis of science as a social system, historical factor, and worldview that can detect and correct unwarranted assumptions, errors, problematic effects, ethical conflicts, and dogmatic scientistic ideologies.

I have identified the third and last family of arguments against OSR as *antitotality arguments*. They are directed against the idea of the single real external world. More specifically, they aim to argue either against the possibility of a privileged totality of this ontological domain or against its causally fixed and/or law-of-nature-governed eternal structure. When considering Goodman's, Gabriel's and Badiou's arguments, it became clear that they cause difficulties for many variants of OSR, which had already been foreshadowed in the problem of pseudo entities, among others, and which point to the natural scientific indescribability of some domains of reality.

However, variants of OSR also reach their limits in the face of these arguments due to the errors of a naïve ontology of simple objects and metaphysical realism.

In the subsequent analysis of Gabriel's main ontological arguments, it became clear that the conception of the external world as the single real one amounts to a totality, through which every OSR ends up in a problematic regress. Moreover, the problem of pseudo entities, which naturalisms cannot solve, remains. Rovane's argument against unimundialism cannot be rejected easily in OSR either. With this, Rovane supports the conclusion that nature can contain multiple ontologically isolated spacetimes (or other domains), as well as that there are facts in nature or existing natural entities that are not epistemologically accessible to us from a single scientific standpoint, not even as a community. Finally, I have reconstructed Cartwright's argument against fundamental laws of nature as a fixed structure of the external world in OSR. Cartwright convincingly shows that the validity of laws of nature is local and that there is no comprehensive and unified structure of nature in the form of fundamental laws of nature. Thus, Gabriel's, Rovane's and Cartwright's arguments are important contributions to NSR and speak against OSR.

I introduced the antitotality arguments into the debate about realism and anti-realism in philosophy of science. Only through the minimal criteria of my concept of OR does it become clear why this family of arguments has a decisive role to play in this dispute alongside historical and underdetermination arguments. This focus and the definition of the debate as metalocal shows clearly that these are by no means arguments from other debates—a global ontological debate or a local, unconnected debate about laws of nature—but that they rightly take their place here. All options proposed by OSR to avoid antitotality arguments either amount to leaving the debate (e.g. by presupposing an OR in metaphysics, which would face even more dire problems) or lead to becoming entangled in contradictions that eventually emerge in the interplay with the other two families of arguments. The reconstructed arguments may refute OSR, but they do not provide good independent grounds for ontological or epistemological anti-realism; on the contrary, they leave the motivations for realism intact. Consequently, the first stalemate arises.

I have conceded that there is a variant of philosophy of science that has many of the classic problems of the debate in mind: structural realism. In the following step, I identified French's ontic structural realism (OntSR) as the best proposal of a structural realism and subsequently labelled it as an OSR. In this way, I was able to demonstrate that, like Chakravartty's semirealism, it cannot withstand the combination of counterarguments. My analysis of epistemic structural realism has shown that the assumption of unknowable objects and their structure as abstract runs into unsolvable difficulties, of which the Newman problem is only one. Ladyman does not repeat these errors in his structural realism, but cannot (so far) give a convincing explanation for what physical structure is as a fundamental metaphysical entity without positing a metaphysics of the single real external world. French goes beyond Ladyman at this point by postulating a theory of metaphysical primitives, namely simples, which he identifies as aspects of structure. I have established that they can only be introduced because there is a comprehensive overall structure of the single real external world that results from fundamental laws of nature and physical principles of symmetry. This proved that French propounds an OSR. I then highlighted several ways to refute this OSR—including Wolff's argument against French's reductionism and the argument against French's eliminativism of objects. Since I was thus able to expound that French's variant is not superior to Chakravartty's, the thesis of a second stalemate was established.

This proof at the end of the first part gave me the opportunity to argue in favour of a new philosophy of science without OSR. Such a prospect raised the question of whether a NSR or anti-realism is the best candidate. With regard to the first stalemate, neither ontological nor epistemological anti-realisms seemed to be sufficiently motivated in philosophy of science from my perspective. There is, however, a more important reason against known attempts to develop a better version of these philosophies: their tendency towards systematic correlationism. I illustrated this in the second part of the book using Meillassoux's conceptual tools of correlationism and ancestrality. This part pursued the goal of resolving the double stalemate and outlining a new path in the realism-anti-realism debate in philosophy of science.

In the sixth chapter, I first reconstructed Meillassoux's concept of correlationism, developed it into an analytical tool and applied it to some anti-realist and alternative realist theory options in philosophy of science. It turned out that Meillassoux's definitions of correlate, correlation and, based on them, the multidimensional characterisation of correlationism require interpretative clarification. The central objection, that Meillassoux was creating a straw man, no longer seemed entirely accurate, as he was able to capture successfully some philosophies using the reasoning schemes of the correlationist circle and two-step, as well as the "decisions" of correlationism. However, this was only possible because the epochal thesis associated with correlationism, namely that all post-Kantian philosophy is correlationist at heart, was separated from a systematic core.

The subsequent discussion of further points of criticism targeted at Meillassoux's proposed concept led to a refinement of it, which resulted in the newly forged analytical tool of *systematic correlationism*. Positions that qualify as such derive from the reasoning schemes of circle and two-step, employing in some way the comparison problem, and which on this basis advocate the positive ontological or epistemological correlationist tenets: either they defend an ontological reciprocal dependence between the natural external world and subjects, or they conclude that knowledge of the natural external world is dependent on subjective practices, interests or an intersubjective consensus. Following the elaboration of this notion, I have argued that systematic correlationism as a categorisation of philosophical positions runs transversely to the dichotomy of *realism—anti-realism*. The two categorisations cannot be reduced to or projected onto each other. This means that there are realist, ontological anti-realist and epistemological anti-realist systematic correlationisms. Examples from the field of philosophy of science were delineated to substantiate this thesis: Stanford's instrumentalism for correlationist ontological anti-realism about scientific entities, Wray's and van Fraassen's constructive empiricism for correlationist epistemological anti-realism and the radical Sellars reading, as well as Koch's hermeneutic realism for correlationist realism. The use of the analytical tool of systematic correlationism in philosophy of science thus reveals similarities between all these positions. In conclusion, I was able to demonstrate that NSR is as distinct from systematic correlationism as it is from OSR and provides a third path between these tested-out options.

In the seventh chapter, I answered the question of how new scientific realism refutes the approaches of systematic correlationism and analysed the *problem of ancestrality*. My reinterpretation of Meillassoux's proposal offers the opportunity to avoid the objections of his numerous critics. To this end, I have distinguished between a *scientific-discourse* and an *ontological level* of the problem: On the first level, a dilemma arises for any systematic correlationism that defends an epistemological anti-realism. It must either run into self-contradictions or offer a revisionist interpretation of scientific discourse, according to which statements about, for example, ancestral events cannot be considered full knowledge claims. I argued that the ancestral problem on this level concerns two recent interpretations of Kant, put forward by Chiba and Allais, as well as Wray's and van Fraassen's constructive empiricism. At this point, the latter is guilty of a hidden (and possibly unintended) revisionism and an untenable complete distinction of the scientific level of empirical concerns from the interpretative meta-level of philosophical reflection.

For proponents of systematic correlationism who advocate ontological anti-realism or realism, on the other hand, it is difficult to provide information about the ontological conditions and the natural-historical genesis of their respective correlations. This is the core of the challenge from ancestrality on the ontological level. The critic has a right to ask what enables and realises the correlation; after all, it is presented as inevitable. An answer to this question must either be denied, and thus a form of local quietism advocated, and the systematic correlationist's response leads her to irredeemable self-contradictions or she must pursue the retrojection strategy of the ancestral. I have argued that local quietism cannot be convincingly defended in the context of philosophy of science, as otherwise the scientific-discourse level of the problem rears its ugly head again. Since only the retrojection strategy remains, I first tested this against Stanford's instrumentalism. He needs to argue that ancestral statements, as tools for predictions in the present, do not apply to anything, and do not even need to apply to anything. However, this makes the origin, the conditions for and the possibility of scientific discourse itself a mystery. I have argued that Stanford therefore finally runs into self-contradictions. In contrast, Koch develops an elaborate ontological-metaphysical variant of the retrojection strategy in the form of his subjectivity thesis, according to which only spacetimes in which a subject in a discourse community that indexically refers to natural entities exists sometimes and somewhere are possible. I have established that he not only makes weighty metaphysical assumptions but, more importantly, that his argument is begging the question against the ancestrality argument, such that his retrojection strategy cannot be justified either. In doing so, I expounded that systematic correlationism is not an appealing option in philosophy of science, thus fully demonstrating the necessity of a NSR to overcome the double stalemate.

In the final chapter, I utilised the lessons from the previous chapters to develop the framework for further research on this new kind of philosophy of science. I have drawn a sketch of possible theory options and determined the neuralgic issues for which any variant of NSR must find answers. Here I emphasised that there can be more than one NSR, and at several points I set out choices that must be made within any single NSR, leading to consequences that could shape a heterogeneous set of philosophical views.

Both the single real external world of OSR and the natural external world of systematic correlationism enclosed in a correlation are rejected in NSR. I have proposed a reformulation of the independence criterion, according to which natural entities are maximally modally robust, and argued for the necessity of recognising ancestrality as a precondition of our own scientific interrelation with nature. Furthermore, I have established it as indispensable that temporal and modal realism are parts of any NSR, which is why the step beyond systematic correlationism and OR always forces a reorganisation of the metalocal debate in philosophy of science. Against the background of my debate map, this means that NSR amounts to a discursive renewal of the discussion, hopefully rousing OSR and systematic correlationism, by which it is to be hoped that philosophy will remedy the sluggishness of the debate. At the very least, it goes beyond the double stalemate and envisages a genuinely new proposal that can incorporate the most defensible and strongest elements of OSR and systematic correlationisms. My goal in this concluding chapter on NSR was to demonstrate that such a philosophy is possible, and has already been developed in many of its foundational features through my analysis and recent innovative proposals in the debate.

The first genuinely new element of any NSR was established with the need for a *new philosophy of nature*. I made the case that nature is conceived in NSR as *radically plural*, the respective plurality of natural contexts as *irreducible* and nature as *one ontological domain among many*. Furthermore, nature is scientifically inexhaustible, neither made to be known nor fundamentally eluding our knowledge. I went on to shown that the plural natural contexts can be differentiated from one another in various ways and that their interconnections (or lack thereof) can be explained. However, interdisciplinary work is always required that combines philosophical and various scientific insights to make the best of our epistemic situation. I discussed how both Gabriel's, Meillassoux's and other variants of new realism could be utilised for the philosophical work involved and articulated some questions that arise in each case. I also tentatively tried out the option of studying nature bottom-up without enlisting some new realist general ontology. The possibility of combining both approaches struck me as the most appealing path. In what followed, I outlined the importance of new orientation for NSR with regard to laws of nature, causation and complexity. Hence, I discussed Cartwright's metaphysical nomological pluralism about local laws of nature and analysed one option for using it to define the plural natural contexts. I analysed one possible starting point for integrating theories of causality, complexity and emergence, such as those of Ellis and Woodward, and deemed it helpful in the context of elaborating such a new philosophy of nature for NSR.

The second genuine element of any NSR is the interpretation of the *natural-scientific interrelation*. Here I referred to the already established research debates that attempt to permeate philosophically the interrelation of us subjects with the historically developed, disunified natural sciences including their different practices, the plurality of methods and use of models and theories with their bonds to plural nature as such, nature's different natural contexts as well as the scientific ontologies of individual natural-scientific theories. In highlighting a few already existing debates and some of their positions affinity or incompatibility with NSR, such as those regarding models or perspectivism, I attempted to convince the reader that NSR has many points of conjunction with established debates and does by no means imply that we have to begin again with philosophy of science. My considerations of the natural-scientific interrelation lead to the question of whether there is a methodological or other crucial divide between the natural sciences and the humanities, cultural and social sciences. By working through possible arguments, I answered in the negative; instead, I was able to establish the necessity for a large plurality of methods in the sciences. I discussed how selective scepticism with its many possible and heterogeneous selection criteria could be transferred to NSR.

In addition, I argued in favour of not assuming any difference in objectivity between different sciences and of not attributing excellent objectivity to the natural sciences either. This leads to the important consequence that an interpretation of the natural-scientific interrelation always gives rise to the possibility of specific critique of sciences with regard to individual questions, methods and scientific research programmes. This raised the following considerations on the interplay between the sciences. The relations of individual sciences, such as mathematics and the natural sciences, must be evaluated according to the decisions that the respective NSR has previously made, for example according to the position regarding whether abstract local laws of nature are treated as natural entities. In this step, I could establish that a stance in philosophy of mathematics is also a necessary ingredient in any NSR. It turned out that a deeper understanding of NSR's natural-scientific interrelation requires more than considerations of interdisciplinarity. In a final step, I reflected on the need to combine different approaches towards the tasks mentioned, which can never be realised by a single science, not even by philosophy. Therefore, this led me to demand a new theory of transdisciplinarity from NSR, which insists on a plurality of disciplines in the study of reality, to which nature and its natural context are integral.

This closes the analysis of this book. I have undoubtedly left many questions unanswered and philosophical tasks unfinished. As my cartography suggested, it is difficult to cover any metalocal territory. I hope I have been able to convince the reader that new scientific realism is both desirable and possible. But even in case I have not achieved this goal, I hope at least to have invigorated (maybe even enraged) objective realist and correlationist readers enough to rid the fascinating and timely debate about scientific realism and anti-realism of the gridlock that has been created by the double stalemate for too long. My plea for collaborative transdisciplinary research is profoundly serious: if I have succeeded in convincing you, then join the discussion as a new scientific realist. I hope that the framework I have presented is integrative, but also flexible enough to test out and to discover the best positions we can take together.

References

Abbott, B. P., R. Abbott, T. D. Abbott, M. R. Abernathy, F. Acernese, K. B. P. Ackley, et al. (LIGO Scientific Collaboration, and Virgo Collaboration). 2016. Observation of Gravitational Waves from a Binary Black Hole Merger. *Physical Review Letters* 116: 061102.

Adams, Robert. 1986. Time and Thisness. *Midwest Studies in Philosophy* 11 (1): 315–329.

Adlam, Emily. 2022. Operational Theories as Structural Realism. *Studies in History and Philosophy of Science Part A* 94 (C): 99–111.

Aguiriano Benéitez, Mario. 2022. Forget Meillassoux! A Critique of Speculative Materialism. *Hastapenak* 4: 94–103.

Albert, Hans. 1985. *Treatise on Critical Reason*. Princeton: Princeton University Press.

Allais, Lucy. 2015. *Manifest Reality. Kant's Idealism and His Realism*. Oxford: Oxford University Press.

Alloa, Emmanuel. 2023. Korrelation: Phänomenologie als Korrelationsforschung. In *Handbuch Phänomenologie*, ed. Emmanuel Alloa, Thiemo Breyer, and Emanuele Caminada, 148–153. Tübingen: Mohr Siebeck.

Alloa, Emmanuel, and Élie During (eds.). 2018. *Choses en soi. Métaphysique du réalisme*. Paris: Presses Universitaires de France Humensis.

Allzén, Simon. 2021. Scientific Realism and Empirical Confirmation: A Puzzle. *Studies in History and Philosophy of Science Part A* 90: 153–159.

Amerini, Fabrizio, and Laurent Cesalli (eds.). 2017. *Universals in the Fourteenth Century*. Pisa: Edizioni della normale.

Anderson, Don L. 2007. *New Theory of the Earth*, 2nd ed. Cambridge: Cambridge University Press.

Appiah, Kwame Anthony. 2006. How to Decide If Races Exist. *Proceedings of the Aristotelian Society* 106 (3): 363–380.

Aquinas, Thomas. 1920. *Summa Theologica*, trans. by Fathers of the English Dominican Province, 2nd ed. London: Burns Oats & Washbourne.

Aristotle. 1984. *The Complete Works of Aristotle: The Rev. Oxford Translation*, ed. Jonathan Barnes, vol. 2. Princeton: Princeton University Press.

Armstrong, David M. 1978a. *Universals and Scientific Realism, Volume I: Nominalism and Realism*. Cambridge: Cambridge University Press.

Armstrong, David M. 1978b. *Universals and Scientific Realism, Volume II: A Theory of Universals*. Cambridge: Cambridge University Press.

Armstrong, David M. 1986a. In Defense of Structural Universals. *Australasian Journal of Philosophy* 64 (1): 85–88.

J. Voosholz, *Towards a New Scientific Realism*, Synthese Library 518,
https://doi.org/10.1007/978-3-031-98886-8

Armstrong, David M. 1986b. The Nature of Possibility. *Canadian Journal of Philosophy* 16 (4): 575–594.
Armstrong, David M. 1989. *A Combinatorial Theory of Possibility*. Cambridge: Cambridge University Press.
Armstrong, David M. 1997. *A World of States of Affairs*. Cambridge: Cambridge University Press.
Armstrong, David M. 2004a. *Combinatorialism Revisited*. http://eprints.nottingham.ac.uk/716/.
Armstrong, David M. 2004b. *Truth and Truthmakers*. Cambridge: Cambridge University Press.
Ásta. 2011. The Metaphysics of Sex and Gender. In *Feminist Metaphysics. Explorations in the Ontology of Sex, Gender and the Self*, ed. Charlotte Witt, 47–65. Dordrecht: Springer.
Augsburg, Tanya. 2014. Becoming Transdisciplinary: The Emergence of the Transdisciplinary Individual. *World Futures* 70 (3–4): 233–247.
Austin, Michael. 2015. Correlationism: Weak and Strong. In *The Meillassoux Dictionary*, ed. Peter Gratton and Paul J. Ennis, 48–50. Edinburgh: Edinburgh University Press.
Azzouni, Jody. 2004. The Derivation-Indicator View of Mathematical Practice. *Philosophia Mathematica* III (12): 81–105.
Azzouni, Jody. 2009. Why Do Informal Proofs Confirm to Formal Norms. *Foundations of Science* 14: 9–26.
Azzouni, Jody. 2024. A White Horse is Not a Horse: Markus Gabriel's "New Realism". In *Markus Gabriel's New Realism*, ed. Jan Voosholz, Synthese Library, 492, 33–73. Cham: Springer.
Backman, Jussi. 2014. Transcendental Idealism and Strong Correlationism: Meillassoux and the End of Heideggerian Finitude. In *Phenomenology and the Transcendental*, ed. Sara Heinämaa, Mirja Hartimo, and Timo Miettinen, 276–294. London: Routledge.
Backmann, Marius, Andreas Berg, Marie I. Kaiser, Michael Pohl, Raja Rosenhagen, and Christian Suhm. 2005. Pragmatism, Realism, and Science. In *Richard Rorty. His Phiosophy under Discussion*, ed. Andreas Vieth, 65–78. Frankfurt (Main): ontos.
Badiou, Alain. 2005. *Being and Event*. London: continuum.
Badiou, Alain. 2008. Preface. In *After Finitude. An Essay on the Necessity of Contingency*. London: continuum.
Badiou, Alain. 2009. *Logics of Worlds. Being and Event, 2*. London: continuum.
Balaguer, Mark. 1998a. Attitudes Without Propositions. *Philosophy and Phenomenological Research* 58 (4): 805–826.
Balaguer, Mark. 1998b. *Platonism and Anti-Platonism in Mathematics*. Oxford: Oxford University Press.
Barad, Karen. 2007. *Meeting the Universe Halfway: Quantum Physics and the Entanglement of Matter and Meaning*. Durham: Duke University Press.
Barcan Marcus, Ruth. 1993. *Modalities: Philosophical Essays*. Oxford: Oxford University Press.
Barish, Barry C., and Rainer Weiss. 1999. LIGO and the Detection of Gravitational Waves. *Physics Today* 52 (10): 44–50.
Bark, Joachim, and Hans-Christoph Graf von Nayhauss. 2015. (eds.) *Profile deutscher Kulturepochen. Vom Realismus in die Moderne 1849–1918*. Stuttgart: Kröner.
Barnes, Barry, David Bloor, and John Henry. 1996. *Scientific Knowledge. A Sociological Analysis*. Chicago: University of Chicago Press.
Baron, Samuel, and Kristie Miller. 2013. Characterizing Eternalism. In *New Papers on the Present*, ed. Roberto Ciuni, Kristie Miller, and Giuliano Torrengo, 31–65. München: Philosophia.
Baron, Samuel, Kristie Miller, and Jonathan Tallant. 2021. Temporal Fictionalism for a Timeless World. *Philosophy and Phenomenological Research* 102: 2.
Beall, Jc. 2004. True and False – As If. In *The Law of Non-Contradiction*, ed. Graham Priest, Jc Beall, and Bradley Armour-Garb, 197–216. Oxford: Clarendon Press.
Beall, Jc. 2009. *Spandrels of Truth*. Oxford: Oxford University Press.
Beall, Jc. 2014. Finding Tolerance Without Gluts. *Mind* 123 (491): 829–849.
Beall, Jc, and Greg Restall. 2006. *Logical Pluralism*. Oxford: Oxford University Press.
Beierwaltes, Werner. 1994. *Eriugena. Grundzüge seines Denkens*. Frankfurt (Main): Klostermann.
Benacerraf, Paul. 1965. What Numbers Could not Be. *Philosophical Review* 74 (1): 47–73.
Benacerraf, Paul. 1973. Mathematical Truth. *Journal of Philosophy* 70: 661–679.

Bengochea, Gabriel R., Gabriel León, Elias Okon, and Daniel Sudarsky. 2020. Can the Quantum Vacuum Fluctuations Really Solve the Cosmological Constant Problem? *Nature Astronomy* 80: 18.

Benitez, Federico. 2022. Barren Worlds: The Scientific Image of Ontic Structural Realism. *Disputatio* 14 (65): 65–90.

Benitez, Federico. 2023. Structural Realism and Theory Classification. *Theoria* 89 (5), 734–747.

Bennett, Karen. 2004. Global Supervenience and Dependence. *Philosophy and Phenomenological Research* 68 (3): 501–529.

Bennett, Karen, and Brian McLaughlin. 2018. Supervenience. *Stanford Encyclopedia of Philosophy* (Winter 2018 Edition), ed. Edward N. Zalta.

Benoist, Jocelyn. 2011. *Éléments de philosophie réaliste: réflexions sur ce que l'on a.* Paris: Vrin.

Benoist, Jocelyn. 2017. *L'adresse du réel.* Paris: Vrin.

Benoist, Jocelyn (ed.). 2018. *Réalismes anciens et nouveaux.* Paris: Vrin.

Benoist, Jocelyn. 2021. *Toward a Contextual Realism.* Cambridge: Harvard University Press.

Benoist, Jocelyn. 2023. No Limit: On What Thought Can Actually Do. In *Limits of Intelligibility: Issues from Kant and Wittgenstein*, ed. Jens Pier, 262–279. London: Routledge.

Benoist, Jocelyn. 2024. I Can't See the "Sense" in This. In *Markus Gabriel's New Realism*, ed. Jan Voosholz, Synthese Library, 492, 129–145. Cham: Springer.

Benovsky, Jiri. 2021. Modal Realism and the Possibility of Island Universes: Why There Are No Possible Worlds. *Metaphysica* 22 (1): 1–13.

Berenstain, Nora, and James Ladyman. 2012. Ontic Structural Realism and Modality. In *Structural Realism. Structure, Object, and Causality*, ed. Elaine Landry and Dean Rickles, 149–168. Dordrecht: Springer.

Berghofer, Philipp. 2018. Ontic Structural Realism and Quantum Field Theory: Are There Intrinsic Properties at the Most Fundamental Level of Reality? *Studies in History and Philosophy of Science Part B: Studies in History and Philosophy of Modern Physics* 62: 176–188.

Bergson, Henri. 1991. *Matter and Memory*, trans. by Nancy Margaret Paul and W. Scott Palmer. New York: Zone Books.

Bergson, Henri. 2013. *Time and Free Will. An Essay on the Immediate Data of Consciousness*, trans. by F.L. Pogson, Muirhead Library of Philosophy. London: Routledge.

Bernecker, Sven. 2003. Skeptizismus, Naturalismus und Quine. *Philosophisches Jahrbuch* 110 (1): 46–58.

Berto, Francesco, and Mark Jago. 2019. Ersatz Modal Realism. In *Impossible Worlds*, ed. dens., 73–92. Oxford: Oxford University Press.

Bertram, Georg W. 2017. *Hegels "Phänomenologie des Geistes". Ein systematischer Kommentar.* Stuttgart: Reclam.

Bhogal, Harjit. 2020. Humeanism about laws of nature. *Philosophy Compass* 15 (8): e12696.

Bigelow, John. 1986. Towards structural universals. *Australasian Journal of Philosophy* 64 (1): 94–96.

Bigelow, John, and Robert Pargetter. 1989. A theory of structural universals. *Australasian Journal of Philosophy* 67 (1): 1–11.

Bitbol, Michel. 2019. *Maintenant la finitude. Peut-on penser l'absolu?* Paris: Flammarion.

Black, Max. 1952. The Identity of Indiscernibles. *Mind* 61 (242): 153–164.

Blackburn, Simon. 2002. Realism: Deconstructing the Debate. *Ratio* 15 (2): 111–133.

Blair, Ann. 2006. Natural Philosophy. In *The Cambridge History of Science: Early Modern Science*, ed. Katherine Park and Lorraine Daston, vol. 3, 363–406. The Cambridge History of Science. Cambridge: Cambridge University Press.

Blok, Vincent. 2017. Realism without Speculation: Heidegger, Meillassoux and the Question of Philosophical Method. *Studia Phaenomenologica* 17: 381–404.

Blumenberg, Hans. 1965. *Die kopernikanische Wende*. Frankfurt (Main): Suhrkamp.

Blumenberg, Hans. 2000. *The Genesis of the Copernican World*, trans. by Robert M. Wallace, 2nd ed. Cambridge: MIT Press.

Boghossian, Paul. 2007. *Fear of Knowledge. Against Relativism and Constructivism.* Oxford: Oxford University Press.

Bonk, Thomas. 2006. Quine zum Problem des Realismus. *Zeitschrift für philosophische Forschung* 60 (2): 190–212.
Booth, Robert. 2018. Merleau-Ponty, Correlationism, and Alterity. *PhænEx* 12 (2): 37–58.
Bös, Gregor E. 2024. A Logical Challenge to Correlationism: The Church-Fitch Paradox in Husserl's Account of Fulfilment, Truth, and Meaning. *Synthese* 203 (211): 1–25.
Bourne, Craig. 2006. *A Future for Presentism*. Oxford: Oxford University Press.
Bouvier, Audrey, and Meenakshi Wadhwa. 2010. The Age of the Solar System redefined by the Oldest Pb–Pb Age of a Meteoritic Inclusion. *Nature Geoscience* 3: 637–641.
Boyd, Richard. 1983. On the Current Status of the Issue of Scientific Realism. *Erkenntnis* 19 (1): 45–90.
Brading, Katherine, and Elena Castellani. (eds.). 2002. *Symmetries in Physics: Philosophical Reflections*. Cambridge: Cambridge University Press.
Brandom, Robert B. 2009. Pragmatism, Inferentialism, and Modality in Sellars's Arguments against Empiricism. In *Empiricism, Perceptual Knowledge, Normativity, and Realism. Essays on Wilfrid Sellars*, ed. Willem A. deVries, 33–61. Oxford: Oxford University Press.
Brassier, Ray. 2007. *Nihil Unbound. Enlightenment and Extinction*. New York: Palgrave Macmillan.
Briadotti, Rosi, and Rick Dolphijn. (eds.). 2017. *Philosophy After Nature*. London: Rowman & Littlefield.
Brock, Stuart, and Edwin Mares. 2007. *Realism and Anti-Realism*. Montreal: McGill-Queen's University Press.
Brogaard, Berit. 2000. Presentist Four-Dimensionalism. *Monist* 83 (3): 341–356.
Brogaard, Berit. 2006. Tensed Relations. *Analysis* 66 (3): 194–202.
Brouwer, L.E.J. 1908. De onbetrouwbaarheid der logische principes. *Tijdschrift voor Wijsbegeerte* 2: 152–158.
Brouwer, L.E.J. 1919. Intuitionistische Mengenlehre. *Jahresbericht der DMV* 28: 203–208.
Brouwer, L.E.J. 1981. *Brouwer's Cambridge Lectures on Intuitionism*, ed. Dirk van Dalen. Cambridge: Cambridge University Press.
Brown, Harold I. 2001. Incommensurability and Reality. In *Incommensurability and Related Matters*, ed. Paul Hoyningen-Huene and Howard Sankey, 123–142. Dordrecht: Kluwer.
Brown, James Robert. 2001. *Who Rules in Science?: An Opinionated Guide to the Wars*. Cambridge: Harvard University Press.
Brown, Nathan. 2011. The Speculative and the Specific. On Hallward and Meillassoux. In *The Speculative Turn. Continental Materialism and Realism*, ed. Levi Bryant, Nick Srnicek, and Graham Harman, 142–163. Melbourne: re.press.
Brukner, Časlav. 2018. A No-Go Theorem for Observer-Independent Facts. *Entropy* 20: 350.
Bruno, G. Anthony. 2018. Meillassoux, Correlationism, and the Ontological Difference. *PhænEx* 12 (2): 1–12.
Bryant, Levi. 2015. Correlationism. In *The Meillassoux Dictionary*, ed. Peter Gratton, and Paul J. Ennis, 46–48. Edinburgh: Edinburgh University Press.
Bryant, Amanda. 2024. Naturalized Metaphysics Without Scientific Realism. *Argumenta* 19: 13–33.
Bryant, Levi, Nick Srnicek, and Graham Harman. 2011. Towards a Speculative Philosophy. In *The Speculative Turn. Continental Materialism and Realism*, ed. Levi Bryant, Nick Srnicek, and Graham Harman, 1–18. Melbourne: re.press.
Bueno, Otávio. 1999a. Empiricism, Conservativeness, and Quasi-Truth. *Philosophy of Science* 66 (3): 485.
Bueno, Otávio. 1999b. What Is Structural Empiricism? Scientific Change in an Empiricist Setting. *Erkenntnis* 50 (1), 55–81.
Bueno, Otávio. 2000a. Empiricism, Mathematical Truth and Mathematical Knowledge. *Poznan Studies in the Philosophy of the Sciences and the Humanities* 71: 219–242.
Bueno, Otávio. 2000b. Empiricism, Scientific Change and Mathematical Change. *Studies in History and Philosophy of Science Part A* 31 (2): 269–296.
Bueno, Otávio. 2008a. Scientific Representation and Nominalism: An Empiricist View. *Principia* 12 (2): 177–192.

Bueno, Otávio. 2008b. Structural Realism, Scientific Change, and Partial Structures. *Studia Logica* 89 (2): 213–235.
Bueno, Otávio. 2009. Mathematical Fictionalism. In *New Waves in Philosophy of Mathematics*, ed. Otávio Bueno and Øystein Linnebo, 59–79. London: Palgrave.
Bueno, Otávio. 2011. Structural Empiricism, Again. In *Scientific Structuralism*, ed. Alisa Bokulich and Peter Bokulich, 81–103. Dordrecht: Springer.
Bueno, Otávio. 2015. Realism and Anti-Realism About Science. *International Journal for the Study of Skepticism* 5 (2): 145–167.
Bueno, Otávio. 2016. An Anti-Realist Account of the Application of Mathematics. *Philosophical Studies* 173 (10): 2591–2604.
Bueno, Otávio. 2021. Mathematics and Measurement: Causation and the Mind. In *Top-Down Causation and Emergence*, ed. Jan Voosholz and Markus Gabriel, Synthese Library, 439, 65–75. Cham: Springer.
Bueno, Otávio. 2024. Set-Theoretic Ontology and Fields of Sense: Limits and Interpretations. In *Markus Gabriel's New Realism*, ed. Jan Voosholz, Synthese Library, 492, 117–128. Cham: Springer.
Bueno, Otávio, and Décio Krause 2007. Scientific Theories, Models and the Semantic Approach. *Principia* 11 (2): 187–201.
Bueno, Otávio, and Steven French. 2011. How Theories Represent. *British Journal for the Philosophy of Science* 62 (4): 857–894.
Bueno, Otávio, Steven French, and James Ladyman. 2012. Models and Structures: Phenomenological and Partial. *Studies in History and Philosophy of Science Part B: Studies in History and Philosophy of Modern Physics* 43 (1): 43–46.
Burgess, John P., and Gideon Rosen. 1997. *A Subject with No Object: Strategies for Nominalistic Interpretation of Mathematics*. Oxford: Clarendon Press.
Busch, Jacob. 2003. What Structures Could Not Be. *International Studies in the Philosophy of Science* 17 (3): 211–225.
Butler, Judith. 1990. *Gender Trouble. Feminism and the Subversion of Identity*. London: Routledge.
Butler, Judith. 2011. *Bodies That Matter. On the Discursive Limits of "Sex"*. London: Routledge.
Button, Tim. 2013. *The Limits of Realism*. Oxford: Oxford University Press.
Callender, Craig. 2012. Time's Ontic Voltage. In *The Future of the Philosophy of Time*, ed. Adrian Bardon, 73–98. New York: Routledge.
Calosi, Claudio, and Patrick M. Duerr. 2021. The General-Relativistic Case for Super-Substantivalism. *Synthese* 199 (5–6): 13789–13822.
Campbell, Keith. 1990. *Abstract Particulars*. Oxford: Blackwell.
Carabine, Deirdre. 2000. *John Scottus Eriugena*. Oxford: Oxford University Press.
Carnap, Rudolf. 1931. Die physikalische Sprache als Universalsprache der Wissenschaft. *Erkenntnis* 2: 432–465.
Carnap, Rudolf. 1956. *Meaning and Necessity: A Study in Semantics and Modal Logic*, 2nd ed. Chicago: University of Chicago Press.
Carnap, Rudolf. 1998. *Der logische Aufbau der Welt*. Hamburg: Meiner.
Carnap, Rudolf. 2003. *The Logical Structure of the Worlds. And Pseudoproblems in Philosophy*, trans. by Rolf A. George. Chicago: Open Court Press.
Caro, Mario De. 2025. Putnam's Liberal Naturalism. In *The Theoretical and the Moral Face of Realist Philosophy: Themes from Putnam*, ed. Michael Frauchiger. Berlin: De Gruyter.
Carroll, John W. 2020. Laws of Nature. *Stanford Encyclopedia of Philosophy* (Winter 2020 Edition), ed. Edward N. Zalta.
Carter, Jessica. 2008. Structuralism as a Philosophy of Mathematical Practice. *Synthese* 163 (2): 119–131.
Cartwright, Nancy. 1983. *How the Laws of Physics Lie*. Oxford: Clarendon.
Cartwright, Nancy. 1999. *The Dappled World. A Study of the Boundaries of Science*. Cambridge: Cambridge University Press.
Chakravartty, Anjan. 2003. The Structuralist Conception of Objects. *Philosophy of Science* 70 (5): 867–878.

Chakravartty, Anjan. 2007. *A Metaphysics for Scientific Realism. Knowing the Unobservable*. Cambridge: Cambridge University Press.
Chakravartty, Anjan. 2008. What You Don't Know Can't Hurt You. Realism and the Unconceived. *Philosophical Studies* 137: 149–158.
Chakravartty, Anjan. 2013. Realism in the Desert and in the Jungle: Reply to French, Ghins, and Psillos. *Erkenntnis* 78 (1): 39–58.
Chakravartty, Anjan. 2017a. Reflections on New Thinking About Scientific Realism. *Synthese* 194 (9): 3379–3392.
Chakravartty, Anjan. 2017b. *Scientific Ontology. Integrating Naturalized Metaphysics and Voluntarist Epistemology*. Oxford: Oxford Universtiy Press.
Chakravartty, Anjan. 2017c. Scientific Realism. *Stanford Encyclopedia of Philosophy* (Summer 2017 Edition), ed. Edward N. Zalta.
Chakravartty, Anjan. 2018. Realism, Antirealism, Epistemic Stances, and Voluntarism. In *The Routledge Handbook of Scientific Realism*, ed. Juha Saatsi, 225–236. London: Routledge.
Chakravartty, Anjan. 2023. Last Chance Saloons for Natural Kind Realism. *American Philosophical Quarterly* 60 (1): 63–81.
Chakravartty, Anjan. 2024. On Semirealism, Realism More Generally, and Underlying Epistemic Stances. *Journal for General Philosophy of Science / Zeitschrift für Allgemeine Wissenschaftstheorie* 55 (2): 269–288.
Chakravartty, Anjan, and Bas van Fraassen. 2018. What is Scientific Realism? *Spontaneous Generations* 9 (1): 12–25.
Chalmers, David J. 1995. Facing Up to the Problem of Consciousness. *Journal of Consciousness Studies* 2 (3): 200–219.
Chalmers, David J. 2009. Ontological Anti-Realism. In *Metametaphysics. New Essays on the Foundations of Ontology*, ed. David Manley, David J. Chalmers, and Ryan Wasserman, 77–129. Oxford: Oxford University Press.
Chalmers, David J. 2020. Debunking Arguments for Illusionism about Consciousness. *Journal of Consciousness Studies* 27 (5/6): 258–281.
Chang, Hasok. 2001. How to Take Realism Beyond Foot-Stamping. *Philosophy* 76 (1): 5–30.
Chang, Hasok. 2012. *Is Water H2O? Evidence, Realism and Pluralism*. Boston Studies in the Philosophy and History of Science.
Chang, Hasok. 2017. Is Pluralism Compatible with Scientific Realism? In *The Routledge Handbook of Scientific Realism*, ed. Juha Saatsi, 176–186. Milton Park: Routledge.
Chang, Hasok. 2018. Realism for Realistic People. *Spontaneous Generations* 9 (1): 31–34.
Chang, Hasok. 2021. The Coherence of Feyerabend's Pluralist Realism. In *Interpreting Feyerabend: Critical Essays*, ed. Karim Bschir, and Jamie Shaw, 40–56. Cambridge: Cambridge University Press.
Chang, Hasok. 2022. *Realism for Realistic People: A New Pragmatist Philosophy of Science*. Cambridge: Cambridge University Press.
Chen, Eddy Keming. 2024. Density Matrix Realism. In *Open Systems: Physics, Metaphysics, and Methodology*, ed. Michael E. Cuffaro, and Stephan Hartmann. Oxford: Oxford University Press.
Chiba, Kiyoshi. 2012. *Kants Ontologie der raumzeitlichen Wirklichkeit. Versuch einer antirealistischen Interpretation der Kritik der reinen Vernunft*. Berlin: de Gruyter.
Chihara, Charles S. 2001. *The Worlds of Possibility: Modal Realism and the Semantics of Modal Logic*. Oxford: Oxford University Press.
Chisholm, Roderick M. 1990. Referrring to Things That No Longer Exist. *Philosophical Perspectives* 4: 545–556.
Chomsky, Noam. 2000. *New Horizons in the Study of Language*. Cambridge: Cambridge University Press.
Churchland, Paul. 1981. Eliminative Materialism and the Propositional Attitudes. *Journal of Philosophy* 78 (2): 67–90.
Churchland, Patricia. 1986. *Neurophilosophy: Toward a Unified Science of the Mind/Brain*. Cambridge: MIT Press.

Clark, Stewart J., and Tom Lancaster. 2017. The Use of Downward Causation in Condensed Matter Physics. In *Philosophical and Scientific Perspectives on Downward Causation*, ed. Michele Paolini Paoletti, and Francesco Orilia, 131–145. New York: Routledge.

Coggins, Geraldine. 2003. World and Object: Metaphysical Nihilism and Three Accounts of Worlds. *Proceedings of the Aristotelian Society* 103 (3): 353–360.

Colyvan, Mark. 2001. *The Indispensability of Mathematics*. Oxford: Oxford University Press.

Colyvan, Mark. 2010. There is No Easy Road to Nominalism. *Mind* 119 (474): 285–306.

Cordovil, João L., Gil C. Santos, and John Symons. 2023. Reconciling Ontic Structural Realism and Ontological Emergence. *Foundations of Science* 28 (1): 1–20.

Corns, Jennifer. 2016. Pain Eliminativism: Scientific and Traditional. *Synthese* 193 (9): 2949–2971.

Craig, Edward. 2002. *Philosophy. A Very Short Introduction*. Oxford: Oxford University Press.

Craver, Carl F. 2007. *Explaining the Brain: Mechanisms and the Mosaic Unity of Neuroscience*. Oxford: Oxford University Press.

Crisp, Thomas M. 2007. Presentism and the Grounding Objection. *Noûs* 41 (1): 90–109.

Crisp, Thomas M. 2008. Presentism, Eternalism and Relativity Physics. In *Einstein, Relativity and Absolute Simultaneity*, ed. William Lane Craig and Quentin Smith, 262–278. London: Routledge.

Cross, Richard. 1999. *Duns Scotus*. Oxford: Oxford University Press.

Crull, Elise. 2024. Realism with Quantum Faces: The Leggett-Garg Inequalities as a Case Study for Feyerabend's Views. *International Studies in the Philosophy of Science* 37 (4): 195–217.

Cumpa, Javier. 2023. The Regress Argument Against Realism About Structure. *Inquiry* 66 (5): 726–737.

Curry, Haskell. 1951. *Outlines of Formalist Philosophy of Mathematics*. Amsterdam: Nort Holland.

da Costa, Newton. 1974. On the Theory of Inconsistent Formal Systems. *Notre Dame Journal of Formal Logic* 15 (4): 497–510.

da Costa, Newton, and Elias H. Alves. 1977. Semantical Analysis of the Calculi $\mathbf{C}_n$. *Notre Dame Journal of Formal Logic* 18 (4): 621–630.

da Costa, Newton, and Steven French. 2000. Models, Theories, and Structures: Thirty Years on. *Philosophy of Science, Supplement* 67: S116–S127.

Dalton, Drew M. 2022. The Metaphysics of Speculative Materialism. *Philosophy Today* 66 (4): 687–705.

Danowski, Déborah, and Eduardo Viveiros de Castro. 2017. *The Ends of the World*. Cambridge: Polity.

Dasgupta, Shamik. 2015. Substantivalism Vs Relationalism About Space in Classical Physics. *Philosophy Compass* 10 (9): 601–624.

Daston, Lorraine. 2008. On Scientific Observation. *Isis* 99: 97–110.

Daston, Lorraine, and Peter Galison. 2007. *Objectivity*. New York: Zone Books.

Davidson, Donald. 1973. On the Very Idea of a Conceptual Scheme. *Proceedings and Addresses of the American Philosophical Association* 47: 5–20.

Davidson, Donald. 1984. *Inquiries into Truth and Interpretation: Philosophical Essays Volume 2*. Oxford: Oxford University Press.

Davidson, Donald. 2001. *Subjective, Intersubjective, Objective: Philosophical Essays Volume 3*. Oxford: Clarendon Press.

Davies, C.J. 2018. Nietzsche Beyond Correlationism: Meillassoux's History of Modern Philosophy. *Continental Philosophy Review* 51: 81–93.

De, Michael. 2018. On the Humphrey Objection to Modal Realism. *Grazer Philosophische Studien* 95 (2): 159–179.

de Freitas, Elizabeth. 2021. Mathematics in the Middle: The Relationship Between Measurement and Metamorphic Matter. *Matter* 2 (2): 1–24.

de Ridder, Jeroen, Rik Peels, and René van Woudenberg (eds.). 2018. *Scientism: Prospects and Problems*. Oxford: Oxford University Press.

De Sanctis, Sarah, and Anna Longo (eds.). 2015. *Breaking the Spell. Contemporary Realism under Discussion*. Mailand: Mimesis International.

Decelles, Peter, Dolores Robinson, and George Zandt. 2002. Implications of Shortening in the Himalayan Fold-Thrust Belt for Uplift of the Tibetan Plateau. *Tectonics* 21 (6): 1062.

Deleuze, Gilles. 1994. *Difference and Repetition*, trans. by Paul Patton. New York: Columbia University Press.

Deleuze, Gilles, and Félix Guattari. 1987. *A Thousand Plateaus*, trans. by Brian Massumi, vol. II. Capitalism and Schizophrenia. Minneapolis: University of Minnesota Press.

Demeter, Tamás. 2009. Two Kinds of Mental Realism. *Journal for General Philosophy of Science* 40: 59–71.

Demmerling, Christoph. 2002. *Sinn, Bedeutung, Verstehen. Untersuchungen zu Sprachphilosophie und Hermeneutik*. Paderborn: mentis.

Demmerling, Christoph. 2004. Realismus und Antirealismus. Zur Anatomie einer Debatte. In *Was ist wirklich? Neuere Beiträge zur Realismusdebatte in der Philosophie*, ed. Christoph Halbig, and Christian Suhm, 29–48. Frankfurt (Main): ontos.

Demmerling, Christoph. 2008. Welcher Naturalismus? Von der Naturwissenschaft zum Pragmatismus. In *Naturalismus und Menschenbild*, ed. Peter Janich, 240–256. Hamburg: Meiner.

Demmerling, Christoph. 2014. Philosophie als Therapie. Die Auflösung philosophischer Dualismen. In *Die Philosophie John McDowells*, ed. David Lauer, and Christian Barth, 16–36. Münster: mentis.

Demtröder, Wolfgang. 2017. *Experimentalphysik 2. Elektrizität und Optik*, 7th ed. Berlin: Springer.

Dennett, Daniel C. 1991. Real Patterns. *Journal of Philosophy* 88 (1): 27–51.

Dennett, Daniel C. 2016. Illusionism as the Obvious Default Theory of Consciousness. *Journal of Consciousness Studies* 23 (11/12): 65–72.

Dennett, Daniel C. 2019. Welcome to Strong Illusionism. *Journal of Consciousness Studies* 26 (9/10): 48–58.

Derrida, Jacques. 1982. Différance. In *Margins of Philosophy*, trans. by Alan Bass, 1–28. Chicago: The University of Chicago Press.

Derrida, Jacques. 2016. *Of Grammatology*, trans. by Gayatri Chakravorty Spivak, Fortieth Anniversary Edition. Baltimore: Johns Hopkins University Press.

Descartes, René. 1996a. *Meditationes de prima philosophia*, ed. Charles Adam, and Paul Tannery, Oeuvres de Descartes, VII. Paris: Vrin.

Descartes, René. 1996b. *Principia philosophiae*, ed. Charles Adam and Paul Tannery, Oeuvres de Descartes, VIII. Paris: Vrin.

de Spinoza, Baruch. 2002a. Principles of Cartesian Philosophy and Metaphysical Thoughts. In *Spinoza: Complete Works*, ed. Michael L. Morgan, trans. by Samuel Shirley, 108–212. Indianapolis: Hackett Publishing.

de Spinoza, Baruch. 2002b. Treatise on the Emendation of the Intellect. In *Spinoza: Complete Works*, ed. Michael L. Morgan, trans. by Samuel Shirley, 1–30. Indianapolis: Hackett Publishing.

de Spinoza, Baruch. 2018. *Ethics: Proved in Geometrical Order*, ed. Matthew J. Kisner, trans. by Michael Silverstone, Cambridge Texts in the History of Philosophy. Cambridge: Cambridge University Press.

Devitt, Michael. 1983. Dummett's Anti-Realism. *Journal of Philosophy* 80 (2): 73–99.

Devitt, Michael. 1991. *Realism and Truth*, 2nd ed. Oxford: Blackwell.

deVries, Willem A. 2005. *Wilfrid Sellars*. Chesham: Acumen.

deVries, Willem A. (ed.) 2009. *Empiricism, Perceptual Knowledge, Normativity, and Realism. Essays on Wilfrid Sellars*. Oxford: Oxford University Press.

Dewan, Lawrence. 2006. *Form and Being: Studies in Thomistic Metaphysics*. Washington: Catholic University of America Press.

Dickie, Imogen. 2015. *Fixing Reference*. Oxford: Oxford University Press.

Dieks, Dennis. 2000. Space-Time Relationism in Newtonian and Relativistic Physics. *International Studies in the Philosophy of Science* 15 (1): 5–17.

Dieks, Dennis. 2014. Weak Discernibility and the Identity of Spacetime Points. In *Space and Time: A Priori and a Posteriori Studies*, ed. Vincenzo Fano, Francesco Orilia, and Giovanni Macchia, 43–62. Berlin: De Gruyter.

Divers, John. 2002. *Possible Worlds*. London: Routledge.

Dolev, Yuval. 2007. *Time and Realism. Metyphysical and Antimetaphysical Perspectives*. Cambridge: MIT Press.

Drossel, Barbara. 2021. Strong Emergence in Condensed Matter Physics. In *Top-Down Causation and Emergence*, ed. Jan Voosholz and Markus Gabriel, Synthese Library, 439, 79–99. Cham: Springer.

Duhem, Pierre. 1914. *The Aim and Structure of Physical Theory*, 2nd ed. Princeton: Princeton University Press.

Dummett, Michael. 1976. What Is a Theory of Meaning? (II). In *Truth and Meaning. Essays in Semantics*, ed. Gareth Evans and John McDowell. Oxford: Clarendon Press.

Dummett, Michael. 1978. Oxford. *Truth and Other Enigmas*, 67–137. London: Duckworth.

Dummett, Michael. 2004. *Truth and the Past*. New York: Columbia University Press.

Dupré, John. 1983. The Disunity of Science. *Mind* 92 (367): 321–346.

Dupré, John. 1993. *The Disorder of Things. Metaphysical Foundations of the Disunity of Science*. Cambridge: Harvard University Press.

Dvořák, Petr. 2023. Emergence in Ontic Structural Realism. *Pro-Fil* 24 (2): 1–17.

Dyke, Heather. 2012. On Methodology in the Metaphysics of Time. In *The Future of the Philosophy of Time*, ed. Adrian Bardon, 169–187. New York: Routledge.

Dyke, Heather. 2013. Time and Tense. In *A Companion to the Philosophy of Time*, ed. Heather Dyke and Adrian Bardon, 328–344. Malden: Wiley-Blackwell.

Dyke, Heather. 2023. Neutral Realism: A New Metaphysical Approach to Representation. *Philosophies* 8 (2): 19.

Eckholdt, Matthias, and Markus Gabriel. 2019. *Die ewige Wahrheit und der Neue Realismus. Gespräche über (fast) alles, was der Fall ist*. Heidelberg: Carl-Auer.

Egg, Matthias. 2018. Entity Realism. In *The Routledge Handbook of Scientific Realism*, ed. Juha Saatsi, 120–132. London: Routledge.

Egg, Matthias, and Juha Saatsi. 2021. Scientific Realism and Underdetermination in Quantum Theory. *Philosophy Compass* 16 (11): e12773.

Ehring, Douglas. 2011. *Tropes. Properties, Objects, and Mental Causation*. Oxford: Oxford University Press.

Eker, Bahadir. 2023. Realism About Tense and Atemporality. *Synthese* 202 (5): 1–25.

Elgin, Mehmet, and Elliott Sober. 2002. Cartwright on Explanation and Idealization. *Erkenntnis* 57 (3): 441–450.

Elliott-Graves, Alkistis. 2020. What Is a Target System? *Biology and Philosophy* 35 (2): 1–22.

Ellis, Brian. 2005. Physical Realism. *Ratio* 18 (4): 371–384.

Ellis, Fiona. 2014. *God, Value, and Nature*. Oxford: Oxford University Press.

Ellis, George F.R. 2006. On the Nature of Emergent Reality. In *The Re-Emergence of Emergence: the Emergentist Hypothesis from Science to Religion*, ed. Philip Clayton and Paul Davies. Oxford: Oxford University Press.

Ellis, George F.R. 2008. On the Nature of Causation in Complex Systems. *Transactions of the Royal Society of South Africa* 63: 69–84.

Ellis, George F.R. 2012. Top-Down Causation and Emergence: Some Comments on Mechanisms. *Interface Focus* 2: 126–140.

Ellis, George F.R. 2016. *How Can Physics Underlie the Mind? Top-Down Causation in the Human Context*. Berlin: Springer.

Ellis, George F.R., and Markus Gabriel. 2021. Physical, Logical, and Mental Top-Down Effects. In *Top-Down Causation and Emergence*, ed. Jan Voosholz and Markus Gabriel, Synthese Library, 439, 3–37. Cham: Springer.

Ellis, George F.R., Ulrich Kirchner, and William R. Stoeger. 2004. Multiverses and Physical Cosmology. *Monthly Notices of the Royal Astronomical Society* 347 (3): 921–936.

Ellis, George F.R., Denis Noble, and Timothy O'Connor. 2012. Top-Down Causation: An Integrating Theme Within and Across the Sciences? *Interface Focus* 2: 1–3.

Emundts, Dina, and Rolf-Peter Horstmann. 2002. *Georg Wilhelm Friedrich Hegel. Eine Einführung*. Stuttgart: Reclam.

Endicott, Ronald. 2016. Functionalism, Superduperfunctionalism, and Physicalism: Lessons from Supervenience. *Synthese* 193 (7): 2205–2235.

English, Jane. 1973. Underdetermination: Craig and Ramsey. *Journal of Philosophy* 70 (14): 453–462.

Ennis, Paul J. 2011. The Transcendental Core of Correlationism. *Cosmos and History* 7 (1): 37–48.

Esfeld, Michael. 2004. Quantum Entanglement and a Metaphysics of Relations. *Studies in History and Philosophy of Science Part B* 35 (4): 601–617.

Esfeld, Michael. 2009. The Modal Nature of Structures in Ontic Structural Realism. *International Studies in the Philosophy of Science* 23 (2): 179–194.

Esfeld, Michael. 2017. How to Account for Quantum Non-locality: Ontic Structural Realism and the Primitive Ontology of Quantum Physics. *Synthese* 194 (7): 2329–2344.

Esfeld, Michael, and Vincent Lam. 2008. Moderate Structural Realism About Space-Time. *Synthese* 160 (1), 27–46.

Esfeld, Michael, and Vincent Lam. 2011. Ontic Structural Realism as a Metaphysics of Objects. In *Scientific Structuralism*, ed. Alisa Bokulich, and Peter Bokulich, 143–159. Dordrecht: Springer.

Ferraris, Maurizio. 2012. *Documentality. Why It Is Necessary to Leave Traces*. New York: Fordham University Press.

Ferraris, Maurizio. 2014. *Manifesto of New Realism*, ed. Silvia Benso, and Brian Schroeder, trans. by Sarah De Sanctis, SUNY series in Contemporary Italian Philosophy. Albany: SUNY Press.

Feyerabend, Paul. 1958. An Attempt at a Realistic Interpretation of Experience. *Proceedings of the Aristotelian Society* 58 (1): 143–170.

Feyerabend, Paul. 1962. Explanation, Reduction and Empiricism. In *Scientific Explanation, Space, and Time*, ed. Herbert Feigl, and Grover Maxwell, 28–97. Minneapolis: University of Minneapolis Press.

Feyerabend, Paul. 1975. *Against Method. Outline of an Anarchistic Theory of Knowledge*. London: Verso.

Field, Hartry. 1980. *Science Without Numbers: A Defense of Nominalism*. Princeton: Princeton University Press.

Field, Hartry. 1982. Realism and Relativism. *Jorunal of Philosophy* 79 (10): 553–567.

Field, Hartry. 1989. *Realism, Mathematics and Modality*. Oxford: Blackwell.

Field, Hartry. 1994. Disquotational Truth and Factually Defective Discourse. *Philosophical Review* 103 (3): 405–452.

Fine, Arthur. 1986a. Realist and Instrumentalist Attachments to Science. *Mind* 95 (378): 149–179.

Fine, Arthur. 1986b. *The Shaky Game. Einstein, Realism and the Quantum Theory*. Chicago: University of Chicago Press.

Fine, Arthur. 2008. Epistemic Instrumentalism, Exceeding our Grasp. *Philosophical Studies* 137 (1): 135–139.

Fine, Kit. 1995. Ontological Dependence. *Proceedings of the Aristotelian Society* 95 (1): 169–290.

Fitch, G.W. 1994. Singular Propositions in Time. *Philosophical Studies* 73 (2/3): 181–187.

Floridi, Luciano. 2008. A Defence of Informational Structural Realism. *Synthese* 161 (2): 219–253.

Fodor, Jerry A. 1974. Special Sciences (Or: The Disunity of Science as a Working Hypothesis). *Synthese* 28 (2): 97–115.

Fogelin, Robert. 1997. Quine's Limited Naturalism. *Journal of Philosophy* 94 (11): 543–563.

Forbes, Graeme. 1985. *The Metaphysics of Modality*. Oxford: Clarendon Press.

Forbes, Graeme. 1989. *Languages of Possibility. An Essay in Philosophical Logic*. Oxford: Blackwell.

Forrest, Peter. 1982. Occam's Razor and Possible Worlds. *Monist* 65 4: 456–464.

Forrest, Barbara. 2000. Methodological Naturalism and Philosophical Naturalism: Clarifying the Connection. *Philo* 3 (2): 7–29.

Forrest, Peter, and David M. Armstrong. 1984. An Argument Against David Lewis's Theory of Possible Worlds. *Australasian Journal of Philosophy* 62 (2): 164–168.

Foucault, Michel. 1972. *The Archaeology of Knowledge. And the Discourse on Language*, trans. by Alan M. Sheridan Smith and Rupert Sawyer. New York: Pantheon Books.

Foucault, Michel. 1977. *Discipline and Punishment. The Birth of the Prison*, trans. by Alan M. Sheridan Smith, 2nd ed. New York: Vintage Books.
Foucault, Michel. 1978. *Introduction*, trans. by Robert Hurley, 2nd ed., vol. 1. The History of Sexuality. New York: Pantheon Books.
Francescotti, Robert. 1998. The Non-Reductionist's Troubles with Supervenience. *Philosophical Studies* 89 (1): 105–124.
Francescotti, Robert. 2014. *Physicalism and the Mind*. Dordrecht: Springer.
Frankish, Keith. 2016. Illusionism as a Theory of Consciousness. *Journal of Consciousness Studies* 23 (11/12): 11–39.
Franssen, Maarten. 2020. Philosophy of Technology and the Continental and Analytic Traditions. In *The Oxford Handbook of Philosophy of Technology*, ed. Shannon Vallor, 55–77. Oxford: Oxford University Press.
Freedman, Wendy L., Barry F. Madore, In Sung Jang, Taylor J. Hoyt, Abigail J. Lee, and Kayla A. Owens. 2025. *Status Report on the Chicago-Carnegie Hubble Program (CCHP): Measurement of the Hubble Constant Using the Hubble and James Webb Space Telescopes*. arXiv: 2408.06153 [astro-ph.CO], https://arxiv.org/abs/2408.06153.
Frege, Gottlob. 1953. *The Foundations of Arithmetic. A Logico-Mathematical Enquiry into the Concept of Number*, trans. by J.L. Austin, 2nd ed. Oxford: Blackwell.
Frege, Gottlob. 1988. *Die Grundlagen der Arithmetik. Eine logisch mathematische Untersuchung*. Hamburg: Meiner.
Frege, Gottlob. 1998. *Grundgesetze der Arithmetik. Begriffschriftlich abgeleitet. I/II*, Georg Olms, Hildesheim. New York: Zürich.
Frege, Gottlob. 2013. *Basic Laws of Arithmetic. Derived Using Concept-Script*, ed. and trans. by Philip A. Ebert, and Marcus Rossberg. Oxford: Oxford University Press.
French, Steven. 2003. A Model-Theoretic Account of Representation (or, I Don't Know Much About Art... but I Know It Involves Isomorphism). *Philosophy of Science* 70 (5): 1472–1483.
French, Steven. 2010. The Interdependence of Objects, Structure, and Dependence. *Synthese* 175 (S1): 89–109.
French, Steven. 2014. *The Structure of the World. Metaphysics and Representation*. Oxford: Oxford University Press.
French, Steven. 2019. Defending Eliminative Structuralism and a Whole Lot More (or Less). *Studies in History and Philosophy of Science* 74: 22–29.
French, Steven. 2020. *There Are No Such Things as Theories*. Oxford: Oxford University Press.
French, Steven. 2023. Quasi-Structural Realism. In *Non-Reflexive Logics, Non-Individuals, and the Philosophy of Quantum Mechanics: Essays in Honour of the Philosophy of Décio Krause*, ed. Jonas R.B. Arenhart, and Raoni W. Arroyo, Synthese Library, 476, 29–43. Cham: Springer.
French, Steven. 2024a. Does the Claim that There Are No Theories Imply that There Is No History of Theories to be Written?(!). *Journal for General Philosophy of Science/Zeitschrift für Allgemeine Wissenschaftstheorie* 55 (3): 327–346.
French, Steven. 2024b. Phenomenology, Perspectivalism and (Quantum) Physics. *Foundations of Physics* 54 (3): 1–18.
French, Steven, and James Ladyman. 2003a. Between Platonism and Phenomenalism: Reply to Cao. *Synthese* 136 (1): 73–78.
French, Steven, and James Ladyman. 2003b. Remodelling Structural Realism: Quantum Physics and the Metaphysics of Structure. *Synthese* 136 (1): 31–56.
Friedl, Johannes. 2021. Realism and Anti-Realism in Young Carnap. In *Der Junge Carnap in Historischem Kontext: 1918–1935/Young Carnap in an Historical Context: 1918–1935*, ed. Christian Damböck, and Gereon Wolters, 149–168. Berlin: Springer.
Frigg, Roman. 2010a. Fiction and Science. In *Fictions and Models*, ed. John Woods, 247–287. Munich: Philosophia.
Frigg, Roman. 2010b. Models and Fiction. *Synthese* 172 (2): 251–268.
Frigg, Roman. 2023. *Models and Theories. A Philosophical Inquiry*. Oxon: Routledge.
Frigg, Roman, and James Nguyen. 2016. The Fiction View of Models Reloaded. *Monist* 99 (3): 225–242.

Frigg, Roman, and James Nguyen. 2020. *Modelling Nature. An Opinionated Introduction to Scientific Representation*, Synthese Library, 427, Cham.

Frisch, Wolfgang, and Martin Meschede. 2013. *Plattentektonik. Kontinentverschiebung und Gebirgsbildung*, 5th ed. Darmstadt: Wissenschaftliche Buchgesellschaft.

Fulda, Hans Friedrich. 2006. Methode und System bei Hegel. Das Logische, die Natur, der Geist als universale Bestimmungen einer monistischen Philosophie. In *Systemphilosophie als Selbsterkenntnis*, ed. Hans Friedrich Fulda, 25–50. Würzburg: Königshausen & Neumann.

Gabbay, Michael. 2010. A Formalist Philosophy of Mathematics, Part I: Arithmetic. *Studia Logica* 96: 219–238.

Gabovich, Alexander M., and Vladimir Kuznetsov. 2023. Scientific Realism From a Polysystemic View of Physical Theories and Their Functioning. *Global Philosophy* 33 (6): 1–18.

Gabriel, Markus. 2008a. *An den Grenzen der Erkenntnistheorie. Die notwendige Endlichkeit des objektiven Wissens als Lektion des Skeptizismus*, Alber. München: Freiburg.

Gabriel, Markus. 2008b. *Antike und moderne Skepsis zur Einführung*. Hamburg: Junius.

Gabriel, Markus. 2009. *Skeptizismus und Idealismus in der Antike*. Frankfurt (Main): Suhrkamp.

Gabriel, Markus (ed.). 2014. *Der Neue Realismus*. Berlin: Suhrkamp.

Gabriel, Markus. 2015a. *Fields of Sense. A New Realist Ontology*. Edinburgh: Edinburgh University Press.

Gabriel, Markus. 2015b. Neutral Realism. *Monist* 98 (2): 181–196.

Gabriel, Markus. 2015c. *Why the World Does Not Exist*. Cambridge: Polity.

Gabriel, Markus. 2016. *Sinn und Existenz. Eine realistische Ontologie*. Berlin: Suhrkamp.

Gabriel, Markus. 2020. *The Limits of Epistemology*, trans. by Alex Englander and Markus Gabriel. Cambridge: Polity.

Gabriel, Markus. 2021. How Mind Fits into Nature. Mental Realism after Nagel. In *Parallax. The Dialectics of Mind and World*, ed. Dominik Finkelde, Slavoj Žižek, and Christoph Menke, 39–53. London: Bloomsbury Academic.

Gabriel, Markus. 2022. Ancestrality and (in-)Dependence – on Heidegger on Being-in-Itself. *British Journal for the History of Philosophy* 30 (3): 535–546.

Gabriel, Markus. 2023. Facts, Not Fossils: New vs. Speculative realism. In *Against New Materialisms*, ed. Benjamin Jon Boysen, and Jesper Lundsfryd Rasmussen, 73–91. London: Bloomsbury Academic.

Gabriel, Markus. 2024a. *Fictions*. Cambridge: Polity.

Gabriel, Markus. 2024b. Responses. In *Markus Gabriel's New Realism*, ed. Jan Voosholz, Synthese Library, 492, 323–373. Cham: Springer.

Gabriel, Markus. 2024c. *Sense, Nonsense, and Subjectivity*. Cambridge: Harvard University Press.

Galison, Peter. 1998. The Americanization of Unity. *Dædalus* 127 (1): 47–75.

Galison, Peter, and David J. Stump (eds.). 1996. *The Disunity of Science: Boundaries, Contexts, and Power*. Stanford: Stanford University Press.

Gensollen, Mario, and Marc Jiménez-Rolland. 2021. Giere's Scientific Perspectivism as Carte Blanche Realism. *ArtefaCToS* 10 (1): 61–74.

Gentile, Nélida, and Susana Lucero. 2024. On Compatibility Between Realism and Fictionalism: A Response to Suárez' Proposal. *Studies in History and Philosophy of Science Part A* 103 (C): 168–175.

George, Alexander. 2000. On Washing the Fur Witout Wetting It: Quine, Carnap, and Analyticity. *Mind* 109 (433): 1–24.

Geyh, Mebus A. 2005. *Handbuch der physikalischen und chemischen Altersbestimmung*. Darmstadt: Wissenschaftliche Buchgesellschaft.

Ghins, Michel. 2024. *Scientific Realism and Laws of Nature: A Metaphysics of Causal Powers*, Synthese Library, 483. Cham: Springer.

Giere, Ronald N. 1988. *Explaining Science: A Cognitive Approach*. Chicago: University of Chicago Press.

Giere, Ronald N. 1999. *Science Without Laws*. Chicago: University of Chicago Press.

Giere, Ronald N. 2004. How Models Are Used to Represent Reality. *Philosophy of Science* 71 (5): 742–752.

Giere, Ronald N. 2006. *Scientific Perspectivism*. Chicago: University of Chicago Press.
Ginev, Dimitri. 2017. Hermeneutic Realism and the Existence of Theoretical Objects. In *New Perspectives on Realism*, ed. Luca Taddio, 83–107. Mailand: Mimesis International.
Giraud, Thibaut, and Raphaël Millière. 2017. Creeping Up on Things from Behind. The Trouble with Absolute Realism. In *New Perspectives on Realism*, ed. Luca Taddio, 109–132. Mailand: Mimesis International.
Gironi, Fabio. 2011. Meillassoux's Speculative Philosophy of Science: Contingency and Mathematics. *Pli* 22: 25–60.
Gironi, Fabio. 2017. The Return of Natural Philosophy. The Realist Turn as Synoptic Reconciliation of Philosophy and Science. In *New Perspectives on Realism*, ed. Luca Taddio, 133–161. Mailand: Mimesis International.
Gödel, Kurt. 1947. What Is Cantor's Continuum Problem? *American Mathematical Monthly* 54: 515–525.
Gödel, Kurt. 1951. Russell's Mathematical Logic (1944). In *The Philosophy of Bertrand Russel*, ed. P. Schilpp, 119–141. New York: Tudor.
Golumbia, David. 2016. "Correlationism": The Dogma that Never Was. *Boundary 2* 43 (2): 1–25.
Gómez-Torrente, Mario. 2024. Modal Realism and Anthropic Reasoning. *Australasian Journal of Philosophy* 102 (4): 925–938.
Goodman, Nelson. 1978. *Ways of Worldmaking*. Brighton: Harvester Press.
Goodman, Nelson. 1983. *Fact, Fiction, and Forecast*, 4th ed. Cambridge: Harvard University Press.
Goodman, Nelson, and Willard Van Orman Quine. 1947. Steps Toward a Constructive Nominalism. *Journal of Symbolic Logic* 12: 105–122.
Gori, Pietro. 2023. The Perspectival Realist Features of Ernst Mach's Critical Epistemology. *Journal for General Philosophy of Science/Zeitschrift für Allgemeine Wissenschaftstheorie* 54 (1): 99–124.
Goris, Wouter. 2012. De magistro – Thomas Aquinas, Henry of Ghent and John Duns Scotus on Natural Conceptions. *Review of Metaphysics* 66: 435–468.
Grant, Iain Hamilton. 2006. *Philosophies of Nature after Schelling*. London: Continuum International.
Gratton, Peter. 2014. *Speculative Realism. Problems and Prospects*. London: Bloomsbury.
Green, Sara, and Robert W. Batterman. 2021. Making Sense of Top-Down Causation: Universality and Functional Equivalence in Physics and Biology. In *Top-Down Causation and Emergence*, ed. Jan Voosholz and Markus Gabriel, Synthese Library, 439, 39–63. Cham: Springer.
Griesemer, James. 2013. Formalization and the Meaning of "Theory" in the Inexact Biological Sciences. *Biological Theory* 7 (4): 298–310.
Güsten, Rolf, Helmut Wiesemeyer, David Neufeld, Karl M. Menten, Urs U. Graf, Karl Jacobs, Bernd Klein, Oliver Ricken, Christophe Risacher, and Jürgen Stutzki. 2019. Astrophysical Detection of the Helium Hydride Ion HeH^+. *Nature* 568: 357–359.
Haack, Susan. 1976. The Pragmatist Theory of Truth. *British Journal for the Philosophy of Science* 27: 231–249.
Haack, Susan. 1993. The Two Faces of Quine's Naturalism. *Synthese* 94 (3): 335–356.
Habdankaitė, Daina. 2021. *Ontologies of the Future in Contemporary Philosophy: Stiegler and Meillassoux*, Phil. Diss. Vilnius: Vilnius University and University of Turin.
Hacking, Ian. 1982. Experimentation and Scientific Realism. *Philosophical Topics* 13 (1): 71–87.
Hacking, Ian. 1983. *Representing and Intervening*. Cambridge: Cambridge University Press.
Hacking, Ian. 1996. The Disunities of the Sciences. In *The Disunity of Science. Boundaries, Contexts, and Power*, ed. Peter Galison, and David J. Stump, 37–74. Stanford: Stanford University Press.
Hägglund, Martin. 2011. Radical Atheist Materialism. A Critique of Meillassoux. In *The Speculative Turn. Continental Materialism and Realism*, ed. Levi Bryant, Nick Srnicek, and Graham Harman, 114–129. Melbourne: re.press.
Hale, Bob, Crispin Wright, and Alexander Miller (eds.). 2017. *A Companion to the Philosophy of Language*, 2nd ed., vol. I/II. Malden: Wiley Blackwell.

Hallward, Peter. 2011. Anything is Possible: A Reading of Quentin Meillassoux's *After Finitude*. In *The Speculative Turn. Continental Materialism and Realism*, ed. Levi Bryant, Nick Srnicek, and Graham Harman, 130–141. Melbourne: re.press.

Hamilton, T.F., and W.L. Robison. 2004. *Overview of Radiological Conditions on Bikini Atoll*, UCRL-MI-208228, Lawrence Livermore National Laboratory.

Hankinson Nelson, Lynn. 2002. Feminist Philosophy of Science. In *The Blackwell Guide to the Philosophy of Science*, ed. Peter Machamer and Michael Silberstein, 312–331. Malden: Blackwell.

Hankinson Nelson, Lynn. 2017. *Biology and Feminism: A Philosophical Introduction*. Cambridge: Cambridge University Press.

Hanson-Park, Jared. 2023. Structural Realism and Agnosticism About Objects. *Global Philosophy* 33 (2): 1–25.

Haraway, Donna. 1991. *Simians, Cyborgs, and Women. The Reinvention of Nature*. New York: Routledge.

Harman, Graham. 2011. *Quentin Meillassoux. Philosophy in the Making*. Edinburgh: Edinburgh University Press.

Harman, Graham. 2021. Malabou's Political Critique of Speculative Realism. *Open Philosophy* 4 (1): 94–105.

Harrison, Alan K. 2022. A Realistic Theory of Quantum Measurement. *Foundations of Physics* 52 (1): 1–32.

Haslanger, Sally. 2012. *Resisting Reality. Social Construction and Social Critique*. Oxford: Oxford University Press.

Haslanger, Sally, and Roxanne Marie Kurtz (eds.). 2006. *Persistence. Contemporary Readings*. Cambridge: MIT Press.

Hawley, Katherine. 2001. *How Things Persist*. Oxford: Clarendon Press.

Healey, Richard. 2016. Metaphysics in Science. In *The Oxford Handbook of Philosophy of Science*, ed. Paul Humphreys, 356–377. Oxford: Oxford University Press.

Healey, Richard. 2017. *The Quantum Revolution in Philosophy*. Oxford: Oxford University Press.

Healey, Richard. 2021. A Pragmatist Perspective on Causation, Laws and Explanation. In *Top-Down Causation and Emergence*, ed. Jan Voosholz, and Markus Gabriel, Synthese Library, 439, 253–268. Cham: Springer.

Hegel, Georg Wilhelm Friedrich. 1979. *Enzyklopädie der philosophischen Wissenschaften im Grundrisse (1830)*, 2nd ed., vol. 8–10. Werke. Frankfurt (Main): Suhrkamp.

Hegel, Georg Wilhelm Friedrich. 1989. *Phänomenologie des Geistes*, 2nd ed., vol. 3. Frankfurt (Main): Werke, Suhrkamp.

Hegel, Georg Wilhelm Friedrich. 2010. *Encyclopedia of the Philosophical Sciences in Basic Outline*, ed. and trans. by Klaus Brinkmann, and Daniel O. Dahlstrom, Cambridge Hegel Translations. Cambridge: Cambridge University Press.

Heidegger, Martin. 1976a. *Was ist Metaphysik?*, vol. 9, 103–122. Frankfurt (Main): Gesamtausgabe, Vittorio Klostermann.

Heidegger, Martin. 1976b. *Zur Seinsfrage*, vol. 9, 385–426. Frankfurt (Main): Gesamtausgabe, Vittorio Klostermann.

Heidegger, Martin. 1997. *Der Satz vom Grund*, vol. 10. Frankfurt (Main): Gesamtausgabe, Vittorio Klostermann.

Heidegger, Martin. 2006a. *Identität und Differenz*, vol. 11. Frankfurt (Main): Gesamtausgabe, Vittorio Klostermann.

Heidegger, Martin. 2006b. *Sein und Zeit*, 19th ed. Tübingen: Max Niemeyer.

Heil, John. 1998. Skepticism and Realism. *American Philosophical Quarterly* 35 (1): 57–72.

Heller, Joshua, and Jon Cogburn. 2017. Meillassoux's Dilemma: Paradoxes of Totality after the Speculative Turn. In *New Perspectives on Realism*, ed. Luca Taddio, 163–189. Mailand: Mimesis International.

Hellman, Geoffrey. 1989. *Mathematics without Numbers: Towards a Modal-Structural Interpretation*, Oxford University Press, Oxford.

Hellman, Geoffrey. 2001. Structuralism Without Structures. *Philosophia Mathematics* 9 (2): 184–211.

Hempel, Carl Gustav. 1950. Problems and Changes in the Empiricist Criterion of Meaning. *Revue Internationale de Philosophie* 4 (11): 41–63.

Henne, Céline, Hannah Tomczyk, and Christoph Sperber. 2024. Physicists' Views on Scientific Realism. *European Journal for Philosophy of Science* 14 (1): 1–27.

Hilbert, David. 1990. *Foundations of Geometry*, trans. by Leo Unger, 2nd ed. La Salle: Open Court Press.

Hilbert, David. 2015. *Grundlagen der Geometrie (Festschrift 1899)*, ed. Klaus Volkert. Berlin: Springer.

Hinchliff, Mark. 1996. The Puzzle of Change. *Philosophical Perspectives* 10: 119–136.

Hinchliff, Mark. 2000. A Defense of Presentism in a Relativistic Setting. *Philosophy of Science* 67 (3): 575–586.

Hinchliff, Mark. 2010. The Identity of the Past. In *Time and Identity*, ed. Joseph Keim Campbell, Michael O'Rourke, and Harry S. Silverstein, 95–110. Cambridge: MIT Press.

Hirsch, Eli. 2010. *Quantifier Variance and Realism: Essays in Metaontology*. Oxford: Oxford University Press.

Hitchcock, Christopher. 1992. Causal Explanation and Scientific Realism. *Erkenntnis* 37 (2): 151–178.

Holbrad, Martin, and Morten A. Pedersen. 2017. *The Ontological Turn. An Anthropological Exposition*. Cambridge: Cambridge University Press.

Hollis, Morgan. 2024. Hints of a Physical Origin for the S_8 Tension. *Nature Astronomy* 405: 8.

Horan, Barbara L. 1988. Theoretical Models, Biological Complexity and the Semantic View of Theories. *Proceedings of the Biennial Meeting of the Philosophy of Science Association* 265–277.

Horgan, Terry. 1993. From Supervenience to Superdupervenience: Meeting the Demands of a Material World. *Mind* 102 (408): 555–586.

Horgan, Terry. 2006. Materialism: Matters of Definition, Defense, and Deconstruction. *Philosophical Studies* 131 (1): 157–183.

Hoyningen-Huene, Paul. 2012. Santa's Gift of Structural Realism. In *Probabilities, Laws, and Structures*, ed. Dennis Dieks, Wenceslao J. Gonzalez, Stephan Hartmann, Michael Stöltzner, and Marcel Weber, 183–186. Berlin: Springer.

Hoyningen-Huene, Paul, and Eric Oberheim. 2009. Reference, Ontological Replacement and Neo-Kantianism: A Reply to Sankey. *Studies in History and Philosophy of Science Part A* 40 (2): 203–209.

Hubler, James N. 1995. *Creatio ex Nihilo. Matter, Creation, and the Body in Classical and Christian Philosophy Through Aquinas*, Phil. Diss, University of Pennsylvania.

Huneman, Philippe, and Estelle Kulich. 1997. *Introduction a' la phénoménologie*. Paris: Armand Colin.

Hylton, Peter, and Gary Kemp. 2020. Willard Van Orman Quine. *Stanford Encyclopedia of Philosophy* (Spring 2020 Edition), ed. Edward N. Zalta.

Ingham, Mary Beth, and Mechthild Dreyer. 2004. *The Philosophical Vision of John Duns Scotus: An Introduction*. Washington: Catholic University of America Press.

Ismael, Jenann. 2012. Devision and the Open Future. In *The Future of the Philosophy of Time*, ed. Adrian Bardon, 149–168. New York: Routledge.

Ismael, Jenann, and Jonathan Schaffer. 2020. Quantum Holism: Nonseparability as Common Ground. *Synthese* 197 (10): 4131–4160.

Jackson, John P., and Nadine M. Weidman. 2004. *Race, Racism, and Science: Social Impact and Interaction*, ABC-Clio. Santa Barbara: Johns, Charles William.

John P., and Nadine M. Weidman. 2023. *Hegel and Speculative Realism*. Cham: Palgrave Macmillan.

Kaiser, Marie I. 2012. Why It Is Time to Move beyond Negalian Reduction. In *Probabilities, Laws, and Structures. The Philosophy of Science in a European Perspective*, ed. Dennis Dieks, Wenceslao J. Gonzalez, Stephan Hartmann, Michael Stöltzner, and Marcel Weber, vol. 3, 255–272. Heidelberg: Springer.

Kanev, Alexander. 2017. New Realism, Pluralism and Science. In *New Perspectives on Realism*, ed. Luca Taddio, 191–213. Mailand: Mimesis International.

Kant, Immanuel. 1998a. *Critique of Pure Reason*, ed. Paul Guyer, and Allen W. Wood. The Cambridge Edition of the Works of Immanuel Kant. Cambridge: Cambridge University Press.

Kant, Immanuel. 1998b. *Kritik der reinen Vernunft*, ed. Jens Timmerberg. Hamburg: Meiner.

Kant, Immanuel. 2000. *Critique of the Power of Judgment*, ed. Paul Guyer, trans. by Eric Matthews, The Cambridge Edition of the Works of Immanuel Kant. Cambridge: Cambridge University Press.

Kant, Immanuel. 2009. *Kritik der Urteilskraft*, ed. Heiner F. Klemme. Hamburg: Meiner.

Kauffman, Stuart A., Dávid Péter Jelenfia, and Gábor Vattay. 2020. Theory of Chemical Evolution of Molecule Compositions in the Universe, in the Miller–Urey Experiment and the Mass Distribution of Interstellar and Intergalactic Molecules. *Journal for Theoretical Biology* 486: 110097.

Keele, Rondo. 2010. *Ockham Explained. From Razor to Rebellion*. Chicago: Open Court.

Keil, Geert. 1993. *Kritik des Naturalismus*. Berlin: de Gruyter.

Khalifa, Kareem. 2010. Default Privilege and Bad Lots: Underconsideration and Explanatory Inference. *International Studies in the Philosophy of Science* 24 (1): 91–105.

Khalili, Mahdi. 2022. From Phenomenological-Hermeneutical Approaches to Realist Perspectivism. *European Journal for Philosophy of Science* 12 (4): 1–26.

Khalili, Mahdi. 2024. Causation and Realism: The Role of Instrumentally Mediated Empirical Evidence. In *The Routledge Handbook of Causality and Causal Methods*, ed. Federica Russo and Phyllis Illari. New York: Routledge.

Khamis, Diana. 2023. Correlationist Sterility: A Critique of the Absolutization of Contingency in Meillassoux. In *Against New Materialisms*, ed. Benjamin Jon Boysen and Jesper Lundsfryd Rasmussen, 59–72. London: Bloomsbury Academic.

Kiefer, Claus. 2007. *Quantum Gravity*, 2nd ed. Oxford: Oxford University Press.

Kim, Jaegwon. 1993a. "Strong" and "Global" Supervenience Revisited. In *Supervenience and Mind: Selected Philosophical Essays*, ed. Jaegwon Kim, 53–78. Cambridge: Cambridge University Press.

Kim, Jaegwon. 1993b. Supervenience as a Philosophical Concept. In *Supervenience and Mind: Selected Philosophical Essays*, ed. Jaegwon Kim, 131–160. Cambridge: Cambridge University Press.

Kim, Jaegwon. 2011. From Naturalism to Physicalism: Supervenience Redux. *Proceedings and Addresses of the American Philosophical Association* 85 (2): 109–134.

Kind, Rainer, and Xiaohui Yuan. 2003. Kollidierende Kontinente. *Physik unserer Zeit* 34 (5): 213–217.

King, Peter. 2002. Scotus on Metaphysics. In *The Cambridge Companion to Duns Scotus*, ed. Thomas Williams, 15–68. Cambridge: Cambridge University Press.

Kitcher, Philip. 2001. Real Realism: the Galilean Strategy. *Philosophical Review* 110 (2): 151–197.

Klima, Gyula. 2017. The Medieval Problem of Universals. *Stanford Encyclopedia of Philosophy* (Winter 2017 Edition), ed. Edward N. Zalta.

Knuuttila, Tarja. 2011. Modelling and Representing: An Artefactual Approach to Model-Based Representation. *Studies in History and Philosophy of Science Part A* 42 (2): 262–271.

Koch, Anton Friedrich. 1990. *Subjektivität in Raum und Zeit.* , Frankfurt (Main): Klostermann.

Koch, Anton Friedrich. 2006a. The Subjectivity Thesis and its Corollaries. *Philosophical Inquiry* 28 (3–4): 9–20.

Koch, Anton Friedrich. 2006b. *Versuch über Wahrheit und Zeit*. Paderborn: mentis.

Koch, Anton Friedrich. 2016. *Hermeneutischer Realismus*. Tübingen: Mohr Siebeck.

Koch, Anton Friedrich. 2020. Is Hermeneutic Realism a Dialectical Materialism? Thinking With and Beyond Heidegger and Adorno. In *Idealism, Relativism, and Realism. New Essays on Objectivity Beyond the Analytic-Continental Divide*, ed. Dominik Finkelde, and Paul M. Livingston, 81–96. Berlin: De Gruyter.

Koch, Anton Friedrich. 2023. Our Common Extended Consciousness and the Readability of Things. Two Theses of Hermeneutic Realism. *Critical Hermeneutics* 6 (2): 35–59.

Koch, Anton Friedrich. 2024. The Intersubjective Perceptual Field as the Privileged Field of Sense. In *Markus Gabriel's New Realism*, ed. Jan Voosholz, Synthese Library, 492, 223–239. Cham: Springer.

Kollmeier, Juna A., David H. Weinberg, Benjamin D. Oppenheimer, Francesco Haardt, Neal Katz, Romeel Davé, Mark Fardal, Piero Madau, Charles Danforth, Amanda B. Ford, Molly S. Peeples, and Joseph McEwen. 2014. The Photon Underproduction Crisis. *The Astrophysical Journal Letters* 789 (2): L32.

Kolozova, Katerina. 2024. A Critique of the Subjectivity-Centered Epistemo-Ontology: A Triangular Glance Combining Marx, Laruelle, Irigaray in Dialogue with Gabriel. In *Markus Gabriel's New Realism*, ed. Jan Voosholz, Synthese Library, 492, 271–284. Cham: Springer.

Korab-Karpowicz, W. Julian. 2023. Political Realism in International Relations. *Stanford Encyclopedia of Philosophy* (Winter 2023 Edition), ed. Edward N. Zalta.

Koskinen, Inkeri, and Uskali Mäki. 2016. Extra-Academic Transdisciplinarity and Scientific Pluralism: What Might They Learn from One Another? *European Journal for Philosophy of Science* 6 (3): 419–444.

Krebs, Sebastian. 2019. *Kripkes Metaphysik Möglicher Welten*. Berlin: de Gruyter.

Kreis, Guido. 2023. Kant on Why We Cannot Even Judge about Things in Themselves. In *Limits of Intelligibility: Issues from Kant and Wittgenstein*, ed. Jens Pier, 57–86. London: Routledge.

Kripke, Saul. 1980. *Naming and Necessity*. Cambridge: Harvard University Press.

Krockert, Florian. 2016. *Im Sumpf des Grauens. Michael Williams' Kontextualismus als Ausweg aus dem Münchhausen-Trilemma*, MA thesis. Berlin: Freie Universität Berlin.

Kubiak, Adam P. 2024. Perspectival Realism and Frequentist Statistics: The Case of Jerzy Neyman? Methodology and Philosophy. *Synthese* 205 (1): 1–29.

Kuhn, Thomas S. 2000. *The Road Since Structure*, ed. James Conant and John Haugeland. Chicago: University of Chicago Press.

Kuhn, Thomas S. 2003. *The Copernican Revolution. Planetary Astronomy in the Development of Western Thought*, 24th ed. Cambridge: Harvard University Press.

Kuhn, Thomas S. 2008. *Die Struktur wissenschaftlicher Revolutionen*, 23rd ed. Frankfurt (Main): Suhrkamp.

Kukla, André. 1996. Does Every Theory Have Empirically Equivalent Rivals? *Erkenntnis* 44: 137–166.

Kusch, Martin. 1991. *Foucault's Strata and Fields. An Investigation into Archaeological and Genealogical Science Studies*, Synthese Library, 278. Dordrecht: Kluwer.

Kusch, Martin. 2002. *Knowledge by Agreement. The Programme of Communitarian Epistemology*. Oxford: Oxford University Press.

Kusch, Martin. 2010. Hacking's Historical Epistemology: A Critique of Styles of Reasoning. *Studies in History and Philosophy of Science Part A* 41 (2): 158–173.

Kusch, Martin. 2017. Epistemic Relativism, Scepticism, Pluralism. *Synthese* 194 (12): 4687–4703.

Kusch, Martin. 2018. Scientific Realism and Social Epistemology. In *The Routledge Handbook of Scientific Realism*, ed. Juha Saatsi, 261–275. London: Routledge.

Kusch, Martin. 2020. Stances, Voluntarism, Relativism. In *Idealism, Relativism, and Realism: New Essays on Objectivity Beyond the Analytic-Continental Divide*, ed. Dominik Finkelde and Paul M. Livingston, 131–154. Berlin: De Gruyter.

Ladyman, James. 1998. What Is Structural Realism? *Studies in History and Philosophy of Science* 29 (3): 409–424.

Ladyman, James. 2007. On the Identity and Diversity of Objects in a Structure. *Aristotelian Society Supplementary* 81 (1): 23–43.

Ladyman, James. 2011. Structural Realism Versus Standard Scientific Realism: The Case of Phlogiston and Dephlogisticated Air. *Synthese* 180 (2): 87–101.

Ladyman, James. 2018. Scientific Realism Again. *Spontaneous Generations* 9 (1): 99–107.

Ladyman, James. 2020. Structural Realism. *Stanford Encyclopedia of Philosophy* (Winter 2020 Edition), ed. Edward N. Zalta.

Ladyman, James, and Lorenzo Lorenzetti. 2023. Effective Ontic Structural Realism. *British Journal for the Philosophy of Science.*

Ladyman, James, and Don Ross. 2007. *Every Thing Must Go. Metaphysics Naturalized*, with David Spurrett and John Collier. Oxford: Oxford University Press.

Lakatos, Imre. 1970. Falsification and the Methodology of Scientific Research Programmes. In *Criticism and the Growth of Knowledge*, ed. Imre Lakatos and Alan Musgrave, 91–196. Cambridge: Cambridge University Press.

Lam, Vincent. 2017. Structuralism in the Philosophy of Physics. *Philosophy Compass* 12 (6): e12421.

Lam, Vincent, and Michael Esfeld. 2012. The Structural Metaphysics of Quantum Theory and General Relativity. *Journal for General Philosophy of Science* 43 (2): 243–258.

Landry, Elaine. 2007. Shared Structure Need not be Shared Set-Structure. *Synthese* 158 (1): 1–17.

Landry, Elaine. 2011. How to Be a Structuralist All the Way Down. *Synthese* 179 (3): 435–454.

Langhe, Rogier De. 2010. The Division of Labour in Science: The Tradeoff Between Specialisation and Diversity. *Journal of Economic Methodology* 17 (1): 37–51.

Langton, Rae. 1998. *Kantian Humility. Our Ignorance of Things in Themselves*. Oxford: Oxford University Press.

Latour, Bruno. 2005. *Reassembling the Social: An Introduction to the Actor-Network Theory*. Oxford: Oxford University Press.

Latour, Bruno, and Steve Woolgar. 1986. *Laboratory Life. The Construction of Scientific Facts*, 2nd ed. Princeton: Princeton University Press.

Laudan, Larry. 1981. A Confutation of Convergent Realism. *Philosophy of Science* 48 (1): 19–49.

Laudan, Larry. 1990. Demystifying Underdetermination. In *Scientific Theories*, ed. C. Wade Savage, 267–297. Minneapolis: University of Minnesota Press.

Lavocat, Françoise. 2016. *Fait et fiction. Pour une frontie're*. Paris: Seuil.

Le Poidevin, Robin (ed.). 1998. *Questions of Time and Tense*. Oxford: Oxford University Press.

Leahey, Erin, and Ryan C. Reikowsky. 2008. Research Specialization and Collaboration Patterns in Sociology. *Social Studies of Science* 38 (3): 425–440.

Lehmkuhl, Dennis. 2011. Mass-Energy-Momentum: Only There Because of Spacetime. *British Journal for the Philosophy of Science* 62 (3): 453–488.

Lehmkuhl, Dennis. 2012. Super-Substanzialismus in der Philosophie der Raumzeit. In *Philosophie der Physik*, ed. Michael Esfeld, 2nd ed., 50–67. Berlin.

Lehmkuhl, Dennis. 2018. The Metaphysics of Super-Substantivalism. *Noûs* 52 (1): 24–46.

Leibniz, Gottfried Wilhelm. 1965. *Monadology and Other Philosophical Essays*, ed. and trans. by Paul Schrecker and Anne Martin Schrecker, Library of liberal arts. Indianapolis: Bobbs-Merrill Company.

Leidlmair, Karl. 2024. Meillassoux and Heidegger. How to Deal with Things-in-Themselves? *Open Philosophy* 7 (1): 117–132.

Leng, Mary. 2010. *Mathematics and Reality*. Oxford: Oxford University Press.

Levinas, Emmanuel. 1987. *Time and the Other. And Additional Essays*, trans. by Richard A. Cohen. Pittsburgh: Duquesne University Press.

Levinas, Emmanuel. 1991. *Totality and Infinity. An Essay on Exteriority*, trans. by Alphonso Lingis, 4th ed., vol. 1. Martinus Nijhoff Philosophy Texts. Dordrecht: Kluwer Academic Publishers.

Levy, Arnon. 2018. Modeling and Realism. Strange bedfellows? In *The Routledge Handbook of Scientific Realism*, ed. Juha Saatsi, 237–249. London: Routledge.

Lewis, David. 1973. *Counterfactuals*. Oxford: Blackwell.

Lewis, David. 1986a. Against Structural Universals. *Australasian Journal of Philosophy* 64 (1): 25–46.

Lewis, David. 1986b. Counterfactual Dependence and Time's Arrow. In *Philosophical Papers*, ed. David Lewis, vol. II, 32–52. Oxford: Oxford University Press.

Lewis, David. 1986c. *On the Plurality of Worlds*. Oxford: Blackwell.

Lewis, David. 1986d. The Paradoxes of Time Travel. In *Philosophical Papers*, vol. II, 67–80. Oxford: Oxford University Press.

Lewis, David. 1992. Critical Notice: Armstrong, D.M. A Combinatorial Theory of Possibility. *Australasian Journal of Philosophy* 70: 211.

Linsky, Bernard, and Edward N. Zalta. 1995. Naturalized Platonism Versus Platonized Naturalism. *Journal of Philosophy* 92 (10): 525–555.

Lipton, Peter. 2004. *Inference to the Best Explanation*, 2nd ed. London: Routledge.

Liptow, Jasper. 2014. Externalismus. Die Auflösung des Dualismus von Begriff und Welt. In *Die Philosophie John McDowells*, ed. David Lauer and Christian Barth, 63–83. Münster: mentis.

Lloyd, Elisabeth. 1993. Pre-Theoretical Assumptions in Evolutionary Explanations of Female Sexuality. *Philosophical Studies* 69 (2/3): 139–153.

Locke, John. 1975. *An Essay Concerning Human Understanding*, ed. Peter H. Nidditch, The Clarendon Edition of the Works of John Locke. Oxford: Clarendon Press.

Loewer, Barry, and Jonathan Schaffer (eds.). 2015. *A Companion to David Lewis*. Malden: Wiley-Blackwell.

Longino, Helen. 1990. *Science as Social Knowledge. Values and Objectivity in Scientific Inquiry*. Princeton: Princeton University Press.

Longino, Helen. 2002. *The Fate of Knowledge*. Princeton: Princeton University Press.

Longino, Helen. 2009. Perilous Thoughts: Comment on van Fraassen. *Philosophical Studies* 143 (1): 25–32.

Longino, Helen. 2019. The Social Dimensions of Scientific Knowledge. *Stanford Encyclopedia of Philosophy* (Summer 2019), ed. Edward N. Zalta.

Longo, Anna. 2014a. Speculative Realism and Other Heresies. *Philosophical Readings* 6 (2): 26–40.

Longo, Anna. 2014b. The Contingent Emergence of Thought. A Comparison Between Meillassoux and Deleuze. In *Time Without Becoming*, ed. Anna Longo, 31–50. Mailand: Mimesis International.

Longo, Anna. 2015a. The Genesis of the Transcendental: How to make a Realist Speculation Out of Absolute Idealism. *Methode* 4 (5): 150–176.

Longo, Anna. 2015b. The Reality of the End of the World: Out-Science-Fiction and Horror Stories. In *Breaking the Spell. Contemporary Realism under Discussion*, ed. Sarah De Sanctis and Anna Longo, 31–48. Mailand: Mimesis International.

Longo, Anna. 2018. La science des conditions transcendantales de la connaissance, ou la "spéculation réaliste. In *Le transcendantal: réceptions et mutations d'une notion kantienne*, ed. Christophe Bouriau, Anthony Feneuil, and Yves Meessen, 47–57. Nancy: Presses universitaires de Nancy.

Longo, Anna. 2020. Infinite Judgements and Transcendental Logic. trans. by Ekin Erkan and Madeleine Collier. *Cosmos and History* 20 (2): 391–415.

Loux, Michael J. 2003. Realism and Anti-Realism : Dummett's Challenge. In *The Oxford Handbook of Metaphysics*, ed. Michael J. Loux, and Dean W. Zimmerman. Oxford: Oxford University Press.

Lovat, Toby. 2017. The Loss of the Great Outdoors: Neither Correlationist Gem nor Kantian Catastrophe. *Perspectives* 7 (1): 14–27.

Lovat, Toby. 2018. *Back to the Great Outdoors? A Kantian Reply to Meillassoux's Argument*, Phil. Diss, Brighton.

Lowe, E. Jonathan. 2005. Ontological Dependence. *Stanford Encyclopedia of Philosophy* (Summer 2005 Edition), ed. Edward N. Zalta.

Luty, Damian. 2024. Eliminative Ontic Structural Realism and the Metaphysical Underdetermination Argument: A Critique. *Studia Philosophiae Christianae* 60 (2): 199–229.

Luu, Thomas, and Ulf-G. Meißner. 2021. On the Topic of Emergence from an Effective Field Theory Perspective. In *Top-Down Causation and Emergence*, ed. Markus Gabriel, and Jan Voosholz, Synthese Library, 439, 101–114. Cham: Springer.

Lynch, Johanna M., Christopher Dowrick, Pamela Meredith, Sue L.T. McGregor, and Mieke Driel. 2021. Transdisciplinary Generalism: Naming the Epistemology and Philosophy of the Generalist. *Journal of Evaluation in Clinical Practice* 27 (3): 638–647.

Lyons, Timothy. 2002. Scientific Realism and the Pessimistic Meta-Modus Tollens. In *Recent Themes in the Philosophy of Science: Scientific Realism and Commonsense*, ed. Steve Clarke and Timothy Lyons, 63–90. Dordrecht: Springer.

Lyons, Timothy. 2013. A Historically Informed Modus Ponens Against Scientific Realism: Articulation, Critique, and Restoration. *International Studies in the Philosophy of Science* 27 (4): 369–392.

Lyons, Timothy. 2017. Epistemic Selectivity, Historical Threats, and The Non-epistemic Tenets of Scientific Realism. *Synthese* 194 (9): 3203–3219.

Lyons, Timothy. 2025. *Scientific Realism*. Cambridge: Cambridge University Press.

Lyons, Timothy, and Peter Vickers. (eds.). 2021. *Contemporary Scientific Realism: The Challenge from the History of Science*. Oxford: Oxford University Press.

Lyons, Timothy W., Christopher T. Reinhard, and Noah J. Planavsky. 2014. The Rise of Oxygen in Earth's Early Ocean and Atmosphere. *Nature* 506: 307–315.

Lyotard, Jean-François. 1984. *The Postmodern Condition. A Report on Knowledge*, ed. Wlad Godzich, and Jochen Schulte-Sasse, trans. by Geoff Bennington and Brian Massumi, Theory and History of Literature, vol. 10. Manchester: Manchester University Press.

Lyre, Holger. 2004. Lokale Symmetrien, Unterbestimmtheit und Strukturenrealismus. *Proceedings der GAP* 5: 97–107.

Maddy, Penelope. 1990. *Realism in Mathematics*. Oxford: Oxford University Press.

Maddy, Penelope. 1997. *Naturalism in Mathematics*. Oxford: Clarendon.

Maddy, Penelope. 2007. *Second Philosophy. A Naturalistic Method*. Oxford: Oxford University Press.

Maddy, Penelope. 2011. Naturalism, Transcendentalism, and Therapy. In *Transcendental Philosophy and Naturalism*, ed. Joel Smith and Peter Sullivan, 120–156. Oxford: Oxford University Press.

Malabou, Catherine. 2020. Le vide politique du réalisme contemporain. In *L'écho du réel*, ed. Cyril Crignon, Wilfried Laforge, and Pauline Nadrigny, 485–498. Sesto San Giovanni: Mimésis.

Marino, Patricia. 2006. What Should a Correspondence Theory be and Do? *Philosophical Studies* 127 (3): 415–457.

Markosian, Ned. 1993. How Fast Does Time Pass? *Philosophy and Phenomenological Research* 53: 829–844.

Markosian, Ned. 1994. The 3D/4D Controversy and Non-Present Objects. *Philosophical Papers* 23: 243–249.

Markosian, Ned. 2004. A Defense of Presentism. In *Oxford Studies in Metaphysics*, ed. Dean W. Zimmerman, vol. 1, 47–82. Oxford: Oxford University Press.

Markosian, Ned. 2012. The Truth About the Past and the Future. In *Around the Tree: Semantic and Metaphysical Issues Concerning Branching and the Open Future*, ed. Fabrice Correia and Andrea Iacona, 127–141. Dordrecht: Springer.

Martín Alcoff, Linda. 2005. Foucault's Philosophy of Science. Structures of Truth/Structures of Power. In *Continental Philosophy of Science*, ed. Gary Gutting, 211–223. Malden: Blackwell.

Massimi, Michela. 2012. Scientific Perspectivism and its Foes. *Philosophica* 84 (1): 25–52.

Massimi, Michela. 2018a. Four Kinds of Perspectival Truth. *Philosophy and Phenomenological Research* 96 (2): 342–359.

Massimi, Michela. 2018b. Perspectival Modeling. *Philosophy of Science* 85 (3): 335–359.

Massimi, Michela. 2019. Realism, Perspectivism, and Disagreement in Science. *Synthese* 198 (Suppl. 25): 6115–6141.

Massimi, Michela. 2022. *Perspectival Realism*. Oxford: Oxford University Press.

Massimi, Michela, and Casey D. Mccoy. 2019. *Understanding Perspectivism: Scientific Challenges and Methodological Prospects*. New York: Routledge.

Mathews, Grant J., Mayukh Gangopadhyay, Nishanth Sasankan, Kiyotomo Ichiki, and Toshitaka Kajino. 2018. Big Bang Nucleosynthesis, the CMB, and the Origin of Matter and Space-Time. *AIP Conference Proceedings* 1947 (1): 020014.

Matta, Corrado. 2014. Actor Network, Ontic Structural Realism and the Ontological Status of Actants. In *Proceedings of the Ninth International Conference on Networked Learning 2014*, ed. Sian Bayne, Chris Jones, Maarten de Laat, Thomas Ryberg, and Christine Sinclair, 195–202.
Maudlin, Tim. 2007. *The Metaphysics Within Physics*. Oxford: Oxford University Press.
Maudlin, Tim. 2014. What Bell Did. *Journal of Physics A: Mathematical and Theoretical* 47: 424010.
May, Gerhard. 1978. *Schöpfung aus dem Nichts. Die Entstehung der Lehre von der Creatio ex Nihilo*. Berlin: de Gruyter.
May, Gerhard. 2004. *Creatio Ex Nihilo. The Doctrine of 'Creation out of Nothing' in Early Christian Thought*, trans. by A. S. Worrall. London: Bloomsbury Publishing.
McAllister, James W. 2023. Empirical Tests of Scientific Realism: A Quantitative Framework. *Metaphilosophy* 54 (4): 507–522.
McDaniel, Kris. 2004. Modal Realism with Overlap. *Australasian Journal of Philosophy* 82 (1): 137.
McDowell, John. 1981. Anti-Realism and the Epistemology of Understanding. In *Meaning and Understanding*, ed. Herman Paret, and Jacques Bouveresse, 225–248. Berlin: de Gruyter.
McDowell, John. 1996. *Mind and World*, 2nd ed. Cambridge: Harvard University Press.
McDowell, John. 1998a. Knowledge and the Internal. In *Meaning, Knowledge and Reality*, ed. John McDowell, 395–413. Cambridge: Harvard University Press.
McDowell, John. 1998b. Non-Cognitivism and Rule-Following. In *Mind, Value, and Reality*, ed. by John McDowell, 198–218. Cambridge: Harvard University Press.
McDowell, John. 1998c. Singular Thought and the Extent of Inner Space. In *Meaning, Knowledge and Reality*, ed. John McDowell, 228–259. Cambridge: Harvard University Press.
McDowell, John. 1998d. Truth-Value Gaps. In *Meaning, Knowledge and Reality*, ed. John McDowell, 199–213. Cambridge: Harvard University Press.
McDowell, John. 1999. Sellars's Transcendental Empiricism. In *Rationality, Realism, Revision*, ed. Julian Nida-Rümelin, 24–51. Berlin: de Gruyter.
McDowell, John. 2009a. Avoiding the Myth of the Given. In *Having the World in View. Essays on Kant, Hegel, and Sellars*, ed. John McDowell, 256–272. Cambridge: Harvard University Press.
McDowell, John. 2009b. Intentionality as a Relation. In *Having the World in View. Essays on Kant, Hegel, and Sellars*, ed. John McDowell, 44–65. Cambridge: Harvard University Press.
McDowell, John. 2009c. Self-Determining Subjectivity and External Constraint. In *Having the World in View. Essays on Kant, Hegel, and Sellars*, ed. John McDowell, 90–107. Cambridge: Harvard University Press.
McDowell, John. 2009d. Sellars on Perceptual Experience. In *Having the World in View. Essays on Kant, Hegel, and Sellars*, ed. John McDowell, 3–22. Harvard University Press, Cambridge.
McDowell, John. 2009e. Sellars's Thomism. In *Having the World in View. Essays on Kant, Hegel, and Sellars*, ed. John McDowell, 239–255. Cambridge: Harvard University Press.
McDowell, John. 2009f. The Constitutive Ideal of Rationality: Davidson and Sellars. In *Having the World in View. Essays on Kant, Hegel, and Sellars*, ed. John McDowell, 207–220. Cambridge: Harvard University Press.
McDowell, John. 2009g. The Logical Form of an Intuition. In *Having the World in View. Essays on Kant, Hegel, and Sellars*, ed. John McDowell, 23–43. Cambridge: Harvard University Press.
McDowell, John. 2009h. Why Is Sellars's Essay Called "*Empiricism* and the Philosophy of Mind"? In *Empiricism, Perceptual Knowledge, Normativity, and Realism. Essays on Wilfrid Sellars*, ed. Willem A. deVries, 9–32. Oxford: Oxford University Press.
McDowell, John. 2009i. Wittgenstein's Quietism. *Common Knowledge* 15: 365–372.
McKinnon, Neil. 2013. Characterizing Presentism. In *New Papers on the Present*, ed. Roberto Ciuni, Kristie Miller, and Giuliano Torrengo, 13–30. München: Philosophia.
McLaughlin, Brian. 1995. Varieties of Supervenience. In *Supervenience: New Essays*, ed. Elias E. Savellos, and Ümit D. Yalçin, 16–59. Cambridge: Cambridge University Press.
McTaggart, J. Ellis. 1908. The Unreality of Time. *Mind* 17 (68): 457–474.

Meillassoux, Quentin. 2006. *Après la finitude. Essai sur la nécessitéde la contingence*. Paris: du Seuil.

Meillassoux, Quentin. 2007. Potentiality and Virtuality. *Collapse* II: 55–81.

Meillassoux, Quentin. 2008. *After Finitude. An Essay on the Necessity of Contingency*. London: continuum.

Meillassoux, Quentin. 2011a. Excerpts from *L'inexistence divine*. In *Quentin Meillassoux. Philosophy in the Making*, ed. Graham Harman. Edinburgh: Edinburgh University Press.

Meillassoux, Quentin. 2011b. Metaphysics, Speculation, Correlation. *Pli* 22: 3–25.

Meillassoux, Quentin. 2012. *The Number and the Siren. A Decipherment of Mallarmé's* coup de dés, trans. by Robin Mackay. Falmouth: Urbanomic and Sequence Press.

Meillassoux, Quentin. 2013a. Metaphysik, Spekulation, Korrelation. In *Realismus Jetzt*, ed. Armen Avanessian, 23–56. Berlin: Merve.

Meillassoux, Quentin. 2013b. Question canonique et facticité. In *Pourquoi y a-t-il quelque chose plutôt que rien?*, ed. Francis Wolff, 137–156. Paris: Presses Universitaires de France.

Meillassoux, Quentin. 2014. Time Without Becoming. In *Time Without Becoming*, ed. Anna Longo, 9–29. Mailand: Mimesis International.

Meillassoux, Quentin. 2016. Iteration, Reiteration, Repetition. A Speculative Analysis of the Sign Devoid of Meaning. In *Genealogies of Speculation. Materialism and Subjectivity Since Structuralism*, ed. Armen Avanessian and Suhail Malik, 117–198. London: Bloomsbury.

Meillassoux, Quentin, Ray Brassier, Iain Hamilton Grant, and Graham Harman. 2007. Speculative Realism. *Collapse* 3: 306–449.

Melia, Joseph. 1992. A Note on Lewis's Ontology. *Analysis* 52: 191–192.

Melia, Joseph. 2008. Ersatz Possible Worlds. In *Contemporary Debates in Metaphysics*, ed. Theodore Sider, John Hawthorne, and Dean W. Zimmerman, 135–151. Oxford: Blackwell.

Mellor, D.H. 1997. Properties and Predicates. In *Properties*, ed. D.H. Mellor and Alex Oliver, 255–267. Oxford: Oxford University Press.

Mellor, D.H. 2012. Nature's Joints: A Realistic Defence of Natural Properties. *Ratio* 25: 387–404.

Merricks, Trenton. 1995. On the Incompatibility of Enduring and Perduring Entities. *Mind* 104 (415): 523–531.

Merricks, Trenton. 1998. Against the Doctrine of Microphysical Supervenience. *Mind* 107 (425): 59–71.

Merritt, David. 2021. Cosmological Realism. *Studies in History and Philosophy of Science Part A* 88 (C): 193–208.

Metaxopoulos, Emilio. 1986. A Critical Consideration of the Lakatosian Concepts: "Mature" and "Immature" Science. In *Imre Lakatos and Theories of Scientific Change*, ed. Kostas Gavroglu, Yorgos Goudaroulis, and Pantelis Nicolacopoulos, 203–214. Dordrecht: Kluwer.

Mets, Ave, Endla Lõhkivi, Peeter Müürsepp, and Jaana Eigi-Watkin (eds.). 2024. *Practical Realist Philosophy of Science: Reflecting on Rein Vihalemm's Ideas*. Lanham: Lexington Books.

Meyerson, Ben. 2023. Clearing Up Correlationism. *Philosophy Today* 67 (3): 605–622.

Michon, Cyrille. 1994. *Nominalisme: La théorie de la signification d'Occam*. Paris: Vrin.

Midgley, Mary. 1992. *Science as Salvation. A Modern Myth and Its Meaning*. London: Routledge.

Mikkola, Mari. 2006. Elizabeth Spelman, Gender Realism, and Women. *Hypathia* 21 (4): 77–96.

Mill, John Stuart. 1867–1900. *A System of Logic, Ratiocinative and Inductive, Being a Connected View of the Principles of Evidence and the Methods of Scientific Investigation*. New York: Longmans, Green, and Co.

Miller, Kristie. 2013. Presentism, Eternalism and the Growing Block. In *A Companion to the Philosophy of Time*, ed. Heather Dyke, and Adrian Bardon, 345–364. Malden: Wiley-Blackwell.

Miller, Kristie. 2018. The New Growing Block Theory vs Presentism. *Inquiry* 61 (3): 223–251.

Miller, Alexander. 2019. Realism. *Stanford Encyclopedia of Philosophy* (Winter 2016 Edition, rev. Winter 2019), ed. Edward N. Zalta.

Millikan, Ruth G. 1984. *Language, Thought, and Other Biological Categories. New Foundations for Realism*. Cambridge: MIT Press.

Mizrahi, Moti. 2013a. The Argument from Underconsideration and Relative Realism. *International Studies in the Philosophy of Science* 27 (4): 393–407.

Mizrahi, Moti. 2013b. The Pessimistic Induction: A Bad Argument Gone too Far. *Synthese* 190 (15): 3209–3226.

Mizrahi, Moti. 2015. Historical Inductions: New Cherries, Same Old Cherry-picking. *International Studies in the Philosophy of Science* 29 (2): 129–148.

Mizrahi, Moti. 2016. The History of Science as a Graveyard of Theories: A Philosophers' Myth? *International Studies in the Philosophy of Science* 30 (3): 263–278.

Mizrahi, Moti. 2018. The "Positive Argument" for Constructive Empiricism and Inference to the Best Explanation. *Journal for General Philosophy of Science* 49 (3): 461–466.

Mizrahi, Moti. 2020. *The Relativity of Theory: Key Positions and Arguments in the Contemporary Scientific Realism/Antirealism Debate*, Synthese Library, 431. Cham: Springer.

Mizrahi, Moti. 2021. In Defense of Relative Realism: A Reply to Park. *Social Epistemology Review and Reply Collective* 10 (1): 1–6.

Monton, Bradley. 2006. Presentism and Quantum Gravity. In *The Ontology of Spacetime*, ed. Dennis Dieks, vol. 1, 263–280. Amsterdam: Elsevier.

Moore, George E. 1939. Proof of an External World. *Proceedings of the British Academy* 25: 273–300.

Moore, Adrian W. 1997. *Points of View*. Oxford: Oxford University Press.

Moore, Adrian W. 2019. *The Infinite*, 3rd ed. London: Routledge.

Morgenthau, Hans J. 1954 *Politics among Nations: The Struggle for Power and Peace*, 2nd ed. New York: Alfred A. Knopf.

Morris, Kevin. 2018. What's Wrong with Brute Supervenience? A Defense of Horgan on Physicalism and Superdupervenience. *Analytic Philosophy* 59 (2): 256–280.

Morrison, Margaret. 1994. Causes and Contexts: The Foundations of Laser Theory. *British Journal for the Philosophy of Science* 45 (1): 127–151.

Moshe, Josef. 2013. The Night in which all Dinosaurs wear Nightcaps. A supplement to Žižek's critique of Meillassoux. *International Journal of Žižek Studies* 7 (3): S. 1–22.

Moss, Sarah. 2012. Four-Dimensionalist Theories of Persistence. *Australasian Journal of Philosophy* 90 (4): 671–686.

Moss, Gregory S. 2024. Absolute Dialetheism. In *Markus Gabriel's New Realism*, ed. Jan Voosholz, Synthese Library, 492, 75–94. Cham: Springer.

Moulin, Isabelle. 2016. *Theologie et Philosophie chez Jean Scot Érigène*. Paris: Vrin.

Mozersky, M. Joshua. 2013. The B-Theory in the Twentieth Century. In *A Companion to the Philosophy of Time*, ed. Heather Dyke, and Adrian Bardon, 167–182. Malden: Wiley-Blackwell.

Murris, Karin, and Vivienne Bozalek (eds.). 2023. *In Conversation with Karen Barad: Doings of Agential Realism*. New York: Routledge, Taylor & Francis Group.

Myrvold, Wayne, Marco Genovese, and Abner Shimony. 2019. Bell's Theorem. *Stanford Encyclopedia of Philosophy* (Spring 2019 Edition), ed. Edward N. Zalta.

Nagel, Ernest. 1961. *The Structure of Science. Problems in the Logic of Scientific Explanation*. London: Routledge.

Nagel, Thomas. 1986. *The View from Nowhere*. Oxford: Oxford University Press.

Neurath, Otto. 1987. United Science and Psychology. In *Unified Science*, ed. Brian F. McGuiness, 1–23. Dordrecht: Reidel.

Newman, Mark. 2005. Ramsey Sentence Realism as an Answer to the Pessimistic Meta-Induction. *Philosophy of Science* 72 (5): 1373–1384.

Ney, Alyssa. 2023. Three Arguments for Wave Function Realism. *European Journal for Philosophy of Science* 13 (4): 1–18.

Nietzsche, Friedrich. 1966. *Beyond Good and Evil. Prelude to a Philosophy of the Future*, trans. by Walter Kaufmann. New York: Vintage Books.

Nietzsche, Friedrich. 1978. *Thus Spoke Zarathustra. A Book for None and All*, trans. by Walter Kaufmann. New York: Penguin Books.

Nietzsche, Friedrich. 1998. *On the Genealogy of Morality. A Polemic*, trans. by Maudemarie Clark, and Alan J. Swensen. Indianapolis: Hackett Publishing.
Nietzsche, Friedrich. 2001. *The Gay Science*, ed. Bernard Williams, trans. by Josefine Nauckhoff, and Adrian Del Caro (poems). Cambridge: Cambridge University Press.
Niiniluoto, Ilkka. 2014. Scientific Progress as Increasing Verisimilitude. *Studies in History and Philosophy of Science Part A* 46: 73–77.
Noble, Denis. 2012. A Theory of Biological Relativity: No Privileged Level of Causation. *Interface Focus* 2: 55–64.
Noble, Denis. 2016. *Dance to the Tune of Life. Biological Relativity*. Cambridge: Cambridge University Press.
Noble, Denis. 2021. The Principle of Biological Relativity: Origins and Current Status. In *Top-Down Causation and Emergence*, ed. Jan Voosholz, and Markus Gabriel, Synthese Library, 439, 117–133. Cham: Springer.
Novak, Kyle. 2020. Deleuze's Transcendental Empiricism Against Speculative Realism: How Deleuze's Hume Avoids the Challenge of Correlationism. *Journal of Speculative Philosophy* 34 (3): 297–308.
Oaklander, L. Nathan. 2004. *The Ontology of Time*. Amherst: Prometheus.
Oaklander, L. Nathan. 2012. A-, B-, and R-Theories of Time. A Debate. In *The Future of the Philosophy of Time*, ed. Adrian Bardon, 1–24. New York: Routledge.
O'Callaghan, John P. 2003. *Thomist Realism and the Linguistic Turn: Toward a More Perfect Form of Existence*. Notre Dame: University of Notre Dame Press.
Ochiai, Hirofumi. 2024. Functional Realism Suggested From the Actualization of Affordances. *Foundations of Chemistry* 26 (3): 399–411.
Olsen, Wendy (ed.). 2010. *Realist Methodology. Bd. II. Realist Methods*. London: Sage.
Oppenheim, Paul, and Hilary Putnam. 1958. Unity of Science as a Working Hypothesis. *Minnesota Studies in the Philosophy of Science* 2: 3–35.
Orensanz, Martin. 2020. Strange Forms of Argumentation: On Meillassoux's Definition of Philosophy. *Cosmos and History* 16 (1): 271–297.
O'Shea, James R. 2009. On the Structure of Sellars's Naturalism with a Normative Turn. In *Empiricism, Perceptual Knowledge, Normativity, and Realism. Essays on Wilfrid Sellars*, ed. Willem A. deVries, 187–210. Oxford: Oxford University Press.
Pachilla, Pablo. 2022. The Eye is in Things: On Deleuze and Speculative Realism. *Comparative and Continental Philosophy* 14 (1): 44–56.
Padui, Raoni. 2011. Realism, Anti-Realism, and Materialism. Rereading the Critical Turn after Meillassoux. *Angelaki* 16 (2): 89–101.
Panaccio, Claude. 2004. *Ockham on Concepts*. London: Routledge.
Panaccio, Claude. 2015. Ockham's Externalism. In *Intentionality, Cognition and Mental Representation in Medieval Philosophy*, ed. Gyula Klima, 166–185. New York: Fordham University Press.
Panagiotatou, Maria, and Stathis Psillos. 2023. Review of Perspectival Realism. *Notre Dame Philosophical Reviews*.
Papineau, David. 2020. Naturalism. *Stanford Encyclopedia of Philosophy* (Summer 2020 Edition), ed. Edward N. Zalta.
Park, Seungbae. 2022a. *Embracing Scientific Realism*, Synthese Library, 445. Cham: Springer.
Park, Seungbae. 2022b. The Realist and Selectionist Explanations for the Success of Science. *Synthese* 200 (3): 1–12.
Parsons, Charles. 2004. Structuralism and Metaphysics. *Philosophical Quarterly* 54 (214): 55–77.
Paul, L.A. 2010. Temporal Experience. *Journal of Philosophy* 107: 333–359.
Paull, R. Cranston, and Theodore Sider. 1992. In Defense of Global Supervenience. *Philosophy and Phenomenological Research* 52 (4): 833–854.
Pelletier, Jenny. 2013. *William of Ockham on Metaphysics. The Science of Being and God*. Leiden: Brill.

Penner, Myron A. 2021. Selective Scientific Realism and Truth-Transfer in Theories of Molecular Structure. In *Contemporary Scientific Realism: The Challenge From the History of Science*, ed. Timothy Lyons, and Peter Vickers, 130–158. Oxford: Oxford University Press.

Perler, Dominik. 2009. Radical Doubts. Theories of Divine Omnipotence and the Emergence of Scepticism in Medieval Philosophy. In *Religio Academici: Essays on Scepticism, Religion and the Pursuit of Knowledge*, ed. Péter Losonczi, 84–107. Budapest: Akadémiai Kiadó.

Pier, Jens (ed.). 2023. *Limits of Intelligibility: Issues From Kant and Wittgenstein*. London: Routledge.

Pierson, Robert, and Richard Reiner. 2008. Explanatory Warrant for Scientific Realism. *Synthese* 161 (2): 271–282.

Pils, Raimund. 2025. The Scientific Realism Debate and Consensus Reporting. *Synthese* 205 (3): 1–23.

Pils, Raimund, and Philipp Schoenegger. 2024. Scientific Realism, Scientific Practice, and Science Communication: An Empirical Investigation of Academics and Science Communicators. *Studies in History and Philosophy of Science Part A* 105 (C): 85–98.

Pincock, Christopher. 2018. Accommodating Explanatory Pluralism. In *Explanation Beyond Causation. Philosophical Perspectives on Non-Causal Explanations*, ed. Alexander Reutlinger, and Juha Saatsi, 39–56. Oxford: Oxford University Press.

Pippin, Robert. 1989. *Hegel's Idealism: The Satisfaction of Self-Consciousness*. Cambridge: Cambridge University Press.

Pippin, Robert. 2018. *Hegel's Realm of Shadows. Logic as Metaphysics in The Science of Logic*. Chicago: University of Chicago Press.

Pitman, Michael M. 2003. Eliminative Materialism and the Integrity of Science. *South African Journal of Philosophy* 22 (3): 207–219.

Planck Collaboration, N. Aghanim, Y. Akrami, M. Ashdown, J. Aumont, C. Baccigalupi, M. Ballardini, A.J. Banday, R.B. Barreiro, N. Bartolo, et al. 2020. Planck 2018 Results – VI. Cosmological Parameters. *Astronomy & Astrophysics* 641: A6.

Plato. 1997a. Phaedrus. In *Plato: Complete Works*, ed. J.M. Cooper and D.S. Hutchinson, trans. by Alexander Nehamas and Paul Woodruff, 506–556. Indianapolis: Hackett Publishing.

Plato. 1997b. *Plato: Complete Works*, ed. John M. Cooper. Indianapolis: Hackett.

Plato. 1997c. Republic. In *Plato: Complete Works*, ed. J.M. Cooper, and D.S. Hutchinson, trans. by G.M.A. Grube and rev. C.D.C. Reeve, 971–1223. Indianapolis: Hackett Publishing.

Plato. 1997d. Timaeus. In *Plato: Complete Works*, ed. J.M. Cooper, and D.S. Hutchinson, trans. by Donald J. Zeyl, 1224–1291. Indianapolis: Hackett Publishing.

Poincaré, Henri. 2018. *Science and Hypothesis. The Complete Text*, ed. Mélanie Frappier, and David J. Stump. London: Bloomsbury.

Pooley, Oliver. 2013. Substantivalist and Relationalist Approaches to Spacetime. In *The Oxford Handbook of Philosophy of Physics*, ed. Robert Batterman, 522–586. Oxford: Oxford University Press.

Popper, Karl R. 1984. *Logik der Forschung*, 8th ed. Tübingen: Mohr Siebeck.

Popper, Karl R. 2002. *The Logic of Scientific Discovery*, 2nd ed. London: Routledge.

Poslajko, Krzysztof. 2019. Eliminativism: The Problem of Representation and Carnapian Metametaphysics. *Acta Analytica* 34: 181–195.

Potochnik, Angela. 2022. Truth and Reality: How to Be a Scientific Realist Without Believing Scientific Theories Should Be True. In *Scientific Understanding and Representation: Modeling in the Physical Sciences*, ed. Insa Lawler, Kareem Khalifa, and Elay Shech. New York: Routledge.

Price, Huw. 1996. *Time's Arrow & Archemedes' Point. New Directions for the Physics of Time*. Oxford: Oxford University Press.

Price, Huw. 1997. Carnap, Quine, and the Fate of Metaphysics. *Electronic Journal of Analytic Philosophy* 5 (1): 1–22.

Price, Huw. 2004. Naturalism Without Representationalism. In *Naturalism in Question*, ed. Mario De Caro, and David Macarthur, 3–21. Cambridge: Harvard University Press.

Price, Huw. 2007. Quining Naturalism. *Journal of Philosophy* 104 (8): 375–402.

Price, Huw. 2011. *Naturalism Without Mirrors*. Oxford: Oxford University Press.

Price, Huw. 2013. *Expressivism, Pragmatism, and Representationalism*. Cambridge: Cambridge University Press.

Price, Huw. 2019. Global Expressivism by the Method of Differences. *Royal Institute of Philosophy Supplement* 86: 133–154.

Priest, Graham. 1995. *Beyond the Limits of Thought*. Cambridge: Cambridge University Press.

Priest, Graham. 1998. What's so Bad about Contradictions? *Journal of Philosophy* 95: 410–426.

Priest, Graham. 2000a. Could Everything Be True? *Australasian Journal of Philosophy* 78: 189–195.

Priest, Graham. 2000b. Truth and Contradiction. *Philosophical Quarterly* 50: 305–319.

Priest, Graham. 2005. *Towards Non-Being: The Logic and Metaphysics of Intentionality*. Oxford: Oxford University Press.

Priest, Graham. 2014. *One: Being an Investigation into the Unity of Reality and of its Parts, including the Singular Object which is Nothingness*. Oxford: Oxford University Press.

Priest, Graham. 2023. On Transcending the Limits of Language. In *Limits of Intelligibility: Issues from Kant and Wittgenstein*, ed. Jens Pier, 219–137. London: Routledge.

Priest, Graham, and Markus Gabriel. 2022. *Everything and Nothing*. Cambridge: Polity.

Prior, A.N. 1957. *Time and Modality*. Oxford: Clarendon Press.

Prior, A.N. 1967. *Past, Present and Future*. Oxford: Clarendon Press.

Prior, A.N. 1968. *Papers on Time and Tense*. Oxford: Clarendon Press.

Prior, A.N. 1972. The Notion of the Present. In *The Study of Time*, ed. J.T. Fraser, F.C. Haber, and G.H. Müller, 320–323. Belin: Springer.

Prior, A.N. 1996. Some Free Thinking about Time (1958) and A Statement of Temporal Realism. In *Logic and Reality. Essays on the Legacy of Arthur Prior*, ed. Jack Copeland, 43–51. Oxford: Clarendon Press.

Psillos, Stathis. 1999. *Scientific Realism*. How science tracks truth. London: Routledge.

Psillos, Stathis. 2006. The Structure, the Whole Structure and Nothing but the Structure? *Philosophy of Science* 73 (5): 560–570.

Psillos, Stathis. 2011. Living with the Abstract: Realism and Models. *Synthese* 180 (1): 3–17.

Pucciarelli, Daniel. 2022. Meaning and Fate of Critique in the Ontological Turn. *Trans/Form/Ação* 45 (1): 95–114.

Putnam, Hilary. 1971. *Philosophy of Logic*. London: Allen & Unwin.

Putnam, Hilary. 1975. Time and Physical Geometry. *Journal of Philosophy* 64 (8): 240–247.

Putnam, Hilary. 1978. *Meaning and the Moral Sciences*. London: Routledge and Kegan Paul.

Putnam, Hilary. 1979. *Mathematics, Matter and Method. Philosophical Papers. I/II*, 2nd ed. Cambridge: Cambridge University Press.

Putnam, Hilary. 1981. *Reason, Truth and History*. Cambridge: Cambridge University Press.

Putnam, Hilary. 1990. *Realism with a Human Face*. Cambridge: Harvard University Press.

Putnam, Hilary. 1994a. Sense, Nonsense, and the Senses: An Inquiry into the Powers of the Human Mind. *Journal of Philosophy* 91 (9): 445–517.

Putnam, Hilary. 1994b. *Words and Life*, ed. James Conant. Cambridge: Harvard University Press.

Putnam, Hilary. 2004. The Content and Appeal of "Naturalism". In *Naturalism in Question*, ed. Mario De Caro, and David Macarthur, 59–70. Cambridge: Harvard University Press.

Quine, Willard Van Orman. 1950. Identity, Ostension, and Hypostasis. *Journal of Philosophy* 47 (22): 621–633.

Quine, Willard Van Orman. 1951. Two Dogmas of Empiricism. *Philosophical Review* 60 (1): 20–43.

Quine, Willard Van Orman. 1953. On What There Is. In *From a Logical Point of View*, ed. Willard Van Orman Quine, 1–19. Cambridge: Harvard University Press.

Quine, Willard Van Orman. 1960. *Word and Object*. Cambridge: MIT Press.

Quine, Willard Van Orman. 1966. Carnap and Logical Truth. In *The Ways of Paradox and Other Essays*, ed. Willard Van Orman Quine, 100–125. New York: Random House.

Quine, Willard Van Orman. 1969a. Epistemology Naturalized. In *Ontological Relativity and Other Essays*, ed. Willard Van Orman Quine, 69–90. New York: Columbia University Press.

Quine, Willard Van Orman. 1969b. Existence and Quantification. In *Ontological Relativity and Other Essays*, ed. Willard Van Orman Quine, 91–113. New York: Columbia University Press.
Quine, Willard Van Orman. 1969c. Ontological Relativity. In *Ontological Relativity and Other Essays*, ed. Willard Van Orman Quine, 26–68. New York: Columbia University Press.
Quine, Willard Van Orman. 1975. On Empirically Equivalent Systems of the World. *Erkenntnis* 9 (3): 313–328.
Quine, Willard Van Orman. 1981a. Five Milestones of Empiricism. In *Theories and Things*, ed. Willard Van Orman Quine, 67–72. Cambridge: Harvard University Press.
Quine, Willard Van Orman. 1981b. Things and Their Place in Theories. In *Theories and Things*, ed. Willard Van Orman Quine, 1–23. Cambridge: Harvard University Press.
Resnik, Michael D. 1985. How Nominalist is Hartry Field's Nominalism? *Philosophical Studies* 47: 163–181.
Resnik, Michael D. 1997. *Mathematics as a Science of Patterns*. Oxford: Oxford University Press.
Rheinberger, Hans-Jörg. 1991. Experiment, Difference, and Writing: I. Tracing Protein Synthesis. *Studies in History and Philosophy of Science Part A* 23 (2): 305–331.
Rheinberger, Hans-Jörg. 1997. *Toward a History of Epistemic Things: Synthesizing Proteins in the Test Tube*. Redwood City: Stanford University Press.
Rheinberger, Hans-Jörg. 2001. *Experimentalsysteme und epistemische Dinge. Eine Geschichte der Proteinsynthese im Reagenzglas*. Göttingen: Wallstein.
Rheinberger, Hans-Jörg. 2011. Infra-Experimentality: From Traces to Data, From Data to Patterning Facts. *History of Science* 49 (3): 337–348.
Rheinberger, Hans-Jörg. 2015. Preparations, Models, and Simulations. *History and Philosophy of the Life Sciences* 36 (3): 321–334.
Richardson, Alan. 1997. Two Dogmas about Logical Empiricism: Carnap and Quine on Logic, Epistemology, and Empiricism. *Philosophical Topics* 25 (2): 145–168.
Riess, Adam G., Stefano Casertano, Wenlong Yuan, Lucas M. Macri, and Dan Scolnic. 2019. Large Magellanic Cloud Cepheid Standards Provide a 1% Foundation for the Determination of the Hubble Constant and Stronger Evidence for Physics beyond CDM. *The Astrophysical Journal* 876 (1): 85.
Riess, Adam G., Gagandeep S. Anand, Wenlong Yuan, Lucas M. Macri, Stefano Casertano, Andrew Dolphin, Louise Breuval, Dan Scolnic, Marshall Perrin, and Richard I. Anderson. 2024. *JWST Observations Reject Unrecognized Crowding of Cepheid Photometry as an Explanation for the Hubble Tension at 8 sigma Confidence*, arXiv: 2401 .04773 [astro-ph.CO], https://arxiv.org/abs/2401.04773.
Robin, E.J. 2021. Kant Walks Meillassoux: Finitude and Correlationism. *Journal of the Indian Council of Philosophical Research* 38 (2): 197–211.
Roche, Michael. 2016. Physicalism and Supervenience: A Case for a New Sense of Physical Duplication. *Erkenntnis* 81 (4): 669–681.
Rock, Tina. 2021. *Dynamic Realism: Uncovering the Reality of Becoming Through Phenomenology and Process Philosophy*. Edinburgh: Edinburgh University Press.
Rodriguez-Pereyra, Gonzalo. 2002. *Resemblance Nominalism. A Solution to the Problem of Universals*. Oxford: Oxford University Press.
Romano, Claude. 2023. Idealism/Realism: A Metaphysical Distinction? *Phainomenon* 36 (1): 95–110.
Rometsch, Jens. 2018. *Freiheit zur Wahrheit. Grundlagen der Erkenntnis bei Descartes und Locke*. Frankfurt (Main): Klostermann.
Rometsch, Jens. 2019. What is a Realistic Concept of Reality? In *New Realism. Problems and Perspectives*, ed. Alexander Kanev, 109–119. Sofia: St. Kliment Ohridski University Press.
Rosen, Gideon. 1990. Modal Fictionalism. *Mind* 99 (395): 327–354.
Rosen, Gideon. 1993. A Problem for Fictionalism about Possible Worlds. *Analysis* 53 (2): 71–81.
Rosen, Gideon. 1995. Modal Fictionalism Fixed. *Analysis* 55 (2): 67–73.
Rosen, Gideon. 2002. A Study in Modal Deviance. In *Conceivability and Possibility*, ed. John Hawthorne and Tamar Szabó Gendler, 283–307. Oxford: Oxford University Press.
Rosenberg, Jay F. 2007. *Fusing the Images*. Oxford: Oxford University Press.

Rosenberg, Jay F. 2009. On Sellars's Two Images of the World. In *Empiricism, Perceptual Knowledge, Normativity, and Realism. Essays on Wilfrid Sellars*, ed. Willem A. deVries, 283–296. Oxford: Oxford University Press.
Rouse, Joseph. 1987. *Knowledge and Power. Toward a Political Philosophy of Science*. Ithaca: Cornell University Press.
Rouse, Joseph. 2002. *How Scientific Practices Matter. Reclaiming Philosophical Naturalism*. Chicago: University of Chicago Press.
Rouse, Joseph. 2015. *Articulating the World. Conceptual Understanding and the Scientific Image*. Chicago: University of Chicago Press.
Rouse, Joseph. 2016. Toward a New Naturalism: Niche Construction, Conceptual Normativity and Scientific Practice. In *Normativity and Naturalism in the Philosophy of the Social Sciences*, ed. Mark Risjord, 28–42. New York: Routledge.
Rouse, Joseph. 2018. Beyond Realism and Anti-Realism—At Last? *Spontaneous Generations* 9 (1): 46–51.
Roush, Sherrilyn. 2010. Optimism About the Pessimistic Induction. In *New Waves in Philosophy of Science*, ed. P.D. Magnus, and Jacob Busch, 29–58. London: Palgrave-Macmillan.
Routley (Sylvan), Richard. 1979. Dialectical Logic, Semantics and Metamathematics. *Erkenntnis* 14: 301–331.
Routley (Sylvan), Richard. 1980. *Exploring Meinong's Jungle and Beyond*. Canberra: Australian National University.
Rovane, Carol. 2013. *The Metaphysics and Ethics of Relativism*. Cambridge: Harvard University Press.
Rowbottom, Darrell P. 2019. *The Instrument of Science. Scientific Anti-Realism Revitalized*. London: Routledge.
Roy, Tony. 1993. Worlds and Modality. *Philosophical Review* 102 (3): 335–361.
Ruphy, Stéphanie. 2016. *Scientific Pluralism Reconsidered. A New Apprpach to the (Dis)Unity of Science*. Pittsburgh: University of Pittsburgh Press.
Russell, Bertrand. 1938. *The Principles of Mathematics*, 2nd ed. New York: W. W. Norton & Co.
Russell, Bertrand. 1976. *Human Knowledge. Its scope and limits*, 6th ed. London: Allen & Unwin.
Rutten, Emanuel. 2024. Prison Break? In Defense of Correlationism. *Atlantika* 2 (1): 1–22.
Ruyant, Quentin. 2021. *Modal Empiricism: Interpreting Science Without Scientific Realism*, Synthese Library, 440. Cham: Springer.
Saatsi, Juha. 2009. Grasping at Realist Straws. *Metascience* 18 (3): 355–390.
Saatsi, Juha. 2018a. Realism and the Limits of Explanatory Reasoning. In *The Routledge Handbook of Scientific Realism*, ed. dems, 200–211. London: Routledge.
Saatsi, Juha (ed.). 2018b. *The Routledge Handbook of Scientific Realism*. London: Routledge.
Saatsi, Juha. 2019a. Historical Inductions, Old and New. *Synthese* 196 (4): 3979–3993.
Saatsi, Juha. 2019b. What Is Theoretical Progress of Science. *Synthese* 196 (2): 611–631.
Sacilotto, Daniel. 2013. Realism and Representation. On the Ontological Turn. *Speculations* 4: 53–62.
Saldanha, Arun. 2009. Back to the Great Outdoors. Speculative Realism as Philosophy of Science. *Cosmos and History: The Journal of Natural and Social Philosophy* 5 (2): 304–321.
Sangeetha, K.S. 2024. A Historical Perspective in Support of Direct Realism. *Topoi* 43 (1): 115–125.
Sankey, Howard. 2016. The Demise of the Incommensurability Thesis. In *The Kuhnian Image of Science: Time for a Decisive Transformation?*, ed. Moti Mizrahi, 75–91. London: Rowman and Littlefield.
Savitt, Steven F. 2006. Presentism and Eternalism in Perspective. In *The Ontology of Spacetime*, ed. Dennis Dieks, vol. 1, 111–128. Amsterdam: Elsevier.
Schäferling, Kristian. 2022. Meillassoux's Reinterpretation of Kant's Transcendental Dialectic. *Open Philosophy* 5 (1): 702–717.
Schaffer, Jonathan. 2010a. Monism: The Priority of the Whole. *Philosophical Review* 87 (1): 67–87.
Schaffer, Jonathan. 2010b. The Internal Relatedness of All Things. *Mind* 119 (474): 341–376.

Schaffer, Jonathan. 2013. The Action of the Whole. *Proceedings of the Aristotelian Society* 180 (1): 3–17.

Schaffer, Jonathan. 2017. The Ground Between the Gaps. *Philosophers' Imprint* 17 (11): 1–26.

Schelling, Friedrich W.J. 2000. *Von der Weltseele. Eine Hypothese der höhern Physik zur Erklärung des allgemeinen Organismus*, ed. Jörg Jantzen, F.W.J. Schelling. Historisch-kritische Ausgabe, hrsg. v. Jörg Jantzen et al., Bd. 6. Stuttgart: Frommann-Holzboog.

Schelling, Friedrich W.J. 2004. *First Outline of a System of the Philosophy of Nature*, ed. Dennis J. Schmidt, trans. by Keith R. Peterson. Albany: State University of New York Press.

Schelling, Friedrich W.J. 2012. On the World-Soul. Extract. trans. by Iain Hamilton Grant. *Collapse* 6: 58–95.

Schlesinger, George. 1980. *Aspects of Time*. Indianapolis: Hackett.

Schlesinger, George. 1982. How Time Flies. *Mind* 91 (364): 501–523.

Schmitt, Dorothee. 2018. *Das Selbstaufhebungsargument. Der Relativismus in der gegenwärtigen philosophischen Debatte*. Berlin: de Gruyter.

Schmitz, Barbara. 2010. Geschaffen aus dem Nichts? Die Funktion der Rede von der Schöpfung im Zweiten Makkabäerbuch. In *Theology of Creation in Early Judaism and Ancient Christianity. In Honour of Hans Klein*, ed. Tobias Nicklas and Korinna Zamfir, 61–79. Berlin: de Gruyter.

Schopenhauer, Arthur. 2010. *The World as Will and Representation, Volume I*, ed. and trans. by Judith Norman, Alistair Welchman, and Christoper Janaway, The Cambridge Edition of the Works of Schopenhauer. Cambridge: Cambridge University Press.

Schopenhauer, Arthur. 2018. *The World as Will and Representation, Volume II*, ed. and trans. by Judith Norman, Alistair Welchman, and Christoper Janaway, The Cambridge Edition of the Works of Schopenhauer. Cambridge: Cambridge University Press.

Schrenk, Markus. 2017. *Metaphysics of Science. A Systematic and Historical Introduction*, London: Routledge.

Schroeren, David. 2022. Wavefunction Realism Does Not 'Privilege Position'. *Synthese* 200 (1): 1–14.

Schulte, Peter. 2010. Plädoyer für einen physikalischen Naturalismus. *Zeitschrift für philosophische Forschung* 64 (2): 165–189.

Schultz, Michael. 2018. The Problem of Underdetermination in Model Selection. *Sociological Methodology* 48 (1): 52–87.

Schweppenhäuser, Gerhard. 2018. *Revisionen des Realismus. Zwischen Sozialporträt und Profilbild*. Stuttgart: Metzler.

Scott, Douglas. 2006. The Standard Cosmological Model. *Canadian Journal of Physics* 84 (6–7): 419–435.

Searle, John. 1995. *The Construction of Social Reality*. London: Penguin.

Secrest, Nathan J., Sebastian von Hausegger, Mohamed Rameez, Roya Mohayaee, Subir Sarkar, and Jacques Colin. 2021. A Test of the Cosmological Principle with Quasars. *The Astrophysical Journal Letters* 908 (2): L51.

Sellars, Wilfrid. 1956. Empiricism and the Philosophy of Mind. *Minnesota Studies in the Philosophy of Science* 1: 253–329.

Sellars, Wilfrid. 1961. The Language of Theories. In *Current Issues in the Philosophy of Science*, ed. Herbert Feigl, and Grover Maxwell, 57–77. New York: Holt, Rinehart and Winston.

Sellars, Wilfrid. 1962. Philosophy and the Scientific Image of Man. In *Frontiers of Science and Philosophy*, ed. Robert G. Colodny, 35–78. Pittsburgh: University of Pittsburgh Press.

Sellars, Wilfrid. 1963. *Science, Perception and Reality*. London: Routledge & Kegan Paul.

Sellars, Wilfrid. 1965. Scientific Realism or Irenic Instrumentalism: A Critique of Nagel and Feyerabend on Theoretical Explanation. *Proceedings of the Boston Colloquium on Philosophy of Science* 2: 171–204.

Sellars, Wilfrid. 1968. *Science and Metaphysics: Variations on Kantian Themes. The John Locke Lectures for 1965–66*. London: Routledge & Kegan Paul.

Sextus Empiricus. 1994. *Outlines of Scepticism*, ed. Julia Annas and Jonathan Barnes. Cambridge: Cambridge University Press.

Shapiro, Stewart. 1997. *Philosophy of Mathematics: Structure and Ontology*. Oxford: Oxford University Press.

Shoemaker, Sydney. 2007. *Physical Realization*. Oxford: Oxford University Press.

Shook, John R. 2021. Abduction, the Logic of Scientific Creativity, and Scientific Realism. In *Abduction in Cognition and Action: Logical Reasoning, Scientific Inquiry, and Social Practice*, ed. John R. Shook, and Sami Paavola, 207–227. Berlin: Springer.

Sider, Theodore. 1997. Four-Dimensionalism. *Philosophical Review* 106: 197–231.

Sider, Theodore. 1999. Presentism and Ontological Commitment. *Journal of Philosophy* 96: 325–347.

Sider, Theodore. 2001. *Four-Dimensionalism. An Ontology of Persistence and Time*. Oxford: Clarendon Press.

Sider, Theodore. 2011. *Writing the Book of the World*. Oxford: Oxford University Press.

Sider, Theodore. 2020. *The Tools of Metaphysics and the Metaphysics of Science*. Oxford: Oxford University Press.

Sismondo, Sergio. 2004. *An Introduction to Science and Technology Studies*. Malden: Blackwell.

Sklar, Larry. 1975. Methodological Conservatism. *Philosophical Review* 84: 374–400.

Skow, Bradford. 2009. Relativity and the Moving Spotlight. *Journal of Philosophy* 106 (12): 666–678.

Skow, Bradford. 2011a. Experience and the Passage of Time. *Philosophical Perspectives* 25 (1): 359–387.

Skow, Bradford. 2011b. On the Meaning of the Question "How Fast Does Time Pass? *Philosophical Studies* 155 (3): 325–344.

Skow, Bradford. 2012a. One Second Per Second. *Philosophy and Phenomenological Research* 85 (2): 377–389.

Skow, Bradford. 2012b. Why Does Time Pass? *Nou^s* 46 (2): 223–242.

Skow, Bradford. 2015. *Objective Becoming*. Oxford: Oxford University Press.

Smart, J.J.C. 1949. The River of Time. *Mind* 58 (232): 483–494.

Smart, J.J.C. 1963a. Is Time Travel Possible? *Journal of Philosophy* 60 (9): 237–241.

Smart, J.J.C. 1963b. *Philosophy and Scientific Realism*. London: Routledge & Kegan Paul.

Smart, J.J.C. 1989. *Our Place in the Universe*. Oxford: Blackwell.

Smart, J.J.C. 2008. The Tenseless Theory of Time. In *Contemporary Debates in Meta-physics*, ed. Theodore Sider, John Hawthorne, and Dean W. Zimmerman, 226–238. Malden: Blackwell.

Solomon, Miriam. 2001. *Social Empiricism*. Cambridge: MIT Press.

Spade, Paul Vincent. 1999. Ockham's Nominalist Metaphysics: Some Main Themes. In *The Cambridge Companion to Ockham*, ed. Paul Vincent Spade, 100–117. Cambridge: Cambridge University Press.

Spehrs, Adriana. 2024. Dispositional Realism, Conflicting Models and Contrastive Explanation. *Journal for General Philosophy of Science / Zeitschrift für Allgemeine Wissenschaftstheorie* 55: 2.

Spiegel, Thomas Jussuf. 2017. *Quietism, Naturalism, and the End of Philosophy*, Phil. Diss. Leipzig: Universität Leipzig.

Spiegel, Thomas Jussuf. 2020. Ist der Naturalismus eine Ideologie? *Deutsche Zeitschrift für Philosophie* 68 (1): 51–71.

Spiegel, Thomas Jussuf. 2021. *Naturalism, Quietism, and the Threat to Philosophy*. Basel: Schwabe.

Srinivasan, Amia. 2016. Philosophy and Ideology. *Theoria* 31 (3): 371–380.

Stachel, John. 2006. Structure, Individuality and Quantum Gravity. In *Structural Foundations of Quantum Gravity*, ed. Dean Rickles, Steven French, and Juha Saatsi, 53–82. Oxford: Oxford University Press.

Stanford, Kyle. 2000. An Antirealist Explanation of the Success of Science. *Philosophy of Science* 67 (2): 226–284.

Stanford, Kyle. 2001. Refusing the Devil's Bargain: What Kind of Underdetermination Should We Take Seriously? *Philosophy of Science* 68 (3): S. S1–S12.

Stanford, Kyle. 2006. *Exceeding Our Grasp. Science, History, and the Problem of Unconceived Alternatives*. New York: Oxford University Press.
Stanford, Kyle. 2017. Underdetermination of Scientific Theory. *Stanford Encyclopedia of Philosophy* (Winter 2017 Edition), ed. Edward N. Zalta.
Stanford, Kyle. 2021. Realism, Instrumentalism, Particularism: A Middle Path Forward in the Scientific Realism Debate. In *Contemporary Scientific Realism: The Challenge From the History of Science*, ed. Timothy Lyons, and Peter Vickers. Oxford: Oxford University Press.
Stang, Nicholas. 2024. "No World" Is Not Enough: Markus Gabriel and the Possibility of Metaphysics. In *Markus Gabriel's New Realism*, ed. Jan Voosholz, Synthese Library, 492, 95–114. Cham: Springer.
Sterpetti, Fabio. 2024. Counterfactuals, Models, and Scientific Realism. In *Model-Based Reasoning, Abductive Cognition, Creativity*, ed. Emiliano Ippoliti, Lorenzo Magnani, and Selene Arfini, 89–116. Berlin: Springer.
Stokes, Stephen. 1999. Luminescence Dating Applications in Geomorphological Research. *Geomorphology* 29 (1–2): 153–171.
Stoneham, Tom. 2009. Time and Truth: The Presentism-Eternalism Debate. *Philosophy* 84 (328): 201–218.
Strawson, Peter F. 1959. *Individuals. An Essay in Descriptive Metaphysics*. London: Methuen.
Strawson, Peter F. 1966. *The Bounds of Sense: An Essay on Kant's Critique of Pure Reason*. London: Routledge.
Stroud, Barry. 1984. *The Significance of Philosophical Skepticism*. Oxford: Oxford University Press.
Strowick, Elisabeth. 2019. *Gespenster des Realismus. Zur literarischen Wahrnehmung von Wirklichkeit*. Paderborn: Fink.
Stump, David. 1992. Naturalized Philosophy of Science with a Plurality of Methods. *Philosophy of Science* 59 (3): 456–460.
Stump, David. 2021. Fallibilism versus Relativism in the Philosophy of Science. *Journal for General Philosophy of Science* 52 (2): 1–13.
Suárez, Mauricio. 2008. Experimental Realism Reconsidered: How Inference to the Most Likely Cause Might be Sound. In *Nancy Cartwright's Philosophy of Science*, ed. Luc Bovens, Carl Hoefer, and Stephan Hartmann, 137–163. New York: Routledge.
Suárez, Mauricio. 2010. Fictions, Inference and Realism. In *Fictions and Models: New Essays*, ed. John Woods, 225–246. Munich: Philosophia.
Suárez, Mauricio, and Nancy Cartwright. 2008. Theories: Tools versus models. *Studies in History and Philosophy of Science Part B: Studies in History and Philosophy of Modern Physics* 39 (1): 62–81.
Sullivan, Meghan. 2012a. Problems for Temporary Existence in Tense Logic. *Philosophy Compass* 7: 43–57.
Sullivan, Meghan. 2012b. The Minimal A-Theory. *Philosophical Studies* 158: 149–174.
Suppes, Patrick. 1978. The Plurality of Science. *PSA: Proceedings of the Biennial Meeting of the Philosophy of Science Association* 3–16.
Suppes, Patrick. 1993. *Models and Methods in the Philosophy of Science: Selected Essays*, Synthese Library, 226. Dordrecht: Kluwer.
Surovell, Jonathan. 2017. Language, Ontology, and the Carnap-Quine Debate. *Philosophia* 45: 811–833.
Tahko, Tuomas E., and E. Jonathan Lowe. 2020. Ontological Dependence. *Stanford Encyclopedia of Philosophy* (Fall 2020 Edition), ed. Edward N. Zalta.
Tal, Eran. 2018. Naturalness and convention in the International System of Units. *Measurement* 116: 631–643.
Taylor, Christopher C.W. 1999. The Atomists. In *The Cambridge Companion to Early Greek Philosophy*, ed. Anthony A. Long, 181–204. Cambridge: Cambridge University Press.
Tegmark, Max. 2007. The Mathematical Universe. *Foundations of Physics* 38 (2): 101–150.
Teitel, Trevor. 2022. How to Be a Spacetime Substantivalist. *Journal of Philosophy* 119 (5): 233–278.

Tetens, Holm. 1982. Was ist ein Naturgesetz? *Zeitschrift für allgemeine Wissenschaftstheorie* 13 (1): 70–83.

Tetens, Holm. 2015. Der Gott der Philosophen. Überlegungen zur Natürlichen Theologie. *Neue Zeitschrift für Systematische Theologie und Religionsphilosophie* 57 (1): 1–13.

Tetens, Holm. 2017a. *Gott denken. Versuch über rationale Theologie*, 5th ed. Stuttgart: Reclam.

Tetens, Holm. 2017b. Realismus und Anti-Realismus im Allgemeinen. In *Handbuch Metaphysik*, ed. Markus Schrenk, J.B. Metzler, 230–236. Stuttgart.

Thanner, Veronika, Joseph Vogl, and Dorothea Walzer (eds.). 2018. *Die Wirklichkeit des Realismus*. Paderborn: Fink.

Thomas-Fogiel, Isabelle. 2022. Beyond Realism and Correlationism, the Idealist Path. In *Ideas and Idealism in Philosophy*, ed. Jure Simoniti, and Gregor Kroupa, 209–230. Berlin: De Gruyter.

Thomasson, Amie. 2007. *Ordinary Objects*. New York: Oxford University Press.

Thomasson, Amie. 2015. *Ontology Made Easy*. Oxford: Oxford University Press.

Thomasson, Amie. 2020a. If Models Were Fictions, Then What Would They Be? In *The Scientific Imagination. Philosophical and Psychological Perspectives*, ed. Arnon Levy, and Peter Godfrey-Smith, 51–74. Oxford: Oxford University Press.

Thomasson, Amie. 2020b. *Norms and Necessity*. Oxford: Oxford University Press.

Thompson, Naomi. 2023. Realism, Deflationism, and Metaphysical Explanation. In *Thomasson on Ontology*, ed. Miguel Garcia-Godinez, 61–83. Berlin: Springer.

Thorne, Christian. 2012. Outward Bound. On Quentin Meillassoux's *After finitude*. *Speculations* 3: 273–289.

Tomasello, Michael. 2014. *A Natural History of Human Thinking*. Cambridge: Harvard University Press.

Tooley, Michael. 1997. *Time, Tense and Causation*. Oxford: Clarendon Press.

Tooley, Michael. 2012. Against Presentism. Two Very Different Types of Objection. In *The Future of the Philosophy of Time*, ed. Adrian Bardon, 25–40. New York: Routledge.

Travis, Charles. 2024. The Meaning of Existence. In *Markus Gabriel's New Realism*, ed. Jan Voosholz, Synthese Library, 492, 13–31. Cham: Springer.

Tugendhat, Ernst, and Ursula Wolf. 1986. *Logisch-semantische Propädeutik*. Stuttgart: Reclam.

Vaidman, Lev. 2018. Many-Worlds Interpretation of Quantum Mechanics. *Stanford Encyclopedia of Philosophy* (Fall 2018 Edition), ed. Edward N. Zalta.

Vallor, Shannon (ed.). 2020. *The Oxford Handbook of Philosophy of Technology*. Oxford: Oxford University Press.

van den Hoonaard, Jordan. 2023. Defending Heidegger Against the Charge of Correlationism: A Prolegomena to a Phenomenology of Nature. *Southwest Philosophy Review* 39 (1): 41–48.

van Fraassen, Bas. 1980. *The Scientific Image*. Oxford: Oxford University Press.

van Fraassen, Bas. 1989. *Laws and Symmetry*. Oxford: Oxford University Press.

van Fraassen, Bas. 2006. Structure: Its Shadow and Substance. *Journal for the Philosophy of Science* 57 (2): 275–307.

Verhaegh, Sander. 2018. Setting Sail: The Development and Reception of Quine's Naturalism. *Philosophers' Imprint* 18 (19): 1–24.

Vickers, Peter. 2013. A Confrontation of Convergent Realism. *Philosophy of Science* 80 (2): 189–211.

Virtanen, Mikko J., and Antti Silvast. 2023. On Theory-Methods Packages in Science and Technology Studies. *Science, Technology, and Human Values* 48 (1): 167–189.

Voosholz, Jan. 2014. *Der spekulative Realismus als Herausforderung. Quentin Meillassoux'* Nach der Endlichkeit *als Kritik und Fortentwicklung der westlichen Philosophie*, MA thesis. Berlin: Freie Universität.

Voosholz, Jan. 2017. Verifikation. In *Sprachphilosophie*, ed. Christoph Demmerling, and Pirmin Stekeler-Weithofer, Wörterbuch der Sprachund Kommunikationswissenschaften, vol. 15. Berlin: de Gruyter.

Voosholz, Jan. 2019. A Formal Ontological Game. Does Meillassoux's Speculative Realism Need a Correlation? In *New Realism. Problems and Prospects*, ed. Alexander Kanev, 225–234. Sofia: St. Kliment Ohridski University Press.

Voosholz, Jan. 2021a. *Objektiver Realismus und Korrelationismus: Die Realismus-Antirealismus-Debatte der Wissenschaftsphilosophie im Lichte des Neuen Realismus*, Phil. Diss. Bonn: Rheinische Friedrich-Wilhelms-Universität Bonn.

Voosholz, Jan. 2021b. Top-Down Causation Without Levels. In *Top-Down Causation and Emergence*, ed. Jan Voosholz, and Markus Gabriel, Synthese Library, 439, 217–251. Cham: Springer.

Voosholz, Jan. 2024a. Introduction. In *Markus Gabriel's New Realism*, ed. Jan Voosholz, Synthese Library, 492, 1–10. Cham: Springer.

Voosholz, Jan (ed.). 2024b. *Markus Gabriel's New Realism*, Synthese Library, 492. Cham: Springer.

Voosholz, Jan. 2024c. The Concepts of Nature and the Universe in Markus Gabriel's New Realism. In *Markus Gabriel's New Realism*, ed. by Jan Voosholz, Synthese Library, 492, 147–166. Cham: Springer.

Voosholz, Jan, and Markus Gabriel. (eds.). 2021. *Top-Down Causation and Emergence*, Synthese Library, 439. Cham: Springer.

Voosholz, Jan, Debora Eller, and Marlene Sieverdingbeck. 2024. Antirealismus. In *Sprachphilosophie*, ed. Christoph Demmerling, and Pirmin Stekeler-Weithofer, Wörterbuch der Sprach- und Kommunikationswissenschaften, vol. 15. Berlin: de Gruyter.

Voosholz, Jan, Marlene Sieverdingbeck, and Marie Charlotte Wolf. 2025. Realismus. In *Sprachphilosophie*, ed. Christoph Demmerling, and Pirmin Stekeler-Weithofer, Wörterbuch der Sprach- und Kommunikationswissenschaften, vol. 15. Berlin: de Gruyter.

Vora, Kalindi, and Neda Atanasoski. 2019. *Surrogate Humanity: Race, Robots, and the Politics of Technological Futures*. Durham: Duke University Press.

Wagenknecht, Susann, Nancy J. Nersessian, and Hanne Andersen (eds.). 2015. *Empirical Philosophy of Science. Introducing Qualitative Methods into Philosophy of Science*. Studies in Applied Philosophy, Epistemology and Rational Ethics, 21. Cham: Springer.

Wallace, David. 2012. *The Emergent Multiverse: Quantum Theory according to the Everett Interpretation*. Oxford: Oxford University Press.

Wallace, David. 2022. Stating Structural Realism: Mathematics-First Approaches to Physics and Metaphysics. *Philosophical Perspectives* 36 (1): 345–378.

Ward, Thomas M. 2014. *John Duns Scotus on Parts, Wholes, and Hylomorphism*. Leiden: Brill.

Weyer, Jost. 2018. *Geschichte der Chemie. Band 1 Altertum, Mittelalter, 16. bis 18 Jahrhundert und Band 2 19. und 20. Jahrhundert*. Berlin: Springer.

Wheeler, Billy. 2022. How Realist is Informational Structural Realism? *Synthese* 200: 6.

Whittle, Bruno. 2021. Mathematical Anti-Realism and Explanatory Structure. *Synthese* 199 (3–4): 6203–6217.

Wigner, Eugene P. 1939. On Unitary Representations of the Inhomogeneous Lorentz Group. *The Annals of Mathematics, Second Series* 40 (1): 149–204.

Wigner, Eugene P. 1960. The Unreasonable Effectiveness of Mathematics in the Natural Sciences. *Communications in Pure and Applied Mathematics* 13: 1–14.

Willaschek, Marcus. 1997. Der transzendentale Idealismus und die Idealität von Raum und Zeit. Eine 'lückenlose' Interpretation von Kants Beweis in der "Transzendentalen Ästhetik". *Zeitschrift für philosophische Forschung* 51 (4): 537–564.

Williams, Bernard. 1978. *Descartes: The Project of Pure Enquiry*. Hassocks: Penguin Books.

Williams, Michael. 2001. *Problems of Knowledge. A Critical Introduction to Epistemology*. Oxford: Oxford University Press.

Williams, Michael. 2006. Realism. What's Left? In *Truth and Realism*, ed. Patrick Greenough and Michael P. Lynch, 77–99. Oxford: Clarendon Press.

Williamson, Timothy. 2007. *The Philosophy of Philosophy*. Malden: Blackwell.

Wilson, Jessica. 2005. Supervenience-Based Formulations of Physicalism. *Noûs* 39 (3): 426–459.

Wilson, Mark. 2006. *Wandering Significance. An Essay on Conceptual Behavior*. Oxford: Oxford University Press.

Wiltsche, Harald A. 2017. Science, Realism and Correlationism. A Phenomenological Critique of Meillassoux' Argument from Ancestrality. *European Journal of Philosophy* 25 (3): 808–832.

Windley, Brian F. 1995. *The Evolving Continents*, 3rd ed. Chichester: Wiley & Sons.

Winkler, Kenneth P. 1989. *Berkeley: An Interpretation*. Oxford: Oxford University Press.
Wittgenstein, Ludwig. 1989. *Philosophische Grammatik*, ed. G.E.M. Anscombe, Rush Rhees, and Georg H. von Wright, Werkausgabe, vol. 4. Frankfurt (Main): Suhrkamp.
Wittgenstein, Ludwig. 2006a. *Philosophische Untersuchungen*, ed. G.E.M. Anscombe, and Rush Rhees, Werkausgabe, vol. 1, 225–580. Frankfurt (Main): Suhrkamp.
Wittgenstein, Ludwig. 2006b. *Tagebücher 1914–1916*, ed. G.E.M. Anscombe, and Rush Rhees, Werkausgabe, vol. 1, 87–223. Frankfurt (Main): Suhrkamp.
Wittgenstein, Ludwig. 2006c. *Tractatus logico-philosophicus*, Werkausgabe, vol. 1, 7–85. Frankfurt (Main): Suhrkamp.
Wolf, William J. and Patrick M. Duerr. 2024. Promising Stabs in the Dark: Theory Virtues and Pursuit-Worthiness in the Dark Energy Problem. *Synthese* 204 (6): 1–40.
Wolff, J.E. 2012. Do Objects Depend on Structures? *British Journal for the Philosophy of Science* 63 (3): 607–625.
Wolff, J.E. 2023. Coordination and Obsolescence: A Response on Behalf of Measurement Realism. *Synthese* 201 (3): 1–20.
Wood, Adam. 2011. Aquinas, Scotus, and Cajetan on "Horseness is just Horseness". In *Medieval Metaphysics, or is it "Just Semantics"?*, ed. Gyula Klima, and Alexander W. Hall, 69–84. Newcastle upon Tyne: Cambridge Scholars.
Woodward, James. 2003. *Making Things Happen. A Theory of Causal Explanation*. Oxford: Oxford University Press.
Woodward, James. 2007. Causation with a Human Face. In *Causation and the Constitution of Reality*, ed. Huw Price, and Richard Corry, 66–105. Oxford: Oxford University Press.
Woodward, James. 2021. Downward Causation Defended. In *Top-Down Causation and Emergence*, ed. Jan Voosholz and Markus Gabriel, Synthese Library, 439, 217–251. Cham: Springer.
Worrall, John. 1989. Structural Realism: The Best of Both Worlds? *Dialectica* 43 (1–2): 99–124.
Worrall, John. 2007. Miracles and Models: Why Reports of the Death of Structural Realism May Be Exaggerated. In *Philosophy of Science*, ed. Anthony O'Hear, 125–154. Cambridge: Cambridge University Press.
Wray, K. Brad. 2008. The Argument from Underconsideration as Grounds for Anti-Realism: A Defence. *International Studies in the Philosophy of Science* 22: 317–326.
Wray, K. Brad. 2018. *Resisting Scientific Realism*. Cambridge: Cambridge University Press.
Wray, K. Brad. 2021. Kuhn and the Contemporary Realism/Antirealism Debates. *Hopos* 11 (1): 72–92.
Wright, Crispin. 1982. Strict Finitism. *Synthese* 51 (2): 203–282.
Wright, Crispin. 1983. *Frege's Conception of Numbers as Objects*. Aberdeen: Aberdeen Univesity Press.
Wright, Crispin. 1992. *Truth and Objectivity*. Cambridge: Harvard University Press.
Wright, Crispin. 1993. *Realism, Meaning, and Truth*, 2nd ed. Malden: Blackwell.
Yablo, Stephen. 2005. The Myth of the Seven. In *Fictionalism in Metaphysics*, ed. Mark Eli Kalderon, 88–115. Oxford: Oxford University Press.
Yablo, Stephen. 2012. Explanation, Extrapolation, and Existence. *Mind* 121 (484): 1007–1029.
Young, Niki. 2020. On Correlationism and the Philosophy of (Human) Access: Meillassoux and Harman. *Open Philosophy* 3 (1): 42–52.
Zahavi, Dan. 2016. The End of What? Phenomenology vs. Speculative Realism. *International Journal of Philosophical Studies* 24 (3): 289–309.
Zalabardo, José L. 2012. *Scepticism and Reliable Belief*. Oxford: Oxford University Press.
Zimmerman, Dean W. 1996. Persistence and Presentism. *Philosophical Papers* 25 (2): 115–126.
Zimmerman, Dean W. 1998. Temporary Intrinsics and Presentism. In *Metaphysics: The Big Questions*, ed. Peter van Inwagen, and Dean W. Zimmerman, 206–219. Malden: Blackwell.
Zimmerman, Dean W. 2005. The A-theory of Time, the B-theory of Time, and 'Taking Tense Seriously'. *Dialectica* 59: 401–457.
Zimmerman, Dean W. 2008. The Privileged Present: Defending an "A-Theory" of Time. In *Contemporary Debates in Metaphysics*, ed. Theodore Sider, John Hawthorne, and Dean W. Zimmerman, 211–225. Malden: Blackwell.

Zimmerman, Dean W. 2011. Presentism and the Space-Time Manifold. In *The Oxford Handbook of the Philosophy of Time*, ed. Craig Callender, 163–244. Oxford: Oxford University Press.
Žižek, Slavoj. 2012. *Less Than Nothing. Hegel and the Shadow of Dialectical Materialism*. London: Verso.
Zukowski, Marek, and Časlav Brukner. 2014. Quantum Non-locality – It Ain't Necessarily So... *Journal of Physics A: Mathematical and Theoretical* 47: 424009.
Zwick, Martin. 2024. A Systems Theoretic View of Speculative Realism. *Philosophy Today* 68 (2): 263–288.

Index

J. Voosholz, *Towards a New Scientific Realism*, Synthese Library 518,
https://doi.org/10.1007/978-3-031-98886-8

N

Q

R

T

The manufacturer's authorised representative in the EU is Springer
Nature Customer Service Centre GmbH, Europaplatz 3, 69115 Heidelberg,
Germany. If you have any concerns regarding our products, please
contact ProductSafety@springernature.com

Printed and bound by CPI Group (UK) Ltd, Croydon, CR0 4YY
01/10/2026
02225968-0014